Geotechnical, Geological and Earthquake Engineering

Volume 43

More information about this series at http://www.springer.com/series/6011

Susumu Iai
Editor

Developments in Earthquake Geotechnics

Springer

Editor
Susumu Iai
Disaster Prevention Research Institute
Kyoto University
Kyoto, Japan

ISSN 1573-6059 ISSN 1872-4671 (electronic)
Geotechnical, Geological and Earthquake Engineering
ISBN 978-3-319-62068-8 ISBN 978-3-319-62069-5 (eBook)
DOI 10.1007/978-3-319-62069-5

Library of Congress Control Number: 2017954317

Printed on acid-free paper

This Springer imprint is published by Springer Nature
The registered company is Springer International Publishing AG
The registered company address is: Gewerbestrasse 11, 6330 Cham, Switzerland

Preface

The earthquake disasters in Japan and New Zealand in 2011 indicated the urgent need for the latest earthquake geotechnics to be deployed for disaster mitigation. This book aims to provide a timely review and summaries of the most recent advances in earthquake geotechnics. It is unique in its content, including a review of developments in the discipline of earthquake geotechnics over half a century, giving readers of this book a solid grasp of this discipline from its inception up to the most recent developments. The book is based on contributions by 18 leading international experts who met in Kyoto in June 13–15, 2016, to discuss a range of issues related to developments in earthquake geotechnics.

This book addresses comprehensive areas of earthquake geotechnics, including performance-based seismic design, the evolution of geotechnical seismic response analysis from 1964 to 2015, mitigation measures against liquefaction, possible inundations in the Greater Tokyo Area due to tsunamis, and a series of effective stress analyses for the Great East Japan Earthquake in 2011.

The compilation and editing of this book coincided with the second phase of activities of Technical Committee 303 (TC303) "Coastal and River Disaster Mitigation and Rehabilitation" (short name "Floods") of the ISSMGE for the period 2013–2017, within the broader theme "Impact on Society". The publication of this book also coincided with the fifth anniversary of the FLIP Consortium, established in 2011, where academicians and practicing engineers significantly contributed to the development, advancement, and use of effective stress analysis. This book reflects the activities of TC303 and the FLIP Consortium.

This book will be of interest to advanced researchers and practicing engineers in earthquake geotechnics.

Kyoto, Japan

Susumu Iai

Contents

Chapter 1
Performance-Based Seismic Design of Geotechnical Structures

Susumu Iai

Abstract The paper gives an overview of performance-based approach for designing geotechnical structures against earthquakes, including remediation of liquefiable soils. Case histories of implementation and performance of remediation measures during past earthquakes are reviewed. The paper discusses the applicability and limitations of the conventional simplified approach for designing remediation of liquefiable soils and how these limitations can be overcome in the performance-based approach that explicitly considers residual displacements and structural strains beyond elastic limit. The paper also presents recent developments in performance evaluation against extremely long duration earthquake motions such as those during the 2011 East Japan earthquake (M = 9.0).

1.1 Introduction

Seismic damage to geotechnical structures is often associated with significant deformation of a soft or liquefiable soil deposit. Designing appropriate balance between structural components and foundation components, including remediation of liquefiable soils, is essential to improve the seismic performance of geotechnical structures. The conventional simplified approach based on pseudo-static force equilibrium provides reasonable guidelines for designing geotechnical structures against earthquake, including remediation of liquefiable soils against moderate earthquake motions. However, there are cases where the conventional simplified approach is not readily applicable. For example, for intense earthquake motions, acceptable residual displacements and structural strains beyond elastic limit become the primary design criteria parameters, which are not evaluated by the conventional simplified approach. There are a number of cases where site specific constraints pose difficulties in implementing the liquefaction remediation measures as suggested by the conventional simplified approach. Objective of this paper is to discuss the applicability and limitations of the conventional simplified approach for designing geotechnical structures, including remediation of liquefiable soils, and

S. Iai (✉)
Disaster Prevention Research Institute, Kyoto University, Kyoto, Japan
e-mail: iai.susumu@flip.or.jp

S. Iai (ed.), *Developments in Earthquake Geotechnics*, Geotechnical, Geological and Earthquake Engineering 43, DOI 10.1007/978-3-319-62069-5_1

how these limitations can be overcome in the performance-based approach. The paper also presents recent developments in performance evaluation against extremely long duration earthquake motions such as those during the 2011 East Japan earthquake (M = 9.0).

1.2 Standard Procedure

Designing geotechnical structures against earthquakes, including remedial measures against liquefaction, can be classified into two broad categories (Port and Harbour Research Institute 1997; PIANC 2001):

(a) Soil improvement to reduce the potential of liquefaction at foundation soil;
(b) Structural solutions to minimize damage, including in the event of liquefaction.

In practice, a combination of these two measures is often adopted.

A flowchart illustrating a standard design procedure for designing geotechnical structures against earthquakes, including liquefaction remediation, is shown in Fig. 1.1. Once the strategy has been determined, it is common practice to select a method first, and then to compare the advantages and disadvantages of different solutions for the particular project. Typically, a number of different solutions will need to be assessed before a final decision can be reached. Sometimes, solutions will be combined to achieve the optimum design. The procedure illustrated in Fig. 1.1 is intended to be used as a guide only; it may often be more efficient to screen out remediation methods that are clearly unsuitable at an early stage based on a rough assessment of the likely area to be treated and any project specific constraints. A key to the selection of the method is often found in the process to try to find out the most vulnerable portion out of the entire soil-structure system, rather than to go through the routine process of checking and evaluating the structural or foundation components one by one out of the whole system.

A combination of two or more remediation methods is often very effective. A typical example might be the combination of a low noise/low vibration method, such as the use of drain, combined with compaction around the improved area, to constrain or confine the overall site. At present, however, formal procedures for selecting the most efficient combination of methods for a specific site have not been developed and the engineer must use judgment and experience to reach the optimum solution.

1.3 Conventional Approach

The design requirements for a specific improvement in the geotechnical structures, including the ground improvement area, will be based on an assessment of the mechanism of damage likely to be induced in the structural components essential to

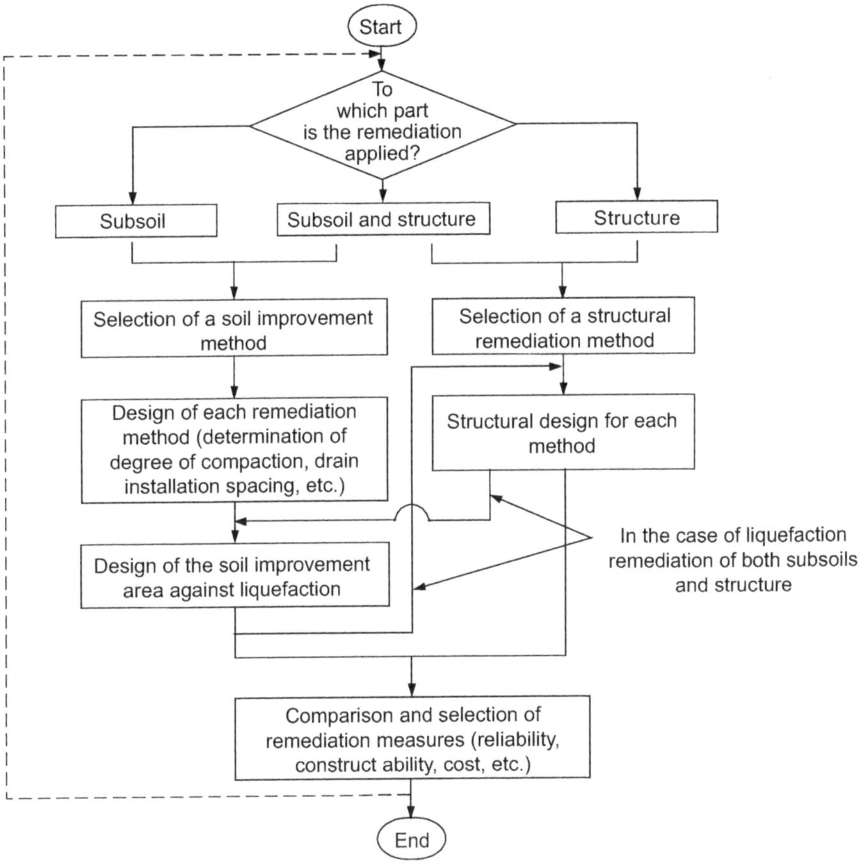

Fig. 1.1 Standard procedure for liquefaction remediation (Port and Harbour Research Institute 1997)

support the soil-structure system or the foundation, often consisting of natural soils by liquefaction. In particular, it is critical to determine whether the damage will be caused primarily by reduction in the shear resistance of the liquefied soil, or by additional external forces due to excessively large displacements. Large displacements of the liquefied subsoil may affect the foundations of existing and new structures, or cause serious damage to lifeline facilities (such as gas, water and telecommunication pipelines and conduits), and special consideration will be necessary in respect of these cases. There are many other cases, however, in which remediation measures against loss of shear resistance will provide benefits against the risk of large displacements; ground compaction being one example.

Generally, even if the soil is predicted to undergo liquefaction over a wide area, it may be possible to limit the area requiring soil improvement to the area that controls the stability of the structure. For example, the zone within the subsoil that contributes predominantly to the stability of spread foundation structures is the part directly below and immediately around the structure; ground far away from the structure does not contribute to the same extent. A key issue is therefore to establish how far the soil improvement needs to extend laterally from the structure. This can be determined by following the procedure summarized below (Iai et al. 1987, 1991).

The design procedure for a soil improvement area presented below is based around use of the compaction method. The procedure may, however, be adapted for other methods with appropriate modification by considering the appropriate characteristics of the improved soil (such as permeability and cyclic strength/ deformation).

1. Propagation of excess pore water pressure into the improved zone

Outside the zone of improved soil, excess pore water pressures will exceed those inside, resulting in a hydraulic gradient driving fluid into the zone of improved soil. This is a complex issue, as the deformation characteristics of dense saturated sand are highly non-linear. As a simplified design procedure, this phenomenon may be addressed as follows:

For excess pore water pressure ratios $u/\sigma_v' < 0.5$, the effect of the excess pore water pressure increase may be ignored because laboratory test data indicate a very small strain generation below this level. For $u/\sigma_v' > 0.5$, however, it is necessary to take into account the effect of excess pore water pressure increase. Shaking table tests and seepage flow analyses suggests that the pore water pressure ratio $u/\sigma_v' > 0.5$ within an area defined by the square ABCD in Fig. 1.2. In this area, the soil shear resistance must be reduced for the purposes of the design. The tests also indicate that an area defined by the triangle ACD exhibited unstable characteristics. This area should therefore be assumed in the design to be liquefiable and treated accordingly.

The exception to this recommendation is when a drain or impermeable sheet or zone has been installed at the perimeter of the improved area in order to shut out the inflow of pore water into the improved area. Under these conditions, the area

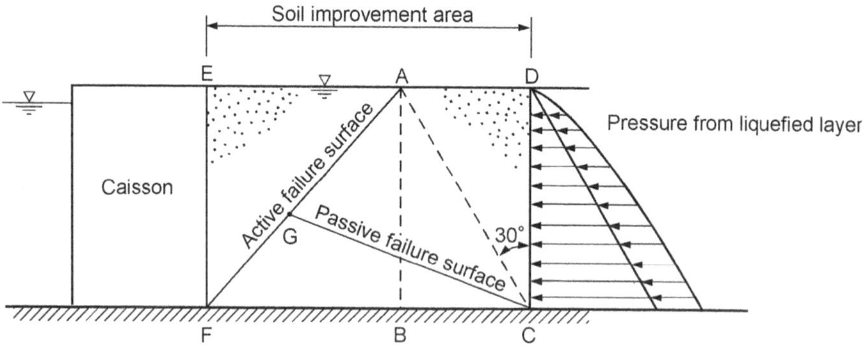

Fig. 1.2 Schematic diagram for investigation of stability with respect to pressures applied from the liquefied sand layer (Iai et al. 1987)

corresponding to the square ABCD need not be included as part of the improvement plan.

2. Pressure applied by the liquefied sand layer

At the boundary between liquefied and non-liquefied ground there is a dynamic force as indicated in Fig. 1.2 and a static pressure corresponding to an earth pressure coefficient $K = 1.0$ which acts on the improved ground due to the liquefaction of the surrounding soils. These forces may greatly exceed the forces acting in the opposite direction from the non-liquefied ground. For a retaining structure backfilled with soil and subject to active static and dynamic lateral earth pressures at EF as shown in Fig. 1.2, the area of soil improvement must be large enough that there is no influence of liquefaction in the active failure zone. To accommodate the net outward force, it is essential to check that sufficient shear resistance can be mobilized along the passive failure surface GC.

3. Loss of shear strength in liquefied sand layer

In the simplified design procedure, unimproved soil of loose or medium relative density should be considered to have negligible shear strength after liquefaction (i.e. the soil is treated as a heavy fluid). Since the shear strength of the improved ground in triangle ACD also cannot be relied upon (see (1) above), then the improvement area should be wide enough to obtain sufficient bearing capacity from the shear resistance along the solid lines EFG and HI in Fig. 1.3.

In practice, lateral pressure from surrounding liquefied sand layers may contribute to the stability of certain structures. Figure 1.3 shows how the dynamic earth pressure may be subtracted from the enhanced static lateral earth pressure (based on an earth pressure coefficient $K = 1.0$) in certain stability calculations. The section on which the pressure from the liquefied sand layer is applied can be assumed to be along the lines GG' or II'.

Although this figure shows only four examples of foundation types, the same principles can be applied to other design arrangements.

4. Design of soil improvement area

Key parameters for any assessment of the area of soil improvement include the angle of internal friction for sands and the anticipated intensity of earthquake shaking. A stability analysis method will also need to be specified. Standard design procedures adopt a simplified, pseudo-static approach to the prediction of earthquake loads and response. Despite this, experience suggests that with appropriate care and selection of parameters, these methods generally provide satisfactory design solutions.

The soil improvement area required for a typical gravity quay wall based on this approach is shown in Fig. 1.2. For anchored sheet pile walls, the zone of improvement in the vicinity of the anchor wall must be carefully considered. For flexible anchor walls, soil improvement must extend to sufficient depth below the tie rod level to ensure that the adequate capacity is maintained to resist the tie rod force.

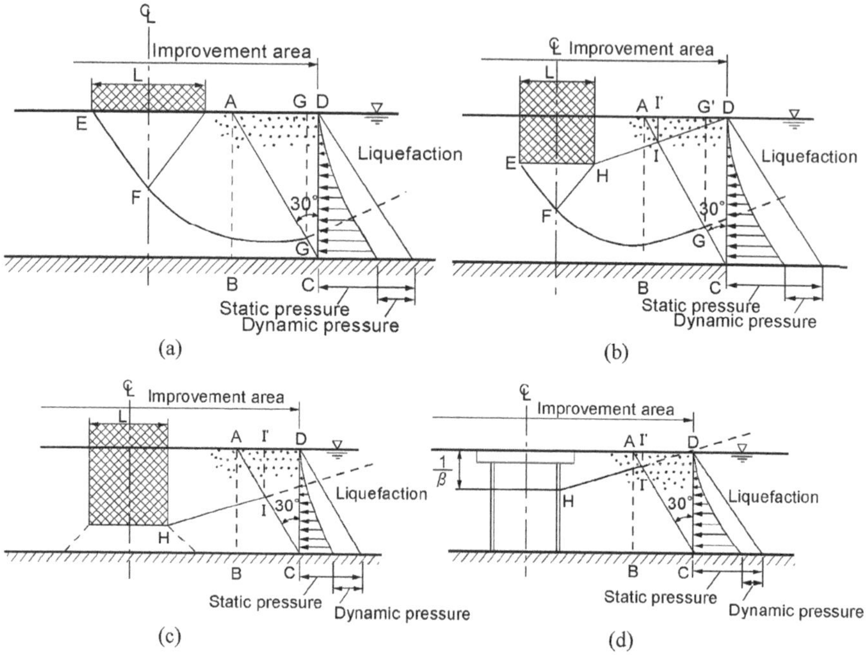

Fig. 1.3 Schematic diagrams for investigation of stability for determining the soil improvement area (Port and Harbour Research Institute 1997; Iai et al. 1987). (**a**) Shallow foundation. (**b**) Foundation with shallow embedment. (**c**) Foundation with deep embedment. (**d**) Pile foundation

1.4 Performance at Moderate Earthquake Motions

The design procedure based on the conventional simplified approach described in the previous chapter provided a basis for implementing the liquefaction remediation in conventional practice. A number of case histories of implementing liquefaction remediation were gradually increasing. The first opportunity to test the effectiveness and adequateness of the remedial measures against liquefaction was provided during 1993 Kushiro-Oki earthquake of magnitude 7.8 (Iai et al. 1994). Kushiro Port located at 15 km from the epicenter was shaken with a peak horizontal acceleration of 0.47 g. If the spiky wave due to cyclic mobility is filtered out, the effective peak ground acceleration was in the order of 0.3 g.

Many of the quay walls at Kushiro port suffered damage due to liquefaction of backfill sand. The most serious damage to the sheet pile quay walls is shown in Fig. 1.4. This quay wall was of a steel sheet pile type anchored by battered steel piles. The ground consists of loosely deposited backfill sand with a thickness of about 10 m, underlain by medium to dense sand deposit that forms the original ground. Liquefaction at the backfill caused serious deformation in the sheet pile wall as shown by the solid line in this figure. Detailed investigation of the sheet pile wall by diving in the sea revealed that cracks opened in the steel sheet pile wall at an elevation of 4 m below the water level.

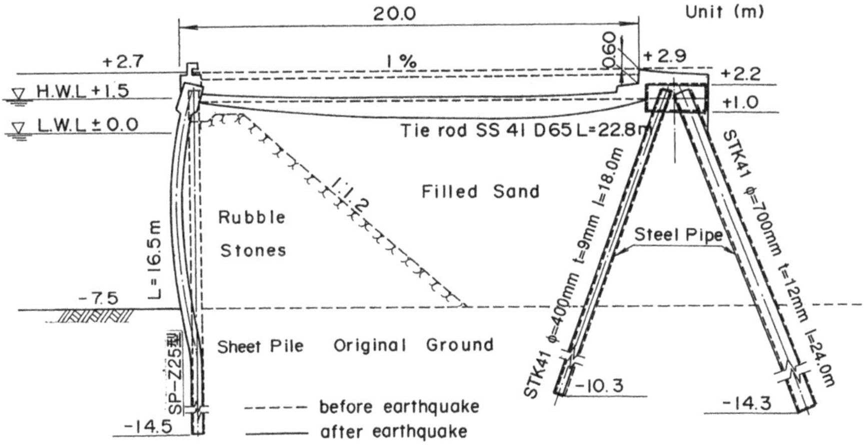

Fig. 1.4 Damaged sheet pile quay wall at South Fishery Wharf, Kushiro Port, during 1993 Kushiro-oki earthquake (Iai et al. 1994)

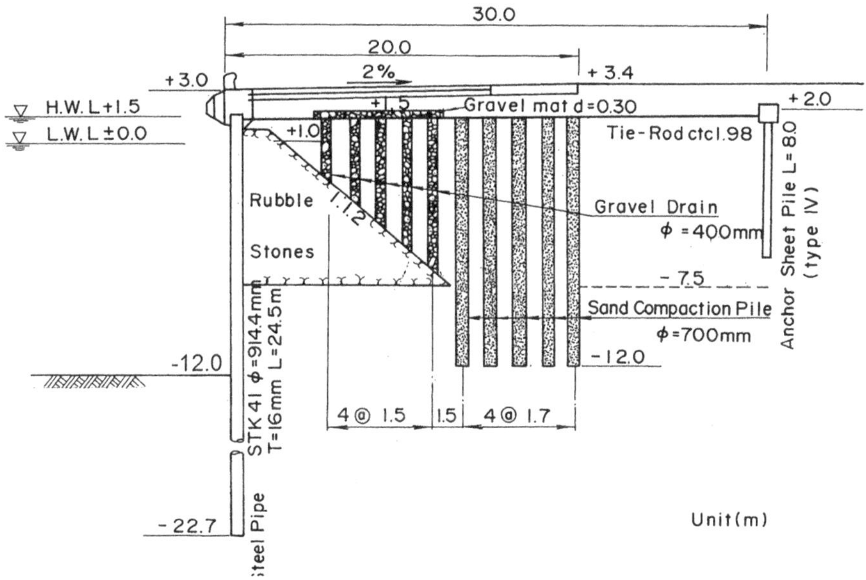

Fig. 1.5 Undamaged sheet pile quay wall at South West Port No.1 Wharf, Kushiro Port, during 1993 Kushiro-oki earthquake (Iai et al. 1994)

In contrast to the damaged sheet pile quay wall, quay walls with compacted backfill sand survived the earthquake without damage. An example, shown in Fig. 1.5, is of a steel pipe pile wall anchored by a steel sheet pile wall with a water depth of 12 m. The ground at this wall originally consisted of fill sand with a thickness of about 10 m, underlain by the original ground of medium to dense sand.

The backfill sand was later treated by a sand compaction pile method as shown in this figure. In order to avoid the affecting the existing steel pipe pile wall during the installation of ground compaction work, gravel drains were installed near the wall as shown in Fig. 1.5.

Despite the earthquake motions with a peak acceleration of 0.3 g level, there was no damage to this quay wall. This case history demonstrated that the measures against liquefaction and design of quay walls according to the procedure based on the conventional simplified approach are sufficient to provide adequate resistance to this level of earthquake motions.

1.5 Performance at Intense Earthquake Motions

1995 Hyogoken-Nambu earthquake with magnitude 7.2 provided a good opportunity to re-evaluate the limitations of the conventional simplified approach of seismic design. Shaken with a peak ground acceleration of 0.5 g, many quay walls in Kobe Port, Japan, suffered serious damage. The damage involves large seaward displacement, settlement and tilt as shown in Fig. 1.6. The damage was cause mainly by deformation in the loosely deposited foundation soil beneath the caisson wall (Inagaki et al. 1996).

In order to evaluate the effects of liquefaction, a series of effective stress analyses were performed. A strain space multiple mechanism model for sand was used for the analysis (Iai et al. 1992). Case-1 is the analysis that simulates the conditions in-situ during the earthquake. In this case, both the foundation soils beneath the caisson wall and the backfill soils were liquefiable decomposed granite.

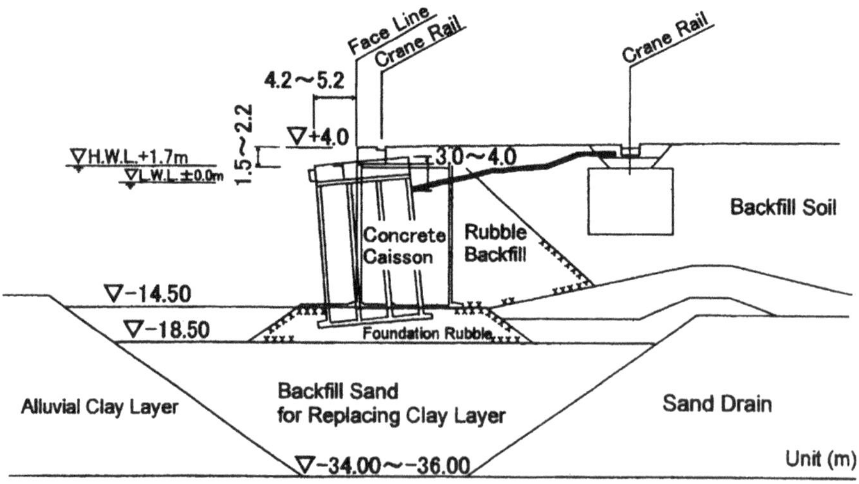

Fig. 1.6 Damage to a caisson quay wall at Kobe Port during 1995 Hyogoken-Nambu earthquake (Inagaki et al. n.d.)

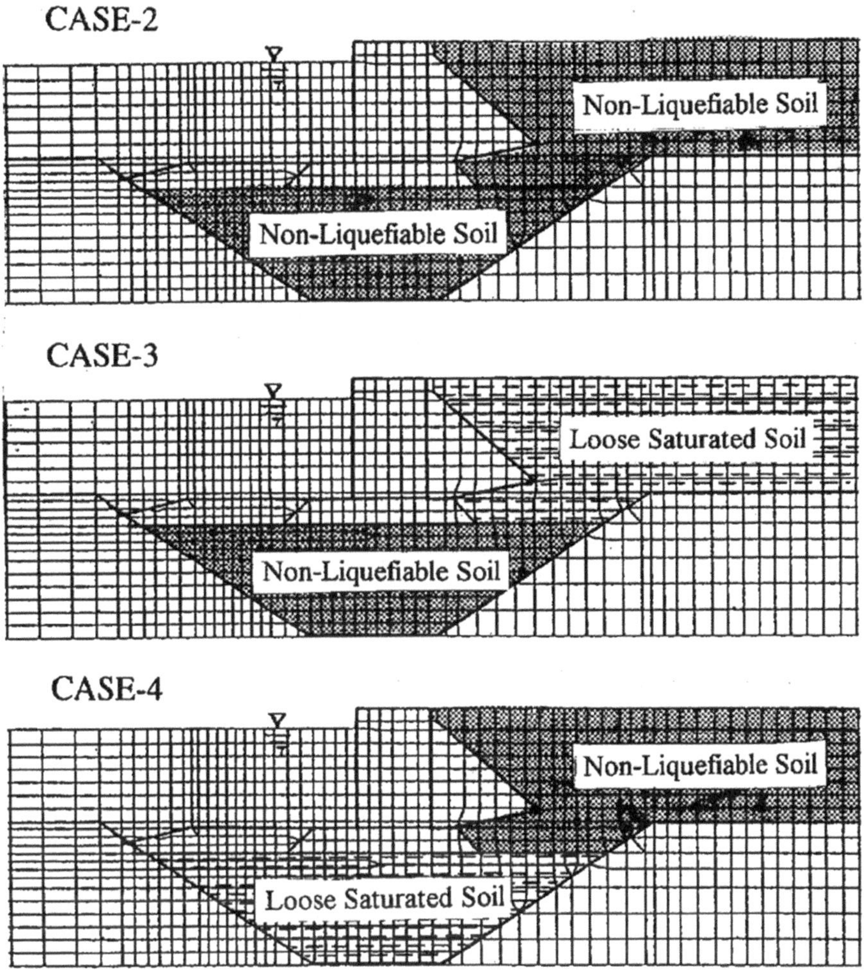

Fig. 1.7 Conditions assumed for parametric study, Cases 2 through 4 (Iai et al. 1992)

Cases-2 through 4 are defined by the extent of the non-liquefiable soil relative to the caisson wall as shown in Fig. 1.7.

Case-1 analysis was successful in simulating the damage to a caisson type quay wall as shown in Fig. 1.8. The major results of the parameter study are summarized in Table 1.1. These results indicate that the deformation of the gravity wall may be reduced up to about one half of that at the earthquake if the excess pore water pressure increase was prevented in the subsoil as in Case-2. In particular, horizontal displacements of Cases-3 and 4 are 0.5(=2.1–1.6) and 0.9(=2.5–1.6)m larger than that of Case-2, suggesting that the effect of the pore water pressure increase in the foundation soil beneath the caisson wall (Case-4) is about twice as that of the backfill (Case-3).

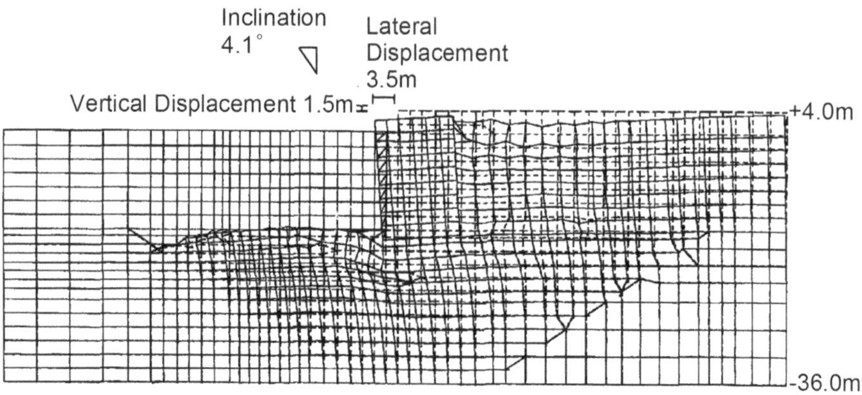

Fig. 1.8 Computed deformation of a gravity quay wall, Case-1 (Iai et al. 1998)

Table 1.1 Major results of parametric study for gravity quay wall (Iai et al. 1998)

Residual displacements of caisson			
Case	Horizontal (m)	Vertical (m)	Tilt (degrees)
Case 1	3.5	1.5	4.1
Case 2	1.6	0.6	2.4
Case 3	2.1	0.7	3.1
Case 4	2.5	1.1	2.2

These results were compared with the performance of the quay walls at Port Island (phase II). One (PC-13) was constructed on a loose deposited foundation similar to that shown in Fig. 1.6, whereas others (PC-14 and -15) were constructed on a foundation improved by the sand compaction pile (SCP) method as shown in Fig. 1.9. Although these quay walls were constructed along a straight face line, displacements of the quay walls PC-14 and -15 were about 2.5 m and 0.3 m in horizontal and vertical directions whereas those at PC-13 were about 3.5 m and 1.5 m in horizontal and vertical directions. The analysis conditions of Case-3, having non-liquefiable foundation, approximate the performance of the quay walls PC-14 and 15. The analysis condition of Case-1, having loose deposited foundation, approximates the performance of the quay wall PC-13. The results for the parameter study of Cases-1 and 3 discussed above can explain the difference between the quay walls constructed on SCP and a loose deposited foundation (Iai et al. 1998).

For variations in the peak acceleration of the earthquake motion input at the base, the horizontal residual displacement at the top of the caisson wall was computed as shown in Fig. 1.10. These response curves constitute the basis for performance-based design described in the next section.

The case history of seismic performance of quay walls during 1995 Hyogoken-Nambu earthquake provided a number of valuable lessons. In particular, the case history and analysis described above suggest that (1) liquefaction remediation is

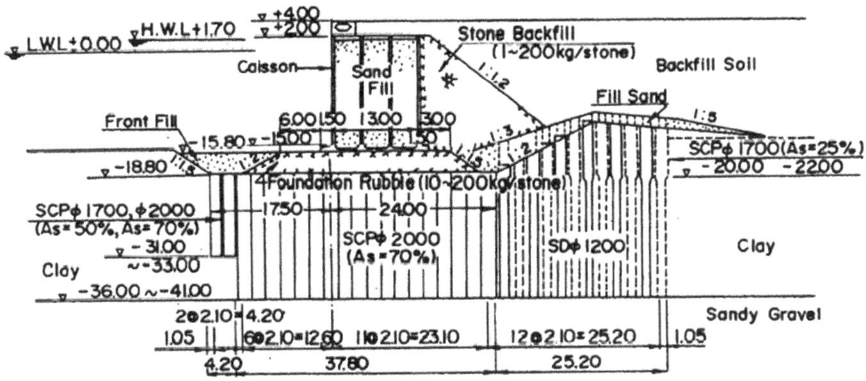

Fig. 1.9 Cross section of a quay wall in Kobe Port improved with sand compaction piles (SCP) (Inagaki et al. n.d.; Iai et al. 1998)

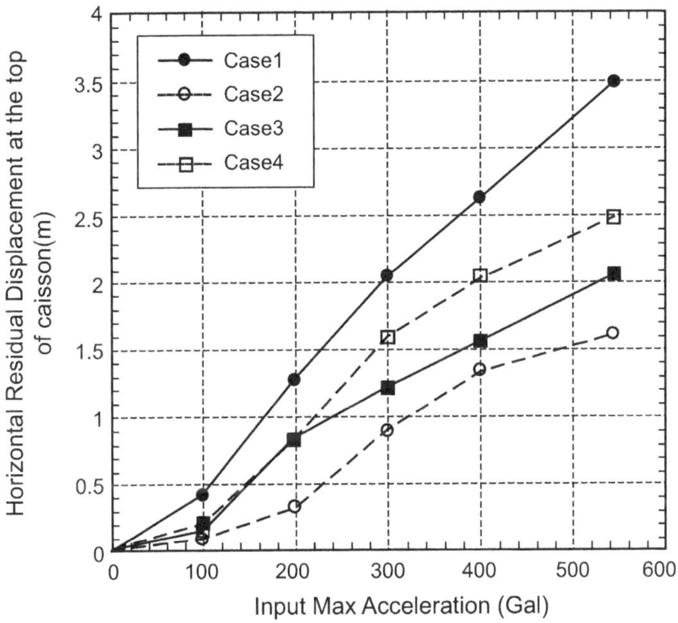

Fig. 1.10 Effects of input acceleration levels on horizontal residual displacement (PIANC 2001)

effective for improving performance of quay walls even for intense earthquake motions, (2) some residual deformation may be acceptable, (3) deformations in ground and foundation soils and the corresponding structural deformation and stress states are key design parameters, and (4) conventional simplified limit equilibrium-based methods are not well suited to evaluating these parameters. All of these issues indicate that there is a strong need to develop performance-based approach for liquefaction remediation (PIANC 2001).

1.6 Toward Performance-Based Design

The goal of the performance-based design is to overcome the limitations present in conventional simplified seismic design. The conventional seismic design is based on providing capacity to resist a design seismic force, but it does not provide information on the performance of a structure when the limit of the force-balance is exceeded. If we demand that limit equilibrium not be exceeded in conventional design for the relatively high intensity ground motions associated with a very rare seismic event, the construction/retrofitting cost will most likely be too high. If force-balance design is based on a more frequent seismic event, then it is difficult to estimate the seismic performance of the structure when subjected to ground motions that are greater than those used in design.

In performance-based design, appropriate levels of design earthquake motions must be defined and corresponding acceptable levels of structural damage must be clearly identified. Two levels of earthquake motions are typically used as design reference motions, defined as follows:

Level 1 (L1): the level of earthquake motions that are likely to occur during the life-span of the structure;
Level 2 (L2): the level of earthquake motions associated with infrequent rare events, that typically involve very strong ground shaking.

The acceptable level of damage is specified according to the specific needs of the users/owners of the facilities and may be defined on the basis of the acceptable level of structural and operational damage given in Table 1.2. The structural damage category in this table is directly related to the amount of work needed to restore the full functional capacity of the structure and is often referred to as direct loss due to earthquakes. The operational damage category is related to the amount of work needed to restore full or partial serviceability. Economic losses associated with the loss of serviceability are often referred to as indirect losses. In addition to the fundamental functions of servicing sea transport, the functions of port structures may include protection of human life and property, functioning as an emergency base for transportation, and as protection from environmental threats from spilling hazardous materials such as oils. If applicable, the effects on these issues should be considered in defining the acceptable level of damage in addition to those shown in Table 1.2.

Once the design earthquake levels and acceptable damage levels have been properly defined, the required performance of a structure may be specified by the appropriate performance grade S, A, B, or C defined in Table 1.3. In performance-based design, a structure is designed to meet these performance grades.

Table 1.2 Acceptable level of damage in performance-based design[a]

Acceptable level of damage	Structural	Operational
Degree I: Serviceable	Minor or no damage	Little or no loss of serviceability
Degree II: Repairable	Controlled damage[b]	Short-term loss of serviceability[c]
Degree III: Near collapse	Extensive damage in near collapse	Long-term or complete loss of serviceability
Degree IV: Collapse[d]	Complete loss of structure	Complete loss of serviceability

[a]Considerations: protection of human life and property, functions as an emergency base for transportation, and protection from environmental threats from spilling hazardous materials such as oils, if applicable, should be considered in defining the damage criteria in addition to those shown in this table
[b]With limited inelastic response and/or residual time for repairs
[c]Structure out of service for short to moderate duration
[d]Without significant effects on surroundings

Table 1.3 Performance grades S, A, B, and C

Performance grade	Design earthquake	
	Level 1(L1)	Level 2(L2)
Grade S	Degree I: Serviceable	Degree I: Serviceable
Grade A	Degree I: Serviceable	Degree II: Repairable
Grade B	Degree I: Serviceable	Degree III: Near collapse
Grade C	Degree II: Repairable	Degree IV: Collapse

The principal steps taken in performance-based design are shown in the flowchart in Fig. 1.11:

1. Choose a performance grade from S, A, B, or C: This step is typically done by referring to Tables 1.2 and 1.3 and selecting the damage level consistent with the needs of the users/owners. Another procedure for choosing a performance grade is to base the grade on the importance of the structure. Degrees of importance are defined in most seismic codes and standards. If applicable, a performance grade other than those of S, A, B, or C may be introduced to meet specific needs of the users/owners.
2. Define damage criteria: Specify the level of acceptable damage in engineering parameters such as displacements, limit stress states, or ductility factors.
3. Evaluate seismic performance of a structure: Evaluation is typically done by comparing the response parameters from a seismic analysis of the structure with the damage criteria. If the results of the analysis do not meet the damage criteria, the proposed design or existing structure should be modified. Soil improvement including remediation measures against liquefaction may be necessary at this stage. Details of liquefaction remediation can be found in the publication of the Port and Harbour Research Institute (Port and Harbour Research Institute 1997).

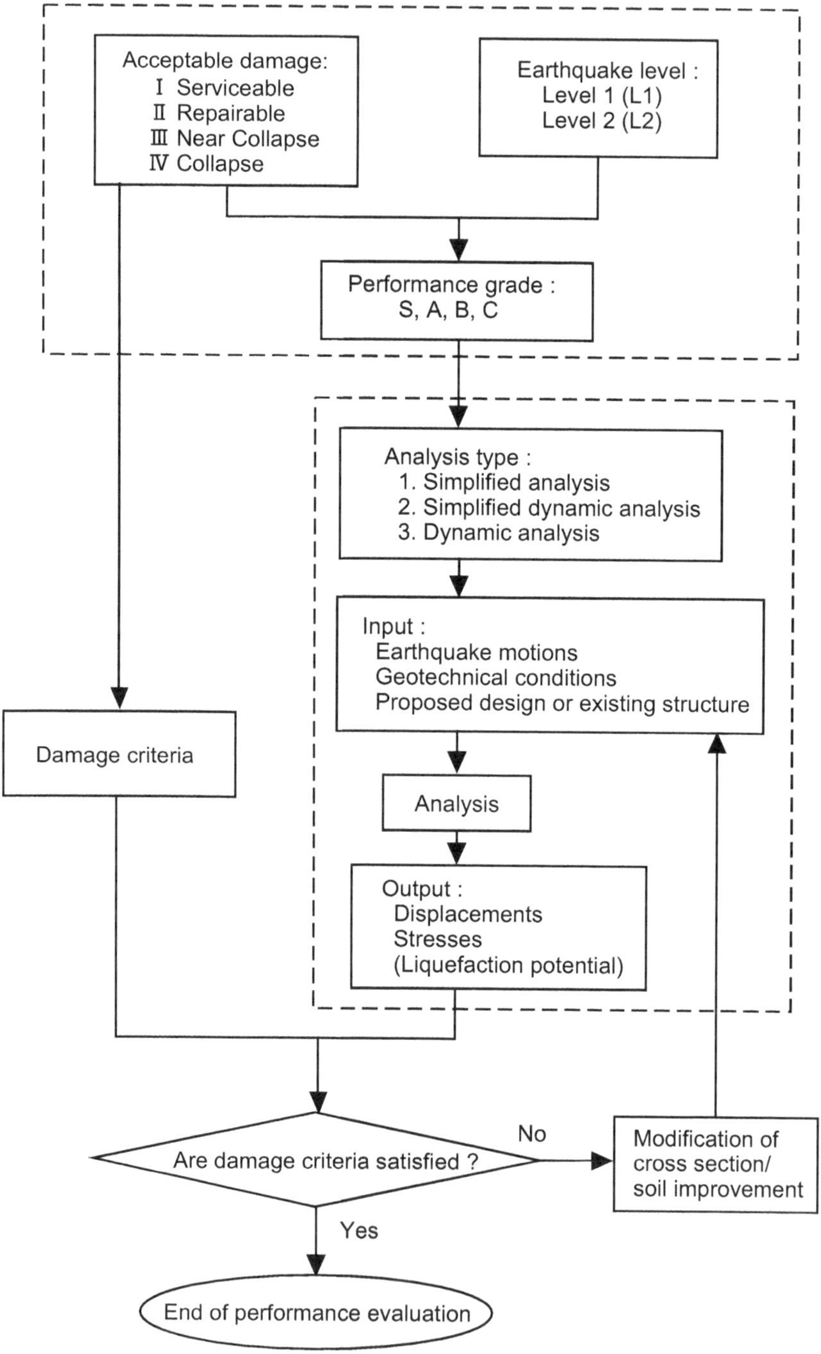

Fig. 1.11 Flow chart for performance evaluation (PIANC 2001)

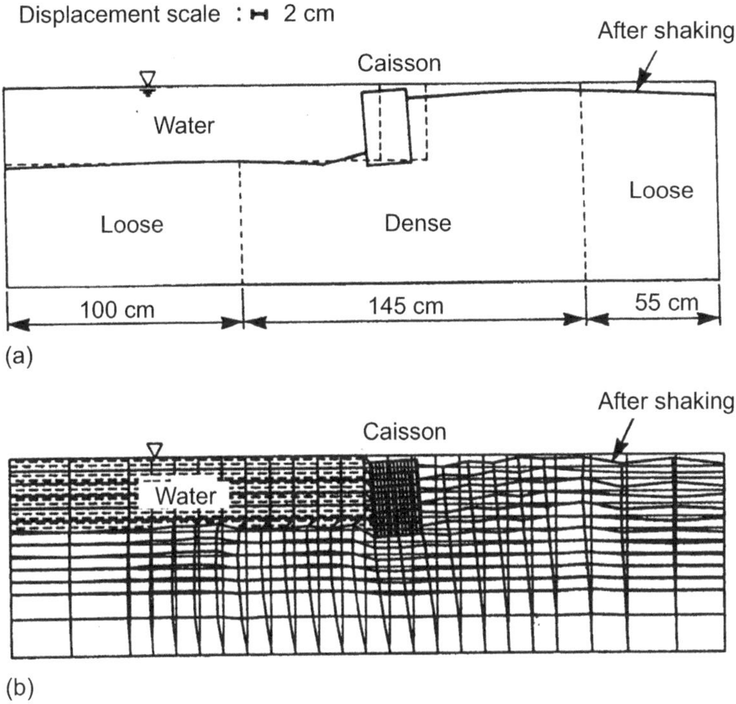

Fig. 1.12 Model test and computed results of residual deformation of a caisson quay wall (Iai 1994). (**a**). Shaking table tests. (**b**) Effective stress analysis

More comprehensive discussions on the performance-based design of port structures can be found in a publication by a working group on seismic effects for port structures, International Navigation Association (PIANC 2001).

Design charts useful for determining the area of ground improvement within the context of performance-based design were obtained in early 1990s. An example for a caisson quay wall is shown in Figs. 1.12 and 1.13, where based on a series of shaking table tests and effective stress analysis (Fig. 1.12) by varying the extent of ground improvement area (Fig. 1.13a), the area of ground improvement is shown in normalized design chart (Fig. 1.13b) (Iai et al. 1998).

Another example is the degree of uplift of a buried structure due to liquefaction as shown in Figs. 1.14 and 1.15 (Iai and Matsunaga 1991). Similar work was also performed on a sheet pile quay wall (Port and Harbour Research Institute 1997; Iai and Matsunaga 1990; McCullough and Dickenson 1998).

In the framework of performance-based design, an example of the liquefaction remediation for a gravity quay wall may be given using the design chart shown in Fig. 1.13 as follows. In this example, let us assume that the quay wall was given the performance grade A with seismic criteria over L1 (PGA = 250 cm/s^2) and L2 (PGA = 400 cm/s^2) earthquake motions specified by the residual displacements (d/H) of less than 1 % for maintaining full operation and 4 % for allowing quick

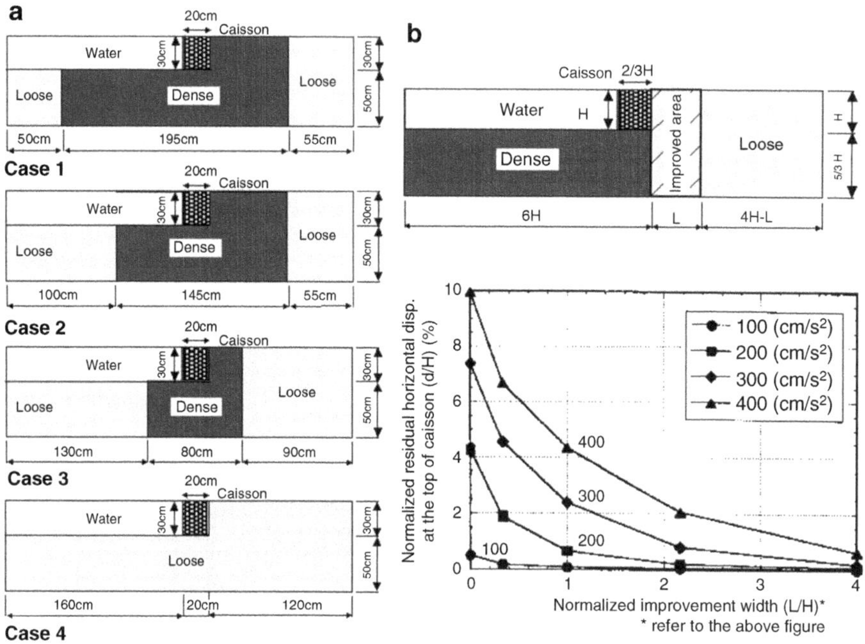

Fig. 1.13 Design chart for ground improvement area (Iai 1994). (a) Variation of cross section. (b) Width of compaction area vs. residual horizontal displacement

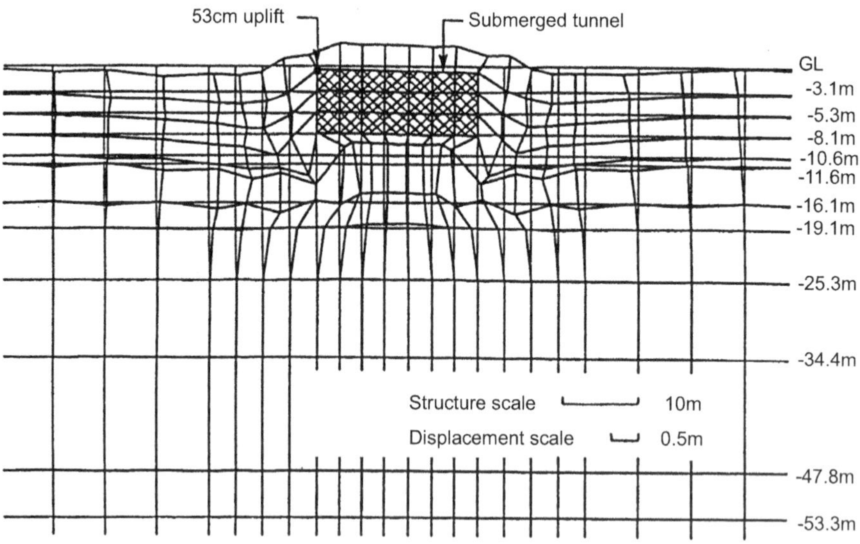

Fig. 1.14 Example of computed uplift of a buried structure in liquefied soil (Iai and Matsunaga 1990)

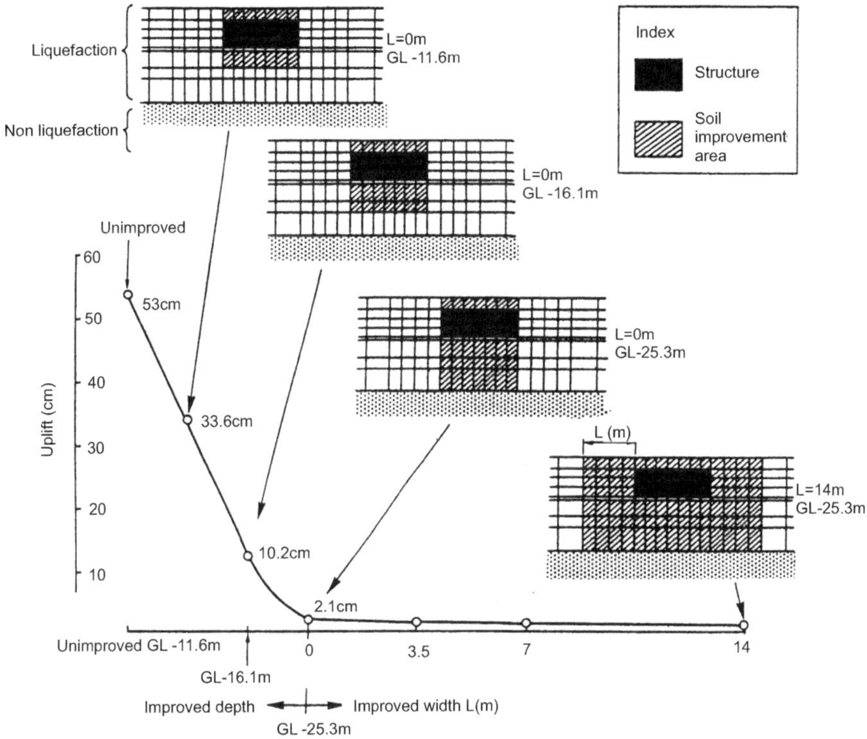

Fig. 1.15 Uplift and improved depth of buried structures (Port and Harbour Research Institute 1997)

recovery of operation. By referring to Fig. 1.13, the L1 criterion gives the area L/H = 1.7 whereas the L2 criterion gives the area of L/H = 1.2. Thus, the improvement area required to meet these performance requirements is L/H = 1.7. In this particular example, stronger L2 excitation does not govern the final design that was highly influenced by a high performance requirement for L1.

Performance-based approach can readily be applicable to other types of structures, including pile-supported wharves (shown in Figs. 1.16 and 1.17), that may require the structural damage criteria with respect to piles and a deck as well as displacements of dykes or retaining structures. Details can be found in the guidelines by PIANC (2001).

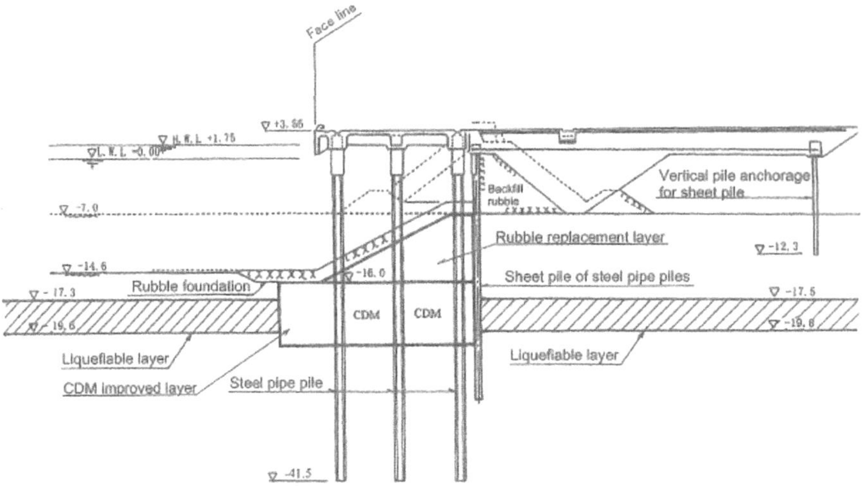

Fig. 1.16 Example of ground improvement for a pile-supported wharf (PIANC 2001)

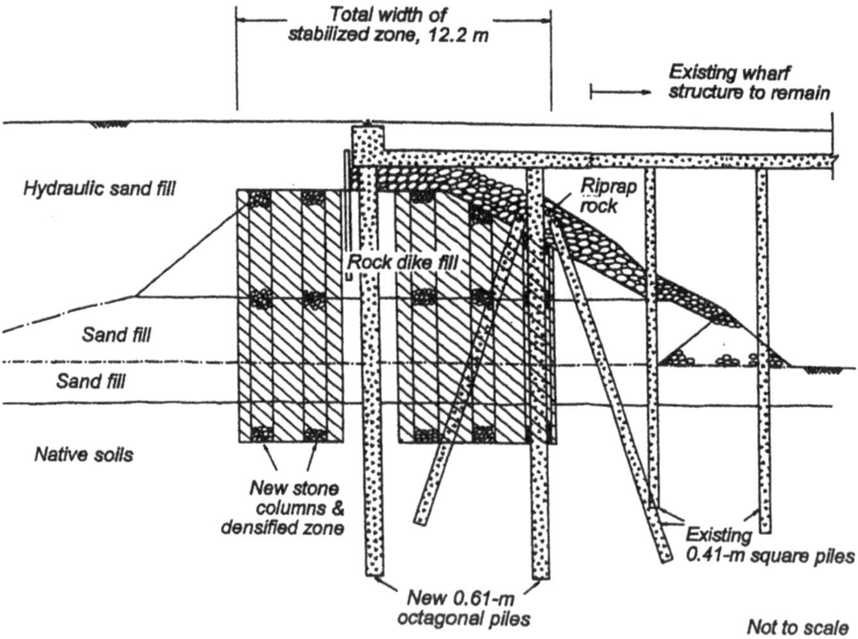

Fig. 1.17 Example of ground improvement for a pile-supported wharf (Egan et al. 1992)

1.7 Seismic Performance with Effects of Partial Drainage During Earthquake Motions of Extremely Long Duration

Duration of the earthquake motions that have been discussed in the preceding sections is about 30 s or less so that undrained behavior of sand provides reasonable representation of the seismic performance of geotechnical structures. For a longer duration of earthquake motions such as those recorded during 2011 East Japan earthquake of magnitude 9.0, the duration of the earthquake motion is in the order of 200 s and the effect of partial drainage during this duration will likely to be significant. Thus, the effective stress analysis allowing dissipation and redistribution of excess pore water pressures during and after earthquake should become the primary analysis procedure to appropriately evaluate the seismic performance of geotechnical structures.

A typical example of the damage to a sheet pile quay wall during 2011 East Japan earthquake is shown in Fig. 1.18. In this example, the sheet pile wall was subject to the earthquake motion with PGA = 0.28 g shown in Fig. 1.19, and displaced 200 mm toward the sea with back fill settlements in 100 mm immediately behind the sheet pile wall. The settlement at back fill extends further inland and resulted in 300 mm differential settlement behind the anchor with coupled piles.

The effective stress analysis of this sheet pile was performed by multiple step analysis of initial gravity analysis by closely following the sequence of construction of the quay wall (Tashiro et al. 2017). After the multiple step analysis of initial

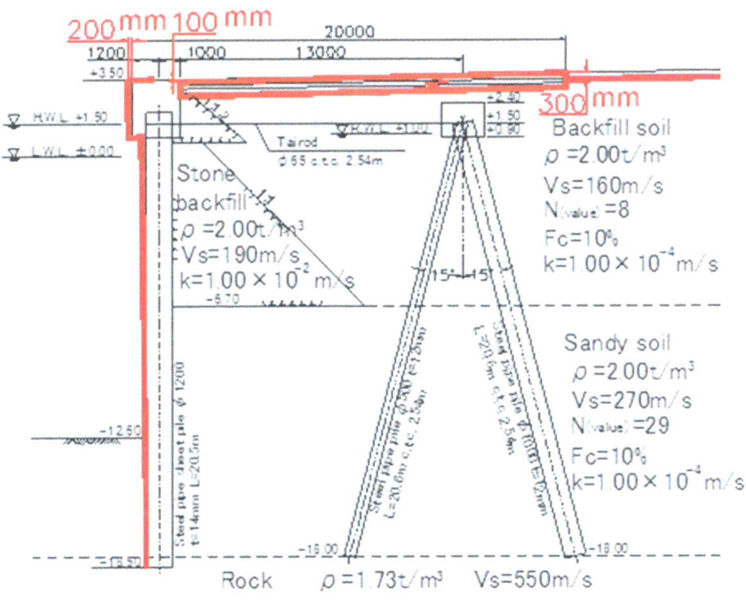

Fig. 1.18 Damage to a sheet pile quay wall subject to earthquake motion of extremely long duration (2011 East Japan Earthquake (M = 9.0), Soma Port, No.2 Wharf (−12 m))

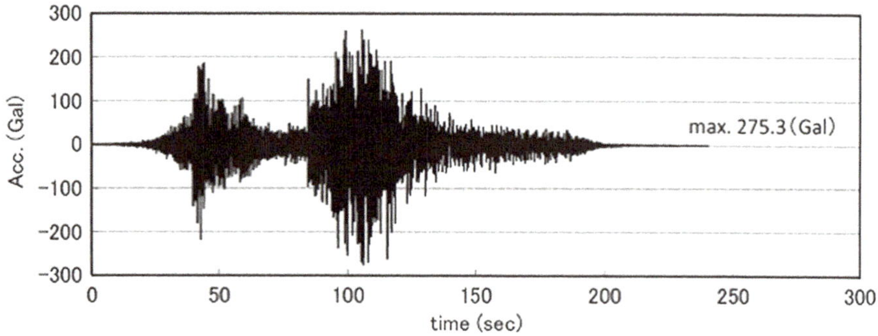

Fig. 1.19 Input earthquake motion at Soma Port during 2011 East Japan Earthquake

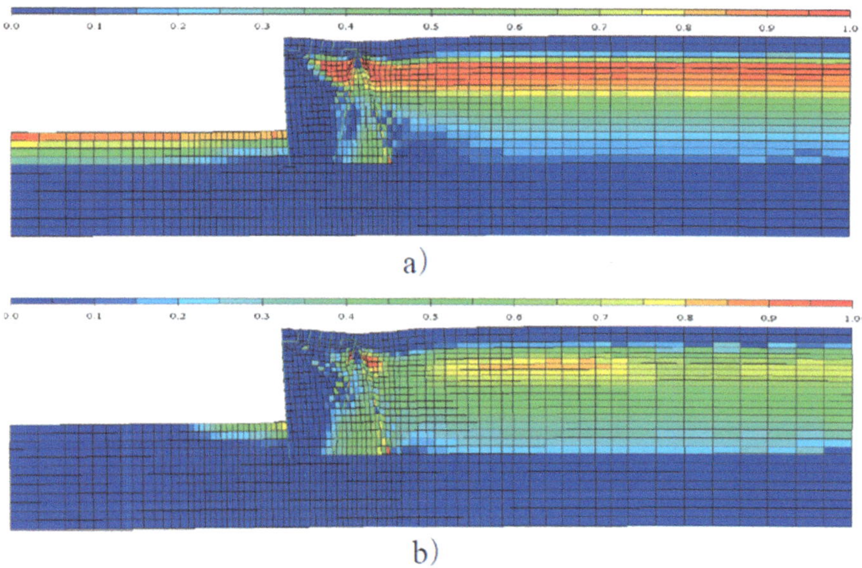

Fig. 1.20 Results of the effective stress analysis of a sheet pile quay wall (Displacement scale: enlarged by five times, Color contour: excess pore water pressure ratio). (**a**) t = 126 s (immediately after primary portion of shaking). (**b**) t = 240 s (partially dissipated state of excess pore water pressure)

gravity analysis, the dynamic effective stress analyses were performed in partially drained and undrained conditions using Cocktail glass model (Iai et al. 2011) for partially drained (Case 1) and undrained analyses (Case 6) and Multiple spring model (Iai et al. 1992) for undrained analysis (Case 5). For further details of the analyses, refer to (Tashiro et al. 2017). The computed results of the sheet pile quay wall in terms of global deformation and excess pore water pressure ratios are shown in Fig. 1.20. The upper half of backfill soil liquefied and the sheet pile wall displaced toward the sea, resulted in settlements at the ground surface inclining toward inland.

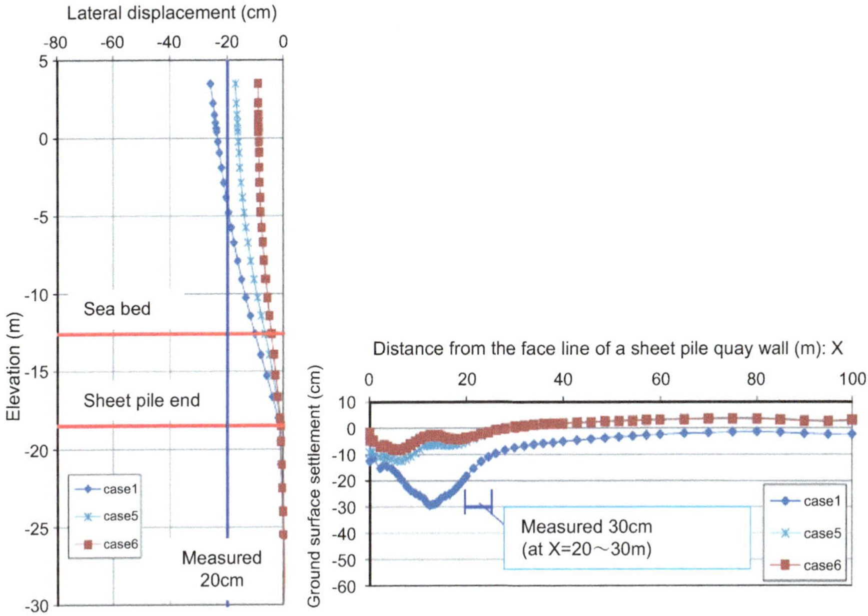

Fig. 1.21 Computed and measured lateral and vertical displacements of a sheet pile quay wall (*Right*: Lateral displacement of a sheet pile wall, *Left*: vertical displacement at the ground surface behind the face line of a sheet pile quay wall) (Case 1: Cocktail glass model (Iai et al. 2011), partial drainage, coefficient of permeability $k = 1 \times 10^{-4}$ m/s; Case 5: Multi spring model (Iai et al. 1992), undrained condition; Case 6: Cocktail glass model (Iai et al. 2011), undrained condition)

The computed and measured displacements at the sheet pile and the ground surface are shown in Fig.1.21. These results indicate that the analyses with partially drained and undrained conditions all capture the order of magnitude lateral displacements of the sheet pile quay wall. However, the partially drained analysis (Case-1) captures the order of magnitude vertical displacements at the ground surface much better than the undrained analyses (Cases-5 and 6).

Consequently, seismic performance of geotechnical structures subject to earthquake motions of long duration can be more precisely evaluated by partially drained analysis, allowing the dissipation and redistribution of excess pore water pressures during and after earthquakes.

1.8 Conclusions

Seismic damage to geotechnical structures is often associated with significant deformation of a soft or liquefiable soil deposit; hence, if the potential for liquefaction exists, implementing appropriate remediation measures against liquefaction is an effective approach to significantly improve seismic performance. The conventional simplified approach based on pseudo-static force equilibrium provides reasonable guidelines for designing remediation of liquefiable soils against moderate earthquake motions. However, there are cases where the conventional simplified approach is not readily applicable. For example, for intense earthquake motions, acceptable displacements and structural stresses become the primary design criteria, which are not evaluated by the conventional simplified approach. There are a number of cases where site specific constraints pose difficulties in implementing the liquefaction remediation measures as suggested by the conventional simplified approach. In order to meet these situations, performance-based approach provides a reasonable framework to work on for adequate liquefaction remediation. Case histories of performance and examples of design described in the paper may be useful to apply these conclusions in practice.

References

Egan JA, Hayden RF, Scheibel LL, Otus M, Serventi GM (1992) Seismic repair at Seventh Street Marine Terminal. Spec Geotech Publ 30., ASCE:867–878

Iai S (1994) Area of ground compaction against soil liquefaction. In: Proceedings of 13th international conference on Soil Mechanics and Foundation Engineering. New Delhi, pp 1075–1078

Iai S, Matsunaga Y (1990) Soil improvement area against liquefaction. In: Proceedings of 8th Japan earthquake engineering symposium, pp 867–872

Iai S, Matsunaga Y (1991) Mechanism of uplift of underground structures due to liquefaction. In: Proceedings of international symposium on Natural Disaster Reduction and Civil Engineering, Osaka, JSCE, pp 297–306

Iai S, Koizumi K, Kurata E (1987) Basic consideration for designing the area of the ground compaction as a remedial measure against liquefaction. Technical Note of Port and Harbour Research Institute, No. 590, pp 1–66 (in Japanese)

Iai S, Koizumi K, Kurata E (1991) Ground compaction area as a remedial measure against liquefaction. Tsuchi-to-Kiso, JSSMFE 39(2):35–40. (in Japanese)

Iai S, Matsunaga Y, Kameoka T (1992) Strain space plasticity model for cyclic mobility. Soils Found 32(2):1–15

Iai S, Matsunaga Y, Morita T, Miyata M, Sakurai H, Oishi H, Ogura H, Ando Y, Tanaka Y, Kato M (1994) Effects of remedial measures against liquefaction at 1993 Kushiro-Oki earthquake. In: Proceedings of 5th U.S.-Japan workshop on Earthquake Resistant Design of Lifeline Facilities and Countermeasures against Soil Liquefaction, NCEER-94-0026, pp 135–152

Iai S, Ichii K, Liu H, Morita T (1998) Effective stress analyses of port structures. In: Special issue on Geotechnical Aspects of the January 17 1995 Hyogoken-Nambu Earthquake, No. 2, Soils and Foundations, pp 97–114

Iai S, Tobita T, Ozutsumi O, Ueda K (2011) Dilatancy of granular materials in a strain space multiple mechanism model. Int J Numer Anal Methods Geomech 35(3):360–392

Inagaki H, Iai S, Sugano T, Yamazaki H, Inatomi T (1996) Performance of caisson type quay walls at Kobe Port. Soils Found Spec Issue, pp 119–136

McCullough NJ, Dickenson SE (1998) Estimation of seismically induced lateral deformations for anchored sheetpile bulkheads. In: Geotechnical Earthquake Engineering and Soil Dynamics III, Geotechnical Special Publication No.75, ASCE, pp 1095–1106

PIANC (2001) International Navigation Association: seismic design guidelines for port structures, Balkema, 474p

Port and Harbour Research Institute (1997) Handbook on liquefaction remediation of reclaimed land, Balkema, 312p

Tashiro S, Sumiya K, Sakakibara T, Kyoku D, Nishiyama S, Miyazawa T, Isayama T, Sato S, Mori A, Kaneko H, Shibata D, Sugihara K, Murakami K, Imono T, Murakami H, Kohama E, Ohya Y (2017) Analysis of quay walls during 2011 East Japan Earthquake. In: Iai (ed) Developments in earthquake geotechnics. Springer

Chapter 2
Evolution of Geotechnical Seismic Response Analysis from 1964 to 2015

W.D. Liam Finn

Abstract The evolution of dynamic response analysis and its role in geotechnical earthquake engineering primarily in North American practice is traced from 1964–2015: from the elementary visco-elastic analyses of site response of the 1960s to the more complex methods of analysis in use today that are based on sophisticated plasticity constitutive models. This review is limited to North American practice because the author is most familiar with this practice, having served it as researcher, professor and professional engineer over the past 50 years. A selection of case histories is presented below that demonstrate the wide-ranging capability of geotechnical engineers today of solving critical and important types of problems in practice since the widespread availability of reliable nonlinear methods of analysis.

2.1 Introduction

The evolution of dynamic response analysis and its role in geotechnical earthquake engineering primarily in North American practice is traced from 1964 to 2015: from the elementary visco-elastic analyses of site response of the 1960s to the more complex methods of analysis in use today that are based on sophisticated plasticity constitutive models. This review is limited to North American practice because the author is most familiar with this practice, having served it as researcher, professor and professional engineer over the past 50 years. A very good picture of the development of dynamic analysis in Japanese practice has been provided by Iai (1998), Iai and Ozutsumi (2005), and Iai et al. (2011, 2013). The research of Iai and his colleagues has resulted in the development of the program FLIP which is a major component of Japanese practice in geotechnical earthquake engineering today. The program can be accessed at http://flip.or.jp/en/flip_english.html.

A comprehensive review of European practice is not available. Potts (2003) in his Rankine Lecture has given a very interesting account of the use of finite element

W.D.L. Finn (✉)
University of British Columbia, Vancouver V6T 1Z4, Canada
e-mail: finn@civil.ubc.ca

S. Iai (ed.), *Developments in Earthquake Geotechnics*, Geotechnical, Geological and Earthquake Engineering 43, DOI 10.1007/978-3-319-62069-5_2

methods for static analysis in European practice. He gives some interesting and important insights into potential pitfalls with the use of such analysis and stresses the importance of understanding thoroughly the fundamental basis of any constitutive model to be used. His cautionary comments are also relevant to nonlinear dynamic analysis.

The first major contributions to seismic response analysis of earth structures were made by Ambraseys (1960a, b, c). He assumed that soil was a visco-elastic material and treated dams as 1-D and 2-D shear beams in his analyses. He demonstrated how the incoming motions were amplified throughout the dam, the contribution of the different modes of vibration of the dam to the global response and how the seismic coefficient varied with depth below the crest of the dam. Ambraseys studied dams in both wide and narrow rectangular valleys and showed that when the ratio of the width of the valley to the height of the dam was less than 3, the seismic response changed significantly. Despite the limitations of the visco-elastic model of soil behavior, analyses based on the model captured many of the important characteristics of seismic response and provided the starting point for some of the subsequent developments. Seed and Martin (1966) expanded the scope of Ambrasey's analyses to a variety of dam sizes and material properties and provided a comprehensive database for selecting appropriate values of the seismic coefficients required for the pseudo-static method of analysis. They also drew attention to the deficiencies in the seismic coefficient method should the materials in the dam lose strength during an earthquake.

Probably one of the more significant events which contributed to the rapid development of geotechnical earthquake engineering in the 1960s was the application of finite element methods to the analysis of embankment dams for the first time by Clough and Chopra in 1966. This was followed by the seismic analysis of slopes by Finn (1966a, b) and the analysis of central and sloping core dams by Finn and Khanna (1966) and Finn (1967).

A major improvement in analysis occurred in 1972 when Seed and his colleagues at the University of California in Berkeley developed the equivalent linear method of analysis for approximating non-linear behavior. This method was incorporated in the program SHAKE by Schnabel, Lysmer and Seed in 1972. The technique was extended to two dimensional finite element analysis by Idriss et al. (1973) in the program Quad-4. and by Lysmer et al. (1975) in the program FLUSH. These programs led to more realistic, total stress analyses of embankment dams and other earth structures under earthquake loading, especially under strong shaking and the equivalent linear constitutive model has remained a major factor in engineering practice to the present day.

While this program development was going on, the capability of testing soils under cyclic loading was also being developed. The cyclic triaxial test was developed by Seed and Lee (1966) and made possible the study of mechanics of liquefaction and seismically induced deformations. At around the same time, the resonant column test was developed for measuring dynamic shear modulus and damping at low strains by Drnevich (1967). In the early 1970s the use of the cyclic simple shear test was pioneered by Seed and Peacock (1971) and Finn et al. (1971).

Therefore by 1975 geotechnical engineers had many of the analytical and laboratory capabilities necessary for more realistic analysis of the seismic response of earth structures. These developments revolutionized the assessment of the safety of embankment dams during earthquakes and the seismic response of sites to seismic excitation.

These new methods were soon put to the test, when Seed et al. (1973, 1975a, b), undertook a comprehensive study of the liquefaction induced flow failure of the lower San Fernando Dam which occurred as a result of the San Fernando, California, earthquake of 1971. The analyses predicted that liquefaction would occur and that the dam would fail by undergoing very large deformations upstream during the earthquake. An extensive region under the upstream slope liquefied during the earthquake and the dam did develop very large deformations but just after, not during, the earthquake. This delayed failure was attributed by Seed (1979) to pore water pressure redistribution.

The equivalent linear method of analysis used in the study of the San Fernando Dam is a total stress analysis and therefore did not track the development and redistribution of seismic pore water pressures during shaking and could not take into account during the dynamic analysis the effects of pore water pressures on soil properties and dynamic response. These effects were taken into account after the analysis by relating the effects of stresses calculated during shaking to pore water pressures and shear strains in laboratory tests under equivalent cyclic loading. The San Fernando case history provided the stimulus for the development of effective stress methods of dynamic analysis which could track the development of pore water pressures during shaking and take the effects of pore water pressures on soil properties into account directly during analysis. The Martin et al. (1975) model for generating pore water pressures during earthquake loading based on the strain response of the soil paved the way for dynamic effective stress analysis.

The first non-linear dynamic effective stress analysis, based on the MFS pore water pressure model was the 1-D program DESRA-2 by Finn et al. (1976, 1996, 2000) and Lee and Finn (1978). The constitutive model is based on hysteretic response. The stress-train relation is hyperbolic, with Masing response on unloading. A comprehensive 2-D version of DESRA, the program TARA-3 was developed by Siddharthan and Finn (1982) and was developed further by Finn et al. (1986). TARA-3 was verified over a 3 year period by centrifuge tests conducted at Cambridge University in the UK on behalf of the United States Nuclear Regulatory Commission (Finn 1988). TARA-3 has the capability to conduct both static and dynamic analysis under total stress or effective stress conditions and can compute permanent deformations directly. It has the option of allowing pore water pressure dissipation and void ratio re-distribution during and after shaking. The program uses properties that are normally measured in connection with important engineering projects.

Since the mid 1980s, many non-linear effective stress programs have been developed, for the most part based on some version of plasticity theory. Detailed presentations of the constitutive models in these programs may be found in Pande and Zienkiewicz (1982), Finn (1988), and Arulanandan and Scott (1994). These programs are mathematically and analytically quite powerful but some use properties that are related to the mechanics of implementing the model rather than characterizing the soil. These parameters usually have to be found by adjusting them to match global response in an element test.

The estimation of large liquefaction displacements is a very important part of assessing the consequences of liquefaction in embankment dams. Evaluating large displacements means that small strain theory is inadequate. Finn and Yogendrakumar (1989) developed the program TARA-3FL to track the post-liquefaction deformations using a Lagrangian formulation. A similar technique is employed in the now widely used program FLAC (Itasca 1996).

One of the most significant developments in fostering the increased use of dynamic analysis in geotechnical engineering, apart from the availability of some reliable constitutive models, is the availability of the well-maintained and documented platform, FLAC (Itasca 1996), for incorporating constitutive models. More recently the program PLAXIS (2001) has become available. The analytical frameworks in FLAC and PLAXIS are different; FLAC is based on an explicit formulation and requires very small time steps, while PLAXIS is based on an implicit formulation. As a result FLAC is much slower than PLAXIS.

A selection of case histories is presented below that demonstrate the wide-ranging capability of geotechnical engineers today of solving critical and important types of problems in practice since the widespread availability of reliable nonlinear methods of analysis.

2.2 Performance Based Design

Performance based design (PBD) is based on tolerable displacement criteria and has become part of practice in geotechnical earthquake engineering. It has been widely used for developing remediation strategies for embankment dams with foundations susceptible to liquefaction under design seismic loadings. There are two crucial requirements for implementing PBD: acceptable performance criteria and a reliable method of analysis. For embankment dams the criterion of acceptable performance is specified by tolerable displacements, usually of the crest. A nonlinear analysis is essential because soil behaves as a nonlinear solid under strong shaking. If significant seismic pore water pressures are developed during shaking, the analysis must be based on effective stresses. Nonlinear dynamic effective stress analysis has many forms and its safe use requires a good technical understanding of the mechanics of the constitutive model selected for use and knowledge of its validation history based on element test data, centrifuge test data and any evidence from case histories. It also requires an adequate understanding of how the computation procedure works. The application of PBD to an embankment dam is illustrated by the process of selecting and designing remediation measures for Sardis Dam.

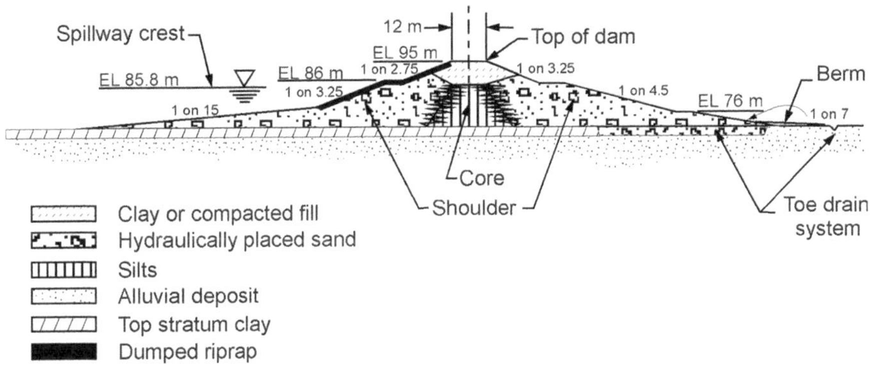

Fig. 2.1 Cross-section of Sardis dam

2.2.1 *Sardis Dam*

Sardis Dam is a hydraulic fill structure located in north-western Mississippi within the zone of influence of the New Madrid seismic zone. A cross-section of the dam is shown in Fig. 2.1.

In situ investigations revealed zones with the potential for liquefaction or significant strength loss which could threaten the upstream stability of the dam; the hydraulically-placed silt core, and a discontinuous layer of weak silty clay or clayey silt ranging in thickness from 1.5 to 4.5 m located in the top stratum clay beneath the upstream slope over a 310 m long portion of the dam. The upper 3 m of sand shell along the lower portion of the upstream slope was also identified as having a potential for liquefaction. However, loss of strength in this zone has a relatively minor effect on the stability of the dam. Stability analyses showed that, although the silt core might liquefy along the entire length of dam, the factor of safety with respect to upstream stability of the dam would still be adequate except in areas where the weak layer of silty clay occurred beneath the upstream slope within 75 m of the centerline. The results of these investigations indicated that remedial measures were necessary to improve the stability of the upstream slope of Sardis Dam during seismic loading.

The studies began in 1972 and the remediation was completed in 1995. During these years, the process for evaluating the seismic safety of embankment dams evolved, driven by advances in earthquake engineering, site investigation techniques, laboratory testing and procedures for seismic response analysis. The last study phase which began in 1989 and incorporated much new technology, finally led to an innovative cost-effective, reliable method of remediation based on performance based design. In this section the aim is to demonstrate how the use of comprehensive analysis helped engineers make informed decisions on complex design decisions. An interesting feature of the Sardis case history is the efforts made to assess the potential reliability of a new method of analysis and to develop some confidence in its use in a situation where a failure would have serious consequences for public safety. Reliability of analysis and guidelines for conducting non-linear analyses similar to those used for Sardis dam are discussed in more detail in Sections 5.0 and 6.0.

Post-liquefaction flow deformation in Sardis Dam were analyzed usingTARA-3FL (Finn and Yogendrakumar 1989) which could cope with large strains and so made it possible to adopt performance displacement criteria for Sardis Dam. The displacement demands on Sardis were evaluated by analyses conducted on the unremediated cross-section of Sardis Dam (Finn 1990). The initial studies were conducted to determine the necessary size and average shearing resistance of a remediated zone and its optimum location to limit deformations to acceptable values in the most economical way. Later it was decided that the most reliable way of supplying the necessary resistance was by driving piles which would nail the upstream slope to the stable foundation soils. This necessitated modifying the program to include piles in the analysis. Some of the early studies on pile behavior were reported in Finn (1993). The full process of reaching the final design for the remediation of Sardis Dam is described in Finn et al. (1998).

To check the performance of TARA-3FL, the post-earthquake deformed shape of Sardis Dam was computed using the specified distribution of residual strengths in the weak layer with a minimum value of 17.5 kPa. The initial and final deformed shapes of the dam for this case are shown in Fig. 2.2. Note different horizontal and vertical scales. Very substantial vertical and horizontal deformations may be noted, together with intense shear straining in the weak thin layer. The static stability of the deformed shape was analyzed using the program UTEXAS2 (USACE 1989) which satisfies both moment and force equilibrium. The factor of safety was found to be close to 1.0. The critical slip surface exited the slope near the location suggested by the finite element analysis.

Many results of this type, for different assumptions about the residual strengths, suggest that the TARA-3FL analysis does indeed achieve equilibrium positions even for large drops in strength due to liquefaction and associated large deformations (Finn 1990).

Analyses of this type can give the loss in freeboard associated with various factors of safety based on the original configuration of the dam. For Sardis Dam, the

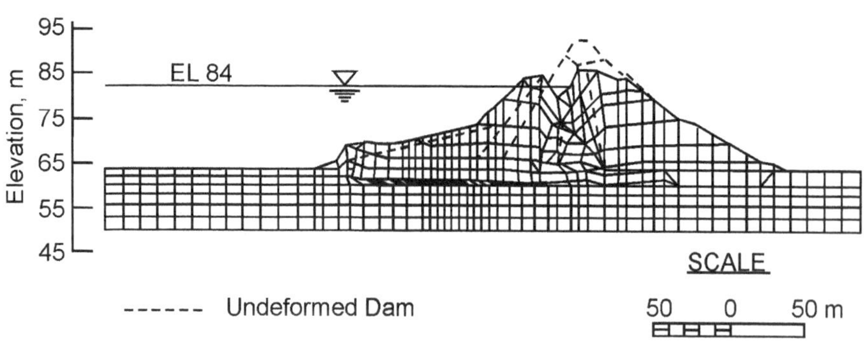

Fig. 2.2 Post-liquefaction deformed shape of Sardis dam. Note different vertical and horizontal scales

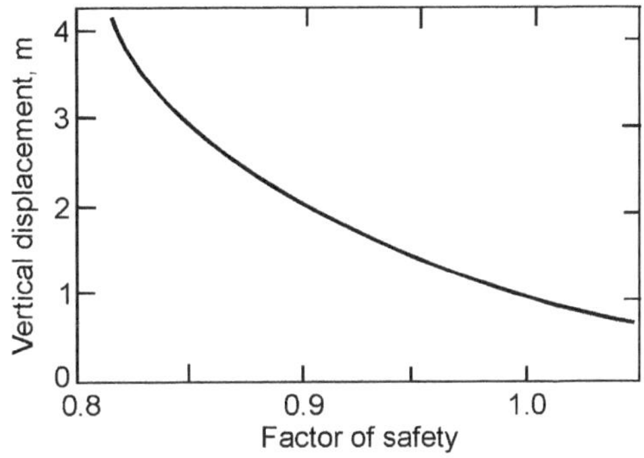

Fig. 2.3 Variation of loss of freeboard with factor of safety of undeformed dam

variation of vertical crest displacement with such factors of safety, corresponding to various values of residual strength, are shown in Fig. 2.3. For the first time, a designer could see the consequences of selecting a particular factor of safety for a dam. This information was helpful to engineers making the transition from a factor of safety based design to displacement based design.

2.2.2 *Remediation Studies*

The selection of remedial measures for Sardis Dam focused on ways of strengthening the weak liquefiable foundation layer. The general idea was to develop a plug of much stronger material across the weak layer which would restrain post-liquefaction deformations as shown in Fig. 2.4. The important properties of this remediation plug were its length, strength and location. Two kinds of analyses were conducted to define the properties of the remediation plug; conventional stability analyses and deformation analyses using the program TARA-3FL.

Various methods of creating the plug were investigated. Because of limitations on lowering the reservoir level, any remediation would have to be placed under water. In these circumstances, nailing the upstream slope to the stable foundation layer by piles was found to be the most cost-effective and reliable method of remediation. The layout of the piles is shown in elevation and in plan in Figs. 2.5 and 2.6 respectively.

2.2.3 *Load Transfer Mechanism in the Pile Remediated Section*

The displacement analysis of the remediated section giving the moments, shears and displacements of the piles are described in Finn et al. (1998). The distribution of

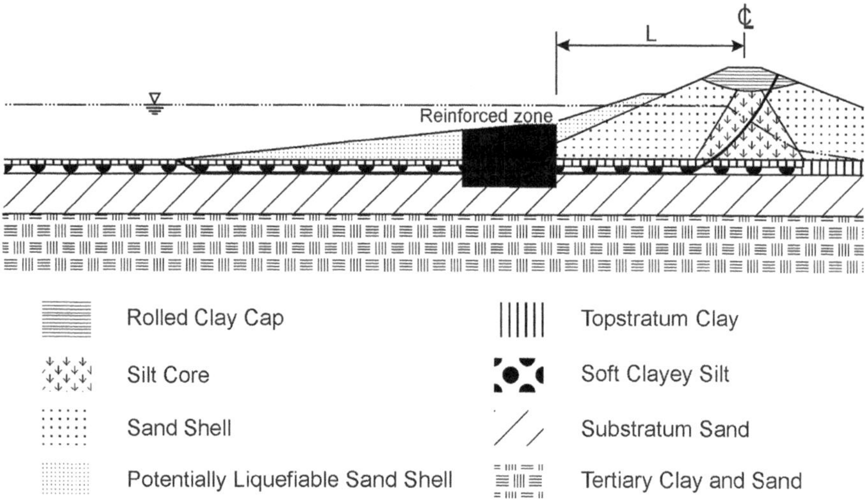

Fig. 2.4 Location of remediation plug to stabilize upstream slope of Sardis dam

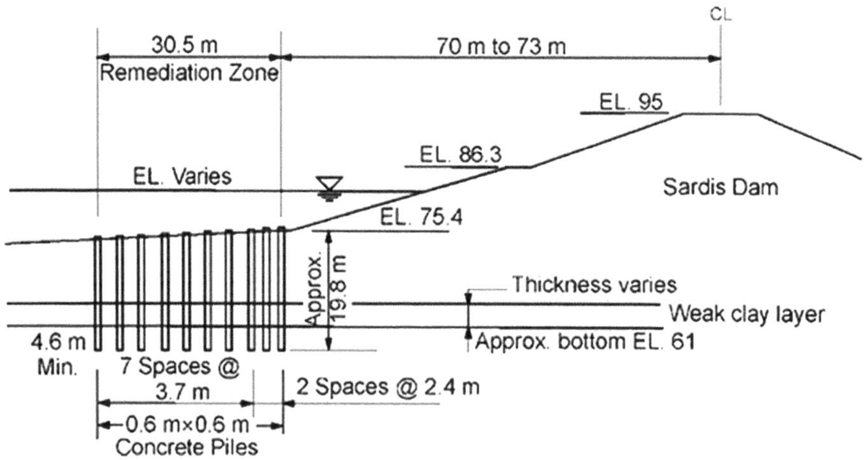

Fig. 2.5 Elevation of pile remediation of Sardis dam (After Stacy et al. 1994)

pile shears between the pile rows is shown in Fig. 2.7 and the distribution of moments in Fig. 2.8.

The distributions of post-liquefaction shears and moments controlled the placement of the piles. Since the piles were not capped, the only load transfer mechanism between the pile rows was the pressures exerted by the soil between them in response to the deflection of the pile rows. The designers were concerned by the possibility of progressive failure if the pile rows were too far apart. Therefore the spacing of the piles in the first three rows was kept small. The spacing was only one third of the spacing between the remaining rows of the piles. Maximum dynamic

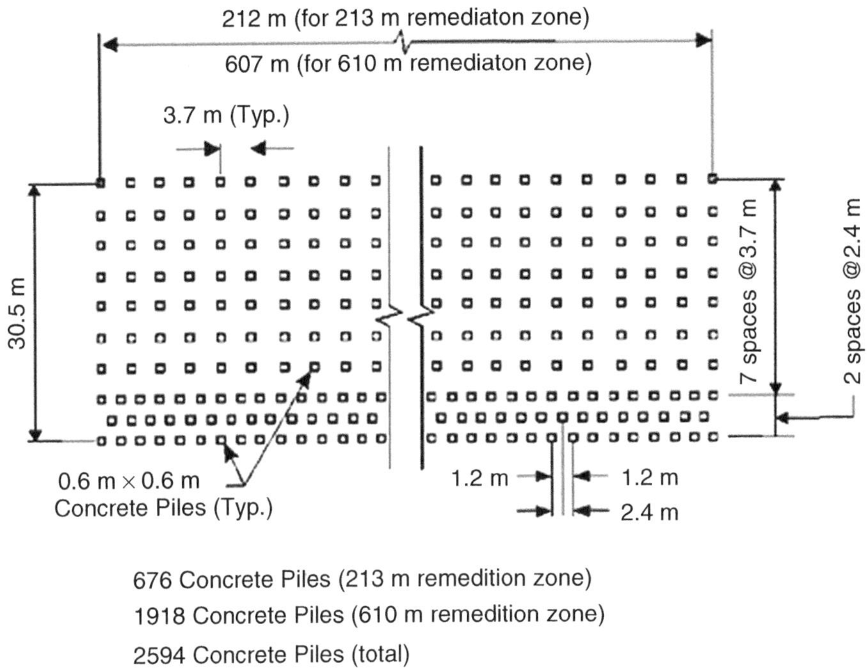

Fig. 2.6 Plan view of pile remediation of Sardis dam (After Stacy et al. 1994)

Fig. 2.7 Distribution of pile moments between pile rows

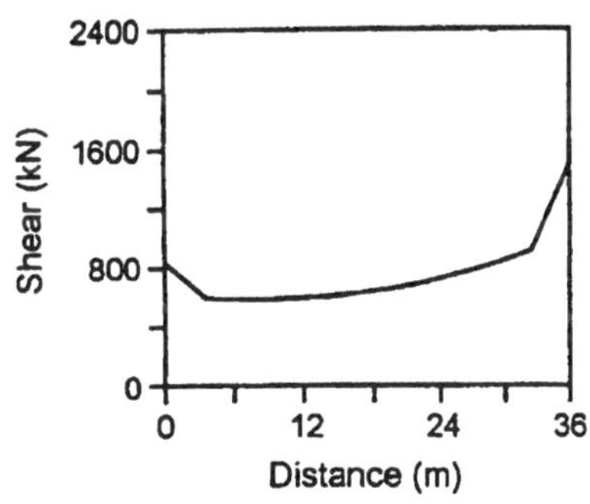

moment occurred in the leading row of upstream piles and is shown in Fig. 2.9. The design moment was the sum of 67 % of the peak dynamic moment and the moments due to the static thrust of the embankment along the weak plane.

The use of displacement criteria for evaluating the post-liquefaction stability of Sardis dam and determining the design and placement of the remedial measures resulted in huge savings over estimated costs for other procedures. This type of displacement analysis has become very much a part of practice today.

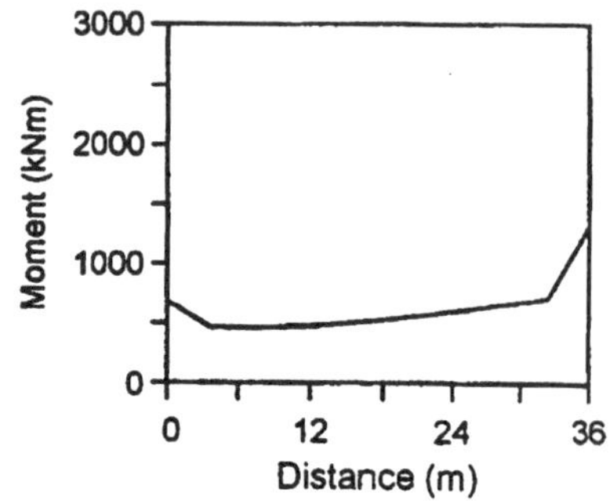

Fig. 2.8 Distribution of pile moments between pile rows

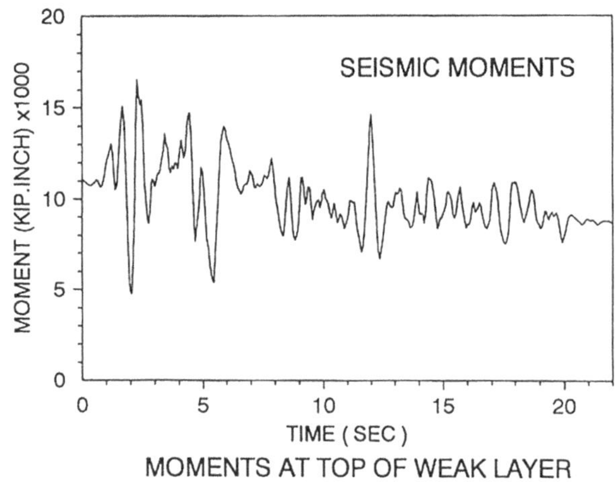

Fig. 2.9 Dynamic building moments in the leading upstream piles

2.3 Development of Seismic Stability Screening Relations

Parametric studies of the response of a particular type of soil structure such as a flood protection dike using nonlinear dynamic analysis give a data base for the potential development of simple screening methods for preliminary assessment of the behavior of a similar type of structure under strong shaking and criteria for prioritizing remediation. Two examples are given: preliminary screening of the seismic performance of slopes for residential development which has been approved by the Professional Engineers and Geologists of British Columbia (APEGBC) as good engineering practice and a criterion for prioritizing remediation intervention for flood protection dikes in Hokkaido, Japan.

2.3.1 Screening Seismic Performance of Slopes for Residential Development

The British Columbia Building Code (BCBC 2006) adopted the ground motions for seismic design proposed in the 2005 National Building Code for Canada (NBCC 1995). These ground motions have a probability of exceedance of 2% in 50 years (annual rate of exceedance of 1/2475), whereas the previously adopted ground motions for seismic design (NBCC 1995; BCBC 1998) had a probability of exceedance of 10% in 50 years (annual rate of exceedance of 1/475). The effect of this change was to greatly increase the number of slopes that could be considered unstable during an earthquake, based on current pseudo-static methods of analysis and therefore potentially not suitable for residential development.

APEGBC, with support from the Provincial Government, established a task force on seismic slope stability (TFSSS) to study this issue and to make recommendations for dealing with it. During its deliberations, the TFSSS reviewed current engineering practice and recent developments in seismic analysis of slopes and recommended analysis based on the concept of acceptable earthquake-induced slope displacements.

In BC, the most common method currently used to carry out seismic slope stability analysis is the pseudo-static limit equilibrium method. In this method, earthquake loading is represented by a constant horizontal force, $F = kW$, applied to the center of gravity of the potential sliding mass, where W is the weight of the sliding mass and k is a seismic coefficient. There was, however, no generally accepted method in BC practice for selecting seismic coefficients for slopes. From a limited survey of BC practice, the TFSSS found seismic coefficients in the range $0.5(\mathrm{PGA}) \leq k \leq 1.0(\mathrm{PGA})$, where PGA is the peak ground acceleration. The TFSSS reviewed recent developments in methods for assessing the potential performance of slopes during earthquakes and selected a new approach based on the work of Bray and Travasarou (2007).

Bray and Travasarou (2007) conducted approximately 55,000 Newmark-type slope displacement analyses involving eight different soil slope configurations, ten different yield accelerations for each slope configuration, and 688 different recorded ground motions from a database compiled by the Pacific Earthquake Engineering Research Center (PEER 2005). From a regression analysis of the resulting slope displacements, they developed an equation to estimate the magnitude of slope displacement due to shearing of the soil along a slip surface.

Bray and Travasarou's equation for estimating median slope displacements, D, greater than 1 cm, is expressed as:

$$\begin{aligned}\ln(\mathrm{D}) = &-1.10 - 2.83\ \ln(\mathrm{k_y}) - 0.333(\ln(\mathrm{k_y}))^2 + 0.566\ \ln(\mathrm{k_y})\ln\mathrm{S(T)} \\ &+ 3.04\ \ln\mathrm{S(T)} - 0.244(\ln(\mathrm{S(T)}))^2 + 1.5\mathrm{T_s} + 0.278(\mathrm{M} - 7)\end{aligned} \quad (2.1)$$

where k_y is the yield acceleration, T_s is the period of the potential sliding mass before the earthquake, S(T) is the spectral period of the potential sliding mass at the

Table 2.1 Slope displacements along a slip surface, estimated using Eq. 2.1

					S_a (1.5T_s) BCBC 2006				
Slope location	Height (m)	M	Ts (s)	PGA (g)	0.2	0.5	1.0	k_y	D (cm)
Nanaimo	30	7	0.35	0.50	1.0	0.69	0.35	0.17	13
Duncan	22	7	0.31	0.54	1.1	0.74	0.37	0.49	2
Victoria	13	7	0.23	0.61	1.2	0.82	0.38	0.52	2

Note: The applicable values of S were obtained by interpolation between the values of $S_a(1.5T_s)$ listed in BCBC (1998, 2006) as shown above

degraded period of the sliding mass as a result of shaking and is taken as $T = 1.5T_s$ and M is the moment magnitude of the design earthquake.

This equation is valid for periods, T_s, in the range $0.05\ s < T_s < 2.0\ s$, for values of yield coefficient, k_y, in the range $0.01 < k_y < 0.5$ and for $4.5 < M < 9.0$.

TFSSS recommended a limiting displacement of 15 cm be used in conjunction with Eq. 2.1. This displacement was based on experience with wood frame construction for residential homes.

As examples of the use of Eq. 2.1, slope displacements along slip surfaces were estimated for three actual soil slopes considered for residential development in Nanaimo, Duncan, and Victoria, BC. Table 2.1 shows that the estimated median slope displacements, D, were relatively small (2–13 cm). Using a tolerable slope displacement of 15 cm, these slopes were considered suitable for residential development. Using the prior method of analysis these three slopes were considered unsuitable for development.

The Bray and Travasarou (2007) method has three features that made it attractive to practitioner's in British Columbia as a screening tool for use with an acceptable displacement based performance criterion: it rests on an extensive data base, it is simple to use and it precludes the necessity of conducting non-linear displacement analysis except in special cases such as when liquefaction may be a problem.

2.3.2 *Prioritizing Flood Protection Dikes for Remediation*

The remediation of long-line structures such as flood protection dikes against the consequences of liquefaction is a very long and expensive process. Therefore it is necessary to identify the locations that are most at risk and remediate these first. A simple predictive relation based on easily measured characteristics of the dikes was developed for prioritizing the selection of locations for remediation of flood protection dikes along major rivers in Hokkaido, Japan. This project demonstrates clearly another potential role for nonlinear analysis.

The flood protection dikes along the Kushiro and Tokachi rivers suffered considerable damage during the 1993 Kushiro-oki earthquake off eastern Hokkaido, Japan. Damage included longitudinal and transverse cracks, slope failures and cave-ins. The more severely damaged dike sections were 6–8 m in height, and were constructed of compacted sand fill resting on a comparatively thick peat layer. The dikes were damaged at 18 locations for a total length of about 10 km along the Kushiro river. The severest damage occurred in Kushiro Marsh (Sasaki et al. 1993; Sasaki 1994a, b). Dike sections which failed were reconstructed after the foundation soils had been improved by the installation of sand compaction piles.

In 1994, the Nansei-oki earthquake occurred off the west coast of Hokkaido and caused failures of flood protection dikes along several river basins in western Hokkaido. After these earthquakes, the Hokkaido Development Bureau initiated a program of improving the diking systems. Because of the great length of dikes, they wished to develop a criterion for prioritizing the remediation work. One of the approaches considered was to use potential crest settlements as a criterion. The strategy to meet this objective was to conduct parametric studies on many dikes and develop an equation that linked dike characteristics to crest settlements for repeats of the 1993–1994 earthquakes. This equation would then be used to identify the dikes at greatest risk of failure and to prioritize interventions for remediation. The parametric studies were commissioned by the Hokkaido Development Bureau through the Advanced Construction Technology Center (ACTEC) in Tokyo. The analyses were conducted at the University of British Columbia (UBC), Vancouver, Canada and the resulting equation was validated by ACTEC using data from many damaged dikes not included in the parametric studies.

First simulations of some dike failures during the 1993 Kushiro-oki earthquake were conducted at UBC, based on soil data and input motions provided by ACTEC (1995). After these simulations proved satisfactory, a program of parametric studies were authorized to investigate how the crest settlements of dikes correlated with slope angle, dike height, and the thicknesses of non-liquefiable and liquefiable layers. A typical dike cross-section is shown in Fig. 2.10.

On the basis of the parametric studies, the displacement criterion, given by Eq. 2.2, was developed (Finn 1998), for symmetrical dikes with the most common side slopes, 1:2.5.

$$S = 0.01 \quad \exp \quad \left(0.922 \frac{H_D \ H_L}{H_{NL} H_{NL}}\right) \tag{2.2}$$

where the normalized settlement S is the crest settlement divided by H_D, the height of the dike; H_L and H_{NL} are the thicknesses of the liquefiable and non-liquefiable layers, respectively. This relationship is shown by the curve in Fig. 2.11 in terms of the non-dimensional variables S and β where $\beta = H_D.H_{NL}/H^2_{NL}$. The data points are the settlements computed from the parametric studies.

The measured crest settlements from a wide variety of dikes which underwent noticeable displacements during the Nansei-oki earthquake in 1994 were plotted by ACTEC engineers in Fig. 2.12 against those predicted by the black prediction curve.

The white square data points were plotted by the client after the correlation in Eq. 2.2 had been submitted. The 8 points close to the curve in Fig. 2.12 are for

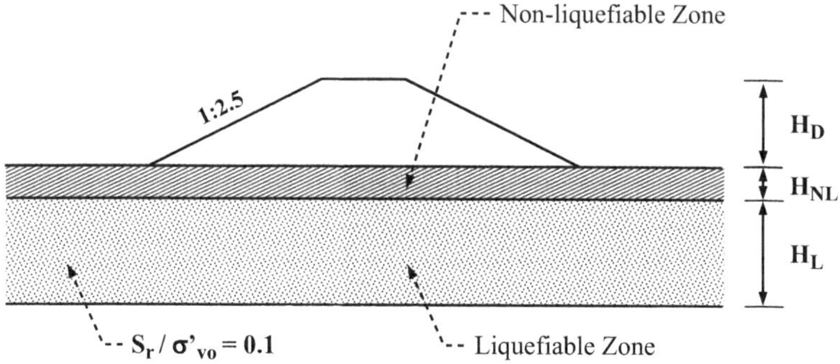

Fig. 2.10 Typical cross-section of dike used in parametric studies

Fig. 2.11 Comparison of computed settlements with the *black* prediction curve

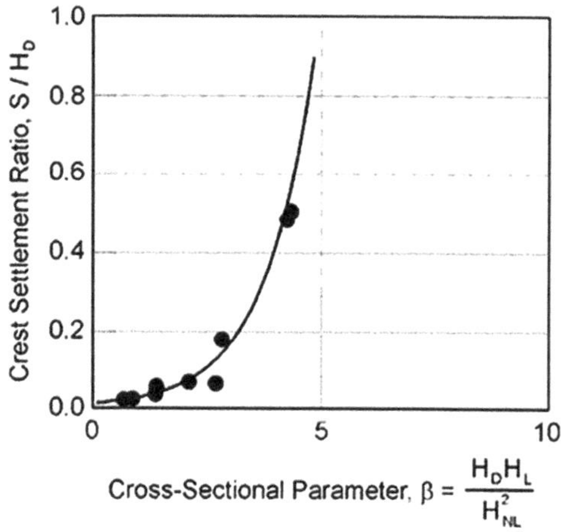

Fig. 2.12 Comparison of observed settlements for all slopes against computed settlements for 1:2.5 slopes (*solid curve*)

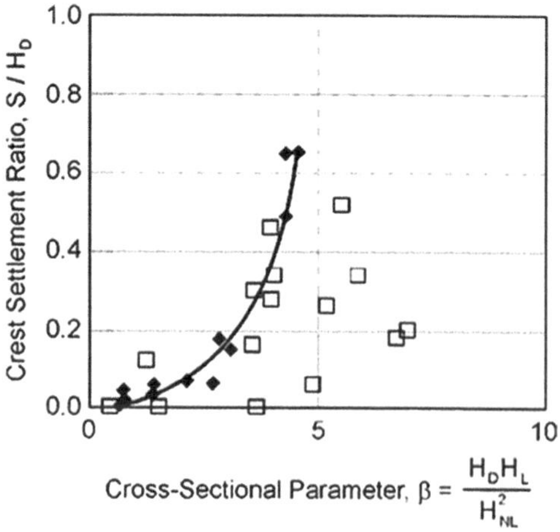

symmetrical dikes with side slopes of 1:2.5. The agreement is very good for these dikes. The remaining points significantly to the right the curve describe the response of dikes with uniform side slopes of 1:5, and unequal slopes, 1:5 and 1:10. These slopes require different correlations.

The predictive equation for dike damage, strongly validated by data from case histories, provides a convenient method for prioritizing dikes with side slopes 1:2.5 in the Hokkaido river basins for remediation. This is another example of the way in which an appropriate analysis can help the engineer formulate a cost effective response to a challenging design problem.

2.4 Soil-Structure Interaction

The seismic safety evaluation of the 44-year-old, 1.3 km long, immersed-tube George Massey Tunnel in Richmond, British Columbia (Fig. 2.13) demonstrates how useful a well-planned seismic response analysis can be in illustrating the consequences of liquefaction and in guiding effective approaches to remediation.

Design criteria were that the retrofitted tunnel should withstand both a 0.25 g M = 7.0 non-subduction earthquake and a 0.15 g M = 8.2 distant subduction earthquake without collapse or loss of life, but with damage to a repairable level including controllable water leakage (Yang et al. 2003).

Soil liquefaction, its consequences, and mitigation, were the key design challenges. Two-dimensional dynamic analyses using the program FLAC were the prime geotechnical analysis and design tool. In the analysis of a cross-section of the tunnel, the modulus of the tunnel section was reduced considerably to speed up the FLAC analyses but the mass was kept unchanged.

In the 2D FLAC analyses transverse and longitudinal sections were studied using total and effective stress constitutive models, UBCTOT (Beaty 2001; Beaty and Byrne 1999) and UBCSAND (Puebla et al. 1997; Beaty and Byrne 1998; Byrne

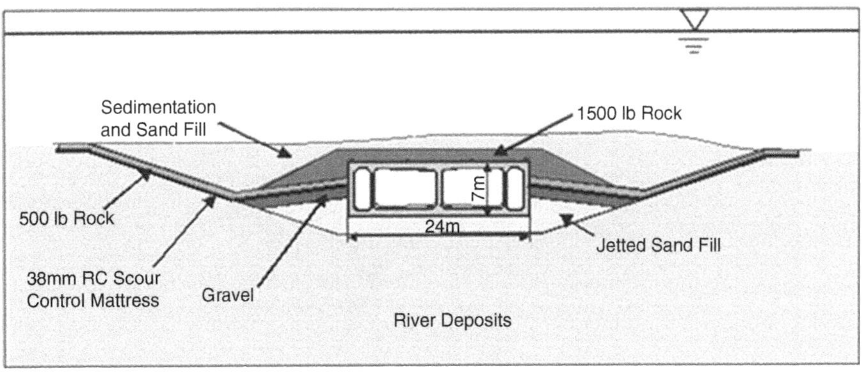

Fig. 2.13 Cross-section of Massey Tunnel

et al. 2004; Naesgaard et al. 2004), developed at the University of British Columbia. Dynamic shaking, liquefaction triggering, consequences of liquefaction and soil-structure interaction were addressed in each of the models. Analyses were carried out with and without retrofit measures. A centrifuge-testing program was carried out to check the numerical model. There was good agreement between numerical and centrifuge test results (Byrne et al. 2004). Field tests of seismic drains were also carried out.

Geotechnical retrofit measures proposed include: ground densification using vibro-replacement stone columns and drain columns on each side of the tunnel. Liquefaction was not mitigated below the tunnel and consequences of post-liquefaction settlement were allowed for in the geotechnical design. Other proposed retrofit measures that include increasing structural reinforcement and adding new emergency pumps will further eliminate seismic vulnerabilities, in particular the consequences of post-liquefaction settlement, by ensuring more uniform ductility, crack and leakage control throughout the tunnel. The combined geotechnical and structural retrofit design will meet the required design criteria.

2.4.1 Results of Dynamic Analyses

The following are some of the key observations from the numerical analyses:

- With level river bottom and no ground improvement the tunnel will heave in the range of 0.5–1.5 m following soil liquefaction (Fig. 2.14). The bulk of the movement occurs during strong shaking following liquefaction.
- With a sloping river bottom and no ground improvement the tunnel heaves less (0.2–0.3 m), but moves laterally significantly in the range of 1–2 m (Fig. 2.15).

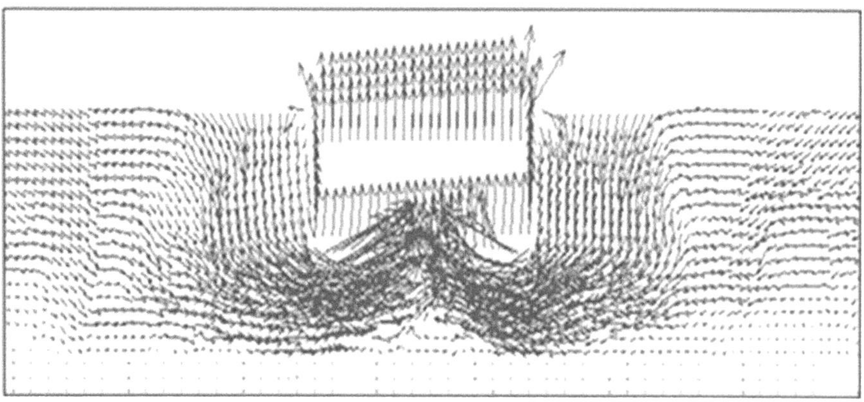

Fig. 2.14 Upward flotation of tunnel when surrounding soil liquefies

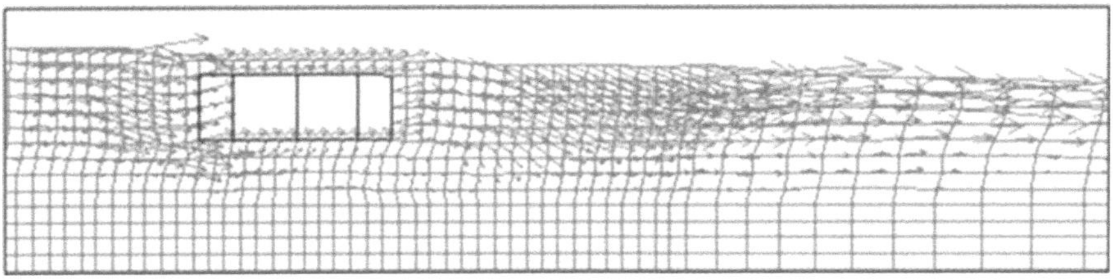

Fig. 2.15 Displacement vectors and displaced grid with sloping river channel

- If horizontal impervious silt layers with large lateral extent are present within the liquefiable sand unit then a slow ongoing heave of the tunnel may continue long after the end of earthquake shaking. This movement is additional to the movement that occurs during earthquake shaking and is due to high pressure groundwater generated by liquefaction in soil adjacent to the tunnel seeping under the tunnel and pushing it up. This movement could continue for many minutes or hours after the end of earthquake shaking. Installation of drainage columns within the proposed densified ground reduces this ongoing movement.
- Sheet piles with large sectional modulus were effective in reducing movement of soil under the tunnel and related tunnel heave. However the sheet piles were more costly than alternative measures. They are also more flexible than the densified ground, and therefore not efficient at reducing lateral movements if the river bed sloped.
- During initial strong shaking the tunnel experiences both compression and tensile stresses, however, with the onset of liquefaction the central river section of the tunnel has a net compressive stress. Lateral movements of the tunnel due to slope in the river bottom and due to differential settlement will give local bending stresses within the tunnel.
- Incoherence (traveling wave effects) and vertical input motion did not have a significant influence on tunnel response compared to the effects of ground liquefaction.
- The total stress model (UBCTOT) and effective stress model (UBCSAND) gave similar displacements during the period of dynamic shaking. However, only the effective stress model could emulate pore pressure migration effects (that could result in tunnel heave after the end of earthquake shaking).
- Based on the numerical analyses the lateral movements of the tunnel due to the design earthquake with the proposed final design remediation scheme should be less than 0.1–0.4 m. Vertical heave should be less than 0.2 m and settlement due to post-liquefaction consolidation should be less than 0.3 m. This includes movements during and following earthquake shaking.

Recommendations for final design for the tunnel are illustrated on Fig. 2.16. The design includes a 15–20 m width of vibro-replacement densification on either side of the tunnel with the outer two rows of columns designed to act as drains by being constructed with 2–5 mm diameter filter (drainage) gravel. Densification was not

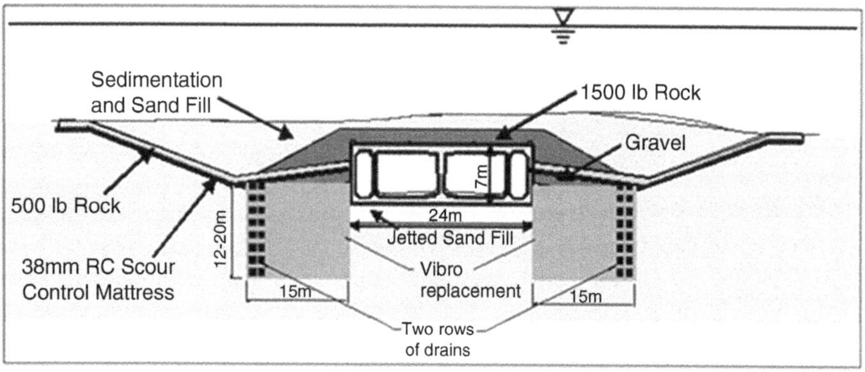

Fig. 2.16 Final remediation scheme with ground densification and drainage

deemed necessary adjacent to the approaches. However, one to two rows of seismic drains on 2 m centers were specified to assist in relieving high pore pressures. The on-land seismic drains are to be at least 200 mm in diameter and include a slotted pipe surrounded with 2–5 mm diameter drainage gravel.

2.5 Reliability of Nonlinear Analysis

Prior to about 1985, the equivalent linear constitutive model incorporated in the total stress programs SHAKE (Schnabel et al. 1972), QUAD-4 (Idriss et al. 1973) and FLUSH (Lysmer et al. 1975) was the industry standard for dynamic response analysis in geotechnical earthquake engineering. The need to calculate directly permanent displacements in earth structures as a result of earthquake shaking and to take into account the effects of seismic pore water pressures during analysis finally spurred a drive to non-linear effective stress analysis. However, despite 35 years of research and development, an industry standard constitutive model has not emerged. Well over a hundred constitutive models have been developed but many have never been used in practice and the remaining models have been used only sporadically. Probably the only widely used non-linear effective stress programs over the last 30 years are DESRA and the various modifications of it such as D-MOD. The advent of the FLAC and PLAXIS computing platforms will facilitate progress towards a standard. They make computing capacity more widely available in standard formats and provide the opportunity for the evaluation of various constitutive models in practice.

The response of earth structures to seismic loading is determined by dynamic analysis using computer programs. The reliability of the computer program used for dynamic analysis, especially for critical structures, is a crucial issue for a designer. The term reliability is not used in this context in its formal mathematical sense as defined, for instance, by a reliability index. Rather it describes the confidence of the designer or analyst that the program can help to bracket the likely range of behaviour of the structure under consideration and provide him with the data on which he can exercise his professional judgement to achieve a safe and cost effective design. How does the analyst develop this feeling of confidence, especially when using a program for the first time and in the absence of an industry standard? This requires a knowledge and understanding of the constitutive model, the computational procedures to implement the model, how a program was tested and validated and whether any verification is available from case histories in the field or in centrifuge tests.

Any constitutive model operates in an ideal environment, typically a continuum. Within this continuum with well–defined properties, all models can perform perfectly. One of the main problems in achieving a reliable response from analysis of an actual structure for a specified seismic input is the challenge of calibrating the model so that the continuum in which the model operates adequately represents the conditions in the actual structure. The difficulty in achieving satisfactory calibration is model dependent. Some programs such as DESRA-2 can operate directly using properties determined by in situ tests during site investigation but most models, especially the advanced plasticity models require laboratory tests. These tests are frequently conducted on reconstituted samples. Comprehensive research studies have defined the conditions to be satisfied by reconstituted test specimens to considered representative of field conditions. The main conditions are: the specimens should be formed by a process that mimics natural deposition in the field and have field density; the stress path in the calibration tests should be comparable to the dominant anticipated stress path in the field.

The process of sample formation has a very significant effect on the stress strain behavior of test specimens. The radically different stress-strain curves for specimens prepared by moist tamping, pluviation in air and pluviation in water are shown in Fig. 2.17 (Vaid et al. 1998). Stress-strain curves on undisturbed test specimens of Holocene sand retrieved from frozen ground and curves from reconstituted test specimens of the same sand formed by pluviation in water are shown in Fig. 2.18. For both types of specimens the results are almost identical. Evidence of this kind clearly indicates the care that should be taken to ensure that as far as possible test specimens should be formed to mimic formation in the field.

The effects of the stress paths followed in loading test specimens has very significant effects on stress-strain behavior also as shown in Fig. 2.19. These stress-strain curves were developed by different loading paths in hollow cylinder torsional cyclic shear tests and demonstrate clearly the sensitivity of stress-strain behavior to stress paths in loading.

A computer program for analysis of geotechnical structures has two main components; a constitutive model and a computational model. The constitutive model is verified using data from element tests, usually triaxial or simple shear tests. The stresses in these tests are defined and the verification does not require a full, dynamic, stress-strain analysis. If the constitutive model is based on material properties each of which can be independently measured, then agreement with element test data is validation that the model can simulate reliably the element

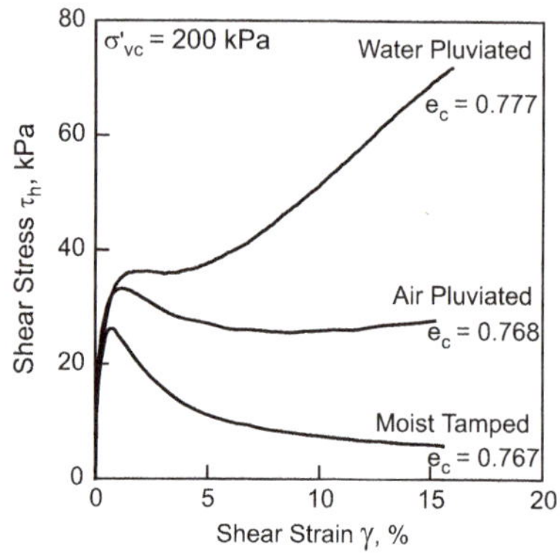

Fig. 2.17 The effect of method of specimen preparation on stress-strain response

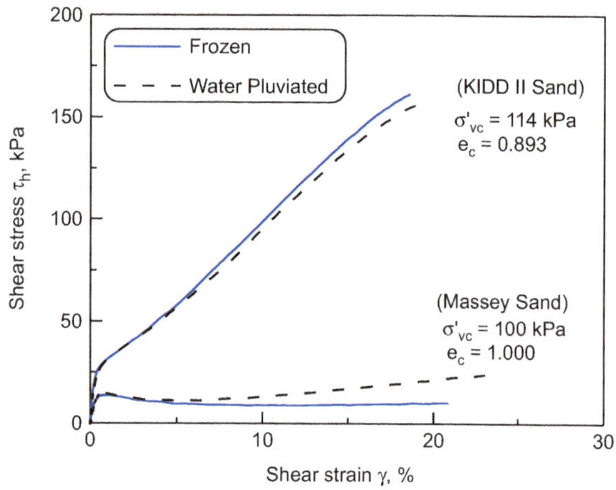

Fig. 2.18 Comparison of stress-strain curves from undisturbed frozen specimens and from water pluviated specimens

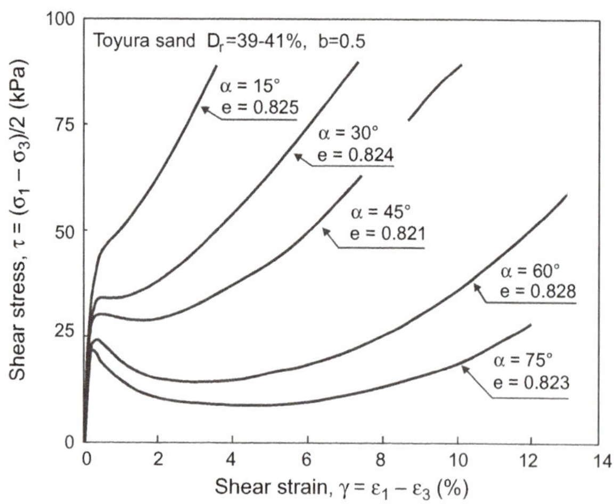

Fig. 2.19 Effect of stress path on stress-strain relations

response. If the model contains constants or parameters which cannot be individually measured but must be globally determined on the basis of the test data, then the validation is much weaker and there is uncertainty whether the global determination of the non-material constants from laboratory data, will be valid for the field application.

There have been three comprehensive prediction studies in North America to assess the capability and reliability of constitutive models and associated computer programs; Case Western Reserve University (Bianchini and Saada 1987), VELACS (Arulanandan and Scott 1994) and Turkey Flat (Shakal et al. 2006). These studies have provided great insight into the factors controlling the capability of constitutive models in practice and the uncertainties associated with computational models that need to be taken into account in practical applications. A key finding from these tests was the strong dependence of the constitutive models on the loading path used in calibration.

2.6 Tentative Guidelines for Analysis

A set of suggested guidelines for dynamic nonlinear analysis is given below

1. The analyst should have a good understanding of the mechanics of the constitutive model and the details of the computational procedures used to implement it.
2. The analyst should critically review the validation history of the constitutive model adopted for analysis, particularly its performance in blind prediction tests.
3. The analyst should test the capability of the model himself using available data bases of centrifuge test data, preferably from tests with significantly non-homogeneous stress states.
4. The parameters of the constitutive model should be determined in a manner that reflects field conditions and is compatible with the predominant type of stress path in the field.
5. The sensitivity of the response to variations in model parameters should be checked by parametric studies.
6. A confirmatory analysis should be conducted using a different constitutive model.
7. The review board on critical projects should include an expert on dynamic response analysis to advise on model selection, calibration and data interpretation.

2.7 Concluding Remarks

Non-linear dynamic effective stress analysis is an essential tool in seismic design in geotechnical earthquake engineering. It is crucial to success in performance based design especially when extensive remediation is involved. But it is a complex process and requires analytical skills and a very thorough knowledge of soil behavior.

The selection of an appropriate constitutive model for the job at hand requires knowledge of the past history of the model; how it fared in blind predictions tests, how it was validated and what its track record is in practice.

The calibration of the model for the job at hand, if it cannot be done on the basis of in situ data, needs to be conducted on test specimens that are representative of field conditions and for stress paths that reflect the dominant stress paths in the field application.

The designer needs to be aware of what the main factors controlling the results of the analysis are and make judgments as to whether any of them, such for instance, dilation, cannot be relied on in the field situation.

It is advisable for high consequence projects to have on the review board an expert on the important aspects of dynamic response analysis and to use this expert in the early stages of the analysis.

Non-linear analysis of critical, high consequence projects is a high risk endeavor to be conducted with the greatest vigilance.

References

ACTEC (1995) Used to refer to data supplied by Advanced Construction Centre, Tokyo, Japan

Ambraseys N (1960a) The seismic stability of earth dams. In: Proceedings of 2nd world conference on earthquake engineering, 2. pp 1345–1363

Ambraseys N (1960b) On the seismic behaviour of earth dams. In: Proceedings of 2nd world conference on earthquake engineering, 1. pp 331–356

Ambraseys N (1960c) On the shear response of a two-dimensional truncated wedge subjected to an arbitrary disturbance. Bull Seismol Soc Am 50:45–46

Arulanandan K, Scott R (ed) (1994) Verification of numerical procedures for the analysis of soil liquefaction problems. Balkema, ISBN 90 54 10 360 4

BCBC (BC Building Code) (1998, 2006) Published by the Office of Housing and Construction Standards, BC Ministry of Forests and Range

Beaty M (2001) A synthesized approach for estimating liquefaction-induced displacements of geotechnical structures. Ph.D. thesis, University of British Columbia

Beaty M, Byrne PM (1998) An effective stress model for predicting liquefaction behaviour of sand. In: Dakoulas P, Yegian M, Holtz R (eds) Geotechnical earthquake engineering and soil dynamics III. ASCE, Geotechnical Special Publication 75(1):766–777

Beaty M, Byrne PM (1999) A simulation of the Upper San Fernando dam using a synthesized approach. In: Proceedings of slope stability & landslides, 13th Annual Vancouver Geotechnical Society Symposium, pp 63–72

Bianchini G, Saada A (1987) Constitutive equations for granular non-cohesive soils. Proc. International Workshop on Constitutive Equations for Granular Non-Cohesive Soils, Cleveland, A.A. Balkema, 733p

Bray JD, Travasarou T (2007) Simplified procedure for estimating earthquake-induced deviatoric slope displacements. J Geotech Environ Eng ASCE 153(4):381–392

Byrne PM, Park SS, Beaty M, Sharp M, Gonzalez L, Abdoun T (2004) Numerical modeling of liquefaction and comparison with centrifuge tests. Can Geotech J 41:193–211

Drnevich VP (1967) Effect of strain on the dynamic properties of sand. Ph.D. thesis, University of Michigan, Ann Arbor, Michigan

Finn WDL (1966a) Static and dynamic stresses in slopes. In: Proceedings of first international congress on rock mechanics, Lisbon, II. pp 167–171

Finn WDL (1966b) Static and seismic analysis of slopes. Rock Mech Eng Geol IV(3):268–277. Springer, Wien/New York

Finn WDL (1967) Behaviour of earth dams during earthquakes. In: Proceedings, 9th International Congress on Large Dams, Istanbul, pp 355–367

Finn WDL (1988) Dynamic analysis in geotechnical earthquake engineering. In: Lawrence von Thun J (ed) Foundation Engineering Division State-of-the-art paper in Earthquake Engineering and Soil Dynamics II – recent advances in ground motion evaluation, Geotechnical Special Publication No. 20

Finn WDL (1990) Analysis of post-liquefaction deformations in soil structures. In: Editor Duncan JM (ed) Invited Paper, Proceedings, H. Bolton Seed Memorial Symposium, University of California, Berkeley, Bi-Tech Publishers, Vancouver, Canada, 2, pp 291–311, May 9–11

Finn WDL (1993) Seismic safety evaluation of embankment dams. In: Proceedings of international workshop on dam safety evaluation, 4. pp 91–135, Grindewald, Switzerland, 26–28 April

Finn WDL (1998) Seismic safety of embankment dams: developments in research and practice 1988–1998. In: Dakoulas P, Yegian M, Holtz RD (eds) Proceedings, geotechnical earthquake engineering and soil dynamics III, Geotechnical Special Publication, vol 75. ASCE, Reston., 2, pp 812–853

Finn WDL, Khanna J (1966) Dynamic response of earth dams. In: Proceedings of third symposium on earthquake engineering, Roorkee, pp 315–3224

Finn WDL, Yogendrakumar M (1989) TARA-3FL: a program for analysis of flow deformations in soil structures with liquefied zones. Soil Dynamics Group, Department of Civil Engineering, University of British Columbia, Vancouver

Finn WDL, Pickering DJ, Bransby PL (1971) Sand liquefaction in triaxial and simple shear tests. J Soil Mech ASCE 639–659

Finn WDL, Lee KW, Martin GR (1976) An effective stress model for liquefaction. J Geotech Eng Div ASCE 103(GT6):517–533. Proc. Paper 13008

Finn WDL, Yogendrakumar M, Yoshida N, Yoshida H (1986) TARA-3: a program for nonlinear static and dynamic effective stress analysis. Soil Dynamics Group, University of B.C., Vancouver

Finn WDL, Lee MK, Yoshida N (1996) DESRA-2C, dynamic effective stress response analysis of soil deposits with energy transmitting boundary, version 4. Department of Civil Engineering, University of British Columbia, Vancouver

Finn WDL, Ledbetter RH, Fleming RL Jr, Forrest TW, Stacy ST (1998) Stabilization of an earth dam using driven prestressed concrete piles. J Earthq Eng 2(2):173–195

Finn WDL, Yoshida N, Lee MK (2000) DESRA-2C, dynamic effective stress analysis of soil deposits with liquefaction beneath an impermeable layer with energy transmitting boundary. Department of Civil Engineering, University of British Columbia, Vancouver

Iai S (1998) Seismic analysis and performance of retaining structures. In: Dakoulas P, Yegian M, Holtz RD (eds) Proceedings, geotechnical earthquake engineering and soil dynamics III, Geotechnical Special Publication, vol 75. ASCE, Reston., 2, pp 1020–1044

Iai S, Ozutsumi O (2005) Yield and cyclic behaviour of a strain space multiple mechanism model for granular materials. Int J Numer Anal Methods Geomech 29(4):417–442

Iai S, Tobita T, Ozutsumi O, Ueda K (2011) Dilatancy of granular materials in a strain space multiple mechanism model. Int J Numer Anal Methods Geomech 35(3):360–392

Iai S, Ueda K, Tobita T, Ozutsumi O (2013) Finite strain formulation of a strain space multiple mechanism model for granular materials. Int J Numer Anal Methods Geomech 37 (9):1189–1212

Idriss IM, Lysmer J, Hwang R, Seed HB (1973) QUAD-4: a computer program for evaluating the seismic response of soil structures by variable damping finite element procedures. Report No. EERC 73-16, University of California at Berkeley

Itasca (1996) FLAC (Version 3.3). Itasca Consulting Group Inc., 708 South 3rd Street, Suite 310, Minneapolis, Minnesota 55415

Lee MK, Finn WDL (1978) DESRA-2, Dynamic effective stress response analysis of soil deposits with energy transmitting boundary including assessment of liquefaction potential, Soil Mechanics Series, vol 38. Department of Civil Engineering, University of British Columbia, Vancouver

Lysmer J, Udaka T, Tsai CF, Seed HB (1975) FLUSH – a computer program for approximate 3-D analysis of soil structure interaction problems. Report No. EERC 75-30, Earthquake Engineering Research Centre, University of California, Berkeley

Martin GR, Finn WDL, Seed HB (1975) Fundamentals of liquefaction under cyclic loading. J Geotech Eng Div ASCE 101(GT5):423–438

Naesgaard E, Yang D, Byrne P, Gohl B (2004) Numerical analyses for the seismic safety retrofit design of the immersed-Tube George Massey Tunnel. In: CD proceedings of 13th world conference on earthquake engineering, Vancouver. Paper 112

NBCC (National Building Code of Canada) (1995, 2005) Published by the National Research Council of Canada, Ottawa

Pacific Earthquake Engineering Research (PEER) Center Database (2005) [Web]

Pande GN, Zienkiewicz OC (eds) (1982) Soil mechanics – transient and cyclic loads. Wiley, Chichester

PLAXIS (2001) PLAXIS: a finite element code for soil and rock analysis, 2d version7, P.O. Box 572, 2600 AN Delft, The Netherlands

Potts DM (2003) 42nd rankine lecture: numerical analysis: a virtual dream or practical reality? Geotechnique 53(6):535–573

Puebla H, Byrne PM, Phillips R (1997) Analysis of CANLEX liquefaction embankments prototype and centrifuge models. Can Geotech J 34:641–657

Sasaki Y (1994a) River Dike Failure Due to the Kushiro-oki Earthquake of January 15, 1993. In: Proceedings of 4th US-Japan workshop on soil liquefaction: remedial treatment of potentially liquefiable soils. Public Works Research Institute, Tsukuba City, July 4–6

Sasaki Y (1994b) Embankment failure caused by the Kushiro-oki Earthquake of January 15, 1993. In: Special volume: performance of ground and soil structures during earthquakes, 13th international conference on soil mechanics and earthquake engineering, New Delhi, pp 61–68

Sasaki Y, Oshiki H, Nishikawa J (1993) Embankment failure caused by the Kushiro-oki Earthquake of January 15, 1993. In: Performance of ground and soil structures during earthquakes. JSSMFE, pp 61–68

Schnabel PL, Lysmer J, Seed HB (1972) SHAKE: a computer program for earthquake response analysis of horizontally layered sites. Report No. EERC 72–12, University of California, Berkeley, California

Seed HB (1979) Considerations in the earthquake-resistant design of earth and rockfill dams. Geotechnique 29(3):215–263

Seed HB, Lee KL (1966) Liquefaction of sands during cyclic loading. J Soil Mech Found Div ASCE 92(SM3):105–134. Proceedings Paper 4824

Seed HB, Martin GR (1966) The seismic coefficient of earth dam design. J Soil Mech Found Div ASCE 92(SM3):25–48. Proceedings Paper 4824

Seed HB, Peacock WH (1971) Test procedures for measuring soil liquefaction characteristics. J Soil Mech Found Eng Div ASCE, 97(SM8) Proceedings Paper 8330
Seed HB, Lee KL, Idriss IM, Makdisi FU (1973) Analyses of the slides in the San Fernando dams during the earthquake of February 9, 1971. Report No. EERC/73-2, University of California, Berkeley, (NTIS No. PB 223 402)
Seed HB, Lee KL, Idriss IM, Makdisi FU (1975a) The slides in the San Fernando dams during the earthquake of February 9, 1971. J Geotech Eng Div ASCE 101(GT7):651–689
Seed HB, Lee KL, Idriss IM, Makdisi FU (1975b) Dynamic analyses of the slide in the lower San Fernando Dam during the earthquake of February 9, 1971. J Geotech Eng Div ASCE 101 (GT9):889–911
Shakal A, Haddadi H, Real C (2006) Recorded data and preliminary review of predictions in the Turkey Flat blind prediction experiment for the September 28, 2004 Parkfield earthquake. In: Proceedings of SMIP06 seminar on utilization of strong-motion data, pp 137–152, Oakland
Siddharthan R, Finn WDL (1982) TARA-2: two dimensional nonlinear static and dynamic response analysis. Soil Dynamics Group, University of British Columbia, Vancouver
Stacy ST, Forrest TW, Ledbetter RH (1994) Sardis dam and others in the lower Mississippi Valley. In: Proceedings of 4th US-Japan workshop on liquefaction, Tsukuba, July
USACE (1989) Users guide: UTEXAS2 slope stability package, vol II. In: Theory by task group on slope stability, Instruction Report GL-87-1, Final Report, U.S. Army Corps of Engineers, Washington, DC
Vaid YP, Sivathayalan S, Stedman D (1998) Influence of specimen reconstituting method on the undrained response of sand. ASTM Geotech Test J
Yang D, Naesgaard E, Gohl B (2003) Geotechnical seismic retrofit design of George Massey Tunnel. In: Proceedings of 12th Pan-American conference on soil mechanics and geotechnical engineering, Cambridge, June 22–26, pp 2567–2574

Chapter 3
Past, Present, and Future Developments in Liquefaction Hazard Analysis

Steven L. Kramer

Abstract Liquefaction of soils has been widely recognized as an important cause of damage in many past earthquakes. Although recognized and named in the early 1950s (Mogami T, Kubo K, The behavior of soil during vibration. In: Proceedings of the 3rd international conference on soil mechanics and foundation engineering, vol 1, Zurich, pp 152–155, 1953; Terzaghi K, Variety of submarine slope failures. In: Proceedings of the 8th Texas conference on soils and foundation engineering, University of Texas, Austin, pp 1–41, 1956), liquefaction sprang to the attention of the geotechnical engineering profession in 1964 following large earthquakes in Niigata, Japan and Alaska, USA. Since that time, a great deal of research on soil liquefaction has been performed, particularly in Japan and the United states but also in other seismically active countries such as Canada, Chile, New Zealand, Taiwan, and Turkey. This research has led to breakthroughs in understanding of the basic mechanics of liquefiable soils, the development of practical, empirical procedures for evaluation of liquefaction potential, and the development of numerical procedures for site-specific analysis of liquefaction and its effects. This paper presents a brief and incomplete review of the history of liquefaction hazard evaluation, assesses its current status, and discusses future developments in this area.

3.1 Past Developments

Following the 1964 earthquakes, experimental investigations of the behavior of liquefiable soils were undertaken in a number of soil mechanics laboratories in both the U.S. and Japan. Early testing programs included static, or monotonic, as well as cyclic loading. An important and influential set of tests was performed in the late 1960s by Gonzalo Castro. Castro's tests (Castro 1969) showed that liquefiable soils would contract or dilate upon undrained monotonic shearing until their effective stresses reached a constant value, after which they would shear with constant

S.L. Kramer (✉)
University of Washington, Seattle, WA, USA
e-mail: kramer@u.washington.edu

S. Iai (ed.), *Developments in Earthquake Geotechnics*, Geotechnical, Geological and Earthquake Engineering 43, DOI 10.1007/978-3-319-62069-5_3

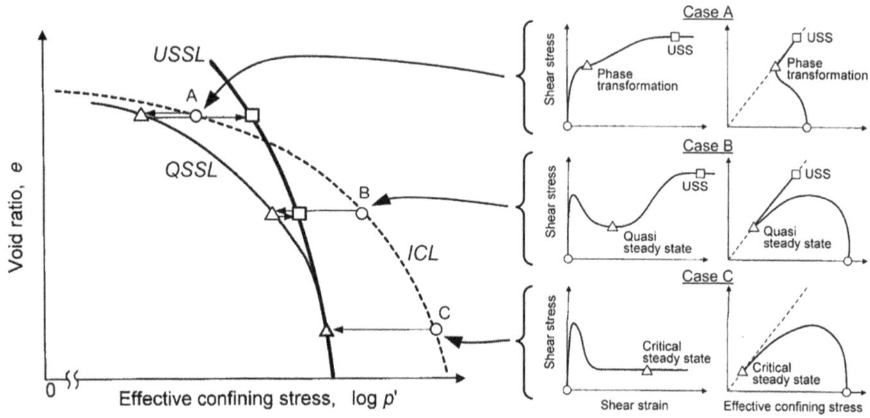

Fig. 3.1 Illustration of ultimate steady state line (*USSL*) and quasi-steady state line (*QSSL*) and their relationship to consolidation curve (*ICL*) and stress-strain behavior (After Yoshimine and Ishihara 1998)

resistance. These results led to the development of steady state concepts, which provided a basic framework for understanding the behavior of liquefiable soils under monotonic and cyclic loading, and also provided guidance for estimating the shear strength of liquefied soil. In those monotonic loading tests on loose test specimens, liquefaction was considered to have been triggered when instability at a threshold stress ratio was reached, after which the soil would strain-soften and flow toward its steady state strength.

In the 1980s, Ishihara and his students identified the existence of a quasi-steady state (Ishihara 1985) at which a contracting soil would begin to dilate at high stress ratios. Yoshimine and Ishihara (1998) improved the steady state framework by adding a quasi-steady state line (Fig. 3.1), which allows understanding of the ranges of contractive and dilative behavior that are observed at different soil densities and effective confining pressures.

While Castro was performing his monotonic loading tests, cyclic loading tests were also being performed on reconstituted test specimens created by pouring or tamping dry, moist, or saturated clean sands into molds of known volume. After saturation and consolidation with no initial shear stress (to simulate level-ground conditions), the specimens were subjected to cyclic loading (Fig. 3.1). The test specimens were instrumented to allow measurement of porewater pressure as well as stress and strain, and stress-controlled loading was applied until the sample liquefied, as evidenced by the development of 100% pore pressure ratio, or "initial liquefaction." Such laboratory testing programs were extremely useful for identifying the soil compositional and environmental characteristics that most strongly affected level-ground liquefaction.

At that point, two different mechanisms had been identified and referred to as "liquefaction." One involved instability of the soil under a static shear stress that would lead the soil to develop large, flow-type deformations. The other involved the

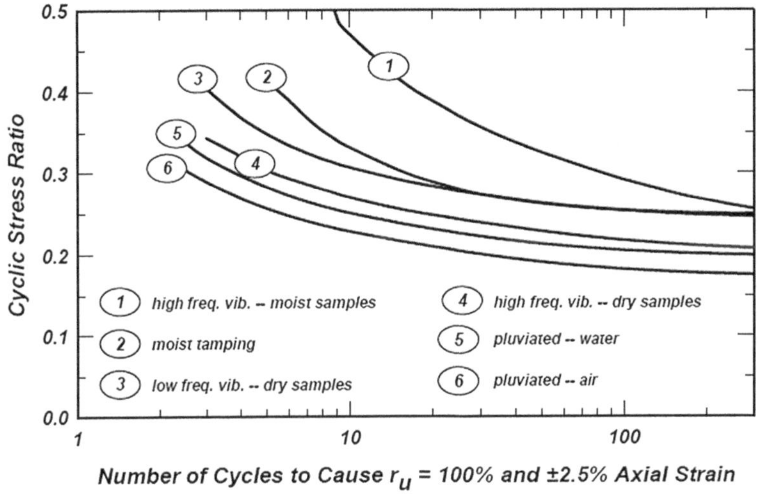

Fig. 3.2 Influence of soil fabric, as related to method of specimen preparation, on liquefaction resistance (After Idriss and Boulanger 2008)

incremental development of pore pressure high enough to cause at least momentary instances of zero effective stress under cyclic loading with no static shear stress. The different uses of the term caused some degree of confusion among U.S. practitioners and led to a period of debate between the Harvard and Berkeley schools of thought regarding the nature of liquefaction.

Liquefaction researchers came to realize, however, that laboratory test specimens reconstituted using different procedures had different soil fabrics that influenced their resistance to the triggering of liquefaction (Fig. 3.2). This resulted in a shift away from laboratory-based procedures to field case history-based procedures for evaluation of liquefaction potential. The case history-based approach relies on documentation of the occurrence, or non-occurrence, of liquefaction at sites subjected to strong earthquake shaking. The documentation includes a measure of earthquake loading at the case history site, typically expressed in terms of a cyclic stress ratio, and a measure of soil density, typically expressed in terms of penetration resistance. Because the types of site response analyses required to estimate loading was difficult at the time, Seed and Idriss (1971) developed a semi-empirical "simplified method" for evaluation of liquefaction potential that remains in use to this day. The simplified procedure underwent several refinements and has been shown to provide a useful and appropriately conservative indication of liquefaction potential when tested against actual liquefaction case histories.

In the 1980 and 1990s, attention shifted toward prediction of the effects of liquefaction. Laboratory investigations had given an indication of how cyclic shear strain amplitudes increased after initiation of liquefaction, and how volumetric strains developed upon dissipation of excess pore pressures. Early procedures for estimating lateral spreading displacements (e.g., Bartlett and Youd 1992) were almost purely empirical but subsequent procedures made use of both laboratory data and lateral spreading case histories. Later, semi-empirical procedures that more accurately represented soil behavior by tying potential shear strains to laboratory data were developed. Most of these procedures (Zhang et al. 2002; Idriss and Boulanger 2008) rely on shear strains levels from the experimental work of Ishihara and Yoshimine (1992).

3.2 Recent Developments

Liquefaction potential is currently evaluated using case history-based cyclic stress procedures, most commonly using updates of the simplified method. A number of researchers and research groups developed their own case history databases, however, and each contained different case histories and different interpretations of common case histories, so a number of different procedures developed over the years. The procedures differed primarily in the manner in which various adjustment factors, such as magnitude scaling factor, depth reduction factor, overburden correction, fines correction, etc., were formulated. Since its inception in the late 1970s, a series of incremental improvements have been made based on additional case history data, experimental data, and improved understanding of the mechanics of liquefaction. The basic procedure, however, remains the same.

Particularly significant developments in empirical liquefaction hazard evaluation included the use of cyclic shear strain (Dobry et al. 1982) and dissipated energy as measures of loading, and the use of CPT resistance (Robertson and Campanella 1985) and shear wave velocity (Andrus and Stokoe 2000) as indicators of soil density and resistance to liquefaction. In the United States, a general consensus on appropriate procedures was developed in the mid-1990s through a workshop organized by Prof. T. Leslie Youd and sponsored by the National Science Foundation (Youd et al. 2001). Since the time of the workshop and subsequent report, however, several earthquakes producing liquefaction have occurred and several investigators have developed newer approaches to liquefaction potential evaluation. An important and comprehensive update on liquefaction hazard evaluation was published in the form of a monograph by Idriss and Boulanger (2008). About 5 years ago, the developer of one of the prominent approaches publically challenged the accuracy of another prominent approach. The ensuing controversy has led to careful review and re-examination of existing procedures and the data upon which they are based. The report of a U.S. National Research Council committee studying the current state of liquefaction knowledge was recently published (NRC 2016).

Improved understanding of the mechanics of liquefiable soils has also led to improvements in the numerical modeling of liquefiable soil profiles. While nonlinear, effective stress analyses were developed as early as the late 1970s by Prof. W.D.L. Finn in Canada and Prof. K. Ishihara in Japan, early soil models degraded effective stress (and, hence, stiffness) monotonically, leading to the

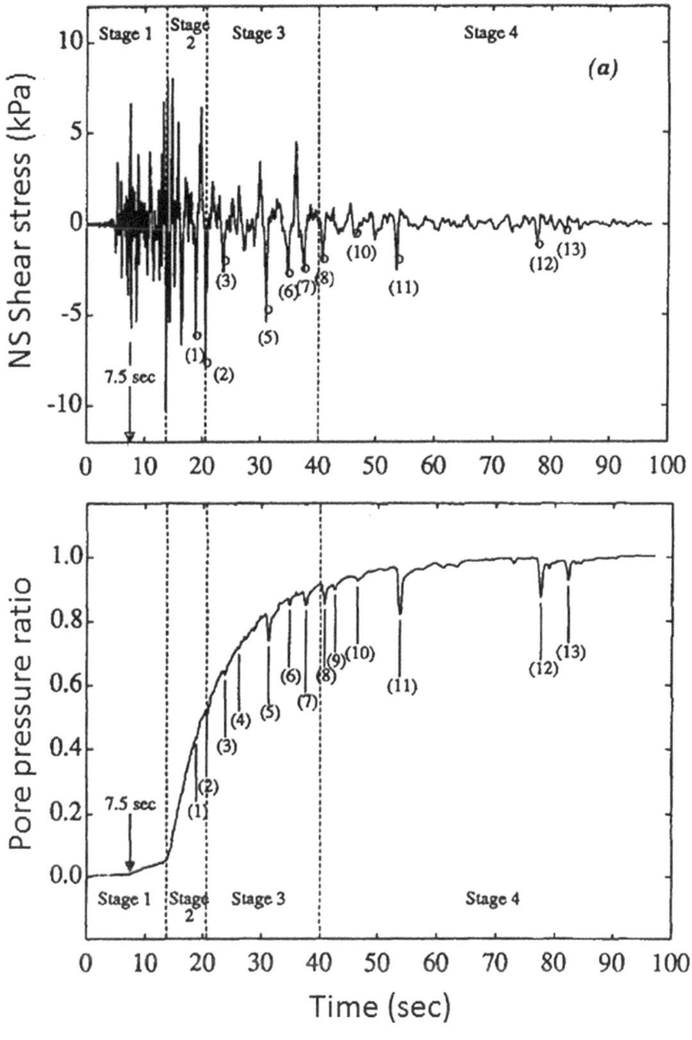

Fig. 3.3 Dilation-induced acceleration pulses coincident with momentary drops in pore pressure (After Zeghal and Elgamal 1994)

development of extremely soft zones of liquefied soil that would effectively "base isolate" overlying soils. With Ishihara's recognition of phase transformation behavior, constitutive models were developed to capture alternating sequences of contractive and dilative behavior that explained the development of sharp pulses of acceleration that were observed to develop even after liquefaction had been triggered. Zeghal and Elgamal (1994) showed that higher-frequency acceleration pulses observed in a ground motion recorded at the Wildlife vertical array in California were associated with dilation-induced drops in pore pressure (Fig. 3.3). These episodes of dilation-induced stiffening are also recognized as having a significant effect on the development of permanent deformations under sloping ground conditions. A number of stress-deformation programs, notably FLIP, OpenSees, and FLAC, have constitutive models capable of representing this important aspect of liquefiable soil behavior. These programs are proving to be quite useful for analysis of many of the complex boundary value problems that geotechnical earthquake engineers encounter in practice.

The potential for pore pressure redistribution during and following earthquake shaking has been recognized for many years; indeed, it was cited as a factor in the failure of Lower San Fernando Dam in the 1971 San Fernando earthquake. A novel set of shaking table experiments by Prof. T. Kokusho (1999, 2000), however, vividly illustrated its particular importance at sites where layers of lenses of impermeable soil may exist. Subsequent research has confirmed this behavior and its significance with respect to flow liquefaction and lateral spreading. Challenges associated with void redistribution, and identification/characterization of soil profiles where its effects are expected to be significant, are likely to be an important focus of future research on soil liquefaction.

Other advances have been made in the manner in which liquefaction hazards are expressed. In the early years, liquefaction hazards were expressed deterministically with respect to some earthquake scenario; a specific earthquake (i.e., an event having a particular magnitude occurring at a particular distance) was assumed, a median ground motion was estimated, and a deterministic factor of safety against triggering is computed. Since the 1970s, ground motions have generally been computed by probabilistic seismic hazard analysis (PSHA). These ground motions are based on all expected magnitudes occurring at all plausible source-site distances. The resulting hazard curves, which were underlain by distributions of contributing magnitudes, are typically used to select a *PGA* value at a single return period (usually 475 or 2,475 years) upon which a deterministic factor of safety against triggering was computed. Recognizing that liquefaction could also be triggered by more frequent, weaker motions or by more rare stronger motions, and explicitly accounting for the uncertainty in liquefaction triggering given some *PGA* level, the idea of fully probabilistic liquefaction hazard analysis (PLHA) was developed. Kramer and Mayfield (2007) considered all levels of shaking with their underlying magnitude distributions, and accounted for uncertainty in liquefaction resistance to develop a PLHA procedure that provides a complete, rational, and objective indication of liquefaction potential. The PLHA process has been extended to allow characterization of lateral spreading and post-liquefaction free-field settlement in terms of deformation hazard curves. These curves account for uncertainties in ground motions, liquefaction potential, and estimated ground deformations.

3.3 Future Developments

Looking to the future, there are certainly improvements that can be made to liquefaction hazard evaluation. These improvements relate to site characterization, case history documentation, and empirical and numerical procedures for evaluation of liquefaction potential and the effects of liquefaction.

Liquefaction hazards remain commonly evaluated by empirical methods and it is expected that they will continue to be for years to come. The databases upon which the most commonly used current empirical models are based, however, have gaps in the ranges of important predictive variables and have not been updated with case histories from recent earthquakes (even as far back as 1999). The recently released National Research Council report (NRC 2016) on liquefaction lists development of a public, empirical liquefaction case history database as its No. 1 recommendation for improvement of liquefaction hazard evaluation. In an effort to improve empirical procedures for evaluation of liquefaction potential, and ultimately the effects of liquefaction, a large collaborative research program called the Next Generation Liquefaction (NGL) project has been initiated by the Pacific Earthquake Engineering Research (PEER) Center. Led by Prof. J. Stewart of UCLA, the NGL program seeks to develop a large, well-vetted, community database of liquefaction case histories (Stewart et al. 2015). It will then support multiple teams of liquefaction researchers to develop empirical liquefaction potential models based on the database. The modelers will meet at regular intervals to compare and discuss interpretations of various case histories and the influence of various factors on liquefaction potential. Modeling teams will be required to use all of the data in the database or justify the reasons for exclusion of any data. Aspects of liquefaction behavior not adequately constrained by empirical data will be investigated by targeted experimental studies. The result of this exercise will be a set of liquefaction potential models based on a consistent set of data; the models are expected to have differences due to different interpretations of case histories and different scaling relationships, but to be generally consistent with each other – certainly more consistent than current procedures are.

The acquisition and documentation of case history data is being revolutionized by new technologies that have been adapted to geotechnical field reconnaissance efforts. These include the use of unmanned aerial vehicles (UAVs), LIDAR, digital photogrammetry, and satellite imagery (Rathje and Franke 2016). Such techniques offer the ability to visualize and measure the three-dimensional geometry of a failed slope or damaged structure more quickly and more accurately than has previously been possible. These measurements can give considerable insight into the complex deformation mechanisms that actually caused observed damage (Fig. 3.4). In the case of satellite imagery, they also often allow comparison and "before and after" geometries from which deformations can be measured.

It is generally understood and accepted that liquefaction potential is strongly affected by soil state, i.e., the in situ density and effective stresses in the soil at the time of an earthquake. In current practice, penetration resistance is used as a

Fig. 3.4 Liquefaction damage at Port of Iqueque in Chile after 2014 earthquake: (**a**) photo taken from ground, and (**b**) photo taken from UAV showing complex three-dimensional deformation pattern (After Rathje and Franke 2016)

substitute for relative density and overburden correction factors are used to account for effective stress. The development of more direct measures of soil density could improve evaluation of liquefaction potential. Nuclear magnetic resonance (NMR) probes are capable of measuring soil water content at some lateral distance beyond the walls of a borehole. For a saturated soil, the water content can be used to produce an accurate measure of soil density, particularly if even disturbed samples of the soil are obtained for measurement of specific gravity.

While most liquefaction case histories involve relatively uniform clean and non-plastic silty sands, there are a number of other liquefaction-susceptible soils for which evaluation of liquefaction potential and the effects of liquefaction are particularly challenging. These include interbedded soil deposits, intermediate plasticity silts and clayey silts, aged soils, and gravelly soils. Additional research is needed to improve liquefaction hazard evaluation in these types of soils.

Empirical procedures for evaluation of liquefaction potential and estimation of permanent ground displacement are almost exclusively based on data from free-field sites, i.e., sites unaffected by the presence of structures. These procedures are therefore of limited use in prediction of ground movements at the location of a structure because they do not account for the additional stresses imposed on the soil by the dynamic response of the structure. In recent years, centrifuge tests and observations of performance in actual earthquakes have provided data upon which new predictive relationship are being built (Dashti and Bray 2013; Bray and Dashti 2014).

As seismically active areas of the world have become more heavily instrumented, a number of ground motions have been recorded by instruments installed at sites containing liquefiable soils. Many of these motions show the effects of liquefaction, principally a sudden and persistent change in frequency content that occurs as the soil softens after initiation of liquefaction. By means of time-frequency analyses, the time at which liquefaction is triggered can be identified and used to compute the level of ground motion at which triggering occurred (Kramer et al. 2015). Knowing the penetration resistance of the critical layer of the soil profile, a data point that plots on, rather than only above or below, the CRR curve can be obtained. Such data can be used to supplement conventional, binary (yes/no) case history data. Knowledge of the timing of liquefaction can also be used to improve empirical estimates of the effects of liquefaction.

Further advances in numerical modeling of liquefaction problems are also expected in the future. These advances are likely to come in three primary areas: (1) improved modeling of the mechanical behavior of liquefiable soils, particularly soils subjected to significant static shear stresses, and improved modeling of fabric effects, (2) improved modeling of hydro-mechanical response, specifically the effects of pore pressure and void ratio redistribution that can lead to increased deformations and even delayed flow failure, and (3) improved modeling of flow sliding using techniques such as the Material Point Method or other procedures originating in computational fluid dynamics.

3.4 Summary

A great deal has been learned about soil liquefaction and its effects in the 52 years since it was so dramatically observed in Niigata and Alaska. Through 50 years of research in Japan, the U.S., Canada, New Zealand, and other countries, great advances in understanding of the basic mechanics of liquefaction have been made, and used to help guide the development of practical, useful empirical procedures. The quality of the empirical procedures has been further advanced by the cooperative interaction of researchers from these countries – it is now common to see teams of researchers from different countries collaborating on post-earthquake reconnaissance, empirical model development, validation of numerical tools, and other activities that improve our collective understanding of liquefaction hazards and their mitigation.

References

Andrus DR, Stokoe KH II (2000) Liquefaction resistance of soils from shear-wave velocity. J Geotech Geoenviron Eng ASCE 126(11):1015–1025

Bartlett SF, Youd TL (1992) Empirical analysis of horizontal ground displacement generated by liquefaction-induced lateral spread, Technical Report NCEER-92-0021. National Center for Earthquake Engineering Research, Buffalo

Bray JD, Dashti S (2014) Liquefaction induced building movement. Bull Earthq Eng Springer 2 (3):1129–1156

Castro G (1969) Liquefaction of sands, Harvard soil mechanics series, vol 87. Harvard University, Cambridge, MA

Dashti S, Bray JD (2013) Numerical simulation of building response on liquefiable sand. J Geotech Geoenviron Eng ASCE 139(8):1235–1249

Dobry R, Ladd RS, Yokel FY, Chung RM, Powell D (1982) Prediction of pore water pressure buildup and liquefaction of sands during earthquakes by the cyclic strain method, NBS building science series, vol 138. National Bureau of Standards, Gaithersburg

Idriss IM, Boulanger RW (2008) Soil liquefaction during earthquakes, Earthquake Engineering Research Institute MNO, vol 12. Earthquake Engineering Research Institute, Oakland

Ishihara, K. (1985). Stability of natural deposits during earthquakes. In: Proceedings of the eleventh international conference on soil mechanics and foundation engineering, vol. 1, pp 321–376

Ishihara K, Yoshimine M (1992) Evaluation of settlements in sand deposits following liquefaction during earthquakes. Soils Found 32(1):173–188

Kokusho T (1999) Formation of water film in liquefied sand and its effect on lateral spread. J Geotech Geoenviron Eng ASCE 125(10):817–826

Kokusho T (2000) Mechanism for water film generation and lateral flow in liquefied sand layer. Soils Found 40(5):99–111

Kramer SL, Mayfield RT (2007) The return period of liquefaction. J Geotech Geoenviron Eng ASCE 133(7):1–12

Kramer, S.L., Sideras, S.S., and Greenfield, M.W. (2015) The timing of liquefaction and its utility in liquefaction hazard evaluation. In: Proceedings of the 6th international conference on geotechnical earthquake engineering, Christchurch, November

Mogami, T. and Kubo, K. (1953). The behavior of soil during vibration. In: Proceedings of the 3rd international conference on soil mechanics and foundation engineering, vol 1, Zurich, pp 152–155

NRC (2016) State of the art and practice in the assessment of earthquake-induced soil liquefaction and its consequences. National Research Council Report, 286 pp

Rathje E, Franke K (2016) Remote sensing for geotechnical earthquake reconnaissance. Soil Dyn Earthq Eng 91:304–316

Robertson PK, Campanella RG (1985) Liquefaction potential of sands using the CPT. J Geotech Eng ASCE 111:384–403

Seed HB, Idriss IM (1971) Simplified procedure for evaluating soil liquefaction potential. J Soil Mech Found Div ASCE 107(SM9):1249–1274

Stewart, J.P., Kramer, S.L., Kwak, D.Y., Kayen, R.E., Tokimatsu, K., Bray, J.D., Cubrinovski, M., Sekigushi, T., Nakai, S., and Bozorgnia, Y. (2015) PEER-NGL project: open source global database and model development for the next-generation of liquefaction assessment procedures. In: Proceedings of the 6th international conference on geotechnical earthquake engineering, Christchurch, November

Terzaghi, K. (1956) Variety of submarine slope failures. In: Proceedings of the 8th Texas conference on soils and foundation engineering, University of Texas, Austin, pp 1–41

Yoshimine M, Ishihara K (1998) Flow potential of sand during liquefaction. Soils Found 38 (3):189–198

Youd TL, Idriss IM, Andrus RD, Arango I, Castro G, Christian JT, Dobry R, Finn WDL, Harder LF, Hynes ME, Ishihara K, Koester JP, Liao SSC, Marcuson WF, Martin GR, Mitchell JK, Moriwaki Y, Power MS, Robertson PK, Seed RB, Stokoe KH (2001) Liquefaction resistance of soils: summary report from the 1996 NCEER and 1998 NCEER/NSF workshops on evaluation of liquefaction resistance of soils. J Geotech Geoenviron Eng ASCE 127(10):817–833

Zeghal M, Elgamal A-W (1994) Analysis of site liquefaction using earthquake records. J Geotech Eng ASCE 120(6):996–1017

Zhang G, Robertson PK, Brachman RWI (2002) Estimating liquefaction-induced ground settlements from CPT for level ground. Can Geotech J 39:1168–1180

Chapter 4
Countermeasures Against Liquefaction

Junichi Koseki

Abstract Within the limited scope of the author's experiences on countermeasures against liquefaction, lattice walls by in-situ cement-mixing, cut-off sheet pile walls and dewatering are picked up. Technical issues on the effectiveness of these countermeasures and their evaluation procedures are discussed based on results from relevant 1-g and centrifuge model tests, numerical analyses and triaxial tests.

4.1 Introduction

The countermeasures against liquefaction that have been and will be adopted in Japan are classified into three major principles as shown in Fig. 4.1.

In this chapter, within the limited scope of the author's experiences on countermeasures against liquefaction as summarized below, attempts are made to pick up the following three topics that are related with the liquefaction countermeasures: lattice walls by in-situ cement-mixing, cut-off sheet pile walls, and dewatering.

1. 1987–1995: Researcher and senior researcher at Public Works Research Institute (PWRI), Ministry of Construction, Japan
2. 1994–2014: Institute of Industrial Science, the University of Tokyo, Japan (1994–1995: Cross-appointment made by PWRI)
3. 2014–current: Department of Civil Engineering, ditto.

4.2 Lattice Walls by In-Situ Cement-Mixing

4.2.1 Background

As schematically shown in Fig. 4.2, the lattice walls reduce cyclic shear deformation and thus prevent full liquefaction of untreated soil layers that are left

J. Koseki (✉)
Department of Civil Engineering, The University of Tokyo, Tokyo, Japan
e-mail: koseki@civil.t.u-tokyo.ac.jp

S. Iai (ed.), *Developments in Earthquake Geotechnics*, Geotechnical, Geological and Earthquake Engineering 43, DOI 10.1007/978-3-319-62069-5_4

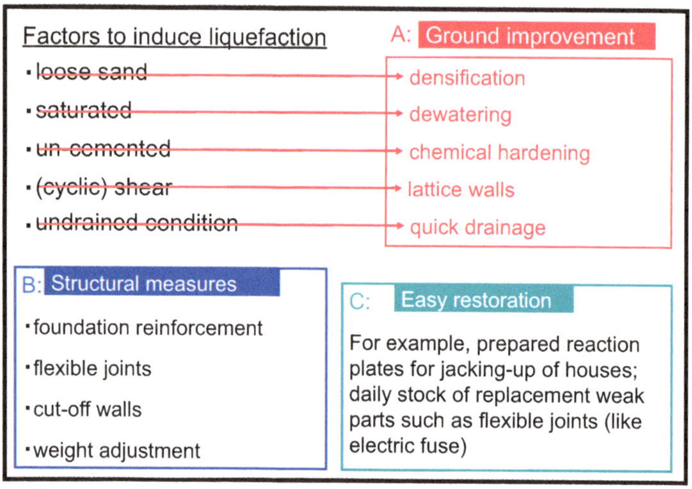

Fig. 4.1 Principles of liquefaction countermeasures

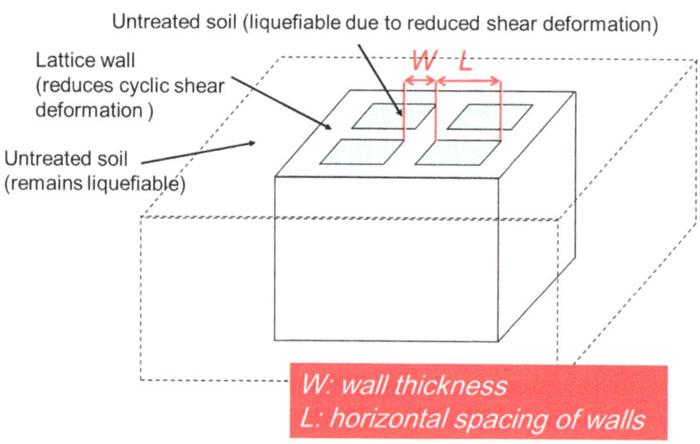

Fig. 4.2 Lattice walls as countermeasure against liquefaction

in-between the walls. On the other hand, the untreated soil layers located outside of the walls remain to be liquefiable.

The design procedure of this particular type of liquefaction countermeasures was initially developed during late 1980s and early 1990s as an outcome from a series of joint researches conducted by Public Works Research Institute, Japan, and private companies (PWRI 1999 among others). At this stage, its application to civil engineering structures, such as river dykes and semi-buried roads were considered. Its application to building foundations has become popular as well, and its good performance could be confirmed during the 1995 Hyogoken-Nanbu (or Kobe) earthquake, as will be introduced briefly in Sect. 2.4 of this chapter.

After the 2011 Great East Japan Earthquake Disaster, its extended application to residential area has become another important issue. In particular, the following questions need to be answered in implementing an effective and economical measure against future earthquakes;

1. How large the horizontal spacing can be?
2. How strong the wall shall be?

4.2.2 *Past Investigations on Horizontal Spacing of Lattice Walls*

The effect of the lattice walls to reduce the cyclic shear deformation of untreated soil layers depends on their horizontal spacing. Smaller the spacing is, more effectively the walls perform. However, the cost would become higher at the same time.

At the initial development stage, the recommended value of the ratio of the horizontal spacing to the thickness of the potential liquefiable layer was 0.8 or less. This value was originally derived from 1 g-field model shaking test results as shown in Fig. 4.3a. It was also confirmed by the results from a series of dynamic centrifuge tests as shown in Fig. 4.3b.

As summarized in Figs. 4.4a, b, it was later reported that the maximum value of the excess pore water pressure ratio depends not only on the shaking intensity but also on the ground water level. If the water level is not at the ground surface but below, then, as shown in Fig. 4.4b, this countermeasure may be effective at L/H equal to 0.9 or even larger.

4.2.3 *Recent Investigations on Horizontal Spacing of Lattice Walls*

After the Great East Japan Earthquake Disaster, another series of dynamic centrifuge tests were conducted by Tsukuni and Uchida (2015) with a base excitation using a strong motion recorded at Urayasu, Chiba Prefecture.

Under this excitation condition, as summarized in Fig. 4.5, it was found that the maximum value of L/H not to induce full liquefaction (i.e., excess pore pressure ratio equal to unity) or excessive settlement of houses may be as large as 1.2. It was also confirmed that the lower the ground water level is, more effective the lattice walls are.

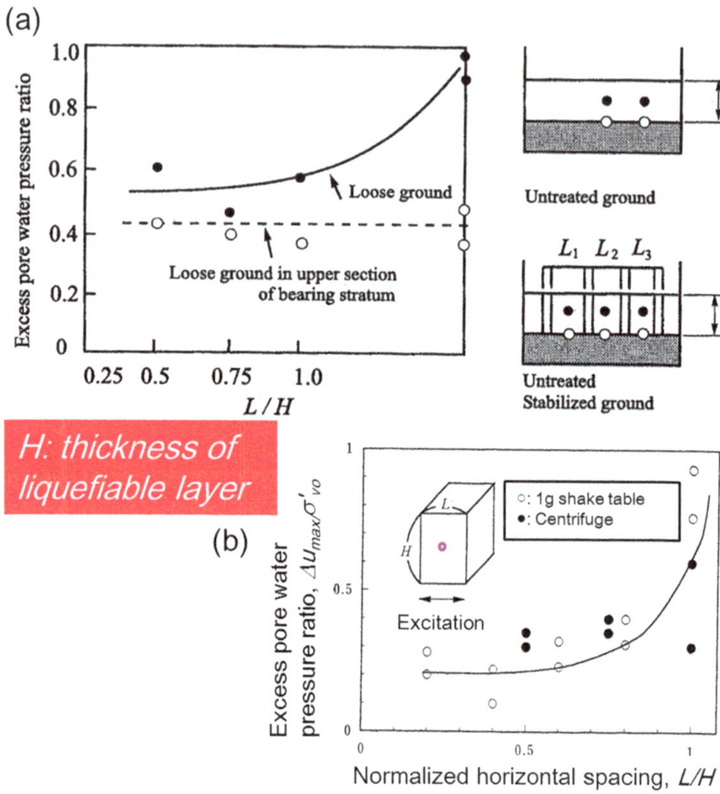

Fig. 4.3 Relationships between normalized horizontal spacing and excess pore water pressure ratio. (**a**) 1 g shake table tests (JGS 1998). (**b**) 1 g shake table and centrifuge tests (PWRI 1999)

4.2.4 *Past Investigations on the Thickness and Strength of Lattice Walls*

Good performance of the lattice walls in preventing the liquefaction of subsoils around pile foundation was demonstrated in the 1995 Hyogoken-Nambu (or Kobe) Earthquake, as reported by Suzuki et al. (1996). It was in contrast to severe damage to adjacent quay wall. After the earthquake, it was confirmed that there is no evidence of liquefaction observed on the surface of untreated sandy ground surrounded by cement-treated soil walls.

However, one issue remained whether the walls themselves were damaged or not. Their performance during future large earthquakes was also discussed, even if the walls could survive the last earthquake load.

As mentioned before, after the 2011 East-Japan Great Earthquake Disaster, their extended application to residential area at L/H>1.0 is being studied, and the

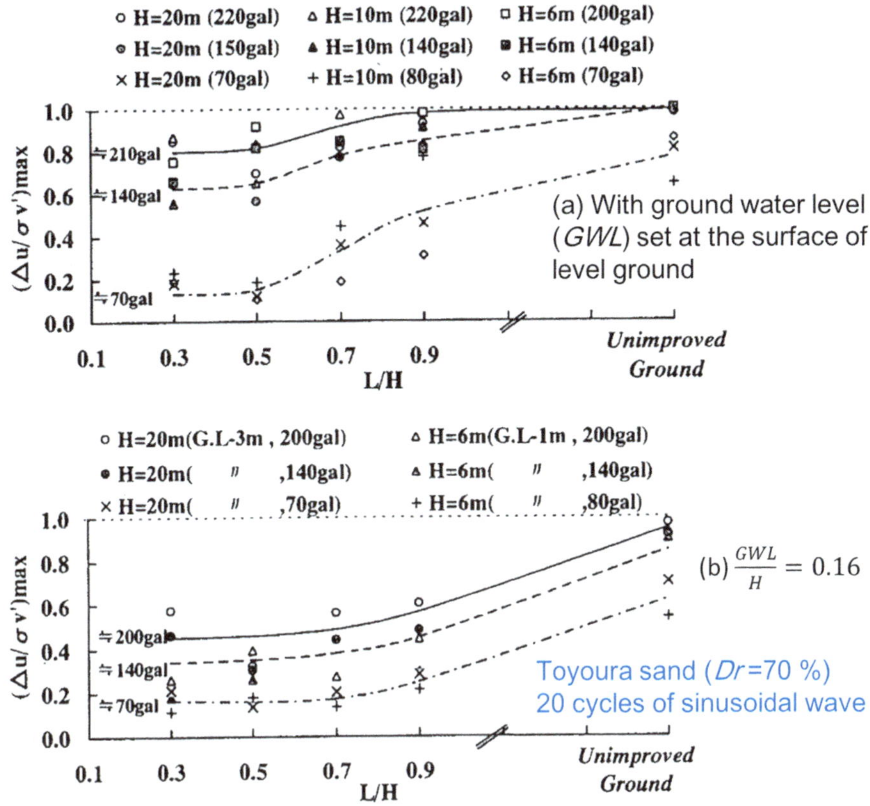

Fig. 4.4 Centrifuge test results on effects of input acceleration and ground water level. (a) with ground water level at ground surface. (b) with ground water level below ground surface (Babasaki et al. 1998)

performance of the walls themselves is of concern in answering the questions; how strong the walls shall be.

In view of these issues, past investigations using numerical analyses on the response of lattice walls under high seismic loads as reported by Namikawa et al. (2007) and Koseki et al. (2008) are summarized herein.

4.2.4.1 Conditions of Numerical Analyses

In order to investigate into the response of the lattice walls under relatively high seismic loads, a three-dimensional numerical analysis was conducted on walls having a center-to-center grid spacing of 8 m and a wall thickness of 0.8 m, as shown in Figs 4.6a, b. The N-S component of the recorded earthquake motion at Kobe Meteorological Observation Station during the 1995 Hyogoken-Nanbu

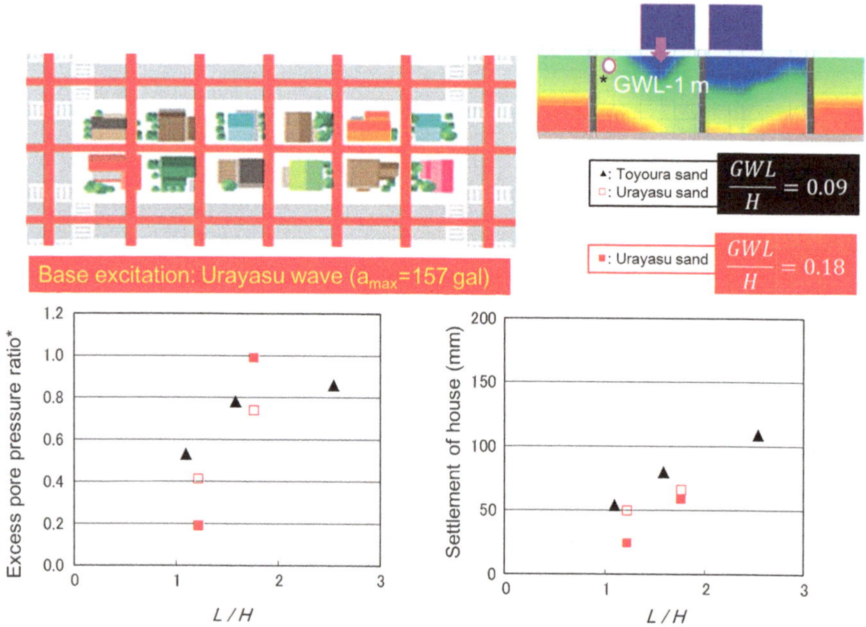

Fig. 4.5 Centrifuge test results on application to residential areas (Modified after Tsukuni and Uchida 2015)

earthquake was modified by re-scaling the maximum acceleration into 300 cm/s^2 and used as an input motion for the analysis. The duration of the excitation was 20 s.

Figure 4.6c shows the profile of the original soil layers and their constitutive models that were employed in the analyses. The liquefaction resistance of the saturated sand layer in terms of the cyclic shear stress ratio to cause liquefaction in 20 cycles of loading with constant amplitude was assumed to increase gradually from 0.18 at the top layer to 0.33 at the bottom layer. The behavior of these saturated sand layers was simulated using the densification model that was developed based on the Mohr-Coulomb type yield criterion (Shiomi et al. 1993).

The improved soil wall was modelled by either the elasto-plastic model considering the post-peak strain-softening properties in shear and tensile regions (Namikawa and Mihira 2007) or a linear elastic model. In using the elasto-plastic model, the unconfined compressive strength of the improved soil, q_u, was assumed to be either 2 or 5 MPa.

4.2.4.2 Results of Numerical Analyses

The distribution of a normal stress of the improved soil walls, which is defined in the horizontal direction that is perpendicular to that of the excitation, is shown in Fig. 4.7a for the case using the elastic modelling for the improved soil with

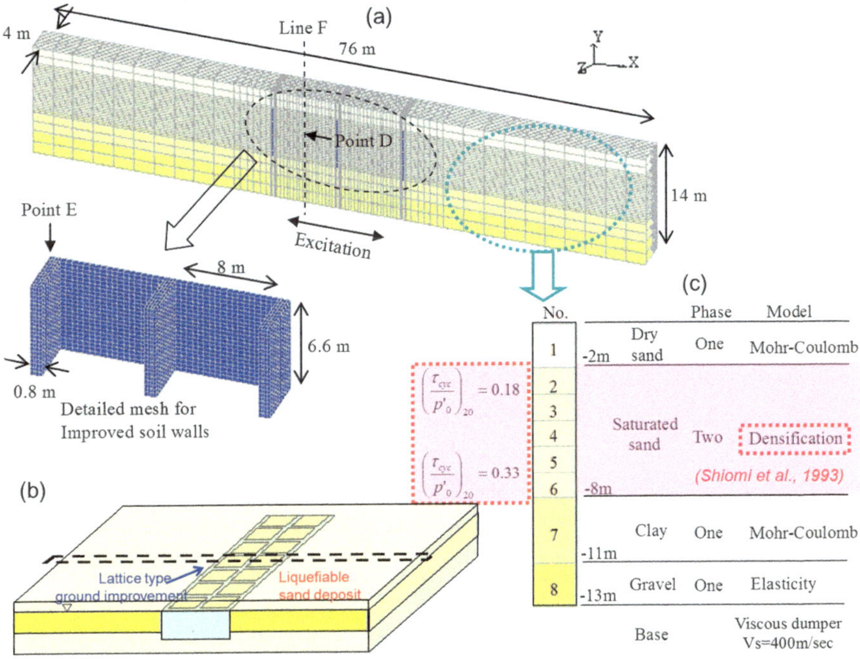

Fig. 4.6 3-D numerical analysis to simulate lattice walls behavior (Namikawa et al. 2007). (**a**) Finite element mesh. (**b**) Assumed improvement of ground. (**c**) Profile of original soil layers

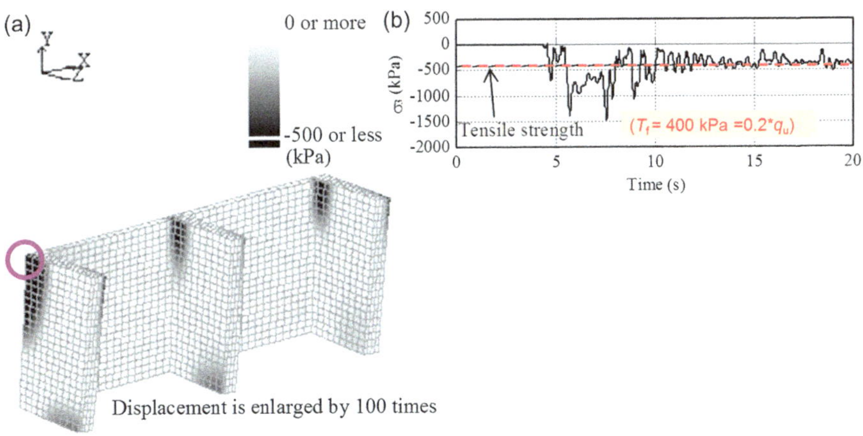

Fig. 4.7 Computed results for case 2a with elastic modeling (Namikawa et al. 2007). (**a**) Distribution of normal horizontal stress σ_z at $t = 5.6$ s. (**b**) Time history of minor principal stress σ_3 at corner of improved grids

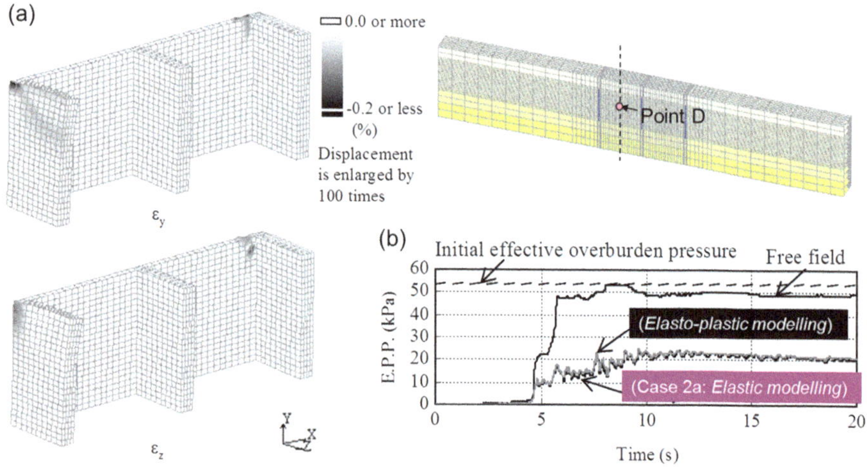

Fig. 4.8 Computed results for case 2b with elasto-plastic modeling (Namikawa et al., 2007). (**a**) Residual strains ($t = 20$ s) (**b**) Time history of excess pore pressure ratio at point in unimproved soil layer

$q_u = 2$ MPa (case 2a). It was drawn based on the results computed at t = 5.6 s, when the input acceleration exhibits nearly the maximum value. Tensile normal stress was concentrated at the corner of the improved soil grids, due to the effect of bending moments that were induced by the external forces. The time history of the tensile stress that was computed at a corner of the improved soil grids is shown in Fig. 4.7b for the same case. For most of the main shaking period, the value of the computed stress exceeded the tensile strength ($T_f = 400$ kPa) that was assigned for the improved soil with $q_u = 2$ MPa, suggesting that the improved soil walls in this case would be partially damaged. With the case using the elastic modelling, however, the extents of the partial damage, such as the resultant strains in the improved soil walls or their permanent deformations, and the effects of the partial failure on the response of the unimproved sand layers that were left in-between the improved soil grids cannot be evaluated properly.

As typical results from the elasto-plasic modelling for the improved soil with $q_u = 2$ MPa (case 2b), the distributions of normal strains of the improved soil walls, which are defined in the vertical direction and the horizontal direction that is perpendicular to that of the excitation, are shown in Fig. 4.8a. They were drawn based on the computed resultant values at the end of the excitation (i.e., t = 20 s). Due to partial failure, tensile strains accumulated at the corner of the improved soil grids and at the upper part of the improved soil walls that faced to the direction of the excitation.

The time histories of the excess pore water pressure in the unimproved sand layers that were left in-between the improved soil grids are compared in Fig. 4.8b between the cases using the elastic modelling (case 2a) and the elasto-plastic

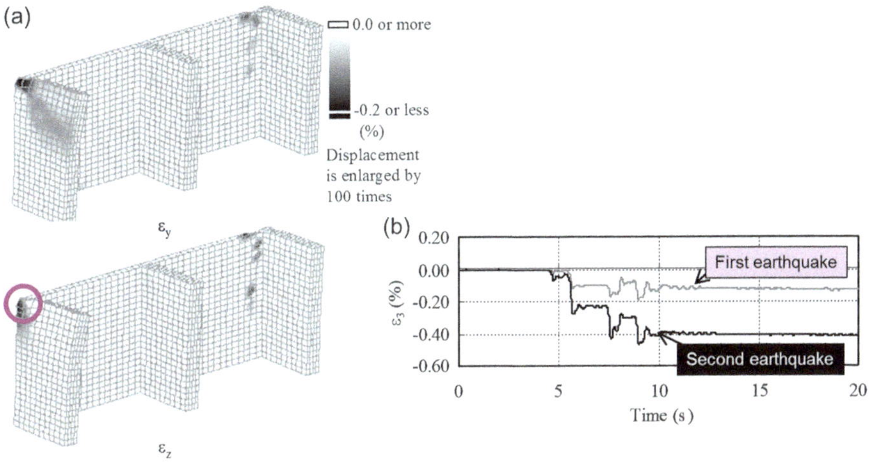

Fig. 4.9 Computed results for cases 2b and 2b-second with elasto-plastic modeling (Namikawa et al., 2007) (**a**) Residual strains during the second earthquake ($t = 20$ s). (**b**) Time histories of minor principal strain ε_3 at corner of improved grids

modelling (case 2b) for the improved soil with $q_u = 2$ MPa. For comparison, the excess pore water pressure in the free field consisting of the original soil deposits that was modelled using two-dimensional plane strain elements was shown in the figure as well. Due to the constraint effect of the improved soil walls, the excess pore water pressure could be reduced as compared to that of the original soil deposits which underwent complete liquefaction. The difference in the amounts of the excess pore water pressures between the results from the elastic modelling and the elasto-plastic modelling for the improved soil was to a limited extent, suggesting that the partial failure of the improved soil walls that was considered in the elasto-plastic modelling did not affect largely the potential of the liquefaction mitigation in these cases.

Once the improved soil walls are damaged partially by a large earthquake, it would become necessary to evaluate their residual resistance against another large earthquake. With the case using the elasto-plastic modelling (case 2b), therefore, an extended analysis (case 2b-second) was conducted by assuming that the second earthquake occurs long after the first one. For simplicity, identical input motions were assigned for both the first and second earthquakes. As can be compared in Figs. 4.8a and 4.9a, the tensile strain increments in the improved soils during the second earthquake were in general larger than those during the first earthquake. The time histories of the minor principal strain that was computed at a corner of the improved soil grids are compared in Fig. 4.9b. The resultant tensile strain in the second earthquake was approximately four times as large as the one in the first earthquake.

Figure 4.10a compares results from the other analyses (cases 3b and 3b-second) using the elasto-plasic modelling for the improved soil walls with

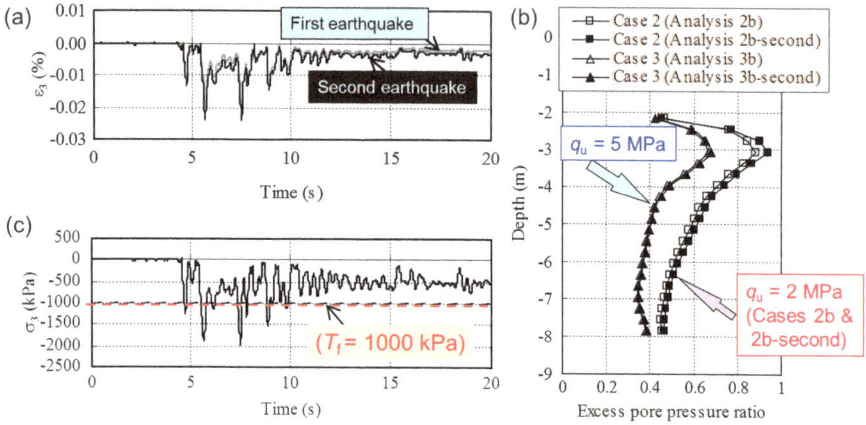

Fig. 4.10 Computed results for cases 3a, 3b and 3b-second with higher strength of improved soil (Namikawa et al. 2007) (**a**) Time histories of minor principal strain ε_3 at corner of improved grids in cases 3b and 3b-second. (**b**) Comparison of vertical distributions of excess pore pressure ratio. (**c**) Time histories of minor principal stress σ_3 at corner of improved grids in case 3a with elastic modeling

$q_u = 5$ MPa. By employing the improved soil with higher strength properties, the resultant tensile strains in the first and second earthquakes became almost equal to each other and could be reduced significantly as compared to those for the improved soil with $q_u = 2$ MPa (Fig. 4.9b). It should be noted that, as shown in Fig. 4.10b, the amount of the excess pore water pressure ratio in the unimproved sand layers that were left in-between the improved soil grids could be also reduced by employing the improved soil with higher strength properties (cases 3b and 3b-second). Even with the cases employing the improved soil with lower strength properties ($q_u = 2$ MPa), on the other hand, the amount of the excess pore water pressure ratio in the second earthquake (case 2b-second) was almost similar to the one during the first earthquake (case 2b), suggesting that the partial failure of the improved soil walls during the first earthquake did not affect largely the potential of the liquefaction mitigation during the second earthquake.

The above comparisons imply that, by using the elasto-plastic modelling for the improved soil walls considering the post-peak strain-softening properties in shear and tensile regions, more rational performance assessment could be made in terms of the extents of the partial damage and the potential of the liquefaction mitigation.

Note also that, as shown in Fig. 4.10c, the tensile stress as obtained from the analysis using the elastic modelling for the improved soil walls with higher strength properties ($q_u = 5$ MPa, case 3a) still exceeds the tensile strength. Thus, when the elastic modelling is adopted in the conventional design procedures based on allowable stress concepts, such performance would not be accepted, resulting into less economical decisions.

4.2.5 Summary and Future Issues

At the initial development of design procedure for civil engineering structures, the recommended horizontal spacing of the lattice wall normalized by the thickness of the liquefiable layer was around 0.8. After the 2011 East-Japan Great Earthquake Disaster, its extended application to residential area is being studied, where the normalized horizontal spacing may be increased to a value exceeding 1.0.

By considering properly the tensile behavior of cement-mixed soils, more rational design in terms of the wall strength may be achieved.

In implementing this method in residential areas, while ensuring the required horizontal spacing, it became necessary to construct the walls not only along municipal roads but also at the boundary between the neighboring housing lots, as has been schematically shown in Fig. 4.5. Within the housing lots, the space available for the additional construction work of lattice walls is not large enough to use the conventional construction equipment. Therefore, development of small-scale construction equipment that would be applicable to such limited space has been made to some extents and is still in progress.

4.3 Cut-Off Sheet Pile Walls

4.3.1 Background

As typically shown in Fig. 4.11, the cut-off sheet pile walls to restrain the lateral deformation of liquefied subsoil layers have been applied to several railway embankments. In applying them to river dykes, on the other hand, the tie rod cannot be installed at the foot of the dyke not to induce any local seepage problem.

It should be reminded that the major function required for river dykes is not only to avoid any failure but also to maintain the crest height to be above the high water level. From the latter point of view, their flood damage risk in Japan seems to be increasing these days, due possibly to effects of global warming. In addition, Japanese populations and properties concentrate at low land areas, which are prone to liquefaction damage during earthquakes.

For example, by the 1995 Hyogoken-Nambu (Kobe) Earthquake, the levee of Yodo river settled down by 2–3 m due to liquefaction of subsoil layers. Reportedly, the high water after the earthquake reached to about 1 m bellow the temporarily rehabilitated embankment. Based on such a case history, we may be able to imagine the worst scenario in which the levees collapse and settle down, due to lateral flow of liquefied subsoil induced by earthquakes; and subsequently, heavy rainfall in the upstream region triggers overflow of high water, leading to levee breakage.

While considering such worst scenario, two series of model shaking tests were conducted for studying the following issues, which will be briefly described herein: (1) how to maintain the crest height? and (2) how to reduce the sheet pile dimension?

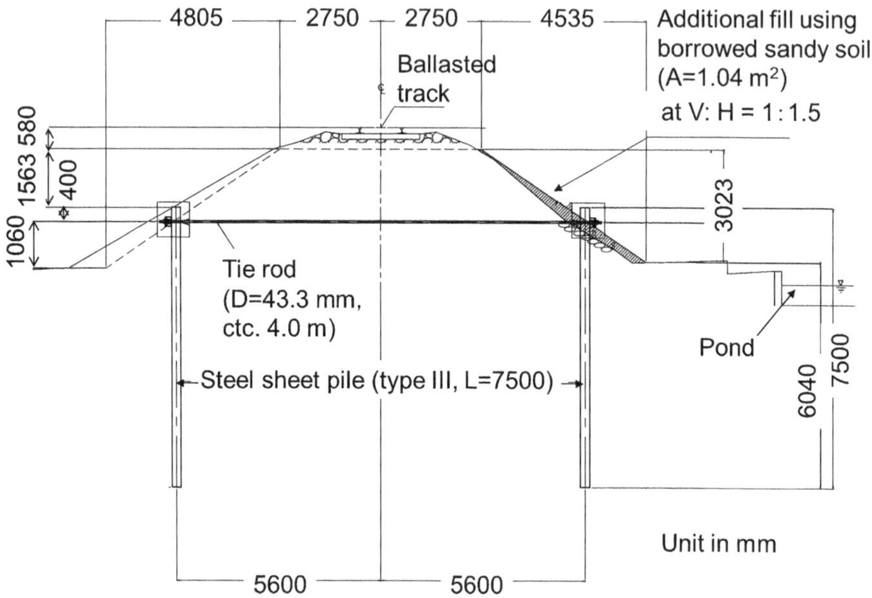

Fig. 4.11 Application of cut-off sheet pile to railway embankment along JR Narita line (Koseki et al. 2012)

4.3.2 First Series of 1-g Model Shaking Tests

4.3.2.1 Test Conditions

In the first series of 1-g model shaking tests (Koseki et al. 2010), three cases of models of an embankment and subsoil layers, as typically shown in Fig. 4.12, were constructed in a rigid soil container with a dimension of 280.0*69.5 cm in plan-view and a depth of 84.5 cm. In order to model non-liquefiable layer to which the sheet piles were embedded, the lower subsoil layer was prepared by compacting Kashima silica #7 sand with a mean diameter of 0.13 mm into a final thickness of about 25 cm. It was overlain by the upper subsoil layer with a thickness of 25 cm, which was prepared by pluviating Kashima #5 sand particles with a mean diameter of 0.34 mm at a relative density of about 30 %. This upper layer liquefied when a large amplitude horizontal excitation was applied.

The embankment had a height of 25 cm. It was made of a mixture of Kashima #5 sand and kaolin clay at a dry mass ratio of 5:1 and a water content of 15 %.

In case 1, as shown in Fig. 4.13a, a pair of steel sheet piles was installed at the foot of the embankment. This type of reinforcement has been widely adapted in Japan as a countermeasure against earthquake-induced liquefaction of subsoil layers (e.g., JGS 1998; Okamura and Matsuo 2002). In case 2, on the other hand, a single sheet pile was installed along the center line of the embankment crest, as shown in Fig. 4.13b. In case 3, as shown in Fig. 4.13c, a pair of sheet piles was installed at the shoulder of the

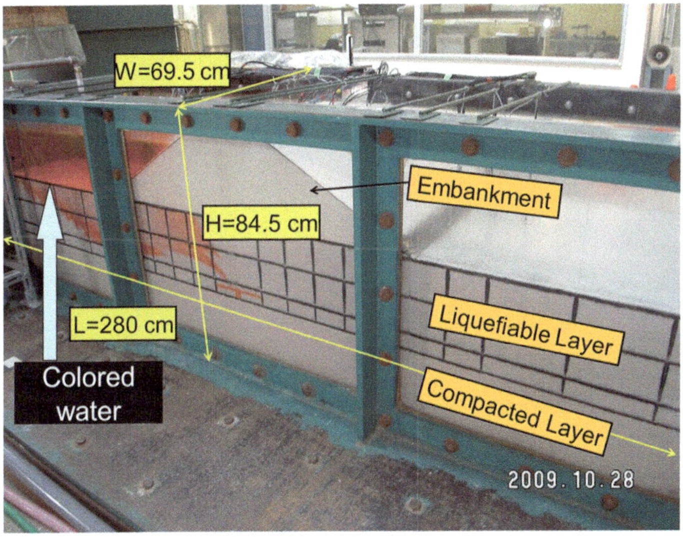

Fig. 4.12 Model of embankment and subsoil layers for first series of shaking tests

embankment, while connecting their top parts with tie-rods. The thickness of the sheet piles was reduced from the one employed for cases 1 and 2, considering the effects of the tie-rod in resisting against the external loads to be applied.

External loads were applied in the following four steps:

Step 1 High water condition without the seismic loading history was simulated by raising the water level on one side of the embankment (herein referred to as "upstream" side), while keeping the water level on the other side (i.e., "downstream" side) at the subsoil surface. The water level on the upstream side was raised up to 20 cm above the subsoil surface at a rate of about 1 cm/min with several intermissions to measure the distributions of pore water pressure and the corresponding steady state flow rate of ground water. For visual observation, colored water was used in this step (except for case 1) as well as in steps 3 and 4.

Step 2 After restoring the water level on the upstream side at the subsoil surface, a horizontal excitation was applied to simulate severe seismic loading history.

Step 3 While keeping the model conditions that were altered by the seismic loading history in step 2, the water level on the upstream side was again raised up to 20 cm (or lower in case the embankment suffered extensive settlement) above the subsoil surface.

Step 4 Following the above step 3, the water level on the upstream side was further raised until overflow was initiated. Then, it was attempted to maintain the water level by providing additional water for the upstream side to the maximum capacity of the available water supply (about 20 liter per min.).

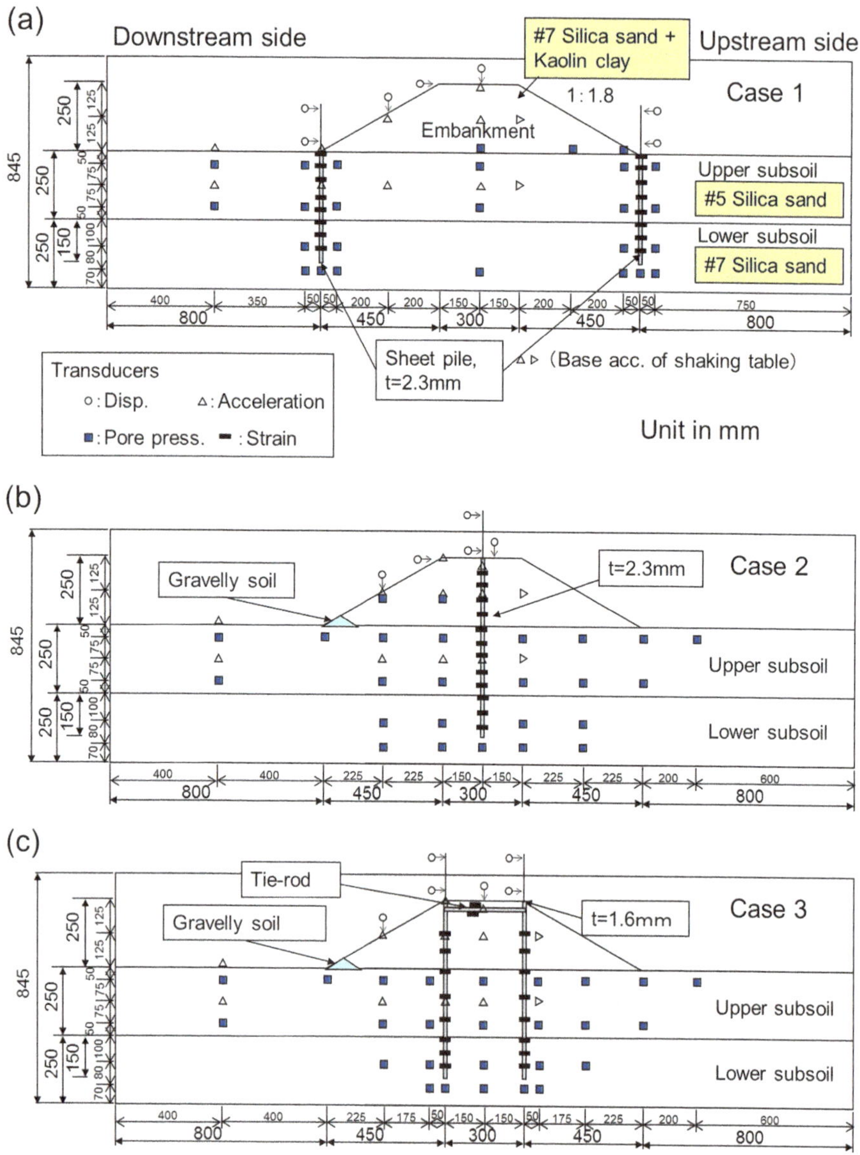

Fig. 4.13 Cross-section of models for first series of shaking tests (Koseki et al. 2010). (**a**) Case 1 with a pair of sheet piles at the front of embankment. (**b**) Case 2 with a single of sheet pile at the crest of embankment. (**c**) Case 3 with a pair of sheet piles at the shoulder of embankment that were connected by tie-rods

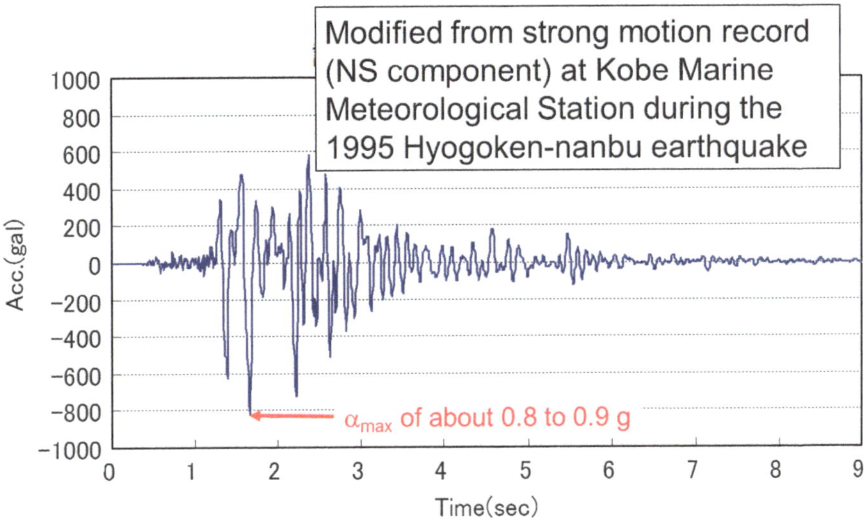

Fig. 4.14 Input wave motion for horizontal excitation in step 2 (Koseki et al. 2010)

In step 2, a horizontal excitation by using an input wave as shown in Fig. 4.14 was applied to simulate severe seismic loading history. This input wave was prepared based on the strong motion record (NS component) observed at Kobe Marine Meteorological Station during the 1995 Hyogoken-nanbu earthquake, where the time scale was modified so that the predominant frequency became 5 Hz, considering the limited size of the model as compared to the prototype one.

4.3.2.2 Test Results

In case 1, local seepage failure was triggered in step 1 at the foot of the embankment on the downstream side, when the water level on the upstream side was raised by 6 cm above the subsoil surface. The failure progressed to the middle height of the embankment when the water level was further raised to 20 cm and kept for 20 min. After lowering the water level on the upstream side, the collapsed part of the embankment was restored to the initial condition. Then the horizontal excitation was conducted in step 2.

The sheet piles installed at the foot of the embankment suffered large residual tilting. Consequently, as illustrated by arrows shown in Fig. 4.15a, the crest of the embankment subsided by about 7 cm, accompanied by re-occurrence of local seepage failure at its foot that was caused by dissipation of excess pore water pressure generated in the upper subsoil layer. In step 3, the water level on the upstream side was raised up to 15 cm above the subsoil surface, while observing the seepage behavior by using colored water. As shown in Fig. 4.15a, the water permeated predominantly through the top part of the upper subsoil layer (i.e., immediately below the embankment) and flowed out from the foot of the

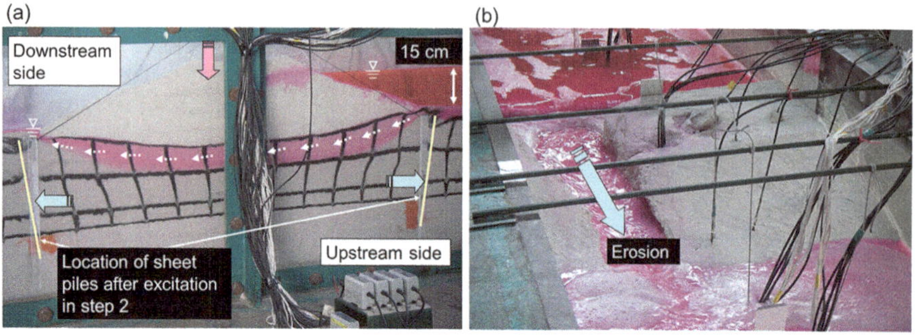

Fig. 4.15 Test results in case 1 (Koseki et al. 2010). (**a**) Step 3 (high water condition after excitation). (**b**) Step 4 (overflow after excitation)

embankment on the downstream side. Subsequently, the water level on the upstream side was further raised to trigger the overflow in step 4. Due to the erosion by the overflowing water, a part of the embankment was completely broken as shown in Fig. 4.15b.

In order to avoid the local seepage failure observed in case 1, in cases 2 and 3, the foot of the embankment on the downstream side was replaced with drainage works using a gravelly material that exhibits much higher permeability than the embankment material (sand and clay mixture).

In step 1 of case 2, as shown in Fig. 4.16a, the water permeated predominantly by passing around the tip of the embedded sheet pile. The photo shown in this figure was taken in about 300 min. After raising the water level on the upstream side to 20 cm above the subsoil surface. The amount of water that flowed out at this state was 0.68 liter per min, which was about half of the corresponding value (about 1.5 liter/min) in case 1. Such change in the water flow rates was affected by the longer seepage distance in case 2 and the reduced permeability of the lower subsoil layer as compared to that of the upper one.

By the excitation in step 2 of case 2, the embankment subsided by about 7 cm. The sheet pile remained at its original position, since it was located along the symmetrical line. In addition, the embankment suffered lateral displacement as well, creating openings with a width of about 3 cm at its interface with the sheet pile as shown in Fig. 4.16b. In these openings, sand particles that were boiled from the liquefied upper subsoil layer deposited to a height of about 6 cm.

In step 3 of case 2, the water level on the upstream side was raised up to 20 cm above the subsoil surface, while observing the seepage behavior by using water that was colored differently from the one employed in step 1. As a result, change in the seepage behavior including the water flow rate from the one observed in step 1 was not significant. Subsequently, the water level on the upstream side was further raised up to 24 cm above the subsoil surface, while no local failure of the embankment or the subsoil layers was observed.

When the water overflowed the sheet pile in step 4 of case 2, the water flowed out while eroding the interface between the embankment and the side wall of the soil

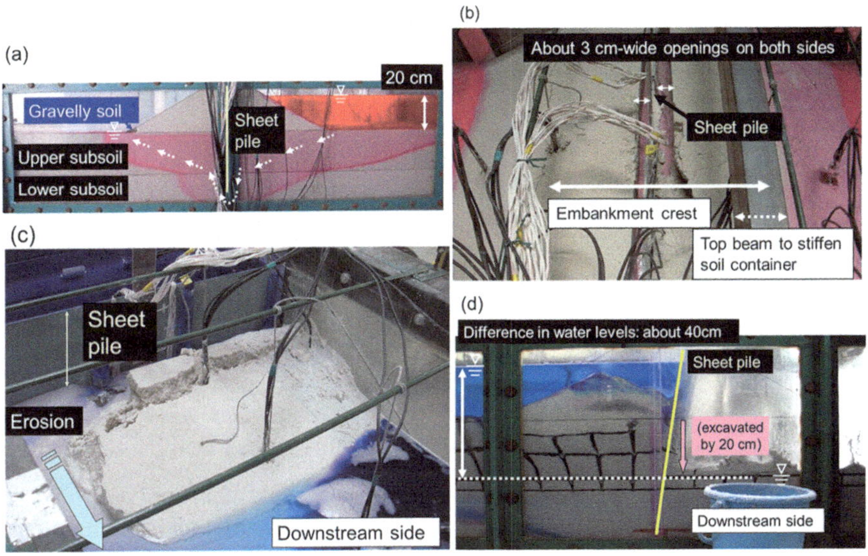

Fig. 4.16 Test results in case 2 (Koseki et al. 2010). (**a**) Step 1 (high water condition before excitation). (**b**) Step 2 (horizontal excitation). (**c**) Step 4 (overflow after excitation). (**d**) Excavation after step

container. As shown in Fig. 4.16c, a part of the embankment on the downstream side was completely eroded, while the sheet pile could retain the water on the upstream side up to its full height.

After step 4 of case 2, it was attempted to induce instability of the sheet pile by removing the embankment on the downstream side and excavating the underlying subsoil layer gradually, while maintaining the water level on the upstream side at 20 cm above the subsoil surface and lowering the water level on the downstream side down to the elevation of the interface between the upper and lower subsoil layers. As shown in Fig. 4.16d, the sheet pile tilted largely when the excavation was made down to the depth of 20 cm below the original subsoil surface. At this state, the embedded depth of the sheet pile on the passive side was 20 cm, which had been 40 cm before the excavation. Therefore, it could be inferred that the sheet pile before the excavation should have been stable against high water condition with an ample safety margin.

In step 1 of case 3 using a pair of sheet piles that are connected by tie rods, as shown in Fig. 4.17a, the water predominantly permeated by passing around the tip of the embedded two sheet piles. The photo shown in this figure was taken in about 420 min. After raising the water level on the upstream side to 20 cm above the subsoil surface. The amount of water that flowed out at this state was 0.41 liter per min, which was even lower than the corresponding value (0.68 liter/min) in case 2. Such change in the water flow rates was affected by the longer seepage distance through the less-permeable lower subsoil layer in case 3.

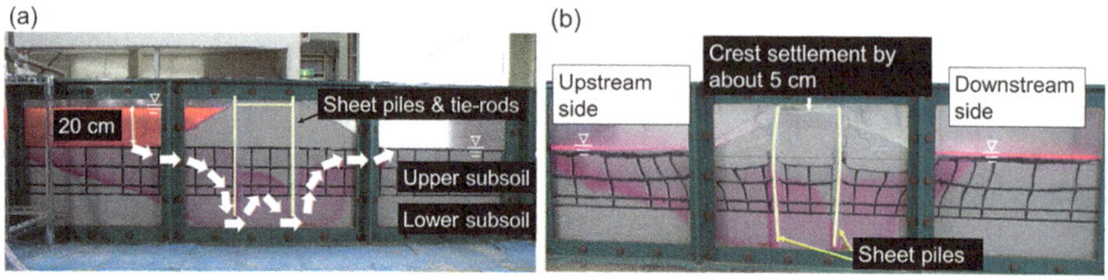

Fig. 4.17 Test results in case 3 (Koseki et al. 2010). (**a**) Step 1 (high water condition before excitation). (**b**) Step 2 (horizontal excitation)

By the excitation in step 2 of case 3, the crest of the embankment, which was surrounded by the pair of sheet piles, subsided by about 5 cm as shown in Fig. 4.17b. On the other hand, the position of the sheet piles was kept almost at their original elevation. Such settlement was possibly caused by the following two mechanisms.

1. The upper subsoil layer that liquefied by the excitation suffered one-dimensional compression due to dissipation of excess pore water pressure.
2. The sheet piles suffered bulging-type deformation as indicated in Fig. 4.17b due to liquefaction-induced reduction of the subgrade reaction of the upper subsoil layer on the passive side.

The behaviors observed in steps 3 and 4 of case 3 were in general similar to those in the corresponding steps in case 2.

Based on results from this first series of model tests, advantages of using sheet pile(s) driven from the crest or shoulder of embankments could be confirmed. It should be noted that recovery of the subgrade reaction in the upper subsoil layer that underwent dissipation of earthquake-induced excess pore water pressure was one of the key features for achieving these advantages.

4.3.3 Second Series of 1-g Model Shaking Tests

In order to reduce the cost of countermeasure, the second series of 1-g model shaking tests were conducted on non-embedded models where a part of the liquefiable subsoil layer was left below the sheet pile wall, as schematically shown in Fig. 4.18a, expecting to mobilize base-isolation effect of the liquefied subsoil layer. Hereafter, the results from case 4 with a pair of sheet piles driven at the embankment shoulder while not embedded to the non-liquefiable subsoil layer (to be called "floating type" for simplicity) will be compared to those from case 3 using a pair of embedded sheet piles (Fig. 4.13c).

Figure 4.18b compares the time histories of the response acceleration measured at the top of the sheet pile at the maximum base acceleration of about 900 gal. As expected, the floating type model exhibited reduced amplitude of the response

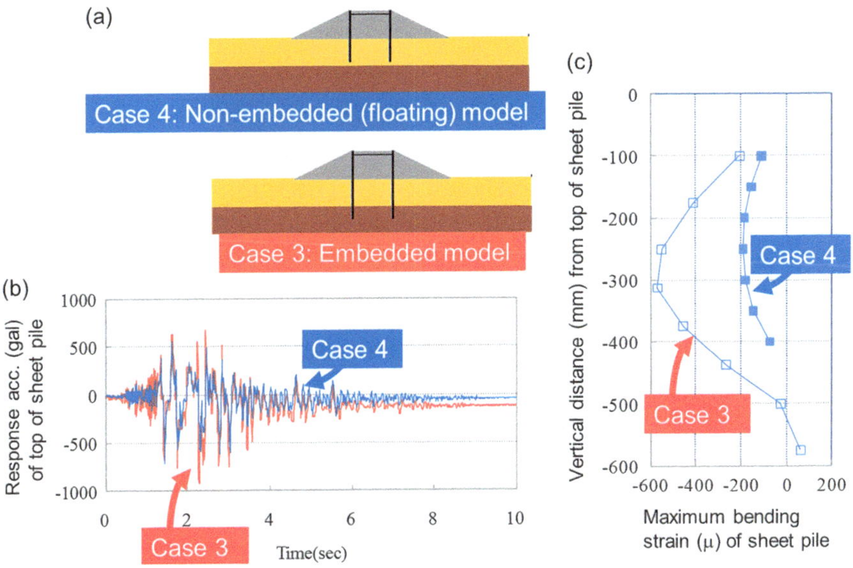

Fig. 4.18 Test results in cases 3 and 4. (**a**) Schematic cross-sections of models. (**b**) Comparison of response acceleration. (**c**) Comparison of maximum bending strain

acceleration. Figure 4.18c compares the profiles of the maximum bending strain of the sheet during excitation. The floating type model underwent much smaller bending strain than the embedded type model, suggesting a possibility of reducing the sheet pile dimension and thus the countermeasure cost.

4.3.4 Summary and Future Issues

In order to maintain the crest height of an embankment after liquefaction of its subsoil layer, a pair of cut-off sheet pile walls installed at the shoulder of the embankment is more effective than sheet pile walls installed near the toe of the embankment. A single sheet pile wall installed at the center-line would be sufficient as well. By leaving a part of the liquefiable subsoil layer below the sheet pile wall (such as case 4), base-isolation may be mobilized, which makes it possible to reduce the sheet pile dimension.

All of the above types exhibited sufficient performance against high water level conditions even after liquefaction. The sheet pile(s) in cases 2 and 3 could retain the water on the upstream side. Thus, the average flow rate of water during overflow was much smaller than the corresponding value of case 1. In these cases, however, the falling water eroded subsoil layer on downstream side, causing a reduction in the embedment depth. Such effects of overflow should be properly considered in future application of this method.

4.4 Dewatering

4.4.1 Background

As schematically shown in Fig. 4.19a, a shallow slope failure was induced in Shiroishi city, Miyagi prefecture by the 2011 East Japan Great Earthquake Disaster. As reported by Kawakami et al. (1978), this slope had suffered from flow-type failure by the 1978 Miyagi-ken-oki Earthquake, exhibiting the maximum residual displacement of about 90 m, where high ground water table was found within the original fill part using volcanic-ash sand. After the failure in 1978, drainage wells had been constructed, as can be seen in Fig. 4.19b, and the fill slope had been made gentle, which seem to have been effective in preventing another large scale-failure in the subsequent large earthquake.

After the 2011 Great East Japan Earthquake Disaster, application of such dewatering method is considered not only to protect slopes but also prevent/reduce liquefaction-induced damage to residential areas developed on level ground, which raised the following questions to be answered;

1. How to apply dewatering method in practice?
2. How to evaluate the liquefaction resistance of partially saturated granular materials?

In view of the above, a case history of liquefaction-induced damage to residential area where the dewatering method will be adopted and results from a series of triaxial liquefaction tests on partially/fully saturated granular materials will be briefly introduced herein.

Fig. 4.19 Performance of slope in Shiroishi city, Miyagi prefecture, during the 2011 East Japan Great Earthquake Disaster. (**a**) Shallow failure at the *upper part*. (**b**) No failure at the *lower part*

4.4.2 Case History of Liquefaction-Induced Damage to Residential Area Where the Dewatering Method Will Be Adopted

A residential area at Minami-Kurihashi, Kuki City, Saitama Pref., Japan suffered from liquefaction-induced damage, as typically shown in Fig. 4.20a during the 2011 East Japan Great Earthquake Disaster (Koseki et al. 2015). In order to construct the housing lot, as shown in Fig. 4.20b, river-origin sandy soils that were dredged from a nearby pond were transported with water through pipes and used for filling works, where temporary cut-off dikes had been constructed in advance.

Figure 4.21 shows typical soil profile at the damaged area. Below the surface sandy soil (F) layer that occasionally includes concrete blocks and cobbles, another sandy soil (Bs) layer that consists of fine sands and silts existed partly. Considering such distribution of Bs layer, it was inferred that this layer corresponds to the reclaimed sandy soil deposits as mentioned above. The ground water level was found to be around the upper surface of Bs layer (i.e., at a depth of about 1 m from the ground surface). Below F and Bs layers, Holocene clayey soil (Ac1 and Ac2) layers deposited to a depth of about 30 m from the ground surface, between which Holocene sandy soil (As1 and As2) layers existed partly.

In selecting countermeasures that are applicable to the damaged residential areas with existing houses, environmental issues, economic aspects and workability were considered. The environmental issues include generation of excessive noise, vibration and fine dusts. On the economic aspects, approximate estimation of the countermeasure costs, including those for keeping the system running, was made.

Fig. 4.20 (**a**) Liquefaction-induced damage to residential area at Minami-Kurihashi, Kuki city, Saitama prefecture, during the 2011 East Japan Great Earthquake Disaster and (**b**) land development works using dredged sandy soils (Koseki et al. 2015)

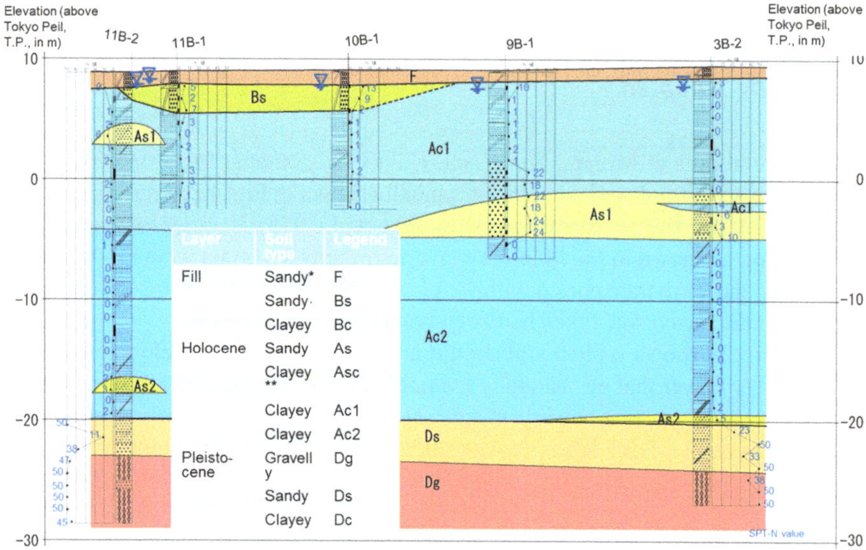

* F layer occasionally includes concrete pieces and cobbles, and clayey soils that originate from partly-filled construction waste soils.
** Asc layer is with a fines content of about 60 %.

Fig. 4.21 Typical soil profile at Minami-Kurihashi, Kuki City (Koseki et al. 2015)

As a result from the above consideration, a method to lower ground water level by about 1.5 m from the original water level has been selected.

After selecting the countermeasure, prototype field tests on the dewatering method were executed as schematically shown in Fig. 4.22, where the test field consists of two sections: drainage ditch section and well section. In the latter section, prefabricated drains had been installed to enhance the consolidation settlement induced by dewatering.

Within 95 days after starting the well operation to lower the ground water level by 2.1 m at maximum, the settlement in the well section accumulated to 15 cm maximum.

Next, the measured ground settlements in the well section were simulated numerically. After verifying the numerical analysis, prediction was made on the long-term settlement without prefabricated drains by eliminating effects of sheet pile friction and extra surcharge that affected the prototype test results. As a result, the maximum settlement in 30 years was predicted to be about 8 cm and was evaluated to be within the allowable range. It should be noted that possible effects of the dewatering method on surrounding unimproved areas were also carefully examined.

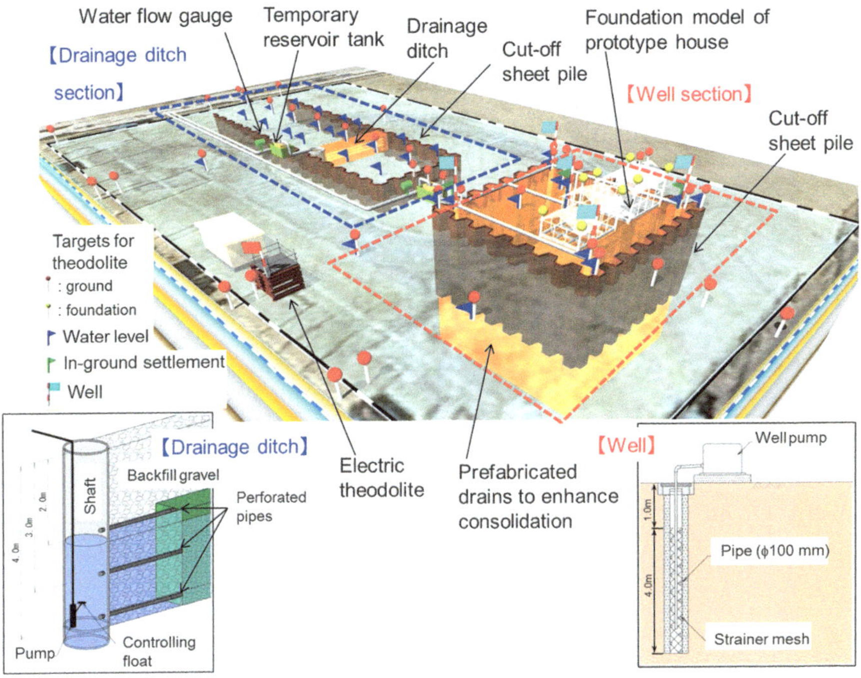

Fig. 4.22 Schematic view of prototype field test on ground water lowering method (Koseki et al. 2015)

4.4.3 Triaxial Liquefaction Tests on Partially/Fully Saturated Granular Materials

Figure 4.23 shows results from triaxial liquefaction tests on partially and fully saturated Toyoura sand conducted under two kinds of density conditions (Wang et al. 2016). Under each of the density conditions, partially saturated specimens exhibited larger liquefaction resistance than the fully saturated ones.

By defining the liquefaction resistance to cause the double amplitude axial strain (DA) of 5 % in 20 cycles and normalizing the resistance of the partially saturated specimens with that of the fully saturated specimens, relationship between the degree of saturation (S_r) and the normalized liquefaction resistance (LRR) is shown in Fig. 4.24a. For comparison, results from past relevant studies are also plotted. No unique relationships between these values could be obtained.

While referring to the study by Okamura and Soga (2006), another index called "potential volumetric strain" is evaluated by applying Boyle's law and used in Fig. 4.24b to represent the pore air compressibility of the partially saturated specimens. The potential volumetric strain is regarded as the volumetric stain of the specimens caused by pore air compression when the excess pore air pressure equals the initial confining pressure. Though the present test result on Toyoura sand

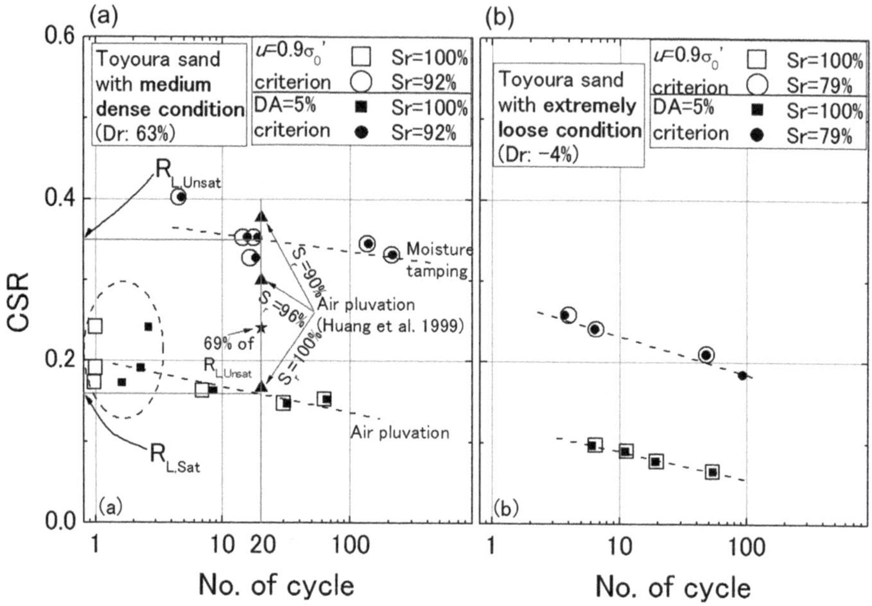

Fig. 4.23 Liquefaction resistance curves of partially and fully saturated Toyoura sand (Wang et al. 2016). (**a**) Medium dense condition. (**b**) Extremely loose condition

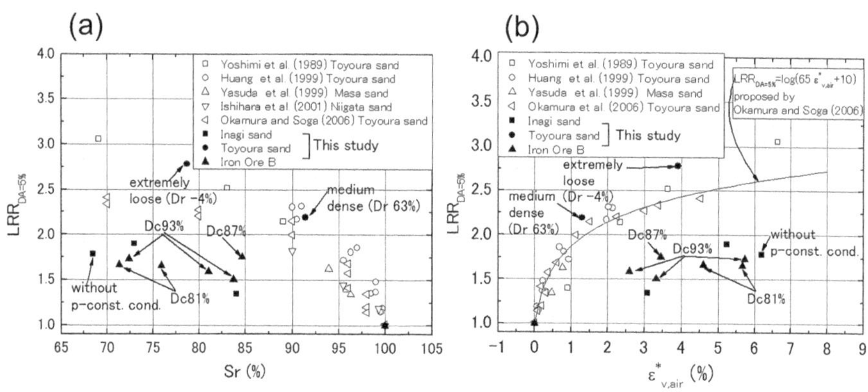

Fig. 4.24 Relationships between liquefaction resistance ratio LRR and (**a**) degree of saturation S_r, (**b**) potential volumetric strain $\varepsilon^*_{v,\,air}$ (Wang et al. 2016)

was consistent with those reported by Okamura and Soga (2006), those on other tested materials (i.e., Inagi sand and one kind of iron ore fines) deviated significantly.

Therefore, while considering the effect of elastic rebounding of the soil particle skeleton as well, a new index R_v called "volumetric strain ratio" was proposed by Wang (2014) as schematically illustrated in Fig. 4.25. If the specimen is fully saturated, its volumetric strain after liquefaction would be zero, where the

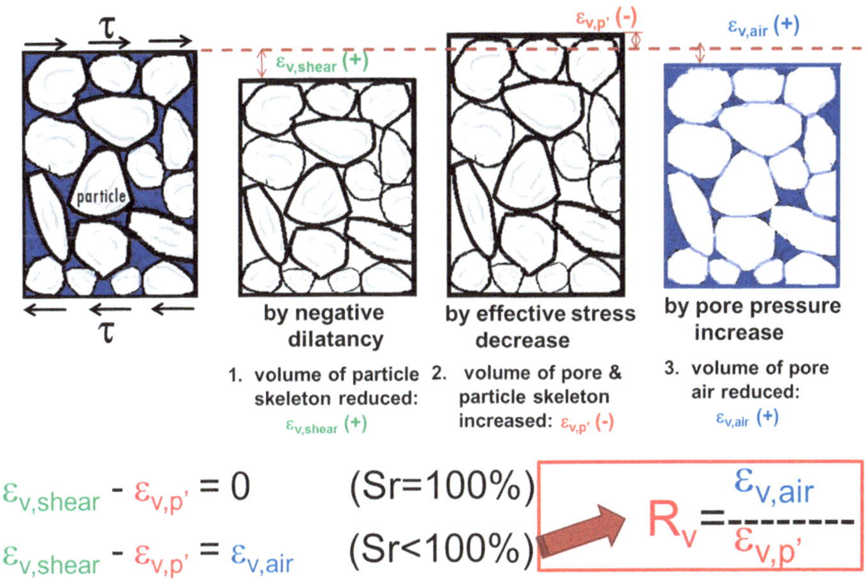

Fig. 4.25 Definition of volumetric strain ratio R_v and its schematic illustration (Wang 2014)

volumetric reduction induced by negative dilatancy would be canceled out by the volumetric increase induced by elastic rebounding due to the decrease in the effective stress. If the specimen is partially saturated, on the other hand, its volumetric strain after liquefaction would be non-zero, because of the pore air compression. Thus, the new index is defined to be the ratio between the volumetric strain due to the pore air compression and that due to the elastic rebound.

As shown in Fig. 4.26a, a special test was conducted to evaluate the amount of elastic rebound experimentally, where the effective stress was reduced isotropically by 90 % from the initial stress state.

Figure 4.26b shows the relationship between R_v, in which the pore air compressibility is represented by the aforementioned "potential volumetric strain" and the normalized liquefaction resistance, LRR of the three types of granular media. Compared with Fig. 4.24b using "potential volumetric strain" only, R_v shows better correlation with LRR.

4.4.4 Summary and Future Issues

After the 2011 Great East Japan Earthquake Disaster, the dewatering method is adopted in some of the residential areas that suffered from liquefaction-induced damage. Prototype field tests were performed to confirm its feasibility from several view points, including the long-term ground settlement and possible effects on surrounding unimproved areas.

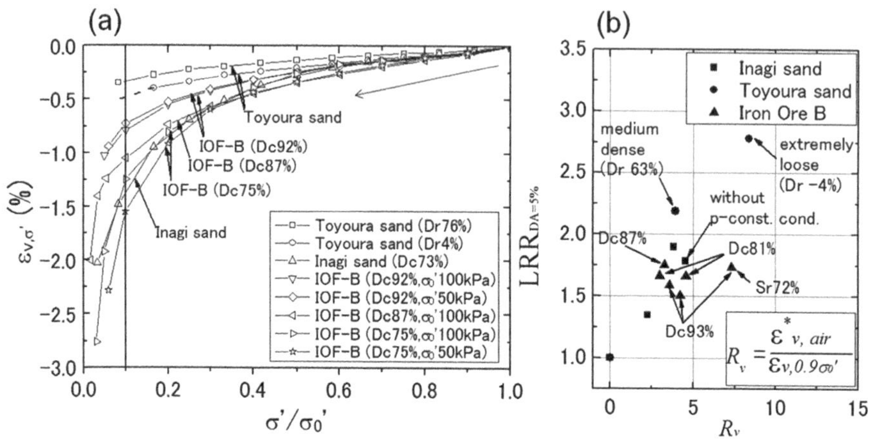

Fig. 4.26 (**a**) Isotropic unloading test results and (**b**) relationship between liquefaction resistance ratio LRR and volumetric strain ratio R_v (Wang et al. 2016)

To consider the effects of pore air compressibility in evaluating the liquefaction resistance of partially saturated granular materials, a new index, volumetric strain ratio (R_v) is proposed.

In order to validate the applicability of the dewatering method, future studies are required to establish the procedures to evaluate reasonably its effect as the countermeasure against liquefaction. In addition, in case of executing ground water lowering, it would be also required to predict accurately the amount of long-term ground settlement to be possibly induced by consolidation.

References

Babasaki R, Suzuki Y, Irie J, Akune M, Ohnuma S (1998) Grid-shaped improved ground by deep cement mixing against soil liquefaction (part 2), Summaries of technical papers of Annual Meeting Architectural Institute of Japan, Paper No. 20297 (in Japanese). http://ci.nii.ac.jp/els/110004215537.pdf?id=ART0006452411&type=pdf&lang=en&host=cinii&order_no=&ppv_type=0&lang_sw=&no=1471500051&cp=

Japanese Geotechnical Society (1998) Remedial measures against soil liquefaction -from investigation and design to implementation-, Balkema

Kawakami F, Asada A, Yanagisawa E (1978) Damage to embankments and earth structures due to Miyagiken-oki Earthquake of 1978. Tsuchi-to-Kiso Jpn Geotech Soc 26(12):25–31. (in Japanese)

Koseki J, Tsutsumi Y, Namikawa T, Mihira S, Salas-Monge R, Sano Y, Nakajima S (2008) Shear and tensile properties of cement-treated sands and their application to liquefaction mitigation, Keynote Lecture. In: Proceedings of 4th international symposium on deformation characteristics of geomaterials (IS-Atlanta 2008), 1, pp 27–50

Koseki J, Tanaka H, Otshushi K, Nagao N, Kaneko M (2010) Model tests on levees reinforced with sheet piles under high water conditions with/without seismic loading history. In: Proceedings of 3rd Korea-Japan geotechnical engineering workshop, pp 49–54

Koseki J, Koda M, Matsuo S, Takasaki H, Fujiwara T (2012) Damage to railway earth structures and foundations caused by the 2011 off the Pacific Coast of Tohoku Earthquake. Soils Found 52(5):872–889

Koseki J, Wakamatsu K, Sawada S, Matsushita K (2015) Liquefaction-induced damage to houses and its countermeasures at Minami-Kurihashi in Kuki city during the 2011 Tohoku Earthquake, Japan. Soil Dyn Earthq Eng 79:391–400.

Namikawa T, Mihira S (2007) Elasto-plastic model for cement-treated sand. Int J Numer Anal Methods Geomech 31:71–107

Namikawa T, Koseki J, Suzuki K (2007) Finite element analysis of lattice-shaped ground improvement by cement-mixing for liquefaction mitigation. Soils Found 47(3):559–576

Okamura M, Matsuo O (2002) Effects of remedial measures for mitigating embankment settlement due to foundation liquefaction. Int J Phys Model Geotech 2(2):1–12

Okamura M, Soga Y (2006) Effects of pore fluid compressibility on liquefaction resistance of partially saturated sand. Soils Found 46(5):695–700

Public Works Research Institute (1999) Design and construction manual of countermeasures against liquefaction, Report of Cooperative Research, No. 186 (in Japanese)

Shiomi T, Shigeno Y, Zienkiewicz OC (1993) Numerical prediction for model No.1, Verification of numerical procedures for the analysis of soil liquefaction problems, Balkema, pp 213–219

Suzuki Y, Saito S, Onimaru S, Kimura T, Uchida A, Okumura R (1996) Grid-shaped stabilized ground improved by deep cement mixing method against liquefaction for building foundation. Tsuchi-to-Kiso Jpn Geotech Soc 44(3):46–48. (in Japanese)

Tsukuni S, Uchida A (2015) Effect of grid-form deep mixing walls to prevent liquefaction damage of small houses. Proc of Deep Mixing 2015:161–170

Wang H (2014) Evaluation of liquefaction potential of partially saturated heap of Iron ore fines during maritime transportation. PhD thesis, Department of Civil Engineering, the University of Tokyo

Wang H, Koseki J, Sato T, Chiaro G, Tan Tian J (2016) Effect of saturation on liquefaction resistance of iron ore fines and two sandy soils. Soils Found 56(4), http://dx.doi.org/10.1016/j.sandf.2016.07.013

Chapter 5
Liquefaction Induced Downdrag and Dragload from Full-Scale Tests

Kyle M. Rollins, Spencer R. Strand, and J. Erick Hollenbaugh

Abstract Frequently, deep foundations extend through potentially liquefiable sand layers near the ground surface and bear on more competent layers at depth. When liquefaction occurs, the skin friction in the liquefied layer would be expected to decrease to some negligible value, but as the liquefiable layer settles, negative skin friction could potentially develop around the pile in this layer as effective stress increases. To investigate the loss of skin friction and the development of negative skin friction, axial load tests were performed on instrumented full-scale piles before and after blast-induced liquefaction at sites in Vancouver, Canada and Christchurch New Zealand. Following blasting, liquefaction developed within sand layers resulting in significant settlement. Skin friction in the liquefied layer initially dropped to essentially zero. However, as the liquefied sand reconsolidated, negative skin friction became equal to about 50% of the pre-blast positive skin friction. Despite significant ground settlement, pile settlement was relatively small. A neutral plane approach for computing pile settlement resulting from negative skin friction provided reasonable agreement with observed behavior.

5.1 Introduction

Deep foundations are typically used to support bridge and high-rise structures when weak or liquefiable soils are encountered. Deep foundations can bypass liquefiable layers and bear in more competent strata at depth. Dead and live loads imposed on the pile foundation are typically resisted by positive skin friction acting on the side of the pile and by end-bearing resistance at the toe of the pile. However, when

K.M. Rollins (✉)
Engineering Department, Brigham Young University, 368 CB, Provo, UT 84602, USA
e-mail: rollinsk@byu.edu

S.R. Strand
US Bureau of Reclamation, 302 E. 1860 S, Provo, UT 84606, USA
e-mail: Sstrand@uc.usbr.gov

J.E. Hollenbaugh
Chatbooks, Inc., 5314 River Run Dr #150, Provo, UT 84604, USA
e-mail: erick.hollenbaugh@gmail.com

S. Iai (ed.), *Developments in Earthquake Geotechnics*, Geotechnical, Geological and Earthquake Engineering 43, DOI 10.1007/978-3-319-62069-5_5

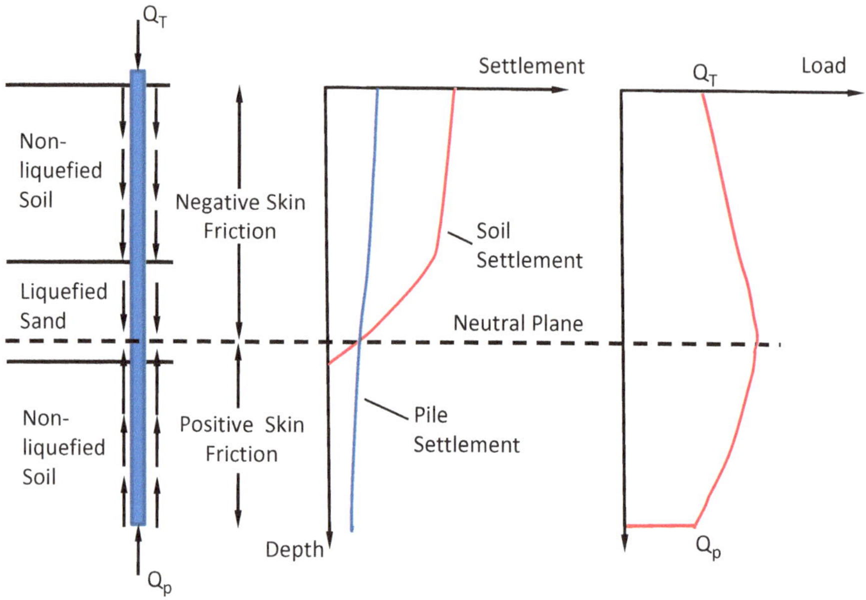

Fig. 5.1 Illustration showing relationships between liquefaction induced settlement, distribution of positive and negative skin friction, and load in the pile

liquefaction occurs in a layer along the pile, settlement of that layer and the soil above it could exceed the settlement of the pile leading to negative skin friction along that length of the pile above the liquefied layer as shown in Fig. 5.1. Negative skin friction acting on the pile creates a "dragload" on the pile in addition to the permanent pile head load above the neutral plane. The neutral plane is the depth where the settlement of the pile equals the settlement of the soil and also where the load in the pile is the greatest. Below the neutral plane, the positive skin friction and end-bearing pressure provide upward resistance which decreases the load in the pile. The end-bearing resistance must be compatible with the pile displacement at the toe to produce static equilibrium. Although these concepts are well known, field measurements that document the magnitude of negative skin friction that would develop in the liquefiable layer have not been available. In contrast to non-liquefiable layers, where the negative skin friction might simply be equivalent to the positive skin friction, the negative skin friction immediately following liquefaction is likely to be a very small fraction of the pre-liquefaction value or perhaps zero. Nevertheless, as the earthquake induced pore pressures dissipate in the liquefiable layer and settlement occurs, the skin friction at the pile-soil interface is likely to increase. Therefore, the negative skin friction which ultimately develops will likely be higher than zero and may depend on the dissipation rate and the increase in effective stress.

In the absence of test results, some investigators have used theoretical concepts to predict the behavior of piles when subjected to liquefaction induced dragloads.

Boulanger and Brandenberg (2004) defined negative skin friction in the liquefied zone in terms of the effective stress during reconsolidation, but concluded that the negative skin friction could be assumed to be zero with little error in the computed pile force or settlement. Fellenius and Siegel (2008) applied their "unified pile design" approach which was developed for downdrag in clays, to the problem of downdrag in liquefied sand, once again assumed that negative skin friction in the liquefied zone would be zero. They conclude that liquefaction above the neutral plane would not increase the load in the pile owing to the development of dragload prior to liquefaction.

To understand better the development of negative skin friction on piles in liquefied sand and the resulting pile response, a full-scale field load tests was conducted using a 324 mm diameter driven steel pipe pile in Vancouver, Canada as well as three augercast piles in Christchurch, New Zealand. Controlled blasting was employed to liquefy a 6-m to 8-m thick layer of sand along the length of the piles. The axial load distribution along the length of the pile was measured and skin friction before and after liquefaction was calculated. This paper describes the testing program, the test results and implications for design practice based on analysis of the test results.

5.2 Vancouver Canada Downdrag Test

The test site was located next to the south portal of the George Massey tunnel south of Vancouver, British Columbia, Canada. Situated in the delta region of the Fraser River, the soil profile consists of naturally deposited channel and alluvial sands.

5.2.1 Geotechnical Site Characterization

Previous site characterization, consisting of cone penetration tests, standard penetration tests, shear wave logging, and undisturbed sampling, was performed at this site for the CANLEX project (Robertson et al. 2000; Monahan et al. 1995) as well as during blast liquefaction studies reported by Rollins and Anderson (2004).

Prior to installation of the test piles, we performed a CPT sounding with a piezocone at the location of the future test pile to define the soil profile. Plots of the CPT test results are provided in Fig. 5.2 along with a generalized soil profile interpreted according to Robertson et al. (1986). In general, the soil profile consists of four major units. The top unit, approximately 2.7 m thick, consists of interlayered sand and silty sand. The second unit, approximately 2.8 m thick (from 2.7 to 5.5 m depth), consists of sandy silts, silts, and clayey silts. The third unit, approximately 9.1 m thick (5.5–14.6 m depth) consists of silty sands and sand. The fourth unit, comprising everything below 14.6 m depth, consists mostly of clean sands with some thin beds of silty sand. The poorly graded fine sands in units 1 and 3 have very

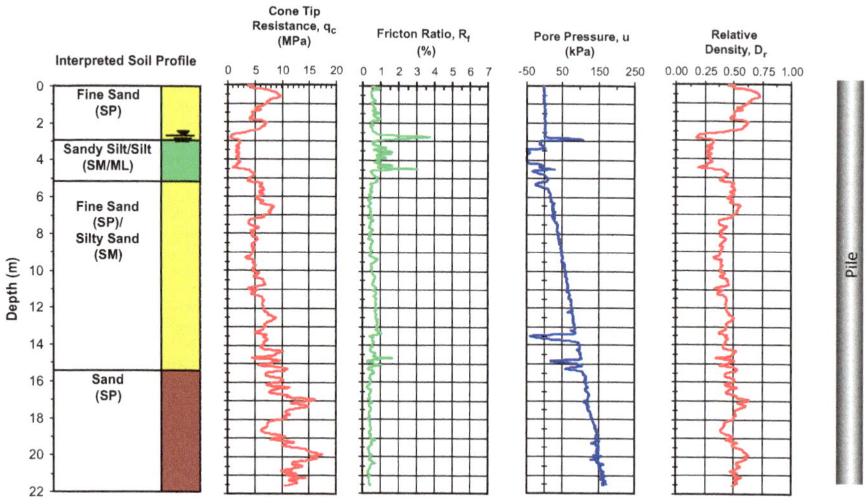

Fig. 5.2 Profiles of CPT cone tip resistance (q_c), friction ratio (R_f), pore pressure, and relative density (D_r) relative to the test pile

similar particle size-distribution curves with a mean grain size of about 0.2 mm and 3–10% fines. The soils in unit 2 contain between 40 and 70 fines with 5–12% clay size particles. Groundwater fluctuated with tides but was typically about 3.6 m below the ground surface during the pile load tests as shown in Fig. 5.2. As part of the CANLEX project, site specific correlations between CPT q_c and $(N_1)_{60}$ values were also developed. For an average q_c of 5.6 MPa in the poorly graded sand layer from 6 to 13 m, this correlation would predict an average $(N_1)_{60}$ value of about 10, while the $(N_1)_{60}$ value below the pile toe would be about 17.

Relative density was also estimated from the cone tip resistance measurements using an equation developed by Kulhawy and Mayne (1990) assuming young normally consolidated, clean sand conditions. The relative density profile is also shown in Fig. 5.2. Within Unit 3, the average relative density is approximately 40% with a standard deviation of about 7%. In contrast, the relative density of unit 1 ranges from 50 to 70%.

5.2.2 Pre-blast Static Pile Load Test

Plan and profile views of the test pile, reaction piles and instrumentation are provided in Fig. 5.3. Two static pile load tests were performed to provide data regarding unit side friction and end-bearing pressures. This data was useful in evaluating the thickness of the sand layer to be liquefied. The pile load tests were performed about 1 day after driving and then again 1-month after driving and 1 day prior to the blast liquefaction test.

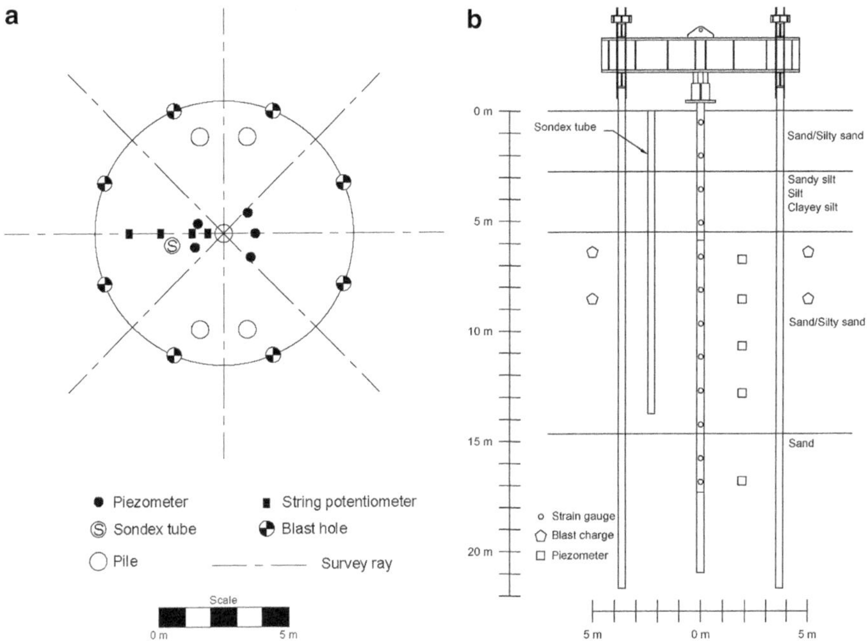

Fig. 5.3 (**a**) Plan view and (**b**) profile view of test pile, blast charges, and instrument layout

5.2.2.1 Test Pile Properties and Instrumentation

The test pile and four reaction piles consisted of steel pipe with an outside diameter of 324 mm and a wall thickness of 9.5 mm. The steel conformed to ASTM A252 Grade 3 specifications and had an average yield strength of 400 MN/m^2 (58,700 psi). The moment of inertia of the piles was 1.16×10^8 mm^4 (279 in.^{4}). The test pile and four reaction piles were driven closed-ended to a depth of 20.9. Prior to driving, the test pile was instrumented with strain gauges at approximately 1.5 m intervals along the pile length to a depth of about 17.3 m. At each depth interval, four strain gauges were applied to the pile at 90° intervals around the circumference of the test pile. The gauges consisted of water-proof electrical resistance type strain gauges. A length of angle iron was placed over the gauges and welded to the pile at regular intervals to protect the gauges and lead wires during driving. During pile driving, strain was measured to determine the stress in the pile and the residual stress was essentially zero at the conclusion of driving.

5.2.3 Test Procedure

The load test was performed using the quick maintained load procedure. Load was applied incrementally and held for 6 min at each increment. Load was applied to the pile using two hydraulic jacks placed between a 100 mm thick plate on the test pile and the reaction beam. As shown in Fig. 5.3, the reaction beam was restrained by four reaction piles which were spaced at about 3.65 m on either side of the test pile (11 pile diameters) which exceeds requirements from ASTM and FHWA standards. Applied load was measured by load cells and displacement of the pile was measured by string potentiometers attached to an independent reference frame.

5.2.4 *Results from Pre-blast Static Load Tests*

An initial pile load test was carried out within 24 h of the completion of pile driving. A plot of the measured pile head load vs. pile head deflection curve for the first test is provided in Fig. 5.4. The curve is relatively linear up to a load of about 550 kN after which the rate of settlement begins to increase. At a load of approximately 725 kN the pile began to settle very rapidly, or plunge, downward. The failure load, interpreted using the Davisson criteria, was 650 kN as shown in Fig. 5.4. This is about 60% lower than predicted by the Eslami and Fellenius (1997) method and about 32% lower than predicted by the LCPC (Bustamante and Gianeselli 1982) method. At the end of the test, the load was released and a residual deflection of 67 mm remained as shown in Fig. 5.4; however, residual stresses in the pile were negligible.

About 1 month after the initial pile load test, a second static load test was performed immediately prior to the blast liquefaction test. The pile head load

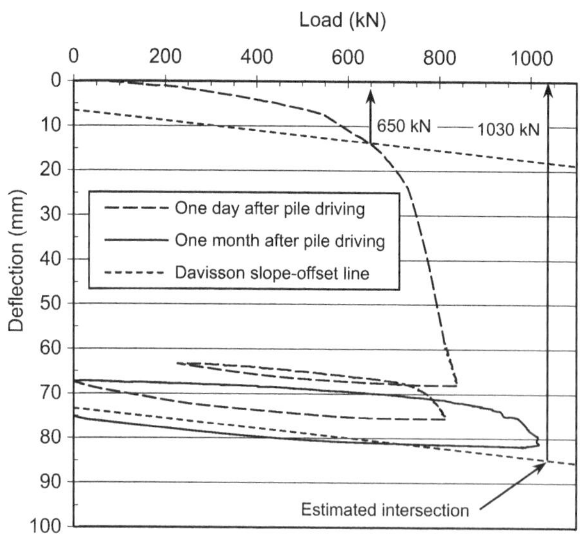

Fig. 5.4 Comparison of the pile head load vs. deflection curves resulting from the two static load tests prior to the blast liquefaction test

vs. deflection curve for the second pile load test is also plotted in Fig. 5.4 along with the curve from the previous load test. Using the Davisson criteria, the failure load was estimated to be 1030 kN, which is about 58% higher than the load from the first test. Most of the increase in axial capacity appears to result from increased side friction. Similar increases in pile resistance with time in sand have been reported by Lim and Lehane (2014). The failure load for the second test is slightly higher than predicted by the LCPC method but still about 30% below the failure load predicted by the Eslami and Fellenius method.

5.2.5 *Pile Load Tests After Blast Liquefaction*

A pile load test was also performed after producing liquefaction using controlled blasting within the target zone from 6 to 13 m. Prior to the blasting, a load equal to about half of the ultimate load resistance based on the Davisson criteria was applied to the pile and maintained throughout the test.

5.2.5.1 Test Layout and Instrumentation

Plan and profile views of the layout of the test pile relative to the blast holes and instrumentation are shown in Fig. 5.3. The test pile was located at the center of a ring of blast holes with a radius of 4.9 m. Eight blast holes were distributed equally around the circumference of the ring as shown in Fig. 5.3a. In each blast hole, 1.35 kg charges were located at depths of 6.4 and 8.5 m below the ground surface as shown in Fig. 5.3b. The eight explosive charges at 8.5 m were detonated sequentially first followed by the eight charges at 6.4 m at 1 s intervals.

The vertical displacement of the pile was measured using two string potentiometers attached to a tensioned cable which was stretched across the site. The cable was anchored at a distance of 19.7 m from the center of test area so that it would not be affected by settlement produced by the blast liquefaction.

Ground surface settlement vs. time caused by the blast tests was monitored using four string potentiometers attached to the tensioned steel cable. The string potentiometers were spaced 1.2 m apart along a line through the center of the test site. The total settlement of the ground surface was also monitored using an array of survey points. Measurements were made at regular intervals along eight rays emanating from the center of the test pile and spaced 45 degrees from each other. The elevation of these survey markers was determined with a level survey before and after blasting. The ground settlement as a function of depth was monitored using a Sondex settlement tube located at 1.83 m from the center of the test area as shown in Fig. 5.3.

The generation and dissipation of excess pore pressure during the blasting process was monitored using five piezometers installed at depths of 6.7, 8.4, 10.7,

12.8 and 16.8 m below the ground surface as shown in Fig. 5.3. The piezometers were typically located about 0.75 m from the center of the test pile.

5.2.6 *Test Procedure for Blast Liquefaction Test*

Initially, a load of about 536 kN was applied to the test pile before blasting and maintained after blasting until pore pressures had dissipated to near pre-blast levels. This represents a load with a factor of safety of about two against ultimate bearing capacity failure. This load was chosen because we estimated that the skin friction would be largely mobilized from the ground surface through a depth of about 13 m. We further estimated that the remaining non-mobilized skin friction below 13 m plus the end bearing resistance would be sufficient to support the pile against the negative skin friction that was expected to develop once blast testing produced liquefaction.

5.2.7 *Results for Blast Liquefaction Test*

5.2.7.1 Excess Pore Pressure Generation and Dissipation

Time histories of excess pore pressure ratio for each of the five piezometers during the blast test are plotted in Fig. 5.5. The pattern of pore pressure generation is non-linear with the greatest increases occurring during the first several blasts followed by smaller increases with each subsequent blast. After about 12 detonations, the residual excess pore pressure ratio for the piezometers at depths of 6.7, 8.4, 10.7, and 12.8 m was above 90% and there was little change in the ratio for the subsequent detonations. However, at a depth of 16.8 m, the excess pore pressure ratio remained quite low and did not exceed 18%. Therefore, the toe of the pile remained in non-liquefied sand.

The pore pressures dissipated more quickly as the depth increased, indicating that the sand reconsolidated from the bottom upward. Excess pore pressure ratios were less than 0.1 at all depths after about 25 min.

5.2.7.2 Blast-Induced Settlement

Level surveys indicated that the maximum settlement was approximately 270 mm near the center of the test area and settlement was greater than 250 mm within 3 m of the center. On average, settlement was less than 3 mm (accuracy of survey) at distances greater than about 12 m from the center. String potentiometers indicate that settlement occurred rapidly during and immediately following blasting. Approximately 80% of the settlement occurred within 5 min after the onset of

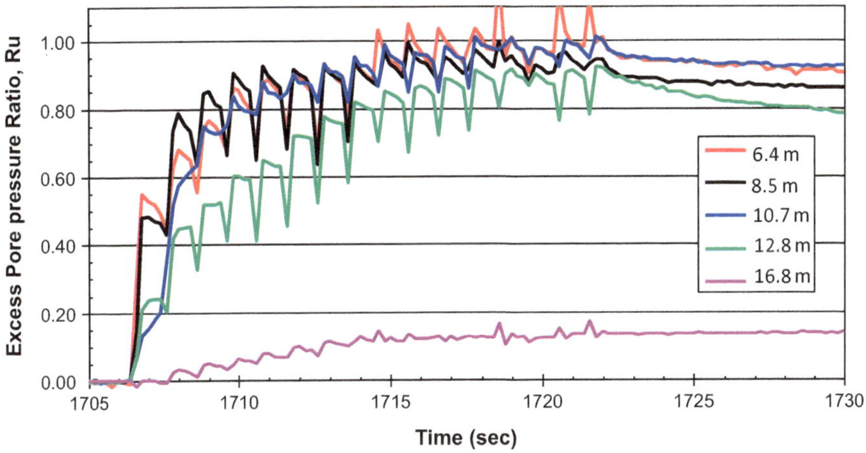

Fig. 5.5 Time histories of excess pore pressure ratios during the first 25 s of the blast liquefaction test at Vancouver, Canada

blasting. At 5 min after blasting, the excess pore pressure ratios in the target zone ranged from 0.2 at 12.8 m depth to 0.52 at 6.7 m depth.

A plot of the settlement vs. depth obtained from the Sondex tube is provided in Fig. 5.6. According to the level survey, the ground surface settlement at the location of the Sondex tube was approximately 270 mm, which is similar to the average settlement recorded by the Sondex tube in the top 3.5 m. The settlement profile was nearly constant from the ground surface to a depth of about 3.5 m and then decreased essentially linearly until it reached zero at a depth of about 13.7 m. This settlement profile indicates that the upper 3.5 m settled as a block on top of an underlying liquefied layer extending from 3.5 m to about 13.6 m. However, the liquefiable sand layer begins at a depth of 5 m and the settlement plot from a previous pilot liquefaction test did not show any appreciable settlement in the silty clay zone. This discrepancy is likely attributable to slippage of the Sondex pipe within the clayey silt layer.

Based on the measured data, the average volumetric strain in the target zone (6–13 m) is 2.3%. However, the dashed line in Fig. 5.11 represents the settlement vs. depth profile assuming that slippage occurred between the Sondex tube and the clayey silt layer. This modified settlement profile indicates that the liquefied layer begins at a depth of about 5 m. This is in reasonable agreement with the generalized soil profile. If settlement is assumed to be negligible in the silty clay layer, then the average volumetric strain in the target zone would be 3.1%. Based on an $(N_1)_{60}$ value of 10 in the liquefied layer, the Tokimatsu and Seed (1987) method would predict a volumetric strain of 2.5% while the Ishihara and Yoshimine (1992) approach would predict about 3.5% strain. Therefore, the reconsolidation settlement for the blast liquefaction test is within the range that would be predicted for an earthquake induced liquefaction event.

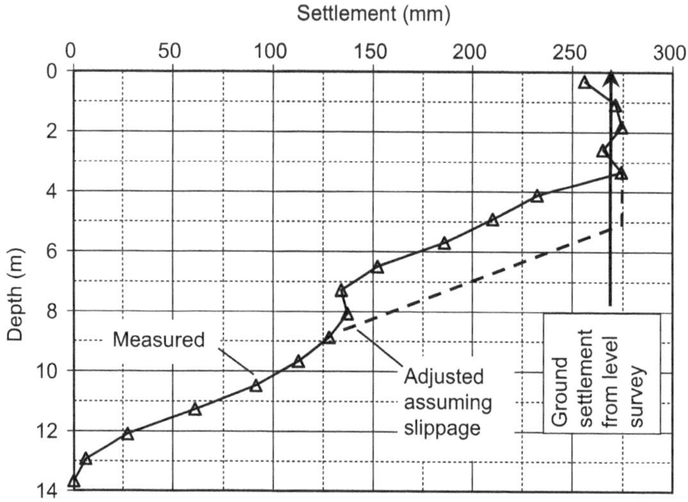

Fig. 5.6 Settlement vs. depth curve measured by the Sondex tube for the blast liquefaction test

5.2.7.3 Load vs. Depth Curves

A load of 536 kN was applied in five increments over a period of 28 min prior to blasting. This load produced a pile head displacement of only 3 mm. At the onset of blasting, the test pile settled slightly so that the load applied by the hydraulic jacks dropped almost immediately to a low of 381 kN, a reduction of 156 kN. Although the hydraulic pump was turned on immediately after the initiation of blasting, the 156 kN load could not be fully re-applied until 18 s after the start of blasting. Within this 18 s period, the ground surface near the pile settled more than 100 mm and the pile head had settled approximately 5 mm.

The load in the pile vs. depth is plotted in Fig. 5.7 for the conditions (a) immediately prior to blasting, (b) immediately after blasting, and (c) at the end of settlement. Based on the curve prior to blasting, the positive skin friction was fully mobilized to a depth of over 13 m, but no end-bearing resistance had yet been developed. The curve immediately after blasting shows positive skin friction to a depth of about 6 m, essentially no skin friction (nearly vertical line) in the liquefied zone from roughly 6 to 13 m, and positive skin friction below 13 m. Skin friction, q_s is given by the equation

$$q_s = K\sigma'_v \tan\phi \tag{5.1}$$

where K is the earth pressure coefficient, σ'_v is the initial vertical effective stress, and ϕ is the friction angle of the soil. As the vertical effective stress approaches zero during liquefaction, the skin friction would also be expected to approach zero.

While the very low skin friction in the liquefied sand layer is expected, the positive friction above this layer requires some explanation. At the onset of blasting, the test pile settled slightly so that the load applied by the hydraulic jacks dropped by 156 kN at the top of the pile. When this 156 kN load was re-applied, this load appears to have developed by positive skin friction from the top downward in the upper section of the pile. It should be noted that the total measured skin friction from the ground surface to a depth of 6 m immediately prior to blasting was approximately 166 kN. Therefore, the redevelopment of positive skin friction due to this applied load appears to be reasonable. The load of 536 kN was maintained throughout the remainder of the test by adding hydraulic fluid to the jack as the pile began to settle and relieved the load. This apparently maintained the positive friction in the upper 6 m of the pile. As noted by Fellenius (2006), very little axial movement is required to produce a shift from negative to positive skin friction.

Following liquefaction, the load originally carried by skin friction in the liquefied zone was transferred to the lower end of the pile where liquefaction had not developed. As a result, at the base of the liquefied zone the load in the pile increased by 130 kN after blast induced liquefaction. The pile settled about 4.5 mm owing to the loss of skin friction in the liquefied sand and the mobilization of skin friction in the underlying sand layer.

Once excess pore pressure had dissipated and settlement had stopped, the load vs. depth curve in the previously liquefied zone developed a negative slope as shown in Fig. 5.7. The negative slope indicates that negative skin friction had developed in this zone and was applying dragload to the pile. As the pore pressures dissipated and effective stresses increased, the skin friction at the pile interface also increased as predicted by Eq. 5.1 and produced a dragload of about 100 kN (22.5 kips). This load was again transferred to the sand below the liquefied zone with a

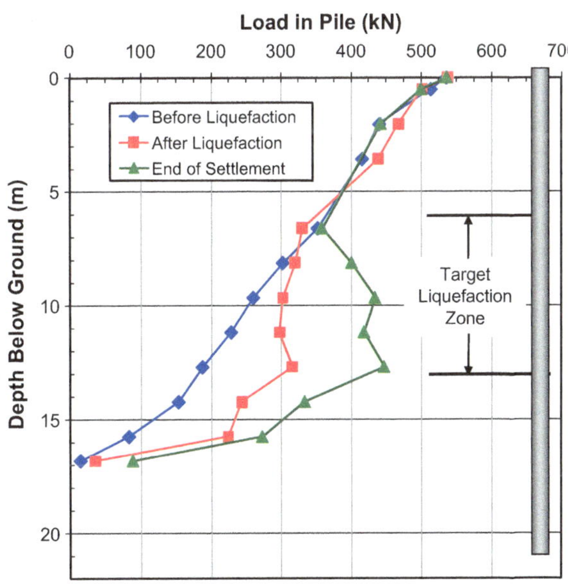

Fig. 5.7 Pile load vs. depth curves before blasting, immediately after blasting and after settlement of the liquefied layer

resulting additional pile settlement of about 2.5 mm or a total pile settlement of 7 mm. Despite the increased dragload, extrapolation of the pile load vs. depth curve in Fig. 5.7 indicates that positive skin friction in the non-liquefied sand below the liquefied zone provided sufficient resistance and relatively little end bearing resistance developed.

The negative skin friction developed in the liquefaction target zone after reconsolidation was approximately 14.3 kN/m^2, not zero as assumed in some approaches. As the liquefied sand reconsolidated and the effective stress increased owing to reduced excess pore pressure, negative skin friction progressively increased on the side of the pile. This negative skin friction at the end of settlement was equal to about 50% of the positive skin friction developed just before blasting. This is consistent with the average excess pore pressure ratio of 50% from the liquefied ($r_u = 100\%$) to the non-liquefied state ($r_u = 0\%$).

5.3 Christchurch New Zealand Downdrag Test

The test site was located in Avondale adjacent to the Avon River in Christchurch, New Zealand. This area experienced significant liquefaction settlement (0.3–1.0 m) during the Christchurch earthquake sequence in 2010–2011 and most homes in the area had been condemned. Test foundations consisted of three nominally 0.6-m diameter augercast piles.

5.3.1 Geotechnical Site Characterization

In connection with this study, site characterization, consisting of cone penetration tests, standard penetration tests, shear wave logging, and undisturbed sampling, was performed by Tonkin and Taylor in association with the Earthquake Commission in New Zealand. Prior to installation of the test piles, two CPT soundings were performed to define the soil profile. An SPT boring was also performed using the sonic drilling system. Corrections to obtain $(N_1)_{60}$ and q_{c1} were made using the procedures outlined by Youd et al. (2001). Plots of the CPT and SPT test results are provided in Fig. 5.8 along with estimates of relative density obtained using correlations proposed by Kulhawy and Mayne (1990).

In general, the soil profile consists of three major units. The top unit, approximately 1.5 m thick, consists of sandy silt. The second unit, approximately 9 m thick (from 1.5 to 10.5 m depth), consists of poorly graded medium-dense clean sand. The third unit, at least 6 m thick (10.5–16 m depth) consists of inter-bedded layers of medium-dense to dense clean sand. The measured fines contents for each unit are also shown in Fig. 5.8. In unit 1, the fines were greater than 50%, but within unit 2 the fines content was typically between 2 and 8.5%. Groundwater fluctuated with tides but was typically about 1.5 m below the ground surface during the pile load

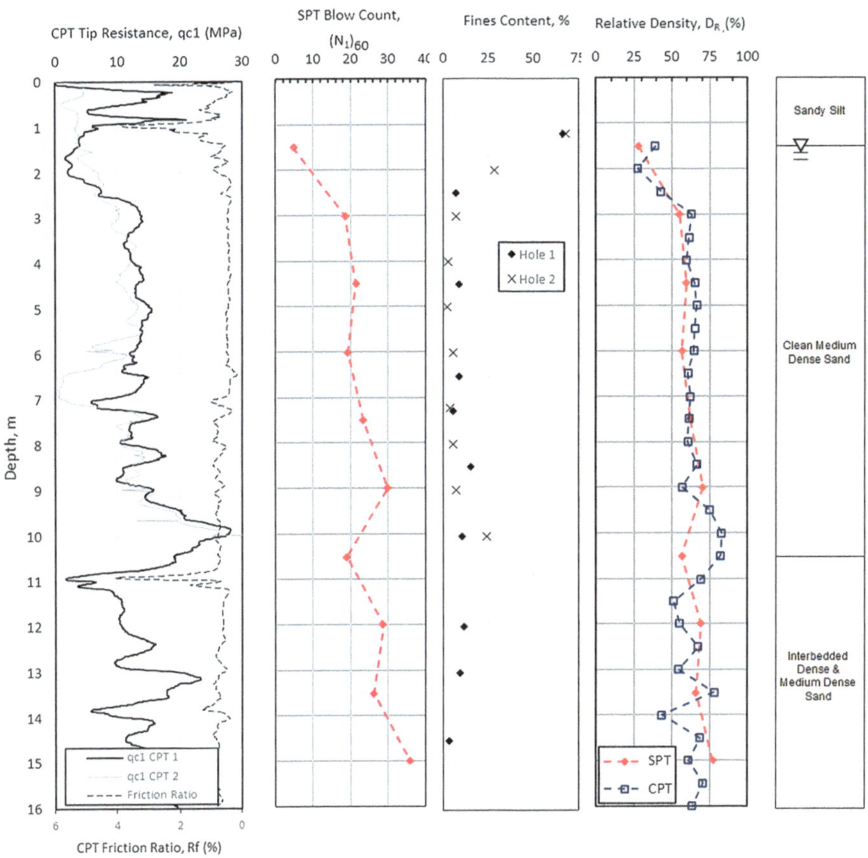

Fig. 5.8 Profile showing CPT cone tip resistance, SPT blow count, fines content, relative density and generalized soil profile for the test site

tests. However, p-wave velocity testing conducted by Cox and Roberts (personal communication) suggests that the soil was not fully saturated until a depth of 2.5–3 m.

5.3.2 Layout of Test Piles and Instrumentation

Plan and profile views of the layout of the test piles relative to the blast holes and instrumentation are shown in Figs. 5.9 and 5.10. The test piles were auger-cast piles constructed with lengths of 8.5 m, 12 m, and 14 m, and consisted of high-strength concrete ($f'_c = 38$ MPa) with a nominal diameter of 610 mm and a steel reinforcing cage extending the full length of the shaft. The reinforcing cages consisted of six 25 mm diameter bars with a spiral bar having a pitch of 30 cm (12 in.). To determine

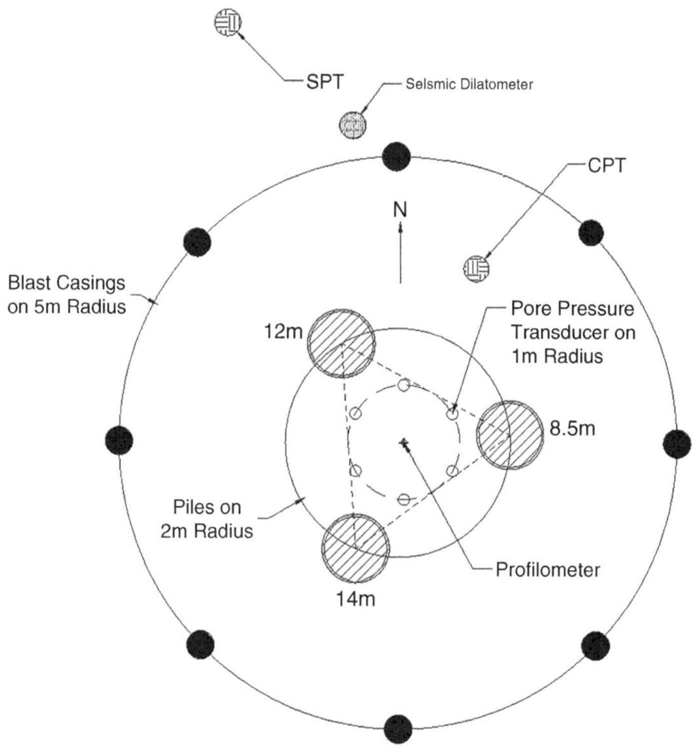

Fig. 5.9 Plan view of test piles, blast holes and instrumentation

load in the pile vs depth, each of the reinforcing cages were instrumented with strain gauges at approximately 1.5 m intervals along the pile length to a depth of about 0.5 m from the bottom of the pile. To determine the cross-sectional area of the shaft versus depth, each test pile was also instrumented with four thermal gauge wires. These gauges were spaced at 30 cm along each wire and used the heat of concrete hydration to indicate the pile diameter throughout the length of the pile.

The test piles were surrounded by a ring of blast holes having a radius of 5.0 m. The piles were 2 m from the center of this ring and 2.3 m from one another. Eight blast holes were distributed equally around the circumference of the ring at 45° as shown in Fig. 5.9. In each blast hole, 1.1 and 2.7 kg charges were located at depths of 4.0 and 8.5 m, respectively, below the ground surface as shown in Fig. 5.10. The eight explosive charges at 8.5 m were detonated sequentially followed by the eight charges at 6.4 m, each at 300 ms intervals. This configuration was selected after a pilot blast liquefaction trial which indicated that these charges would liquefy the soil out to a distance of about 8 m from the center of the blast ring. The ground settlement as a function of depth was monitored using a "Sondex" profilometer located at the center of the test area as shown in Fig. 5.10.

The generation and dissipation of excess pore pressure during the blasting process was monitored using six pore pressure transducers, or piezometers. The

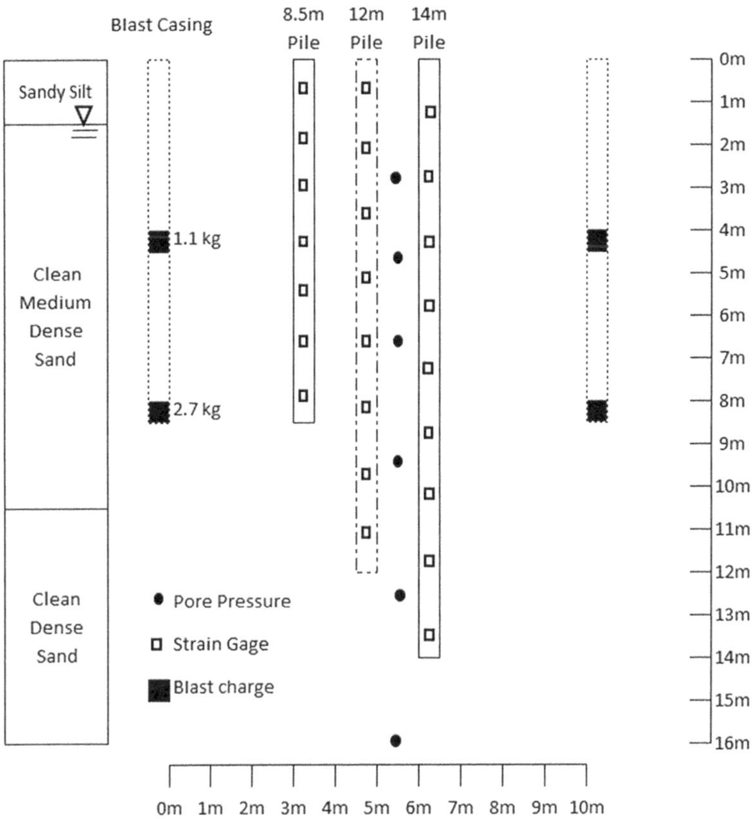

Fig. 5.10 Elevation view of test piles, blast charges, and instrumentation relative to the soil profile

transducers were capable of withstanding maximum blast pressures of 20.7 MPa (3000 psi) associated with the blast but were also capable of recording residual pore pressures to an accuracy of about 1.4 kPa (0.2 psi). Piezometers were installed at depths of 2.75, 4.85, 6.8, 9.7, 12.8, and 15.85 m below the ground surface as shown in Fig. 5.10. The piezometers were typically located about 1 m from the center of the blast ring. Additional details regarding the installation procedures are provided elsewhere (Rollins et al. 2005; Ashford et al. 2004).

5.3.3 *Static Load Testing*

Though presented here for clarity, static load testing was performed on the three test piles about a month after the subsequent blast test to provide information on the actual side resistance and end-bearing in a non-liquefied state. Although the blast liquefaction testing may have affected the test results to some degree, it should be noted that this layer of sand had already been liquefied a number of times in previous earthquakes. As with all deep foundations, there is uncertainty about the axial capacity of auger-cast piles. The piles were loaded using a rectangular steel frame and 272 mt of weights. The rectangular frame rested on the three piles with a load cell installed between the frame and each pile. Pile head settlement for each test pile was monitored by three string potentiometers attached to separate independent reference frames for each pile. Weights were added to the frame and distributed in a manner to concentrate loads first towards the 8.5 m pile and then towards the 12 m and 14 m piles. This testing approach was adopted to maximize load-displacement data regarding all test piles within the constraints of the project budget. A conventional load test on one test pile, with a load frame and reaction piles, would have expended the budget while providing data on only one test pile. In this case, the weights were sufficient to fully mobilize the resistance of the 8.5 m pile, but only provided load-deflection data to about 12 mm deflection for the 12 and 14 m piles which was still sufficient to fully mobilize side friction (Hirany and Kulhawy 2002). End-bearing versus displacement (Q-z) curves proposed by O'Neill and Reese (1999) suggest that mobilized end-bearing resistance was about 50% of ultimate capacity defined by displacement equal to 5% of the pile diameter.

Based on the strain during the static load test, plots of the load in the pile versus depth for both the 12 and 14 m piles were developed as shown in Fig. 5.11. No plots are provided for the 8.5 m pile because lead wires for the strain gauges were damaged during the installation of dowel bars for the support of the load frame. To provide a basis for investigating the measured side friction and end-bearing resistance, the measured values were compared with equations for drilled shaft side friction specified by the US Federal Highway Administration (FHWA) (O'Neill and Reese 1999; Brown et al. 2010). These FHWA equations have often been used as a point of reference in understanding the behavior of auger-cast piles; however, the unit side resistance of auger-cast piles has often been lower than that for drilled shafts in sand. As expected, the FHWA equations predict unit skin friction higher than interpreted from strain in the static load tests.

By trial and error, the unit skin friction predicted by the FHWA method was scaled down to produce improved agreement with the measured skin friction. As shown in Fig. 5.11, the best agreement with measured load versus depth curves was obtained using scaling factors of 70% and 55% of skin friction predicted by the FHWA equations for the 12 and 14 m piles, respectively. Both piles had settled about 12 mm at the time of these measurements so skin friction should have been fully mobilized (Hirany and Kulhawy 2002). Using these reduced skin friction values, the predicted load versus depth curves are in remarkably good agreement with measured curves. Additional analyses to fit the load-settlement curve for the 8.5 m pile suggest that ultimate side friction for that pile was approximately equal to 60% of the value predicted by FHWA equations. These results suggest that on average, for whatever reason, the maximum skin friction on the auger-cast piles is about 60% of the FHWA values.

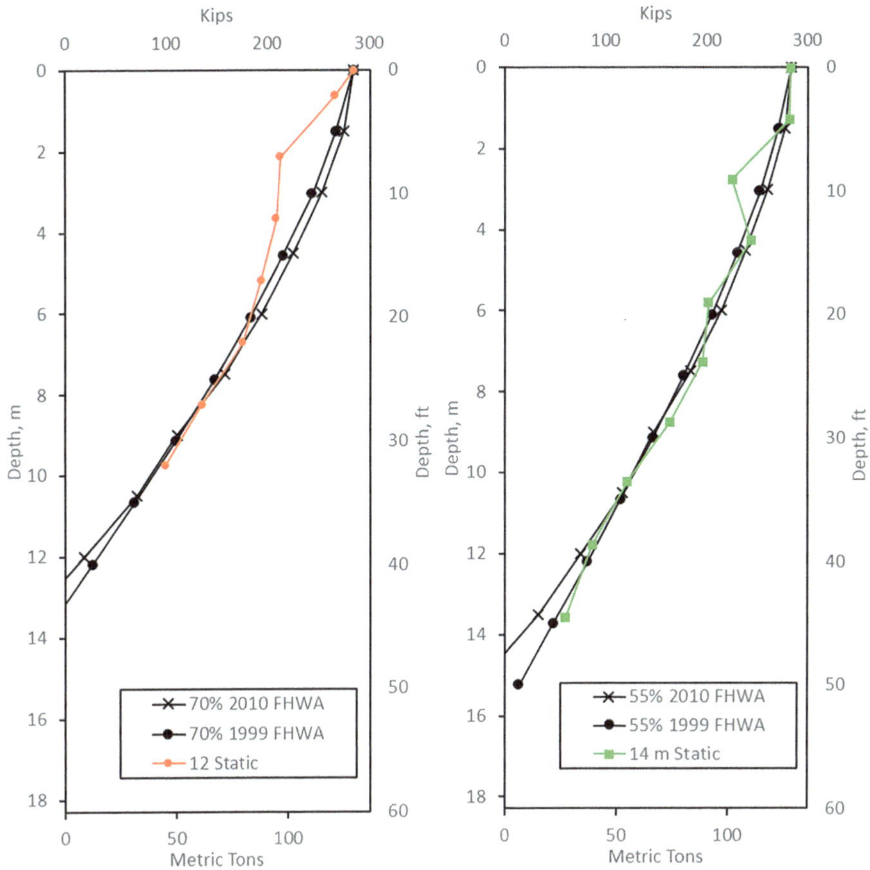

Fig. 5.11 Interpreted pile load versus depth curves from static load tests in comparison with curves as a percentage of load predicted by 1999 and 2010 FHWA equations for drilled shaft design

5.3.4 *Blast Liquefaction Testing*

The explosive charges were detonated one at a time at approximately 0.20 s intervals beginning with the eight 2.7 kg charges at 8.5 m depth followed by the eight 1.1 kg charges at 4.0 m depth. Measurements of the excess pore pressure versus depth relative to the initial vertical effective stress indicate that liquefaction ($r_u = \Delta u/\sigma'_o = 1.0$) occurred between depths of 4 and 14 m which produced settlement in the area around the blast ring of 30–40 mm. Liquefaction was confirmed by the presence of several sand boils around the test area. The settlement around the piles was noticeably less than settlement in adjacent areas subjected to the same blast load and suggests that ground settlement may have been reduced by soil arching around the piles.

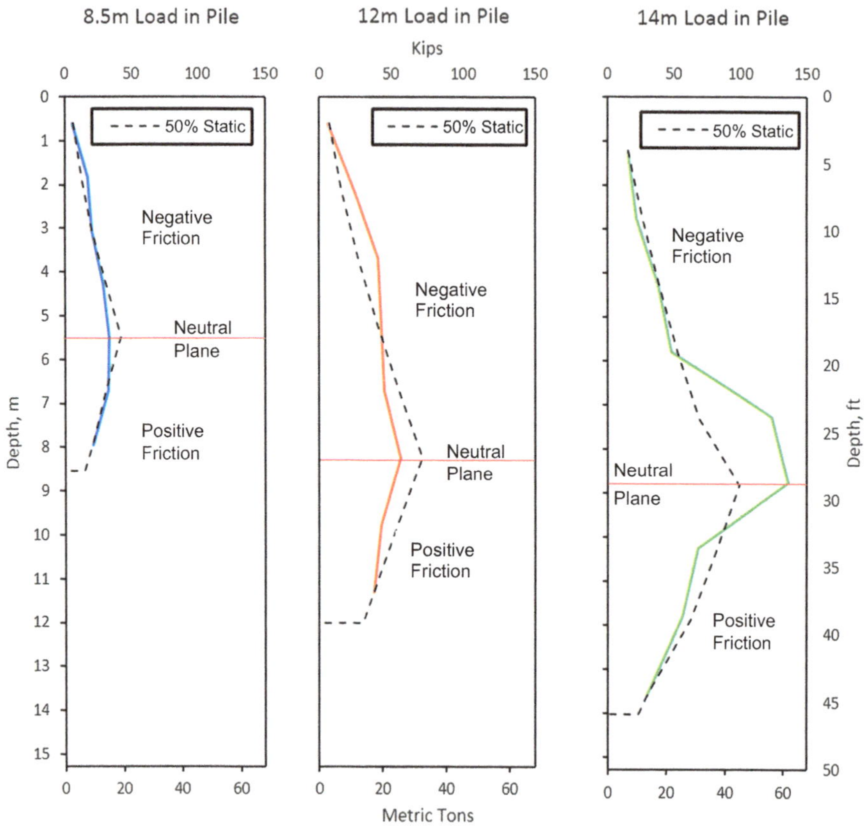

Fig. 5.12 Interpreted pile load versus depth curves (*solid lines*) following blast liquefaction along with predicted curves (*dashed lines*) assuming skin friction equal to 50% of measured average positive skin friction from the static load test. The neutral plane is shown in each plot with a *horizontal line* separating negative skin friction above from positive skin friction below

The test piles settled between 14 and 21 mm. Because the ground settled more than the piles, negative skin friction developed in each case. Plots of the load in each pile as a function of depth interpreted from the strain gauge readings are provided in Fig. 5.12 for the conditions 60 min after blasting when liquefaction induced settlement was completed. Because no pile head load is applied, any load in the piles is induced by negative skin friction or dragload above the neutral plane. Clearly, the negative skin friction is not zero at the end of consolidation. Positive skin friction reduces the load in the pile below the neutral plane. Thus, the neutral plane is visible in each of the plots as the point where the load in the pile begins to decrease. Because the neutral plane in each case is located within the liquefied layer, the positive skin friction below the neutral plane is also reduced relative to the value before liquefaction.

Theoretically, the neutral plane is also the depth where pile and soil settlement are equal. The neutral planes for the 8.5, 12, and 14 m piles are at depths of 5.5, 8.25 and 8.8 m, respectively. The depth to the neutral plane increased as the length of the pile increased suggesting that the pile settlement decreased as the pile length increased. However, the 8.5, 12, and 14 m piles settled 18, 14, and 21 mm, respectively. Variations in pile settlement from this pattern are likely due to differences in soil settlement profiles across the site.

Based on the strain gauges within unit 1 at the top of each pile, the skin friction from the surface down to a depth of about 2.5 m is approximately 24 kPa and is about the same as the non-liquefied value obtained from the static load tests. These results suggest that liquefaction did not occur within unit 1 which is a fine-grained layer or within the zone of partial saturation to a depth of about 2.75 m. This result correlates well with the pore pressure measurements at 2.75 m where the excess pressure ratio was only about 0.4.

Figure 5.12 also presents dashed lines showing the load in the pile assuming 50% of the average positive skin friction found in the static load test developed along the pile length where liquefaction occurred. Because the neutral plane is located within the liquefied zone, both negative and positive skin friction are reduced by 50% in the computations. Agreement with the measured curves is generally very good and confirms the reduced skin friction value obtained from the test in Vancouver. There is a length of the 14 m pile (6–10 m) where the load appears to increase and decrease unexpectedly. The load transfer in this depth interval indicates that skin friction is higher than the static resistance and suggests that the strain gauges may have over-registered the actual applied load. However, even for this pile. the overall agreement between measured and computed load above and below this zone is quite good.

5.4 Conclusions

Blast induced liquefaction led to negative skin friction and pile settlement. Negative friction in non-liquefied layers was approximately the same as positive friction prior to liquefaction. Within liquefied layers, the skin friction immediately after liquefaction may have approached zero. However, as the liquefied layer settled owing to dissipation of excess pore pressures and effective stress increased, negative skin friction also increased at the sand-pile interface. At the end of consolidation, the average negative skin friction was roughly equal to 50% of the positive skin friction before liquefaction obtained from the static load tests in all cases. The consistency of these results strongly suggests that skin friction in liquefied layers is not zero but should be considered to be about 50% of the pre-liquefaction skin friction in assessing dragload and pile settlement following liquefaction. The depth to the neutral plane increased as the pile length increased indicating a tendency for reduced pile settlement for longer piles assuming uniform ground settlement. Although significant ground settlement occurred around the test piles in these field tests, pile settlement was generally within acceptable ranges. For auger-cast piles, the end-bearing and side resistance values can be highly dependent on the construction quality. For the test piles investigated in this study the unit side resistance was typically 50–70% of the unit side resistance predicted by the US FHWA design equations for drilled shafts in sands.

Acknowledgments Funding for the test in Vancouver, Canada was provided by the TRB Ideas Deserving Exploratory Analysis (IDEAS) program and this support is gratefully acknowledged. We also express gratitude to Conetec, Inc. for donating the CPT testing services to the project, and to the British Columbia Ministry of Transportation for allowing use of the test site. Funding for the tests in New Zealand was primarily provided by a grant from the US National Science Foundation (Grant CMMI-1408892) with supplemental funding from the Pacific Earthquake Engineering Research (PEER) Center (Research Agreement Number: 1110-NCTRKR), the Federal Highway Administration and the Utah Department of Transportation Research Division. We also acknowledge the contributions by Michael Olsen from Oregon State University and Tonkin & Taylor in New Zealand in performing this study. This support is gratefully acknowledged; however, the conclusions and recommendations are not necessarily those of the sponsors.

References

Andrus RD, Stokoe KH II (2000) Liquefaction resistance of soils from shear-wave velocity. J Geotech Geoenvir Eng ASCE 126(11):1015–1025

Ashford SA, Rollins KM, Lane JD (2004) Blast-induced liquefaction for full-scale foundation testing. J Geotech Geoenvir Eng ASCE 130(8):798–806

Boulanger RW, Brandenberg SJ (2004) Neutral plane solution for liquefaction-induced drowndrag on vertical piles. In: Proceedings of ASCE geo-trans conference. ASCE, Reston, pp 470–479

Bozozuk M (1981) Bearing capacity of pile preloaded by downdrag. In: Proceedings of 10th international conference on soil mechanics and foundation engineering, vol 2. Stockholm, pp 631–636

Brown DA, Turner JP, Castelli RJ (2010) Drilled shafts: construction procedures and LRFD design methods. Publication No. FHWA-NHI-10-016, US Federal Highway Administration

Bustamante M, Gianeselli L (1982) Pile bearing capacity predictions by means of static penetrometer CPT. In: Proceedings of the 2nd European symposium on penetration testing. Amsterdam, pp 493–500

Eslami A, Fellenius BH (1997) Pile capacity by direct CPT and CPTu methods applied to 102 case histories. Canadian Geotech J NRC Canada 34(6):886–904

Fellenius BH (1996) Basics of foundation design. BiTech Publishers Ltd., Richmond. 134 p

Fellenius BH (2006) Results from long-term measurement in piles of drag load and downdrag. Can Geotech J., April 2006 43(4):409–430

Fellenius BH, Siegel TC (2008) Pile drag load and downdrag in a liquefaction event. J Geotech Geoenvir Eng. ASCE, Reston, Virginia 134(9):1412–1416

Hirany A, Kulhawy FH (2002) On the interpretation of drilled foundation load test results. Deep Foundations GSP116, 2:1018–1028

Ishihara K (1990) Liquefaction and flow failure during earthquakes. Géotechnique, ICE, London 43(3):351–451

Ishihara K, Yoshimine M (1992) Evaluation of settlements in sand deposits following liquefaction during earthquakes. Soils Found 32:173–188

Kulhawy FH, Mayne PW (1990) Manual on estimating soil properties for foundation design. Research Report EERI EL-6800, Electric Power Research Institute, Palo Alto, CA

Lim JK, Lehane BM (2014) Characterisation of the effects of time on the shaft friction of displacement piles in sand. Geotechnique 64:476–485

Monahan PA, Luternauer L, Barrie JV (1995) The geology of the CANLEX phase II sites in Delta and Richmond British Columbia. In: Proceedings of the 48th Canadian geotechnical conference. Vancouver, pp 59–68

O'Neill MW, Reese LC (1999) Drilled shafts: construction procedures and design methods. Publication No. FHWA-IF-99-025, US Federal Highway Administration, 535p

Robertson PK, Campanella RG, Gillespie D, Grieg J (1986) Use of piezometer cone data. In: Proceedings on use of in-situ tests in geotechnical engineering. ASCE, Reston, pp 1263–1280

Robertson PK, Wride CE, List BR, Atukorala U, Biggar KW, Byrne PM, Campanella RG, Cathro DC, Chan DH, Czajewski K, Finn WDL, Gu WH, Hammamji Y, Hofmann BA, Howi JA, Hughes J, Imrie AS, Kinrad JM, Küpper A, Law T, Lord ERF, Monahan PA, Morgenstern NR, Phillips R, Piché R, Plewes HD, Scott D, Sego DC, Sobkowicz J, Stewart RA, Watts BD, Woeller DJ, Youd TL, Zavodni Z (2000) The Canadian liquefaction experiment: an overview. Can Geotech J, NRC Canada 37(3):499–504

Rollins KM, Anderson JKS (2004) Performance of vertical geocomposite drains based on full-scale testing at Massey Tunnel, Vancouver, B.C. Final Report, NCHRP-IDEA Project 94, Transportation Research Board, National Academy Press, Washington, DC, 107 p

Rollins KM, Strand SR (2006) Downdrag forces due to liquefaction surrounding a pile. In: Proceedings of the 8th national conference on earthquake engineering. Earthquake Engineering Research Institute,10 p

Rollins KM, Lane JD, Dibb E, Ashford SA, Mullins AG (2005) Pore pressure measurement in blast-induced liquefaction experiments. In: Transportation research record 1936, "Soil Mechanics 2005." Transportation Research Board, National Academy Press, Washington, DC, pp 210–220

Tokimatsu K, Seed HB (1987) Evaluation of settlements in sands due to eathquake shaking. J Geotech Env Eng 103(8):861–878

Wride, C.E (Fear), Robertson, P.K., Biggar, K.W., Campanella, R.G., Hofman, B.A., Hughes, J.M. O., Küpper, A., Woeller, D.J. (2000) Interpretation of in situ test results from the CANLEX sites. Can Geotech J, NRC Canada, 37(3), 505–529

Youd TL, Idriss IM, Andrus RD, Arango I, Castro G, Christian JT, Dobry R, Finn WDL, Harder LF, Hynes ME, Ishihara K, Koester JP, Liao SSC, Marcuson WF, Martin GR, Mitchell JK, Moriwaki Y, Power MS, Robertson PK, Seed RB, Stokoe KH (2001) Liquefaction resistance of soils: Summary report from the 1996 NCEER and 1998 NCEER/NSF workshops on evaluation of liquefaction resistance of soils. J Geotech Geoenvir Eng, ASCE, Reston 127(10):817–833

Zhang G, Robertson PK, Brachman RWI (2002) Estimating liquefaction-induced ground settlements from CPT for level ground. Can Geotech J, NRC Canada 39(5):1168–1180

Chapter 6
Whole Life Sensing of Infrastructure

Kenichi Soga

Abstract One of the greatest challenges facing civil engineers is the stewardship of ageing infrastructure. At present, it is difficult to inspect the condition of infrastructure while it is being used. The fragility of old infrastructure presents a challenge for new construction in congested urban environments. Little is also known of the long-term performance of such infrastructure. We see many cases around the world of continuous retrofit/renovation of infrastructure during its lifetime. The vision discussed in this chapter is to significantly improve our capability in predicting and managing the life expectancy of large infrastructure. For future proofing, active monitoring of construction and operational processes of civil engineering infrastructure is essential. This implies that structures are instrumented to assess their performance against engineering design parameters or predictive models. In recent years, sensor and communications research has been undergoing a revolution. There are possibilities to use emerging sensor technologies (distributed fiber optics sensing, wireless sensor network, low power miniature sensors, energy harvesting for continuous monitoring, robotic inspections, satellite images, crowd source data, etc.) to address the particular needs to look after our infrastructure.

6.1 Introduction

The economic potential and social wellbeing of cities relies heavily on physical infrastructure. Although many of our physical infrastructure is constructed to last for many years, existing infrastructure is challenged by the change in load and usage (e.g. number of passengers carried, numbers of vehicles or volume of water used). Change in usage is not the sole concern; for example, a demographic change to an older population in certain developed countries will alter the way we use our

---

K. Soga (✉)
University of California, Berkeley, CA, USA
e-mail: soga@berkeley.edu

S. Iai (ed.), *Developments in Earthquake Geotechnics*, Geotechnical, Geological and Earthquake Engineering 43, DOI 10.1007/978-3-319-62069-5_6

infrastructure. We therefore need to assess the resilience of our infrastructure against rapid changes in social networks and policies.

Ideally infrastructure should be designed to be adaptable to satisfy the demands. However, the past engineering design philosophy has been rather rigid based on a demand prediction given at that time. Therefore, we have to embrace the fact that many of our current existing infrastructure in cities will be fixed in space for many decades. The high cost involved in upgrading will lead to a desire to extend their life. These demanding challenges create issues on ageing infrastructure (e.g. UK National Infrastructure Delivery Plan 2016–2021 HM Treasury 2016; State of Nation ICE 2016; Report Card for America's Infrastructure ASCE 2013).

Another challenge for civil engineers and infrastructure owners is to make decisions when a catastrophic event occurs. Infrastructure systems and projects face a multitude of hazards that must be assessed, communicated, and managed appropriately. The Fukushima Power plant event in 2011 in Japan and major flooding and extreme weather events in recent years around the world are good examples of this. In particular, sensing the performance of linear infrastructure systems such as power supply, buried pipelines, railways and flood defense embankments is challenging because any break in a system can potentially disrupt the operation of the whole infrastructure as well as influences the other neighboring infrastructure.

To perform these assessments, there is a need to develop and deploy sensor networks for rapid condition assessment of critical structures in order to operate optimally and to ensure system safety and resilience. There are opportunities to extend the working conditions of the existing sensing devices to the range of extreme civil conditions, such as large deformation, high pressure, extremely high and low temperatures and moisture. These sensor systems provide 'intelligence in extremes'. This effort can be coupled with high-performance computer simulation and advanced visualization tools to conduct risk assessments at the citywide scale, considering multiple hazards such as earthquakes, tsunamis, flooding, and fires.

For future proofing, active monitoring of construction and operational processes of civil engineering infrastructure is essential. This implies that structures are instrumented to assess their performance against engineering design parameters or predictive models. Emphasis on monitoring is not new in civil engineering. In fact, monitoring has been extensively used in civil engineering for decades ever since the use of the observational method where feedbacks from monitoring systems are used to inform and potentially modify the design and construction. In recent years, sensor and communications research has been undergoing a revolution. An effective use of existing and new smart monitoring systems with better understanding of how people use the infrastructure services would lead to the realization of resilient adaptable infrastructure systems. There are possibilities to use other emerging sensor technologies (distributed fiber optics sensing, wireless sensor network, low power miniature sensors, energy harvesting for continuous monitoring, robotic inspections, satellite images, crowd source data, etc.) to address the particular needs to look after our infrastructure.

6.2 Long-Life Infrastructure and Ever Changing Demands

In infrastructure cases with long life cycles, assets face external conditions within much shorter timeframes that deviate from what was known or assumed by the planner or designer at the outset; for instance, population growth/reduction, climate change, energy price change, and transport mode changes. If a designer ignores the possibility that external conditions may evolve unexpectedly over time, then a substantial risk arises that the designed object will become inadequate. We see many cases around the world of continuous retrofit/renovation of infrastructure during its lifetime. An example of this particular issue is given in this section.

At London Bridge station in London, redevelopment work has been ongoing over a period of 5 years since 2013. This station is one of the busiest stations in London and has 250,000 users every day. The new redevelopment work is designed to cope with the prediction that there will be 50% increase in the users in the next 10 years or so. It will have the longest platforms in London with trains departing and arriving every 5 min.

The point to note here is that this station has been continuously renovating from the time it was opened in 1886. Especially, in the 1970s, 40 years ago, a major renovation work took place. This renovation was called the 'Operation London Bridge', which made two stations into one and linked other transportation networks such as buses and subways (see Fig. 6.1a). The following sentence summarizes the features of the renovated station.

> Two old railway stations will be merged into one with a higher capacity, giving easy interchange between buses, tube and trains – and direct access to all service from the spacious concourse with new bars, buffets and shops. (Prouting and Lochner 1975)

The station is now in a situation where it is necessary to redevelop again only 40 years later. The redevelopment work currently undertaken (see Fig. 6.1b) is similar to the previous one as described below.

> The number of platforms will increase and track layout will accommodate higher capacity trains. At the same time, existing bus, train and underground services will be linked with the largest concourse in the UK which will offer a variety of retail services. (Rail 2012)

Figure 6.2 shows the change in the population of London. The trend in the 1970s is a decreasing one, whereas that in the 2000s is an increasing one. Due to the rapid growth of the London financial industry since the 1990s, the population is expected to increase and the current number of railway users in London has doubled compared to 20 years ago. The infrastructure that has a long-life cycle has exceeded its allowable capacity due to the change in users' pattern after 40 years.

As shown in the case of London Bridge station, the utilization of infrastructure continues to change within several decades due to economic growth, industrial structure change and climate change. Timescales in infrastructure and construction are long, with construction time-scales alone stretching from 2 to 10 years, and asset lifetimes in the range of 60–100 years or even more. That is, civil engineering structures are often fixed in space and time (e.g. 120-year design life) and provide

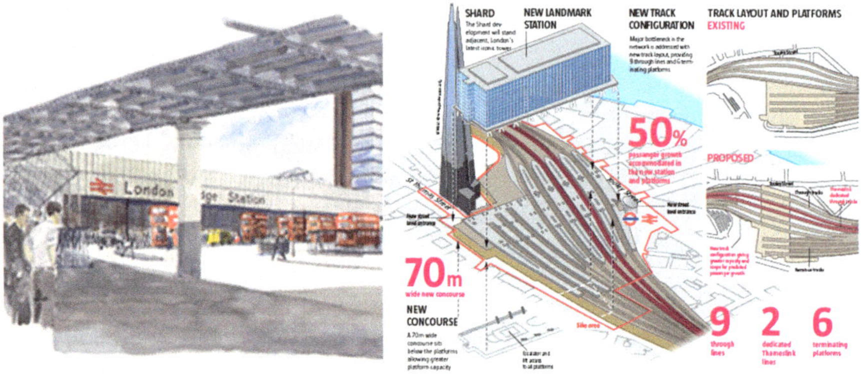

(a) (b)

Fig. 6.1 London bridge station redevelopment projects. (**a**) 1970s (**b**) 2010s (Cole 2010)

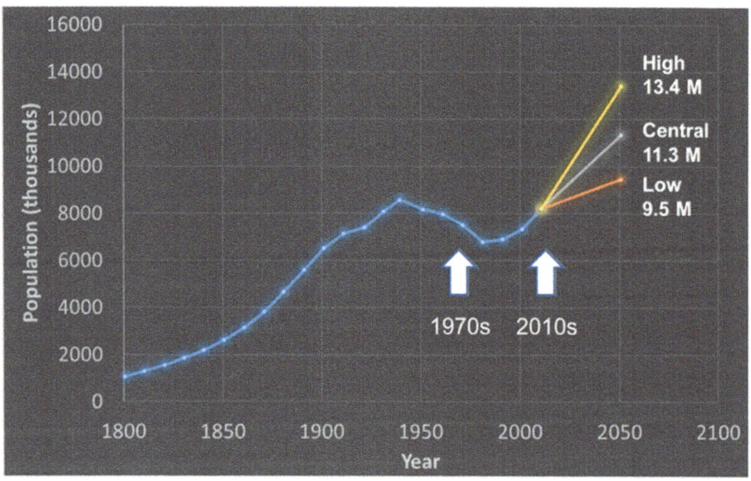

Fig. 6.2 Population change in London

independent services for transportation, energy supply, water, sewage and communication without any appreciable linkage. Each of these elements is operated with different business models, is guided by different performance metrics and deals with systems that involve different degrees of interconnectedness and time scales in terms of ageing and the requirements for repair and maintenance.

Due to different rates of technical development between monitoring system versus infrastructure usage, some data we will be using may be from older sensors, and some of the sensors may be embedded now but data will be used 10, 20 or 50 years later. At present, there is a mismatch between the life span of infrastructure and that of sensor systems, which makes the concept of whole-life-cycle based asset management difficult to achieve. Sensor systems to fulfil this concept ('intelligence for life') need to be either long-life or adaptable for replacement.

6.3 Whole-Life Infrastructure Sensing

The observational method proposed by Peck (1969) for example is to learn from monitoring data and to feedback the findings to design systematically. The intention is to improve the efficiency of construction for future infrastructure. If the infrastructure can proactively be monitored for a long period of time even after construction, we may be able to construct a design concept that considers the lifecycle span, enabling efficient maintenance, remodeling and extension of life.

In the field of civil engineering, there are various monitoring systems such as visual inspection and survey measurement, while introducing a new sensing system is not easy due to dependence on past experiences and customs, safety considerations, etc. In contrast, in other fields (for example, automobiles, aerospace, electronics manufacturing, oil and natural gas), business models have already been established to increase safety and to reduce lifecycle costs by sensing. For example, Rolls-Royce, an aircraft engine manufacturer, has a business model in the form of a comprehensive contract called Power-by-the-hour. A company that purchased their engine increases its safety by paying the expenses on a time-of-flight basis. This is done by installing many sensors on the engine and constantly monitoring it on-line during service and maintenance. Such data can be used to improve their next-generation engines. Rolls-Royce began this business model in the 2000s, but it is now being introduced to other industries.

In recent years, the Internet of Things (IoT) has attracted new technologies related to sensors and communication and they are becoming a part of everyday life. If any sensor technology that can be used for a long period (a period equal to the infrastructure lifetime) is available, it is possible to introduce the whole lifecycle approach in the civil engineering industry. However, to realize this approach, the following questions need to be answered.

- What sensors are needed to measure the performance of infrastructure to make engineering decisions?
- How can the robustness and reliability of sensors be proven?
- Whether data should be collected periodically or automatically?
- How do we analyze the data in order to obtain reliable and valuable results?
- Is it better to measure accurately with a limited number of measurements or to indirectly monitor behavior using data mining?
- Can sensing of general users replace some of the traditional sensor systems?

6.4 Candidate Technologies for Whole-Life Infrastructure Sensing

There are sensing technologies potentially available for monitoring a variety of lifecycle attributes. Some of them are introduced in this section.

6.4.1 Distributed Fiber Optic Sensing

Distributed fiber optics sensing technology utilizes the sensitivity of the optical fiber to environmental changes such as temperature, humidity, strain, vibration and noise (acoustic). In this technology, standard optical fiber becomes a sensor and tens of kilometers of fiber can be sensed at once for continuous distributed measurement of the conditions around the optical fiber such as temperature, strain, acoustic noise, etc. Because of its simple and quick installation, distributed optical fiber sensing can be as equally practical as other conventional measurements.

As shown in Fig. 6.3, when passing light through the fiber, most of it passes but only a small part of light scatters. This scattering phenomenon is related to the nonuniformity of the refractive index of the glass medium. When there is a change in the temperature, strain, vibration, etc. of the fiber, the characteristics of the laser light signal proceeding in the optical fiber and the scattering phenomenon change. The scattering signals include Rayleigh, Brillouin and Raman and their signatures such as amplitude and natural frequency are influenced by changes in temperature and strain.

Different types of analyzer detect different scattering signals in a different way (see Fig. 6.4). For example, Phase-OTDR is capable of quantifying both acoustically induced and dynamic multiple strain perturbations (Lu et al. 2010). It can perform long distance dynamic sensing for vibration and acoustics. The Raman Optical Time Domain Reflectometry (ROTDR) is a classical technology to measure temperature at many points along the length of an optical fiber (Hartog et al. 1985). Distributed temperature sensing systems (DTS) is a Raman scattering based system, which can be used for power cable and transmission line monitoring, fire detection, leakage detection at dikes, dams and sewers. It is widely used in downhole temperature monitoring of oil and gas wells. Phase-OTDR (Distributed Acoustic Sensing (DAS)) can be used for security monitoring for long borders and perimeters at high-value facilities and high-level security locations, or real-time position and speed monitoring of trains. Brillouin scattering based techniques such as time domain techniques called 'Brillouin optical time-domain reflectometery (BOTDR)' and 'Brillouin optical time domain analysis (BOTDA)' are well-established for distributed strain measurement (Horiguchi et al. 1989; Kurashima et al. 1990; Niklès et al. 1996) There are other strain sensing techniques such as BOFDA (Garus et al. 1997) and BOCDA (Hotate and Hasegawa 2000).

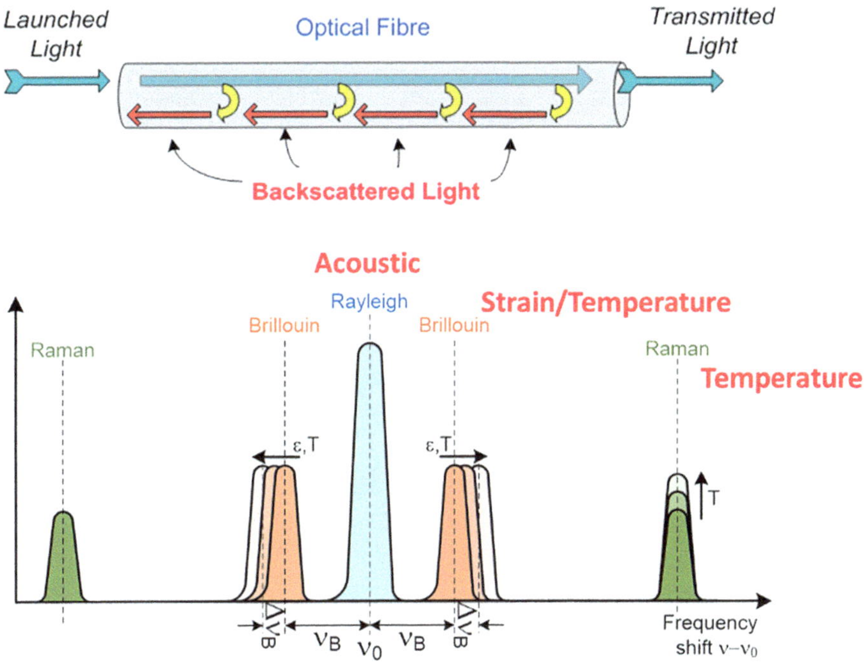

Fig. 6.3 Change in scattered light, amplitude and natural frequency

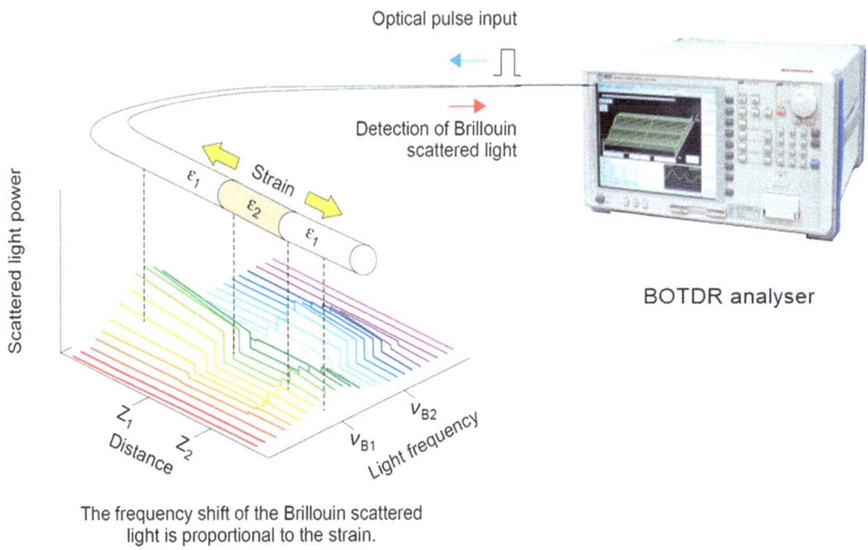

Fig. 6.4 Distributed strain measurement optical fiber system consisting of cable and analyzer

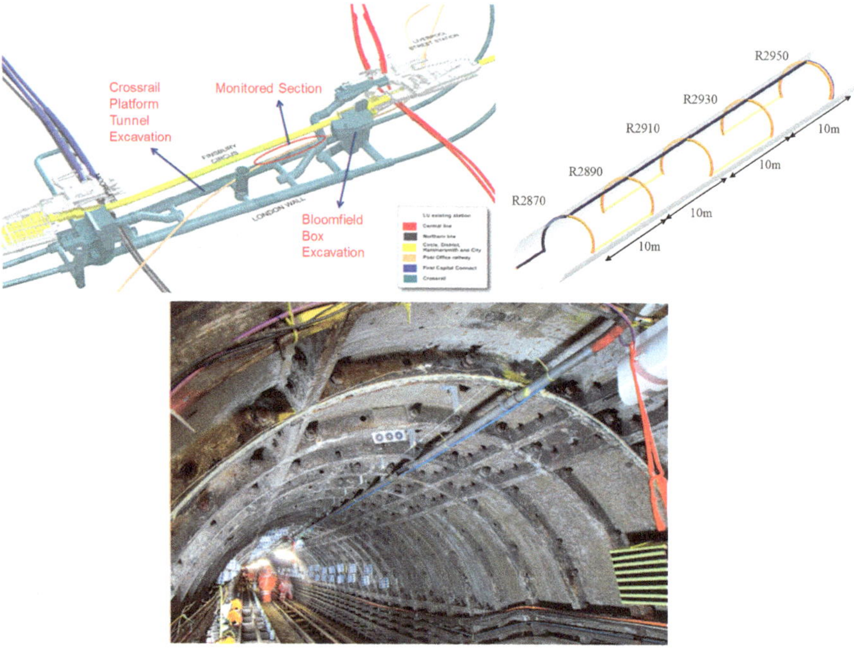

Fig. 6.5 Optical fiber monitoring at the London underground construction

The design of civil engineering structure is often carried out with consideration of limiting values of strain. BOTDR uses spontaneous Brillouin scattering phenomena using incident light to generate Brillouin signals, whereas BOTDA uses the stimulated Brillouin scattering phenomenon with an external excitation signal to amplify the Brillouin signal. The change in the frequency of the Brillouin scattered light is proportional to strain. Strain distribution can be obtained over the entire length of an optical fiber by analyzing the signal of scattered light in space.

In retaining walls, underground structures, tunnels, and pipelines, the external loads (such as soil loads) are often a distributed load. By measuring the distributed strain profile, it is possible to estimate the characteristics of the external loads. Figure 6.5 shows a case where a fiber optic monitoring system was introduced in a 100-year cast iron tunnel (Gue et al. 2015). Here, under the tunnel with a diameter of 2.7 m, an underground station of 11 m in diameter was constructed in parallel to the tunnel. Since the distance between the two tunnels was 2 m or less, optical fibers were installed at five locations in the circumferential direction and the longitudinal direction along the tunnel crown for the purpose of monitoring the degree of distortion during the adjacent construction. Figure 6.6 shows the time-varying data of strain distribution detected by the optical fibers installed in the circumferential direction of ring R2870 (from the side face to the ceiling face and then the opposite side face). At the initial stage (green line group), one side shows tensile strain and the other side shows compressive strain, but over time the compression section moves to the ceiling part and the side experiences tensile strain. This is because an

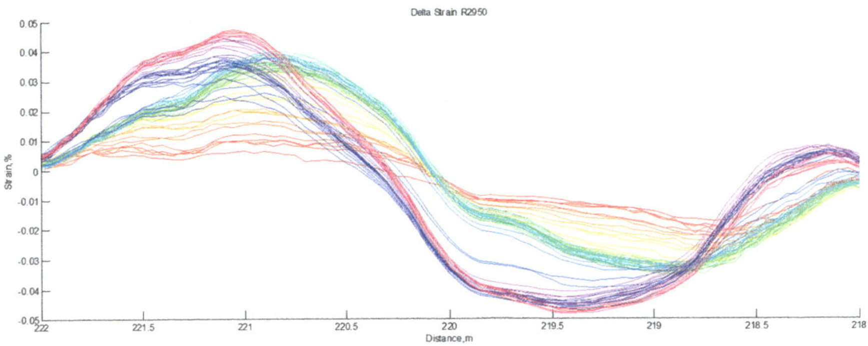

Fig. 6.6 Example of optical fiber data

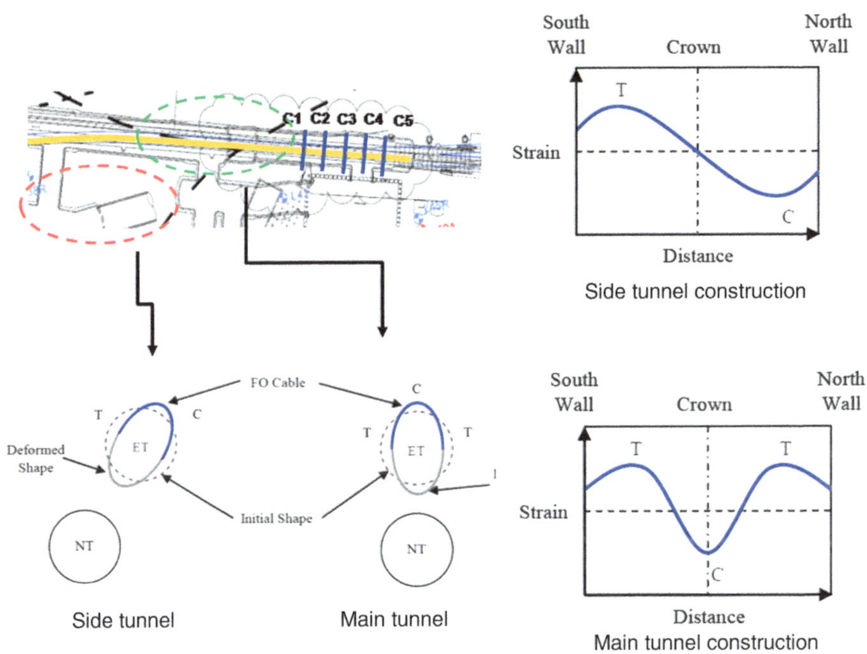

Fig. 6.7 Deformation mechanism of tunnel

obliquely tunnel (the part surrounded by red) was excavated first, and then the platform was built just under it, as shown in Fig. 6.7. The deformation mechanism expected at the time of each construction stage and the strain distribution corresponding to it are shown in the right diagram of Fig. 6.7. The data shows that distributed optical fiber strain measurement can be equivalent to attaching several thousands of "strain gauges". The system can be used not only to verify the amount of strain but also to derive the mechanism of overall deformation.

The characteristic of distributed fiber optics measurement technology is that it can continuously measure distribution of temperature, strain, and acoustic noise of several kilometers or several tens of kilometers using an inexpensive optical fiber. Since the deterioration rate of the glass optical fiber material is slow, it is considered to be an ideal measurement method for long-term monitoring by embedding it in civil engineering structures. Furthermore, the performance of analyzers is likely to be improved in the future, which in turn the accuracy of the data improves with time in the long term. However, techniques for attaching and/or embedding optical fibers to structures are required to be improved in the future. Figure 6.8 shows photos of fiber optics installation in different geotechnical structures. Further details of this technology can be found in (Soga et al. 2015; Kechavarzi et al. 2016).

6.4.2 Computer Vision

Recent advances in digital images have reached to the stage of renewing existing surveying techniques. For example, using the technique called Structure from Motion (SfM), we can automatically detect the same part seen in multiple images and find the location of the camera where the image was taken. The object appearing in the image can then be constructed as a three-dimensional one. For example, Fig. 6.9 shows models of tunnels and rock slopes made from their images. Image data is suitable for long-term record of the state of infrastructure, so it is considered to be very important as a tool to assess our infrastructure in the long term.

Even with an old infrastructure without digital design information, it is possible to compare the past images with the current ones by constructing a three-dimensional model developed from the current images. Visual inspection is one of maintenance work of civil engineering structures in order to detect abnormalities such as cracks. Figure 6.10 shows images of water leaks on a subway tunnel surface in Prague in 2003 and 2007. Comparing the two images can be difficult because the camera positions are different. However, if these images were linked with the three-dimensional model, it is possible to digitally transform the two images as if they are taken from the same location. Figure 6.11 (a) is an image of year 2003, (b) and (c) are images of year 2007. Comparing (a) and (c), it is possible to discriminate color change, spread of water leakage, installation of new cables, etc. As in this example, if images gathered throughout years were linked and organized with the associated three-dimensional model, it is possible to access the image database at the site and compare it with the image record of the past.

Figure 6.12 is a three-dimensional model developed based on images of a new tunnel in London, whereas Fig. 6.13 shows results of three sites using various image analysis algorithms to detect any changes (Stent et al. 2016). Case (a) contains a water splash marker, case (b) has thin lines (like cracks), and case (c) has no change. The ground truths (true solutions) are shown in the third column. The fourth to seventh columns are the results of various image processing methods, and they are

Fig. 6.8 Example of optical fiber strain measurement. (**a**) Piles. (**b**) Tunnels. (**c**) Retaining walls

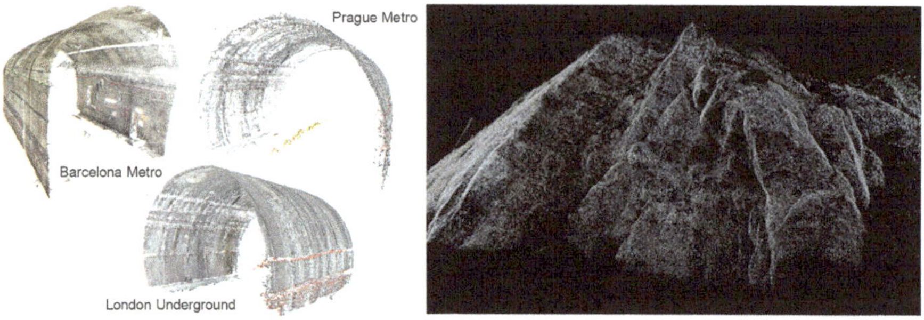

Fig. 6.9 Three-dimensional models of tunnel and rock slope made from digital photograph

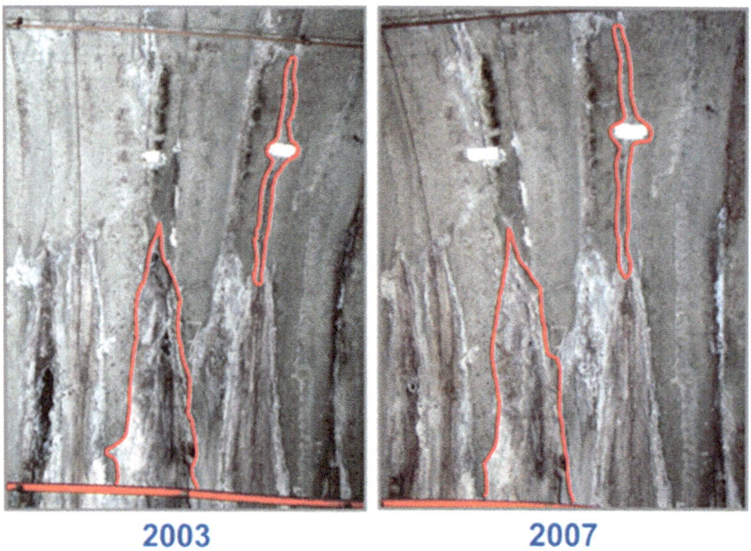

Fig. 6.10 Photo of 2003 and 2007 photos of leakage sites in Prague subway

integrated in the eighth column to show the performance of the change detection algorithm developed for this project. In most cases, the deliberately applied variants were detected without any problems, but there were also false detections. At present, there are no image analysis algorithms that can accurately detect every changes and different algorithms need to be applied appropriately. Further improvement and innovation are needed to utilize computer vision technology for practical use.

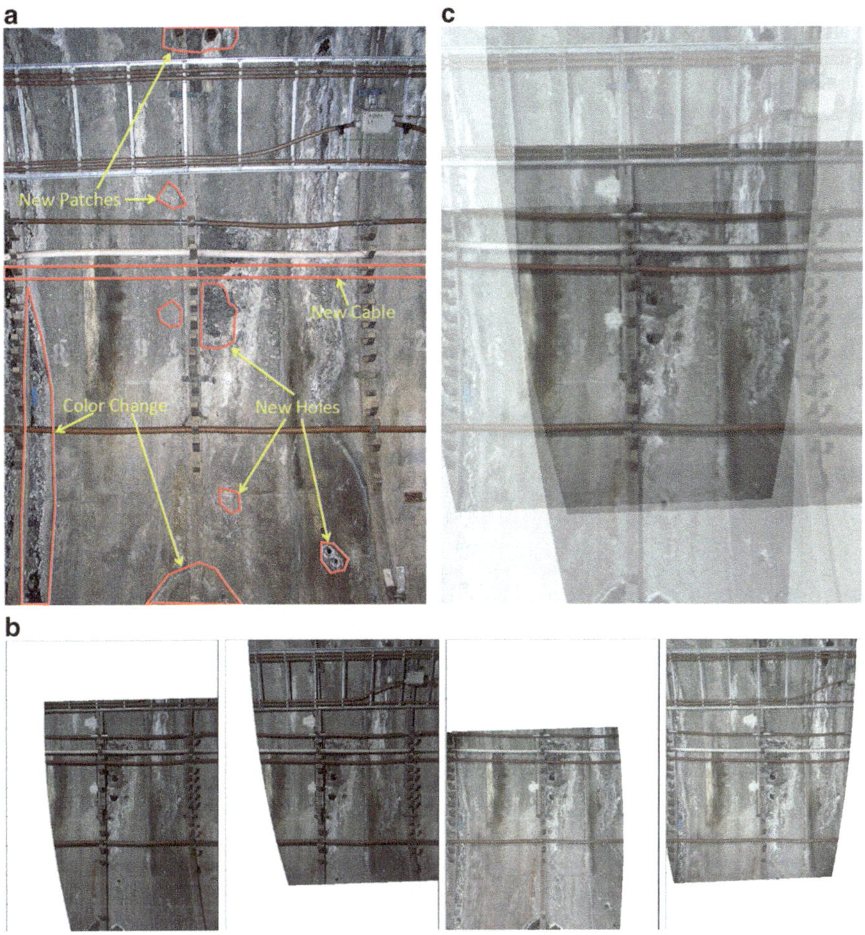

Fig. 6.11 (**a**) Pictures in 2003, (**b**) photographs taken during 2007, matching the ones in 2003 and transforming the pictures as taken from the same place, (**c**) photographs of 2007 Synthesized (Chaiyasarn 2011)

6.4.3 Wireless Sensor Network

For monitoring large-scale infrastructure, introducing a wired monitoring system takes a great deal of time and cost. On the other hand, wireless sensor network (WSN) is a very attractive technology in which cable installation is not necessary. By combining it with high precision and low cost sensors, it enables large scale monitoring which was not possible in the past. The advantages of WSN to monitor the behavior of civil infrastructure will be as follows. (i) multiple sensors can be deployed without installing cables, (ii) sensing in areas difficult to connect is

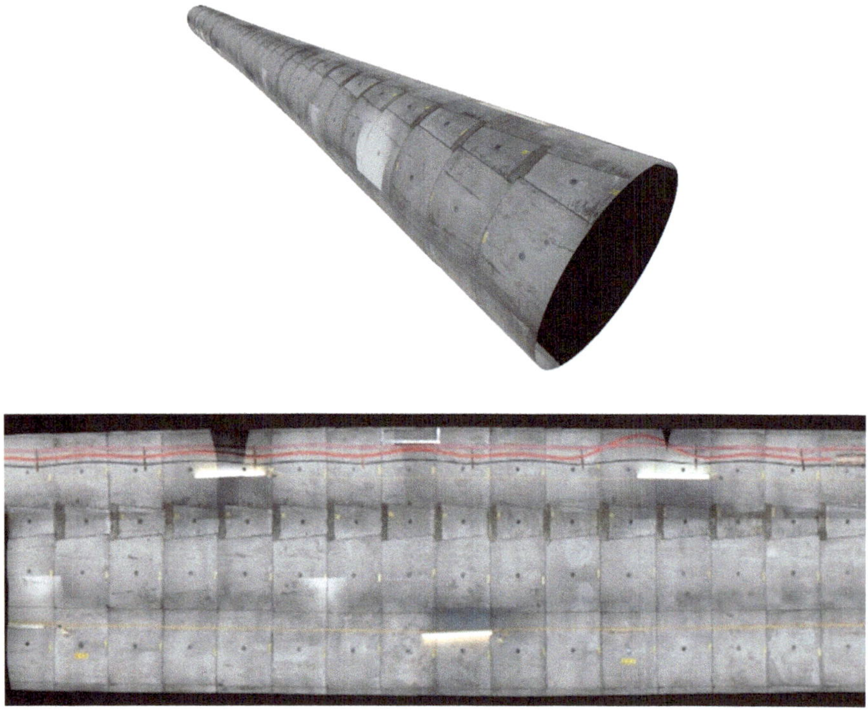

Fig. 6.12 A three-dimensional model of a tunnel constructed from images

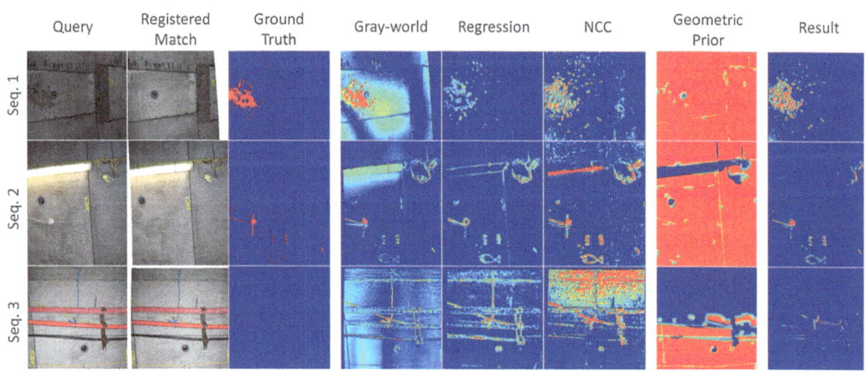

Fig. 6.13 Example of detection of variations using various image analysis algorithms (Stent et al. 2016)

possible, and (iii) since it is possible to obtain a wide range of data in real time, decision making can be made quickly.

Thanks to the Internet of Things (IoT) initiative in recent years, the performance of WSN hardware and software has dramatically improved, not only in data

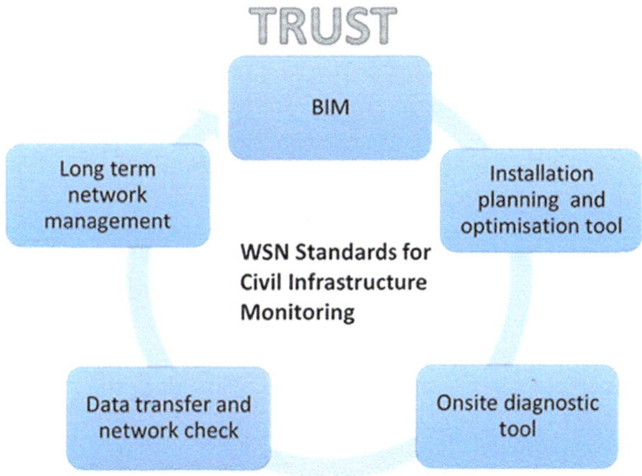

Fig. 6.14 Reliable WSN establishment in infrastructure monitoring

collection and abnormal signal detection, but also in data compression processing, data analysis, network processing etc. Thanks to low power CPUs, it is now possible to perform data processing at the sensing level, which in turn saves energy by sending only the necessary data.

Although WSN's application in the civil engineering field is very attractive, it is necessary to recognize that the current wireless sensor system does not have the robustness as reliable as the wired system (Stajano et al. 2010; Rodenas-Herráiz et al. 2016). In particular, WSN installation still remains as a trial and error exercise and depends greatly on personal experience. It is also difficult to manage large-scale WSNs that end users can trust while maintaining cost performance. Wireless communication, power supply for long-term monitoring and expandability of the network are also issues to be solved. Figure 6.14 shows an integrated framework for designing, installing, and managing of the WSN, which can be used when WSN deployment is considered.

Figure 6.15 shows the latest WSN system developed at the University of Cambridge (Bevan 2016). The key features are ultra low power, small size, and artificial intelligence. Readings and evaluations of acceleration, inclination and displacement are conducted by on-board calculation. Data communication is optimized and power consumption is suppressed. Since it is compact and lightweight, it can be installed relatively easily in places where installation is dangerous or difficult.

6.4.4 MEMS and Energy Harvesting

MEMS (Micro-Electro-Mechanical Systems) is a compact sensor system of micron and millimeter scale. It is characterized by a compact integrated device system combining with mechanical sensing and electrical data processing. Because it uses the semiconductor integrated circuit fabrication technology, it is suitable for mass

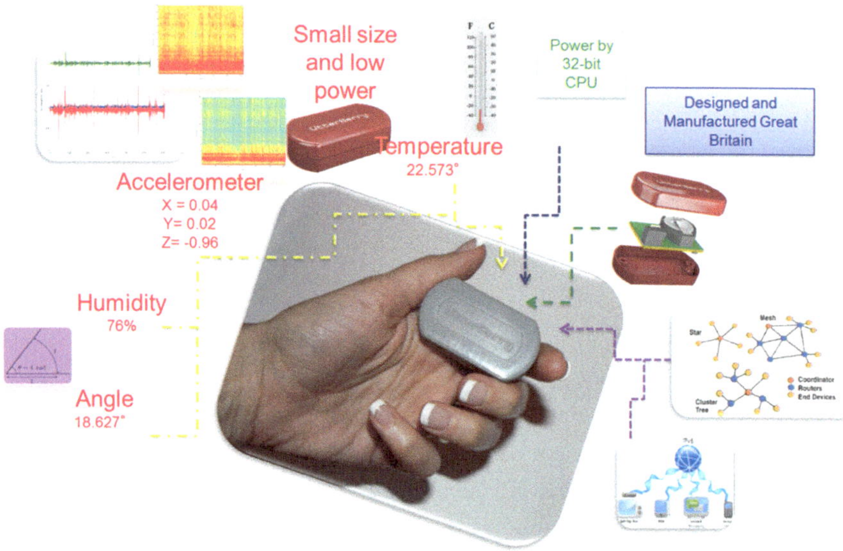

Fig. 6.15 WSN system developed at Cambridge University (Stajano et al. 2010)

production and there is a possibility to provide a low cost sensor. MEMS sensors that have already been commercialized can measure acceleration, inclination, temperature and pressure.

When monitoring civil engineering structures, strain data are useful to check the design. Foil type strain gauges or vibrating wire strain gauges are often used for on-site monitoring, but because of their high power consumption, they are not suitable for WSN. Therefore, a high resolution and low power strain sensor is necessary. There is recent development in a MEMS type strain sensor as shown in Fig. 6.16. The prototype MEM strain sensor was deployed in the field in order to check the progress of cracks in the subway tunnel wall in Prague. A thin steel plate with a newly developed multi-directional MEMS strain sensor attached was placed on the cracks. The details are reported by Belsito et al. (2013, 2016) (Fig. 6.16).

Using a low power consumption MEMS sensor enables continuous long-term monitoring even with small batteries. In some cases, energy present in the environment surrounding the sensor (e.g., sunlight, wind power, vibration, temperature change) can be used to supply power to the MEMS device. There is a possibility that even a sensor system without power supply can be constructed (Ye and Soga 2012). For example, as shown in Fig. 6.17, Cambridge University has developed a vibration energy harvesting system (VEH) that is capable of absorbing vibration energy utilizing the concept of parametric excitation (Jia et al. 2014a, b). Such environmental power generation technology enables long-term monitoring of areas where access is limited. This reduces maintenance costs for battery replacement and hazardous waste disposal. Energy harvesting technologies can contribute to realize long-term monitoring of infrastructure.

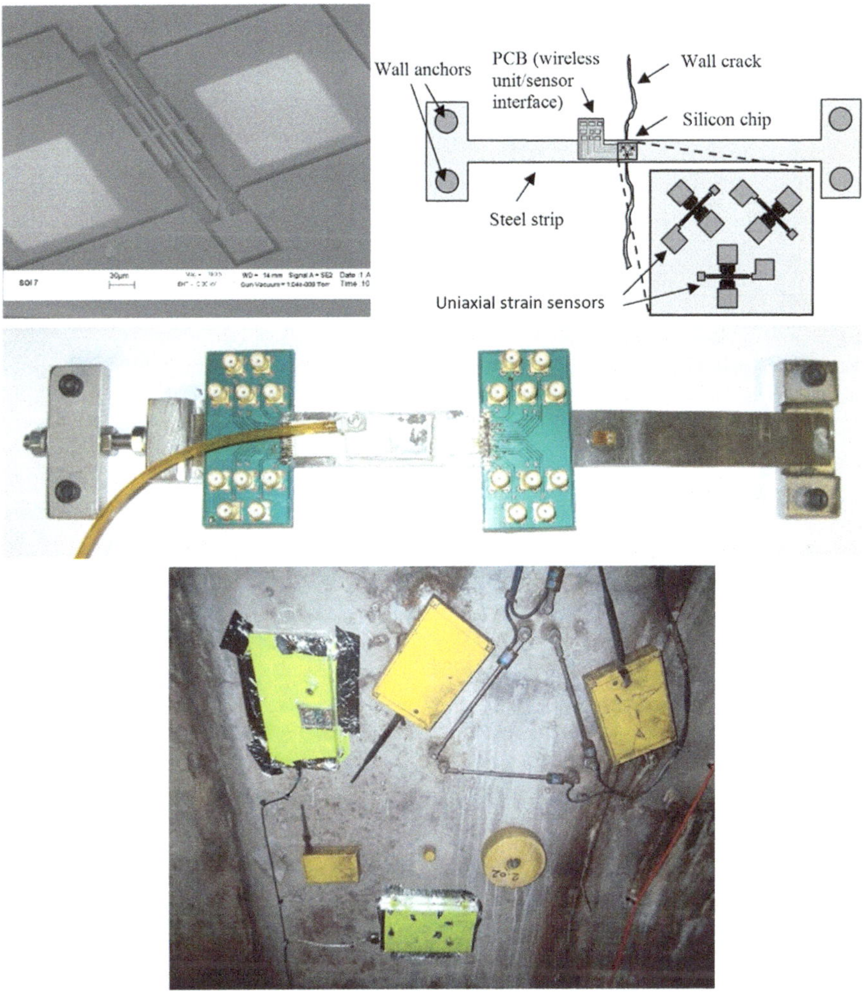

Fig. 6.16 MEMS-type strain sensor developed at the University of Cambridge and its application for crack monitoring in Prague Metro

6.5 Closure

Recent developments in sensor and communication technology are expected to bring about a change in the monitoring method used for civil engineering. This paper introduced the technologies of optical fiber sensing, computer vision, WSN, MEMS, environmental power generation and on-site demonstration that enable large-scale and long-term monitoring. Figure 6.18 shows how monitoring using sensor and

Fig. 6.17 Application example of environmental power generation by vibration using the concept of parametric excitation

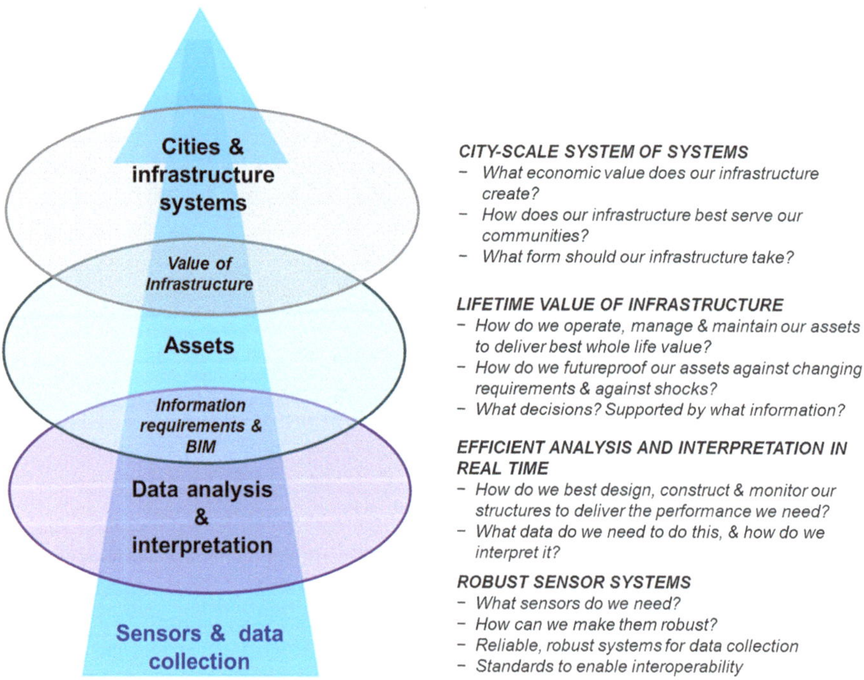

Fig. 6.18 Smart Infrastructure and whole life sensing of infrastructure (Modified from Cambridge Centre for Smart Infrastructure and Construction 2015)

communication technology can contribute to understanding the actual performance of individual civil engineering structures, infrastructure as a whole and urban behavior to build a smart infrastructure. Such an integrated approach may be useful in the future for designing sensor systems, distributing information for decision makers, and managing more efficient infrastructure systems.

References

ASCE (2013) Report card for America's infrastructure, American Society of Civil Engineers. The next report is due on March 2017

Belsito L, Ferri M, Mancarella F, Roncaglia A, Yan J, Seshia AA Soga K (2013) High resolution strain sensing on steel by silicon-on-insulator flexural resonators fabricated with chip-level vacuum packaging, 2013 Transducers and Eurosensors XXVII: the 17th International Conference on Solid-State Sensors, Actuators and Microsystems, TRANSDUCERS and EUROSENSORS 2013, pp 992–995

Belsito L, Ferri M, Mancarella F, Masini L, Yan J, Seshia AA, Soga K Roncaglia A (2016) Fabrication of high-resolution strain sensors based on wafer-level vacuum packaged MEMS resonators, sensors & actuators A

Bevan H (2016) Personal communication

Cambridge Centre for Smart Infrastructure and Construction (2015) Annual review 2016; delivering impact. www-smartinfrastructure.eng.cam.ac.uk

Chaiyasarn K (2011) Detection and monitoring of damage for tunnel inspection based on Computer Vision, PhD dissertation, University of Cambridge

Cole M (2010) Borough Viaduct & London Bridge: the keys to opening capacity. NCE Deliv Thames Link Issue 9:14–15

Garus D, Golgolla T, Krebber K, Schliep F (1997) Brillouin optical frequency-domain analysis for distributed temperature and strain measurements. J Lightwave Technol 15:654–662

Gue CY, Wilcock M, Alhaddad MM, Elshafie MZBE, Soga K, Mair RJ (2015) The monitoring of an existing cast iron tunnel with distributed fibre optic sensing (DFOS). J Civ Struct Heal Monit 5:1–14

Hartog AH, Leach AP, Gold MP (1985) Distributed temperature sensing in solid-core fibre. Electron Lett 21:1061–1062

HM Treasury (2016) National infrastructure delivery plan 2016–2021

Horiguchi T, Kurashima T, Tateda M (1989) Tensile strain dependence of Brillouin frequency shift in silica optical fibres. IEEE Photon Technol Lett 1:107–108

Hotate K, Hasegawa T (2000) Measurement of Brillouin gain spectrum istribution along an optical fibre with a high spatial resolution using a correlation-based technique—proposal, experiment and simulation. IEICE Trans Electron E83-C:405–411

ICE (2016) State of nation. Institution of Civil Engineers, London

Jia Y, Yan J, Soga K, Seshia AA (2014a) A parametrically excited vibration energy harvester. J Intell Mater Syst Struct 25(3):278–289

Jia Y, Yan J, Soga K, Seshia AA (2014b) Parametric resonance for vibration energy harvesting with design techniques to passively reduce the initiation threshold amplitude. Smart Mater Struct 23(6):065011

Kechavarzi C, Soga K, de Battista N, Pelecanos L, Elshafie M, Mair RJ (2016) Distributed optical fibre sensing for monitoring geotechnical infrastructure – a practical guide. ICE Publishing, London

Kurashima T, Horiguchi T, Tateda M (1990) Distributed-temperature sensing using stimulated Brillouin scattering in optical silica fibres. Opt Lett 5:1038–1040

Lu Y, Zhu T, Chen L, Bao X (2010) Distributed vibration sensor based on coherent detection of phase-OTDR. J Lightwave Technol 28:3243–3249

Network Rail (2012) Complex projects procedure for London Bridge and the surrounding area. Report No. K002-NRT-CHG-00001, London

Niklès M, Thévenaz L, Robert PA (1996) Simple distributed fibre sensor based on Brillouin gain spectrum analysis. Opt Lett 21:758–760

Peck RB (1969) Advantages and limitations of the observational method in applied soil mechanics. Geotechnique 19(2):171–187

Prouting N, Lochner D (1975) Operation London Bridge, documentary film

Rodenas-Herráiz D, Soga K, Fidler P de Battista N (2016) Wireless sensor networks for civil infrastructure monitoring – a best practice guide. ICE Publishing, 208 pp

Soga K, Kwan V, Pelecanos L, Rui Y, Schwamb T, Seo H Wilcock M (2015) Skempton lecture – the role of distributed sensing in understanding the engineering performance of geotechnical structures, Geotechnical Engineering for Infrastructure and Development, XVI ECSMGE. ICE Publishing, London, pp 13–48

Stajano F, Hoult N, Wassell I, Bennett P, Middleton C, Soga K (2010) Smart bridges, smart tunnels: transforming wireless sensor net-works from research prototypes into robust engineering infrastructure. Ad Hoc Netw 8:872–888

Stent S, Gherardi R, Stenger B, Soga K, Cipolla R (2016) Visual change detection on tunnel linings. Mach Vision Appl 27(3):319–330

Ye G, Soga K (2012) Energy harvesting from water distribution systems. J Power Eng Am Soc Civil Eng 138(1):7–17

Chapter 7
Recent Efforts to Mitigate the Impacts of Earthquake Hazard in Indonesia from Geotechnical Engineering Perspective

Hendriyawan, M. Irsyam, M. Asrurifak, I. Meilano, D.H. Natawidjaja, S. Widiyantoro, A.D. Nugraha, L.M. Sakti, A. Sabaruddin, L. Faisal, P.T. Simatupang, B.M. Hutapea, and T. Afriansyah

Abstract This paper presents the recent efforts in Indonesia to mitigate the impacts of earthquake hazards. The actions includes as follows: updating of the seismic hazard maps of Indonesia 2010 and 2016; revision and continuous updating of building and infrastructure design codes; development of microzonation maps for big cities in Indonesia; development of academic draft of Indonesian Earthquake Master Plan; development of design guidelines for tsunami vertical evacuation; development of a national design code for geotechnics and earthquake; and establishment of the national center for earthquake studies. Revision of seismic hazard maps for Indonesia 2010 has been developed based upon updated: seismotectonic data, fault models, and GMPEs up to 2010. The updating of infrastructure design codes related to earthquakes activities are being performed for buildings, bridges, dams, harbors and others special structures. The development of microzonation map of seismic risk has been initiated for Jakarta city. The academic draft for

Hendriyawan • M. Asrurifak • B.M. Hutapea
Research Center for Disaster Mitigation, Institute of Technology Bandung, Bandung, Indonesia

Indonesian Society for Geotechnical Engineering (ISGE), Bandung, Indonesia

M. Irsyam (✉)
Research Center for Disaster Mitigation, Institute of Technology Bandung, Bandung, Indonesia

Indonesian Society for Geotechnical Engineering (ISGE), Bandung, Indonesia

Indonesian Academy of Sciences, Bandung, Indonesia
e-mail: masyhur.irsyam@yahoo.co.id

I. Meilano • L.M. Sakti
Research Center for Disaster Mitigation, Institute of Technology Bandung, Bandung, Indonesia

D.H. Natawidjaja
Indonesian Institute of Sciences, Bandung, Indonesia

S. Iai (ed.), *Developments in Earthquake Geotechnics*, Geotechnical, Geological and Earthquake Engineering 43, DOI 10.1007/978-3-319-62069-5_7

Indonesian earthquake master plan has been prepared based on assessment from several aspects, i.e. basic sciences, engineering and risk analysis, and social and legal aspects. The guidelines for tsunami evacuation had been finished and submitted to BNPB in 2013. Lastly, the Indonesia National Research Center for Earthquake has been initiated in June 2016.

Keywords Seismic hazard maps • National seismic code • Microzonation

7.1 Introduction

Indonesia has been well known as one of the most seismically active countries in the world since it is located on the boundaries of four major plates; Eurasia, India-Australia, Pacific and Philippines Sea plates. The India-Australia plate converges northeastward about 50–70 mm/year to the Asian Plate along the Sunda trench (Bock et al. 2003). This tectonically dynamic environment puts most parts of Indonesia prone to earthquake and their secondary hazards such as tsunami, landslides, liquefaction, flood and fire. Considerable numbers of active faults are also identified on offshore and onshore of Indonesia such as Sumatran Fault Zone, Palu-Koro fault, Sorong fault and many others.

Lessons learned from several great earthquakes repeatedly occurred in Indonesia since the 2004 Aceh earthquake have increased the awareness to public and government regarding the impacts of seismic hazard on their structures such as buildings, bridges, dams, and any other critical facilities. Based on Irsyam et al. (2013a), the estimated total losses due to the earthquake from 2004 to 2010 vary

S. Widiyantoro
Research Center for Disaster Mitigation, Institute of Technology Bandung, Bandung, Indonesia

Indonesian Academy of Sciences, Bandung, Indonesia

Global Geophysical Research Group, Faculty of Mining and Peteroleum Engineering, Institute of Technology Bandung, Bandung, Indonesia

A.D. Nugraha
Research Center for Disaster Mitigation, Institute of Technology Bandung, Bandung, Indonesia

Global Geophysical Research Group, Faculty of Mining and Peteroleum Engineering, Institute of Technology Bandung, Bandung, Indonesia

A. Sabaruddin • L. Faisal
Center for Research and Development of Housing and Settlements, The Ministry of Public Works, Bandung, Indonesia

P.T. Simatupang
Indonesian Society for Geotechnical Engineering (ISGE), Bandung, Indonesia

T. Afriansyah
Jakarta Industry and Energy Department, The Government of Jakarta, Bandung, Indonesia

from US$39 million to US$4745 million. In addition, almost more than 200,000 were killed due to the earthquakes activities. Over the last decade, many research agencies, universities, including professional associations have significantly increased the efforts to improve the understanding of earthquake phenomena and the predictive capabilities of earthquake science and to mitigate the impacts of future earthquakes in Indonesia.

The recent efforts in Indonesia to mitigate the impacts of earthquake hazard from geotechnical engineering perspective which involving the Indonesian Society for Geotechnical Engineering (ISGE) and the Research Center for Disaster Mitigation (RCDM) of Institute of Technology Bandung (ITB) include several activities as follows:

- Updating of seismic hazard maps of Indonesia 2010 and 2016
- Revision and continuous updating of building and infrastructure design codes
- Development of microzonation maps for big cities in Indonesia
- Development of academic draft of National Earthquake Master Plan
- Development of design guidelines for tsunami vertical evacuation
- Development of national design code for geotechnics and earthquake
- Preparation for establishment of the National Center for Earthquake Studies

7.2 Updating of Seismic Hazard Maps of Indonesia 2010 and 2016

The need to revise and to update the Indonesian Seismic Hazard Map 2002 (SNI 2002) was driven by several great earthquake occurrences in Indonesia such as the 2004 Aceh Earthquake (M_w9.0–9.3) which was followed by tsunami, the 2005 Nias Earthquake (M_w 8.7), and the 2009 Padang Earthquake (M_W 7.6). In 2009, Ministry of Public Works has assigned eleven (11) researches as a Team for Revision of Seismic Hazard Maps of Indonesia (TRSHMI 2010) to revise the 2002 map. The team consists of several institutions in Indonesia such as: ITB; Institute of Road Engineering Agency for Research and Development and Center for Research and Development of Housing and Settlements of the Ministry of Public Works; Indonesian Institute of Sciences (LIPI); Meteorological, Climatological and Geophysical Agency (BMKG), and Geological Research and Development Center (GSC). The team also collaborated with the United Stated Geological Survey (USGS) and the Australia-Indonesia Facility for Disaster Reduction (AIFDR). Several issues such as recent seismic activities, latest research works regarding fault characteristics in Indonesia, improvements of the method in seismic hazard analysis and latest provisions in International Building Code (ASCE 2010) have been considered in the development of the national hazard maps 2010.

Following TRSHMI 2010 (Irsyam et al. 2010, 2013a, b), seismic sources were classified into three types of source models; fault zone, subduction zone, and background seismicity. Classification was conducted based on seismogenic conditions, focal mechanisms and earthquake catalogs. The seismogenic conditions included geometry and geomorphological of tectonic plate such as faults and subduction zones. Fault sources were modeled as a plane in 3-D space for calculation of distance from a site to a certain point at the plane. Subduction zone model were derived from tomography model conducted by Widiyantoro (2009), historical earthquake catalogs and seismotectonic conditions. Background seismicity was utilized to account for random earthquakes on unmapped faults and smaller earthquakes on mapped faults. A type of background seismicity was gridded models based on spatially smoothed earthquake rates (Frankel 1995).

The fault model and subduction model utilized by TRSHMI 2010 is shown in Fig. 7.1. The hazard maps have been officially published and signed by the Minister of Public Work of Indonesia in 2010 and implemented in a newly revised national code for earthquake resistance building design SNI 1726-2012.

TRSHMI 2010 has also proposed recommendation to the government. The recommendations include: conduct periodic updating of seismic hazard maps and related national design codes every 3–5 years; conduct micro seismicity investigation of active faults that have not been well identified and well quantified, especially near big cities such as Jakarta, Surabaya, Semarang; accelerate the installation of strong-motion accelerometer networks in Indonesia in order to develop database of time histories and attenuation functions; and perform seismic microzonation studies for big cities in Indonesia.

In order to follow up the recommendation from TRSHMI 2010, in 2015 The National Agency for Disaster Management (BNPB) and the Ministry of Public Works started coordinating several institutions in Indonesia to update the national seismic hazard maps 2010. The team consists of experts from ITB, BMKG, Center of Volcanology and Geological Hazard Mitigation (PVMBG), Ministry of Public Works, LIPI, Geospatial Information Agency (BIG), Agency for the Assessment and Application of Technology (BPPT), and professional associations such as Indonesian Disaster Expert Association (IABI). The team is divided into several working groups, such as geology, geodesy, seismology, ground motion prediction equations, and Seismic Hazard Analysis. The organization chart can be seen in Fig. 7.2.

Important information is considered for updating seismic hazard maps such as significant results of recent active-fault studies utilizing trenching, carbon dating, epicenter relocation, strain analysis (GPS) as well as availability of basic data including the SRTM-30, IFSAR, LiDAr and other data that is just recently available. The process in developing seismotectonic model can be seen in Fig. 7.3. It can be expected that this new information will result in more accurate tectonic model and their seismic parameters, such as maximum magnitudes and slip-rates. This updating work is currently on going and it is expected to be completed before end of 2016.

For general mapping of active faults on-land, SRTM with 30 m grid resolutions and IFSAR DSM (Digital Surface Map) with grid data of 5 m spacing were used in analysis (Figs. 7.4 and 7.5). Concurrently, for mapping active fault in the ocean, GEBCO 2009 bathymetry data and some higher resolution bathymetry near the coasts from local surveys were also used in preparing new seismotectonic model for Indonesia.

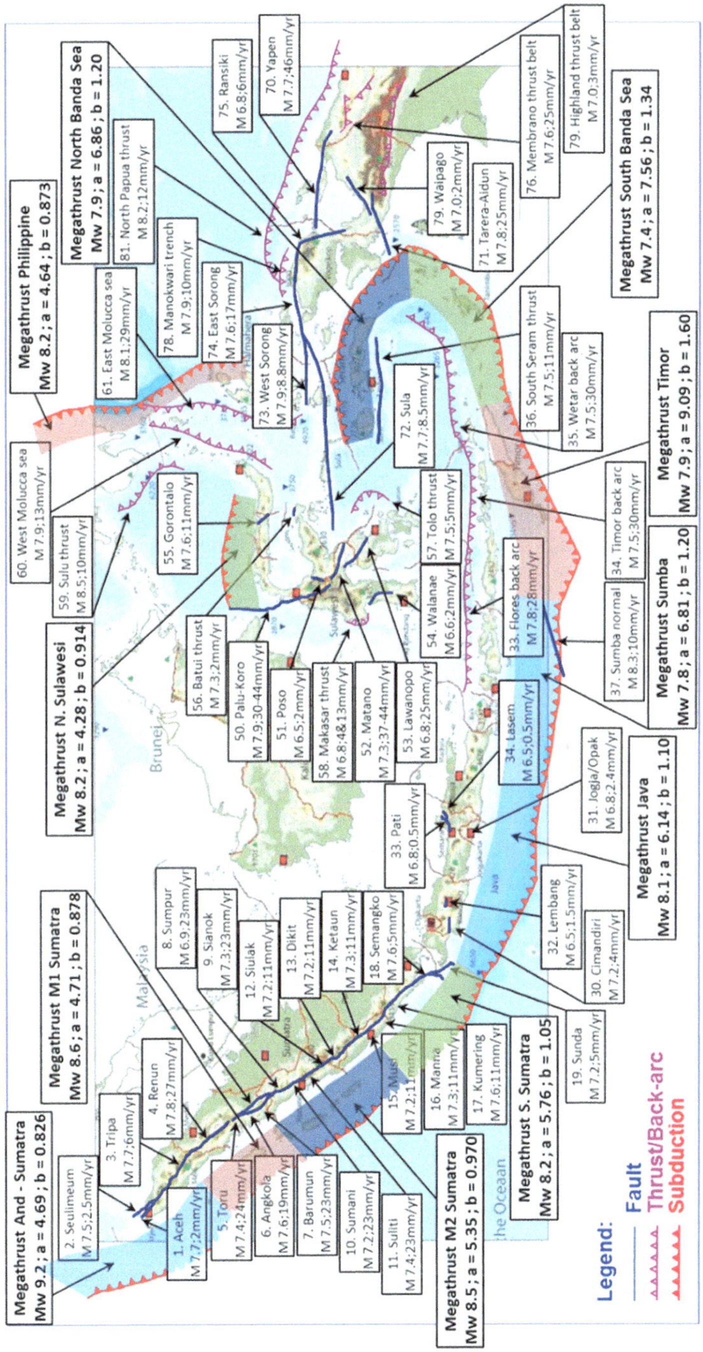

Fig. 7.1 Earthquake sources and their parameters used by TRSHMI 2010 (Irsyam et al. 2010, 2013a)

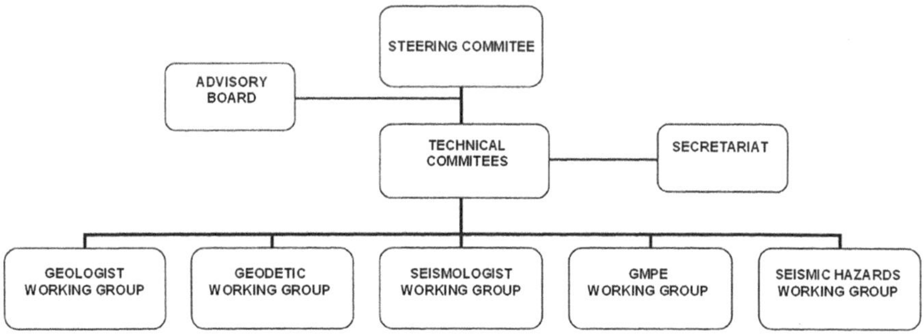

Fig. 7.2 Organization chart of team for updating of Seismic Hazard Maps of Indonesia 2016

To further investigate the fault structures underground and their relations to geological strata, Ground Penetration Radar (GPR) surveys were conducted in selected locations of fault lines that have been identified from the SRTM-30m and IFSAR-5m inspections. In a selected site, more detailed investigations were conducted such as total-station mapping, hand drilling and sample corings to analyze soil and rock strata related to faultings and trenching as shown in Fig. 7.6.

In order to analysis the seismicity, the earthquake databases around Indonesia were compiled from several sources as follows:

- The Preliminary Determination of Epicenters (PDE) from 1901 to 2014 (National Earthquake Information Center – USGS).
- The EHB catalog (International Seismological Comission (ISC) that calculated based on Engdahl et al. (1998).
- The Meteorological, Climatological, and Geophysical Agency (BMKG) of Indonesia earthquake data catalog from April 2009 to June 2014.
- Focal mechanism from ISC database which compile most reliable focal mechanism solution e.g. from Harvard Centroid Moment Tensor.

In order to obtain more precise hypocenter locations, The catalogs were relocated using teleseismic double-difference relocation (teletomoDD) (Pesicek et al. 2010). The teletomoDD code is an improvement from double-difference relocation (DD) (Waldhauser and Ellsworth 2000) which relocated hypocenter based on events pairs. DD method assume that if distance between event pair is shorter than their distance to seismic station then both of events have similar ray paths and waveforms. The example of relocated earthquakes distribution around East Java and Sumatra region by applying 3D seismic velocity from tomographic inversion study (Widiyantoro and Van Der Hilst 1996, 1997) are shown in Figs. 7.7 and 7.8, respectively. These relocated hypocenters data are also used in preparing new seismic hazard maps of Indonesia.

Based on the latest researches and investigations, revised seismotectonic models for Indonesia were developed and to be used in developing new seismic hazard

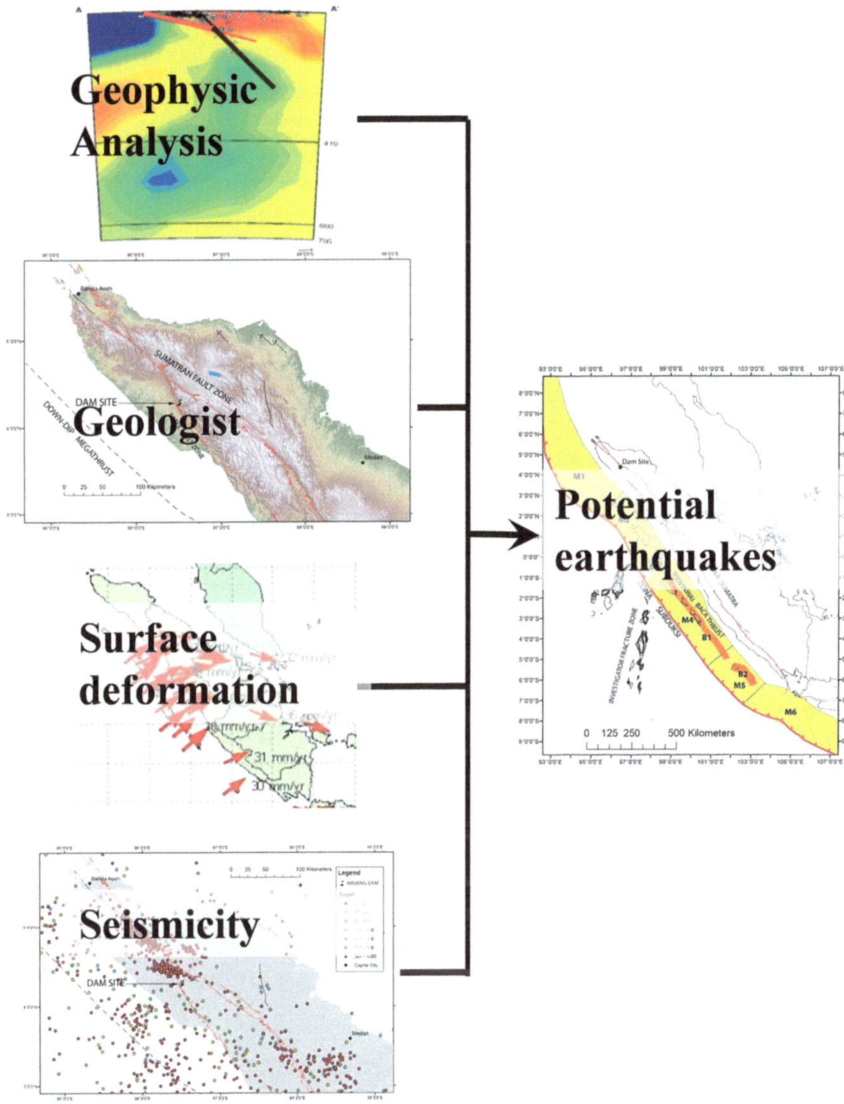

Fig. 7.3 The process in developing seismotectonic model

maps of Indonesia 2016. For example, the detailed of Sumatra Fault Zone (SFZ) segmentations and earthquake-parameter characterizations are revised in the new maps as shown in Fig. 7.9. The locations of some segments in SFZ have been re-mapped more accurately based on SRTM-30 and WORLD-30m digital map. The major different with the previous model (used in SNI 1726-2012) is by adding the Batee Fault Zone in the new model.

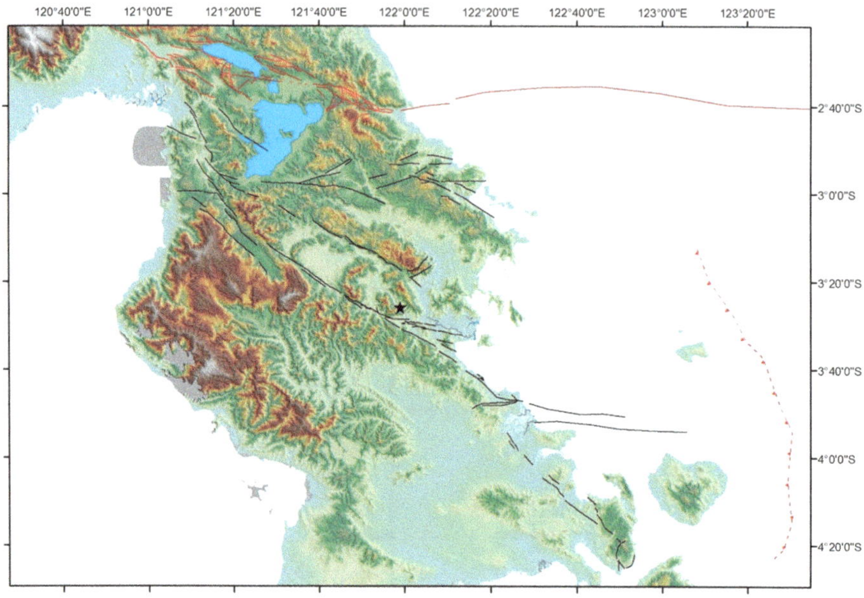

Fig. 7.4 Results of regional fault mapping based on morphotectonic analysis

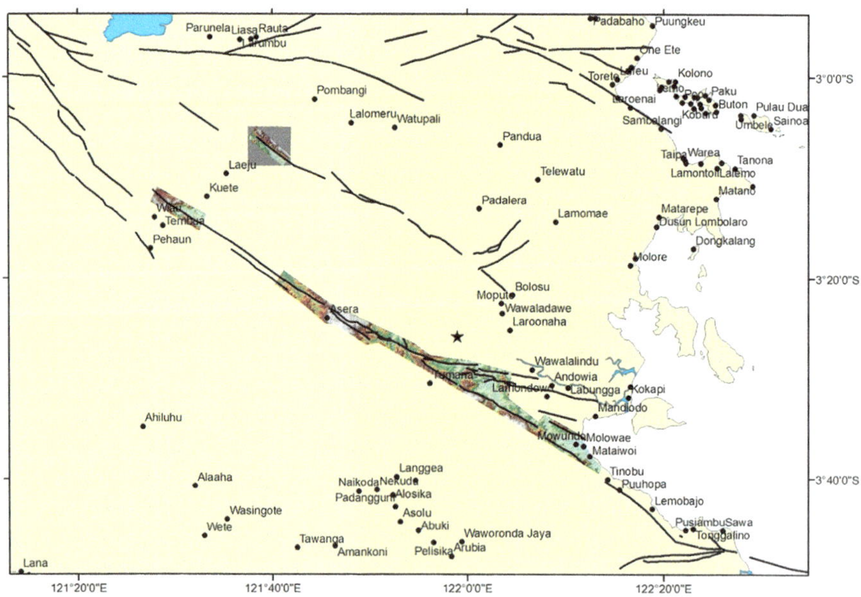

Fig. 7.5 Index map of IFSAR 5m-grid DSM

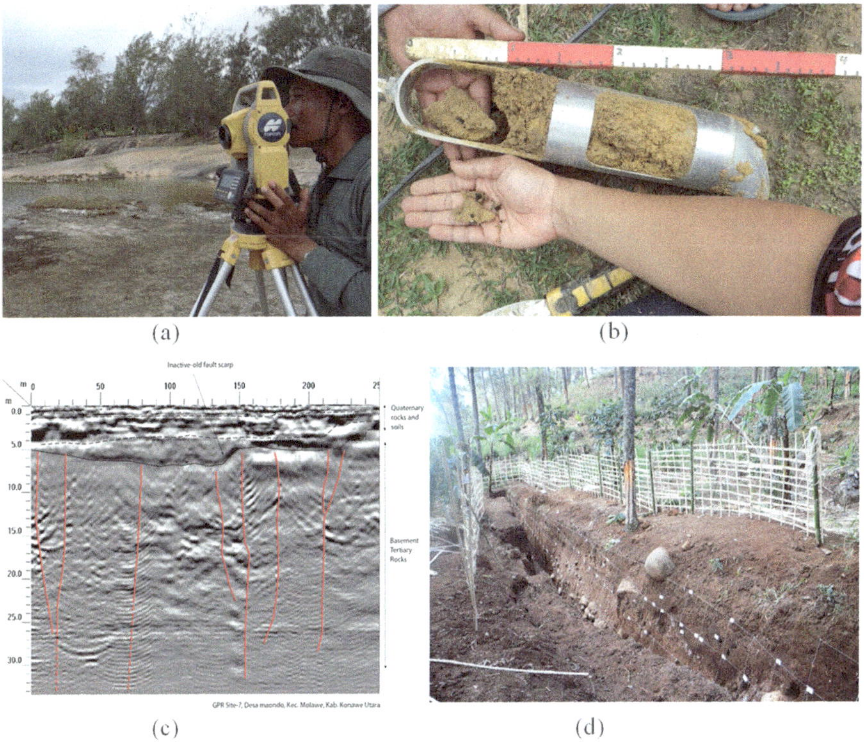

Fig. 7.6 Active fault surveys using: (**a**) total station measurement; (**b**) hand drilling and core sampling; (**c**) ground penetration radar; (**d**) trenching

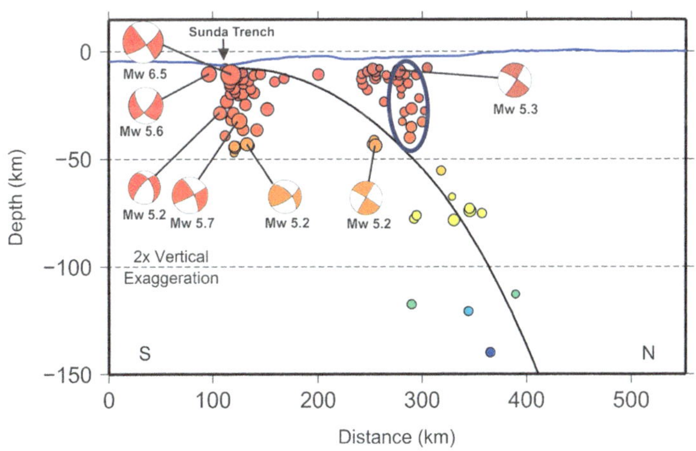

Fig. 7.7 Relocation epicenter for cross-section below East Java using 3D Velocity Model (Nugraha et al. 2015, 2016)

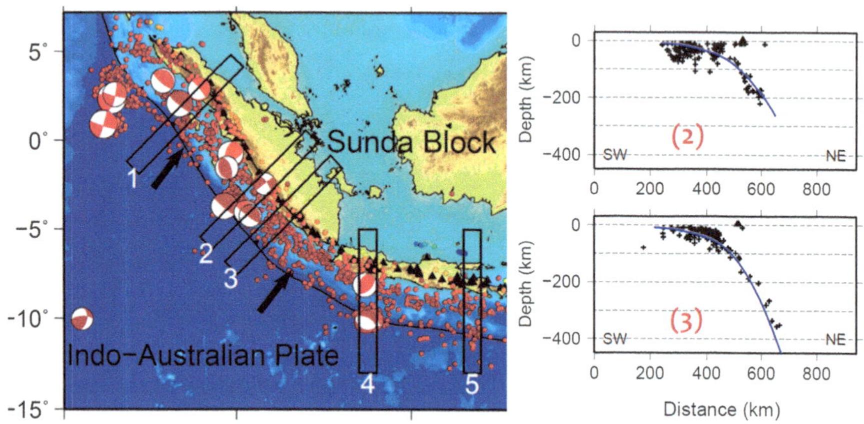

Fig. 7.8 Example of vertical sections of relocated hypocenter by using 3D seismic velocity model through Sumatra regions (No 2 and 3) (Nugraha et al. 2015, 2016)

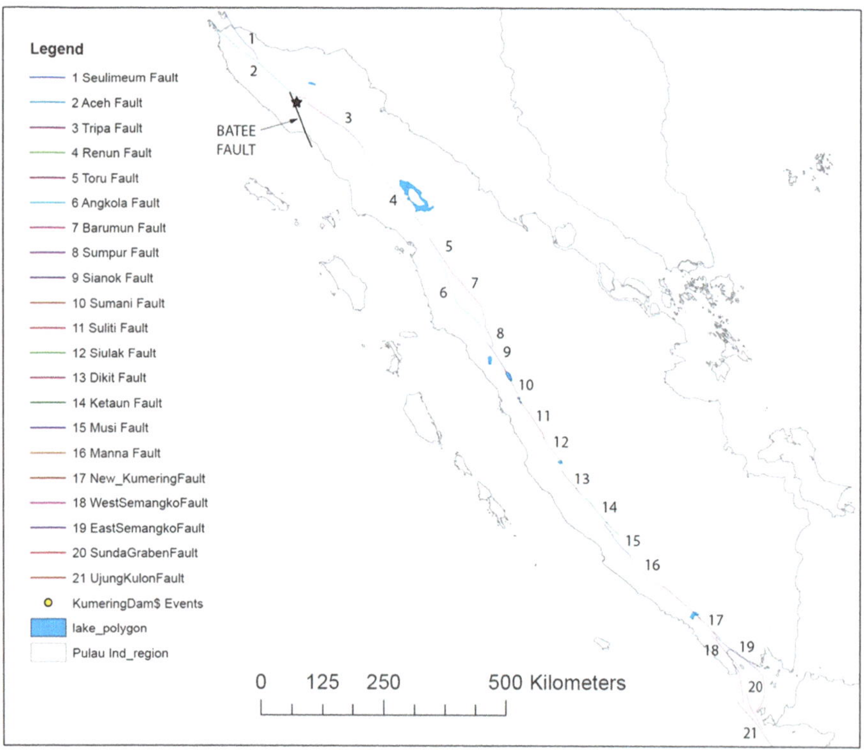

Fig. 7.9 Revised seismic fault segmentation of the SFZ

7.3 Revision and Continuous Updating of Building and Infrastructure Design Codes

The Ministry of Public Works has practically implemented SNI 1726-2012 (SNI 2012) to replace SNI 03-1726-2002 (SNI 2002) as a national code for earthquake resistance building design since 2014. The new code for buildings follows the concept of Maximum Considered Earthquake Geometric Mean (MCE_G) and Risk-Adjusted Maximum Considered Earthquake (MCE_R) used by ASCE 7–10. For the purpose of geotechnical calculation, it combines both the results from probabilistic seismic hazard analysis for 2% probability of exceedance in 50 years (2500 years earthquake) and deterministic seismic hazard analysis for area located near active fault. Both approaches were utilized according to the procedure proposed by Leyendecker et al. (2000) and the result of combining both probabilistic and deterministic analyses is called MCE_G.

In order to evaluate the seimic hazard for low and high risk structure (e.g. building), the spectral acceleration maps are required. The 0.2 s and 1.0 s spectral acceleration maps have been developed by taking into account the probability of collapse for a structure. Probability of collapse of a structure is influenced by the structural capacity that has uncertainty including as site-to-site variability in the shape of hazard curve, material properties, nonstructural components, etc. that will result in a lack of uniformity in structural capacity (Luco et al. 2007). For the new national code, probability of collapse of a structure is targeted to be equal to 1% in 50 years according to ASCE 7–10. Figures 7.10 and 7.11 show the maps of Risk-Adjusted Maximum Considered Earthquake (MCE_R) at 0.2 s and 1.0 s spectral response acceleration in SNI 1726-2012.

Updating of national standards for design of infrastructure is also conducted for bridges, dams, harbours, tunnels, and others special structures. It seems that each

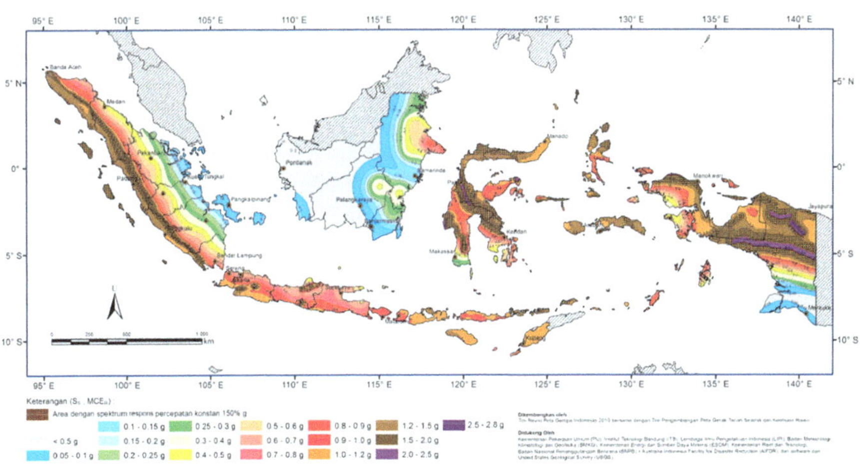

Fig. 7.10 Map of risk-targeted maximum consider earthquake (MCE_R) at bedrock (SB) of Indonesia at 0.2 s spectral response acceleration (SNI 1726-2012)

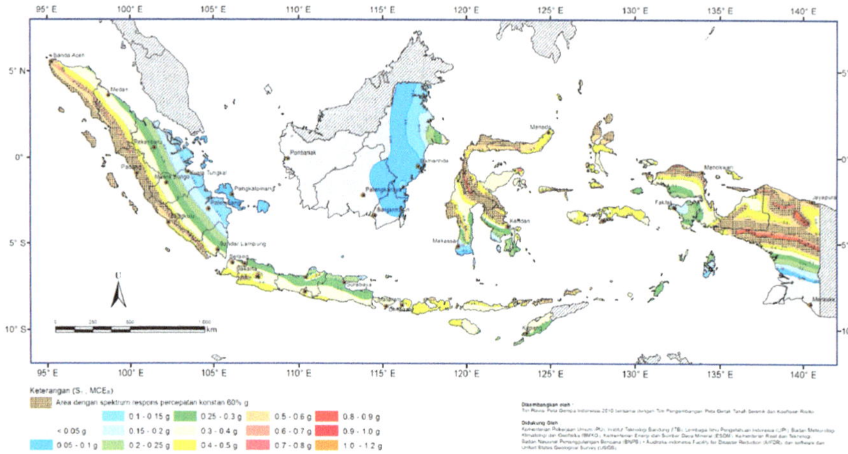

Fig. 7.11 Map of risk-targeted maximum consider earthquake (MCE_R) at bedrock (SB) of Indonesia at 1.0 s spectral response acceleration (SNI 1726-2012)

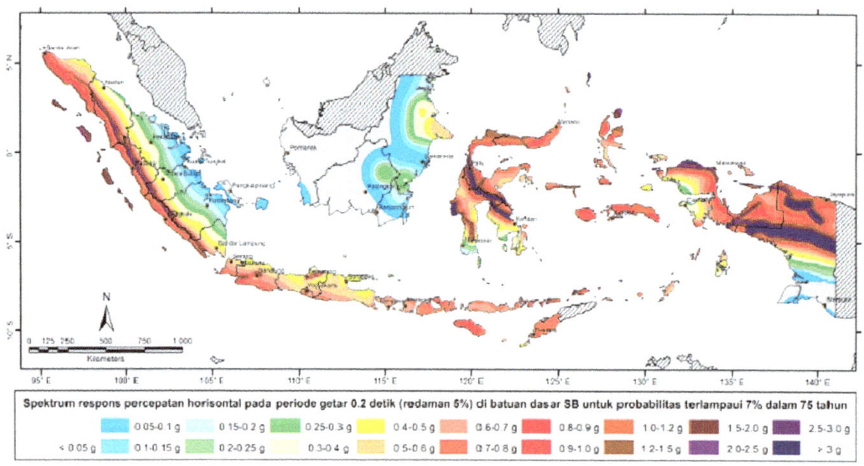

Fig. 7.12 Horizontal response spectra for T = 0.2 s for Site Class S_B (RSNI2 2833:201X) for bridge design

type of infrastructure code will utilize hazard maps having different return period of earthquake. For design of bridge, updating of the code is already completed. The standard RSNI2 2833:201X (RSNI 2015) that follows AASHTO LRFD Bridge Design Specification, 5th Edition, 2012 (AASHTO 2012) utilize return period of earthquake equals to approximately 1000 years that corresponds to 7% probability of exceedance in 75 years. Based on the new code, three maps of spectral response accelerations are required to develop the design response spectra, i.e. spectral acceleration at 0.0 s, 0.2 s, and 1.0 s. The hazard maps for T = 0.2 s for site class S_B can be seen in Fig. 7.12. The code is already in the Indonesian Standardization Body and expected to be launched by the end of 2016.

Fig. 7.13 Several large dams in Indonesia

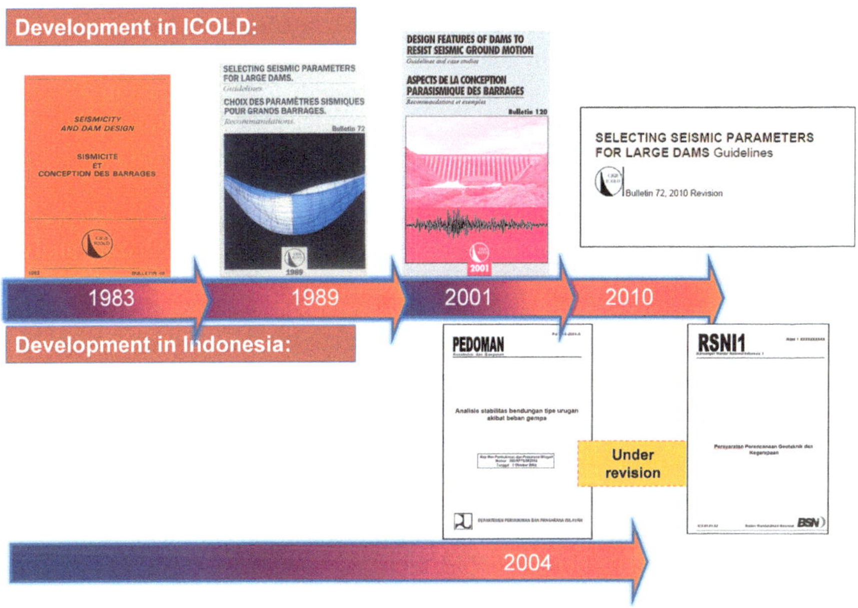

Fig. 7.14 Progress updating seismic design guideline for dam in Indonesia

Indonesia has 203 large dams that most of them were built from 1946 to 2014 (Fig. 7.13). Presently, 16 large dams are still under construction and it is expected in the next 5 years 49 large dams will be built in Indonesia. Presently, Indonesia seismic design guidelines 2004 for dams refer to ICOLD, USBR, USACE, and Japan Guidelines. Currently, the seismic design code for dams will be included in the national standard such as Indonesian Design Code for Geotechnics and Earthquake (Fig. 7.14).

7.4 Development of Seismic Microzonation Maps for Big Cities in Indonesia

In order to enhance disaster preparedness, risk reduction and hazard mitigation, the Coordinating Ministry for People's Welfare in 2011 started coordinating a national team to develop seismic microzonation maps for selected big cities in Indonesia. The team members consist of experts from national agencies and university research center including: RCDM of ITB, the Government of Jakarta; the Ministry of Public Works; BMKG; BNPB; PVMBG; BPPT; and BIG. The team members also collaborate with experts from Australian National University (ANU) through AIFDR.

Currently, Jakarta was selected as prototype for the seismic microzonation work. Seismic risk maps for Jakarta were developed for two hazard levels based on probabilistic approach, i.e. 10% and 2% probability of exceedance in 50 years, and for three scenario earthquakes based on deterministic approach, i.e. subduction Megathrust ($M_w = 8.7$ and $R = 179$ km), subduction Benioff ($M_w = 7.0$ and $R = 145$ km), and shallow crustal ($M_w = 6.1$ and $R = 51$ km).

The work covers estimation of seismic hazard, site characterization, site specific response analysis and risk assessment (Irsyam et al. 2014). Seismic hazard analysis was conducted based on deterministic and probabilistic approaches considering seismic sources influencing Jakarta such as several major fault lines in western Java and subduction zone either Megathrust or deep subduction Benioff sources. Site characterization was carried out by interpreting the results of field measurements including in-situ testing such as standard penetration test (SPT), Dutch cone penetration test (DCPT), shear wave velocity measurement using microtremor array and seismic downhole test and laboratory tests. The site characterization was supported by the Government of Jakarta. Site response analysis were conducted based on the 1-D non-linear wave propagation procedure utilizing the free software NERA (Bardet and Tobita 2001) and by using the constitutive model proposed by Mróz (1967). Peak surface acceleration map of Jakarta due to Megathrust earthquake $M = 8.7$ $R = 179$ km is presented in Fig. 7.15.

The seismic risk microzonation maps were then developed by combining the results of site response analysis with building fragility for two types of low rise buildings that dominate the residential building population, i.e. confined masonry and in-filled frame structures. The procedure in determining the level of seismic risk for two types of low rise buildings in Jakarta can be seen in Fig. 7.16. Seismic risk map of Jakarta for residential building due to Megathrust earthquake $M = 8.7$ $R = 179$ km is presented in Fig. 7.17.

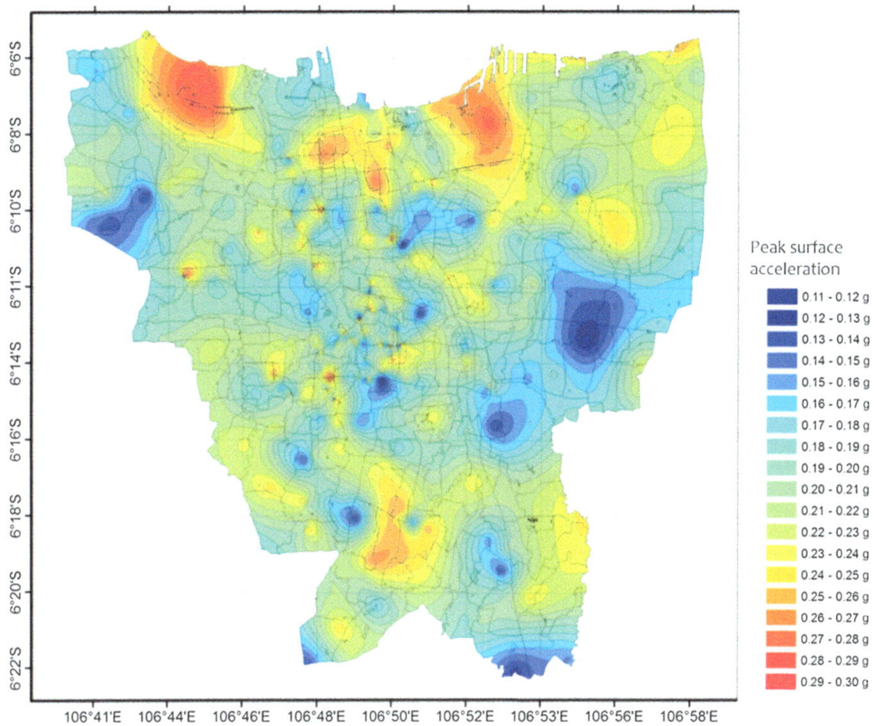

Fig. 7.15 Peak surface acceleration map of Jakarta due to Megathrust earthquake $M = 8.7$ $R = 179$ km (Sakti et al. 2015)

7.5 Development of Academic Draft of the National Earthquake Master Plan

To support disaster risk reduction programs in Indonesia, in 2013 B.P. had selected 12 universities to prepare academic drafts of master plans for 12 hazards. The plans are multiagency programs to reduce the risks of life and property from future hazards in Indonesia. RCDM ITB had been selected by BNPB to lead the establishment of academic draft for Indonesian earthquake master plan. The masterplan is developed with specific purpose: (1) to support the earthquake disaster risk reduction efforts by providing guidance in prevention, mitigation and early warning, (2) to provide a reference related to: program of activities, the priority focus, and an indicative budget for each ministry or agency and all stakeholders in earthquake disaster risk reduction.

The master plan consists of disaster risk reduction program for short, medium and long term based on assessment from several aspects, i.e. basic sciences,

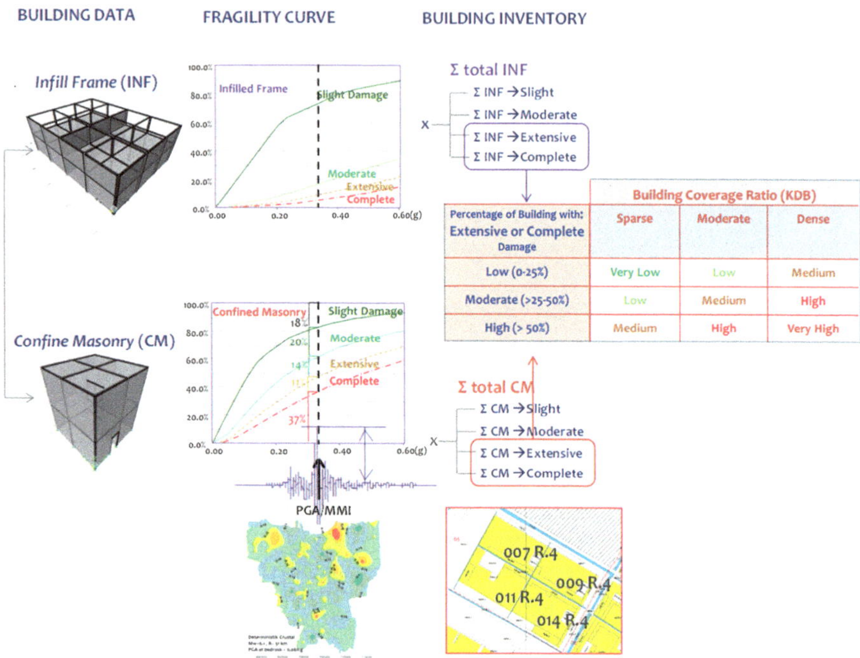

Fig. 7.16 Procedure in determining the level of seismic risk for two types of low rise buildings (INF and CM) in Jakarta city

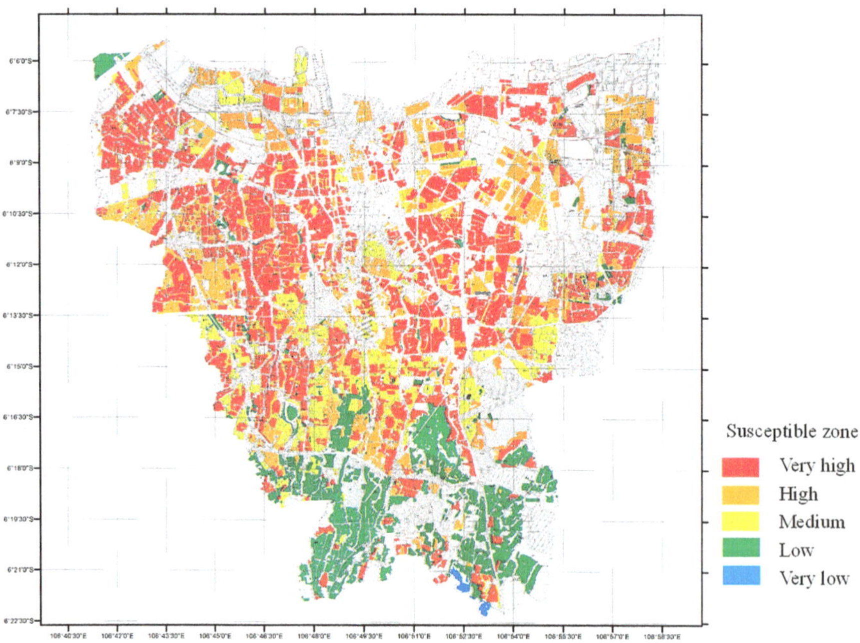

Fig. 7.17 Seismic risk map of Jakarta for residential building due to megathrust earthquake M = 8.7 R = 179 km (Sakti et al. 2015)

engineering and risk analysis, and social and legal aspects. The general outcome of the master plan includes: (1) improving understanding of earthquake processes and impacts, (2) reducing earthquake impacts on building and infrastructures, and (3) improving the earthquake resilience of communities nationwide.

Some important proposed recommendations of the master plan can be found in Irsyam et al. (2013c), i.e. to study seismic sources which urgently need to be identified (such as activity, the maximum magnitude and slip rate), to develop ground motion prediction equations (GMPE) based upon Indonesian data, to establish earthquake national research center, to update Indonesian seismic hazard maps and building-infrastructure codes, and to investigate potential urban mega disaster.

7.6 Development of Design Guidelines for Tsunami Vertical Evacuation

In 2013, the National Agency for Disaster Management (BNPB) requested RCDM ITB to develop guidelines for tsunami vertical evacuation. In order to follow up the request, RCDM ITB established team for developing the guideline. The team consists of several experts from geophysics, geodetic, geotechnics, structures, hydraulics, and construction management. Four (4) books were submitted to BNPB as follows: (1) guideline for developing tsunami hazard maps; (2) guideline for planning of temporary tsunami evacuation sites; (3) guideline for planning and designing structures that serve as tsunami vertical evacuation; and (4) guideline for planning and designing of existing high ground and man-made embankment for temporary tsunami evacuation sites. The covers for each guideline are shown in Fig. 7.18.

Fig. 7.18 Guidelines for tsunami vertical evacuation

7.7 Development of the National Design Code for Geotechnics and Earthquake

Development of the Indonesia's national design code for geotechnics and earthquake has been initiated in 2014 by the Ministry of Public Works. The main purpose of the project is to establish a code as the Indonesian national standard for the design of construction works related to geotechnics and earthquake. The developed code will cover sub structures for buildings, highways, bridges, water resources and settlements. The committee for this work consists of ISGE, Institute of Road Engineering Agency for Research and Development, Research and Development Center for Water Resources, Center for Research and Development of Housing and Settlements, and universities.

The code will provide common structural design rules to be used for design of sub structures including the requirements for insitu and laboratory tests. There are nine (9) topics that are covered by the code as follows: (1) Stability of Slope and Embankment, (2) Deep Excavation, (3) Foundation, (4) Tunnel, (5) Seismicity, (6) Geotechnical Investigation, (7) Retaining Structures, (8) Ground Improvement, and (9) Hydraulic failure. The code are developed by committees through a consensus process such as focus group discussion (FGD) organized by the Ministry of Public Works and ISGE. Development of the code is currently on going and expected for completion by the end of 2016.

7.8 Preparation for Establishment of the National Center for Earthquake Studies

Based upon the academic draft of Indonesia earthquake master plan (Irsyam et al. 2013c), a national center for earthquake research is urgently needed to support disaster risk reduction programs in Indonesia. The center is required to support the enhancement of the science and practice of earthquake science and engineering and to improve the understanding of the impact of earthquakes on the physical, social, and economic. The center will facilitate gathering of scientists and engineers from different institutions at national level to share and integrate all data, resources, knowledge and new scientific findings in earthquake hazards through collaborative researches.

The establishment of the research center was initiated at the end of 2015 by several Indonesian institutions such as: Ministry of Public Works; ITB; BMKG; LIPI; PVMBG; BIG; BNPB; the Ministry of Research, Technology and Higher Education; BPPT; Universities (UNDIP, UGM, and ITS); and professional associations (ISGE, IABI, and AFMI).The research center will cover the following tasks such as: conduct basic and applied researches related to earthquake hazard and risk; support and perform the updating of seismic hazard and risk maps periodically and sustainably; develop standards, manual and guidelines related to seismic hazard and risks; and coordinate and synchronize with the ministry/agency and other institutions regarding all the activities related to monitoring systems, measurement, and seismic characteristics analysis for supporting the updating of national hazard maps and earthquake risk. In order to fulfill the tasks, the research center is supported by eight (8) working groups, i.e. geology, geodesy, seismology, GMPE, seismic hazard analysis, geotechnics, infrastructures and structures, and collateral hazards. The establishment of the center was initiated in June 10, 2016.

7.9 Conclusions

Lessons learned from several great earthquakes repeatedly occurred in Indonesia since the 2004 Aceh earthquake have increased the awareness to public and government regarding seismic activities in Indonesia. Over the last 10 years, many research institutions, universities, including association professions such as Indonesian Society for Geotechnical Engineering (ISGE) have significantly increased the efforts to understand earthquake hazards and to mitigate the impact of future large earthquakes in Indonesia. Some of the recent efforts in Indonesia to mitigate the impacts of earthquake hazards were described briefly in this paper. The actions includes updating of the seismic hazard maps of Indonesia 2010 and 2016, revision and continuous updating of building and infrastructure design codes, development of microzonation maps for big cities in Indonesia, development of academic draft of Indonesian Earthquake Master Plan, development of design guidelines for tsunami vertical evacuation, development of a national design code for geotechnics and earthquake, and establishment of the national center for earthquake studies.

References

AASHTO (2012) LRFD bridge design specification. AASHTO, Washington, DC

ASCE 7 (2010) Minimum design loads for buildings and other structures. American Society of Civil Engineers, Reston

Bardet JP, Tobita T (2001) NERA a computer program for nonlinear earthquake site response analyses of layered soil deposits. Department of Civil Engineering University of Southern California

Bock Y et al (2003) Crustal motion in Indonesia from global positioning system measurements. J Geophys Res 108:2367

Engdahl ER, van der Hilst RD, dan Buland R (1998) Global teleseismic earthquake relocation with improved travel times and procedures for depth determination. Bull Seismol Soc Am 88:722–743

Frankel A (1995) Mapping seismic hazard in the central and Eastern United States. Seismol Res Lett 66(4., 1995):8–21

Irsyam M, Sengara IW, Asrurifak M, Ridwan M, Aldiamar F, Widiyantoro S, Triyoso W, Natawijaya DH, Kertapati E, Meilano I, Suhardjono (2010) Summary: development of seismic hazard maps of Indonesia for revision of seismic hazard map in SNI 03-1726-2002. Research report submitted to the Ministry of Public Works by Team for Revision of Seismic Hazard Maps of Indonesia, July

Irsyam M, Hendriyawan Asrurifak M, Ridwan M, Aldiamar F, Sengara IW, Widiyantoro S, Triyoso W, Natawijaya DH, Kertapati E, Meilano I, Suhardjono, Firmanti A (2013a) Past earthquakes in Indonesia and new seismic hazard maps for earthquake design of buildings and infrastructures. Geotechnical predictions and practice in dealing with geohazards. In: Chu J, Wardani SPR, Lizuka A. Springer. ISBN 978-94-007-5674-8

Irsyam M, Sengara IW, Asrurifak M, Ridwan M, Aldiamar F, Widiyantoro S, Triyoso W, Natawijaya DH, Kertapati E, Meilano I, Suhardjono Hutabarat D, Sidi ID, Merati W (2013b) Development of seismic hazard and risk maps for new seismic building and infrastructure codes in Indonesia. In: Proceedings of the 6th civil engineering conference in Asia Region: embracing the future through sustainability. Jakarta, pp 20–22. ISBN 978-602-8605-08-3

Irsyam M, Widiyantoro S, Imran I, Rahayu H, Hendriyawan Puspito NT, Meilano I, Hutapea BM, Nugraha AD (2013c) Rencana Induk (Master Plan) Penanggulangan Risiko Bencana Gempabumi, Prosiding Seminar Nasional Riset Kebencanaan. Mataram, 8–10 Oktober 2013

Irsyam M, Hutabarat D, Asrurifa M, Imran I, Widiyantoro S, Hendriyawan Afriansyah T, Pindratmo H, Firmanti A, Ridwan M, Haridjono SW, Pandhu R (2014) Development of seismic risk microzonation maps of Jakarta City. In: Iai S (ed) Proceedings of the 4th international conference on Geotechnical Engineering for Disaster Mitigation and Rehabilitation (4th GEDMAR), September 16–18, Kyoto University, Japan. Taylor&Francis Group, London. ISBN 978-1-138-02709-1

Leyendecker EV, Hunt RJ, Frankel AD, Rukstales KS (2000) Development of maximum considered earthquake ground motion maps. Earthquake Spectra 16(1):21–40

Luco N, Ellingwood BR, Hamburger RO, Hooper JD, Kimball JK, Kircher CA (2007) Risk-targeted versus current seismic design maps for the conterminous United States. Structural Engineers Association of California 2007 convention proceedings, pp 163–175

Mróz Z (1967) On the description of anisotropic work hardening. J Mech Phys Solids 15(3):163–175

Nugraha AD, Shiddiqi HA, Widiyantoro S, Ramdhan M, Wandono S, Handayani T, Nugroho H (2015) Preliminary results of teleseismic double-difference relocation of earthquakes around Indonesia archipelago region. AIP Conf Proc 1658:030002. (2015). doi:10.1063/1.4915010

Nugraha AD, Shiddiqi HA, Widiyantoro S, Thurber CH, Pesicek JD, Wiyono S, Ramdhan H, Wandono M, Irsyam M (2016) Hypocenter relocation of the earthquakes along the Sunda Arc, Indonesia using A 3D seismic velocity model, in preparation to be submitted to Bulletin Seismological Society of America

Pesicek JD, Thurber CH, Zhang H, DeShon HR, Engdahl ER, Widiyantoro S (2010) Teleseismic double difference relocation of earthquakes along the Sumatra Andaman subduction zone using a 3D model. J Geophys Res 115:B10303

Rancangan Standar Nasional Indonesia (2015) Perancangan Jembatan terhadap Beban Gempa (RSNI 2833:201X). Badan Standardisasi Nasional

Sakti LM, Irsyam M, Asrurifak M, Imran I, Alfriansyah T (2015) Analysis study of risk map in Jakarta induced by Megathrust Scenario Earthquake. 10th Indonesian geotechnical conference and 19th annual scientific meeting, Jakarta

Standar Nasional Indonesia (2002) Tata Cara Perencanaan Ketahanan Gempa untuk Bangunan Gedung dan Non Gedung (SNI 03 1726-2002), Badan Standardisasi Nasional

Standar Nasional Indonesia (2012) Tata Cara Perencanaan Ketahanan Gempa untuk Bangunan Gedung dan Non Gedung (SNI 1726-2012), Badan Standardisasi Nasional

Waldhauser F, Ellsworth WL (2000) A double-difference earthquake location algorithm: method and application to the northern Hayward fault. Calif Bull Seism Soc Am 90:1353–1368

Widiyantoro S, Van Der Hilst R (1996) Structure and evolution of lithospheric slab beneath the Sunda Arc, Indonesia. Science 271(80):1566–1570

Widiyantoro S, Van Der Hilst R (1997) Mantle structure beneath Indonesia inferred from high-resolution tomographic imaging. Geophys J Int 130:167–182

Widiyantoro S (2009) Seismicity and structure of lithosperic slab beneath the Sunda Arc, Indonesia, SE Asian Gateway Evolution international conference. Royal Holloway University of London, 14–17 September 2009

Chapter 8
Possibility of Four Metre Deep Flooding in Densely Populated Low-Land Area of Tokyo Induced by a Mega-Earthquake

Hideki Ohta, Toshiki Aoyama, Toru Shimizu, Shunichi Hamada, and Masahito Nakamura

Abstract In this chapter, the authors describe the historical evidence of how the ground conditions in Tokyo have changed over the past 150 years and those changes would influence the area if a mega-earthquake hits Tokyo in the near future. In the recent 150 years, Tokyo was developed as a highly industrialized city protected from flooding by dykes, flood-gates and reclaimed land which have been built in progressive manners. Water needed in industrialized low-land area of Tokyo was partly obtained by deep-well pumping of the groundwater during a period of 100 years from 1873 to 1973. The excessive groundwater withdrawal resulted in lowered groundwater level down to a maximum depth of about 60 m below sea water level, which led to serious land subsidence in the low-land area particularly during a period of 1960–1973 when the withdrawal was extremely heavy. The latest big earthquake struck Tokyo area in 1923 when the altitude of Tokyo low-land area was still higher than the sea water level. Since then Tokyo low-land area has experienced the following three major changes: (i) land subsidence (max: about 4.5 m) due to the excessive groundwater pumping, (ii) construction of extremely extensive underground networks of lifelines, railways, roads and shopping areas, and (iii) increasing danger of failure of the riverine levees due to the recent increase in the torrential rainfalls typically accompanied by typhoons that strike Japan several times a year because of increasing typhoon-activation power supplied from the sea water 1 or 2° (Celsius) warmer than before. These changes made the low-land area of Tokyo much more susceptible to flooding than the time of 1923. In case that the sea water of Tokyo Bay flows onto the low-land area of Tokyo through possible breakage of some part of the existing

H. Ohta (✉)
Chuo University, Tokyo, Japan
e-mail: ohtaoffice@kib.biglobe.ne.jp

T. Aoyama • S. Hamada • M. Nakamura
Construction Industry Engineers Center, Tokyo, Japan

T. Shimizu
Research Institute of Construction and Economy, Tokyo, Japan

S. Iai (ed.), *Developments in Earthquake Geotechnics*, Geotechnical, Geological and Earthquake Engineering 43, DOI 10.1007/978-3-319-62069-5_8

seawalls and/or the flood gates, the depth of the water in major part of the area is expected to reach about 4 m at the deepest area and about 1.5 m on average. This may cause unacceptably serious situation in which huge number of people will suffer from the submersion of subways when the sea water flows into the underground facilities. About two million local residents in the low-land area submerged by flood water need to move to higher areas because almost all lifeline systems become out of service in the low-land area. Many of submerged underground facilities will not work anymore even after drying off and therefore will have to be replaced by new ones. About ten million people currently using subways in their daily commuting, business trips etc. every day will have practically no alternative means of transport. In the worst scenario, this will give extremely serious effects on enormous numbers of economic activities not only in Tokyo but also in the entire Japan.

8.1 Introduction

Densely populated Metropolitan area of Tokyo consists of (i) gently sloped hilly area in the western part of Tokyo covered by Pleistocene Kanto loam and (ii) eastern low-land area (Holocene flood plain) which is a basin bordering with another hilly area in the east. Through the basin, two major rivers, the Tone and the Arakawa rivers, flow into the northern part of Tokyo Bay. The two rivers have periodically flooded the area every summer for the past several thousands of years in the earlier days. The low-land area in the eastern part of Tokyo was completely under the sea water of Tokyo Bay about 6000 years ago when the sea water level was about 6 m higher than it is now. It has been formed by repeated flooding of the rivers since then and become a densely populated urbanized area (of about 120 km^2) called by the people as "Zero-metre Area" implying it is lower than the mean high water level (M.H.W.L.:T.P. (Tokyo Peil) 1 m).

Thus born Tokyo low-land area is naturally destined to be susceptible to flooding at every season of raining and typhoons (June to September) that fertilized the land making it suitable for agricultural use. Tokyo low-land area experienced the ceaseless construction of dykes and land reclamation during a period of 200 years from 1660 to 1860. In the middle of nineteenth century, the agricultural fields were converted to residential areas to accommodate workers needed in industrialization of Japan. Those residential areas were inevitably susceptible to flooding.

This susceptibility of Tokyo low-land area to flooding was made even worse by the land subsidence due to heavy deep-well pumping of the groundwater. Huge amount of the groundwater in Tokyo low-land area was continuously pumped up not only for water supply but also for collecting methane gas out of the water for about 100 years. This resulted in lowered groundwater level down to maximum 60 m below the sea level and land subsidence of maximum 4.5 m due to the consolidation of Holocene layers of soft clays and silts (Matsumoto et al. 1988, 2013). In other words, the levels of the river water and the sea water relatively rose maximum 4.5 m above the ground surface of Tokyo low-land area in the past 100 years. This means that Tokyo low-land area would be under water as if it is at the shallow part of the bottom of Tokyo Bay unless the entire area is completely protected by riverine and coastal banks, levees, dykes, flood gates and storm-water pumping stations.

8.2 Major Floods in Tokyo in the Last Century

It is said that the sea water level 6000 years ago when the ocean transgressed most was about 6 m higher than it is now. This means that the lands lower than an altitude of 6 m shown by the green colour in Fig. 8.1 were under the sea water. They emerged above the sea with the progress of coastline regression.

It is interesting to see that all of four great civilizations of the world in Fig. 8.1 are located near/within the green-coloured areas through which meandering big rivers run down before entering the sea. The rivers made the lands even larger by periodical flooding. This regularly repeated process made the deltas fertile, which motivated the people to develop technology needed for flood control and irrigation leading to the formation of the four great civilizations.

In the areas lower than 6 m in Japan (green-coloured areas in Fig. 8.2), similar processes presumably have happened, which turned barren lands first into agricultural fields, and then to urban areas such as Tokyo Metropolitan area, Nagoya Chukyo area and Osaka Kansai area.

Figure 8.3 Shows the coast lines of Tokyo now and 5000 years ago when the sea water level was still much higher than it is now.

Tokyo low-land area densely populated at present was under the water of ancient Tokyo Bay 5000 years ago. Solid circles plotted along the ancient coastlines are kitchen middens indicating that they were located along the old coastlines and remained there without being washed away by tsunamis in the past 5000 years. Kawata (2016) explained that the shallow and narrow mouth of Tokyo

Fig. 8.1 Areas lower than 6 m (shown by *green colour*) (By courtesy of Geospatial Information Authority of Japan)

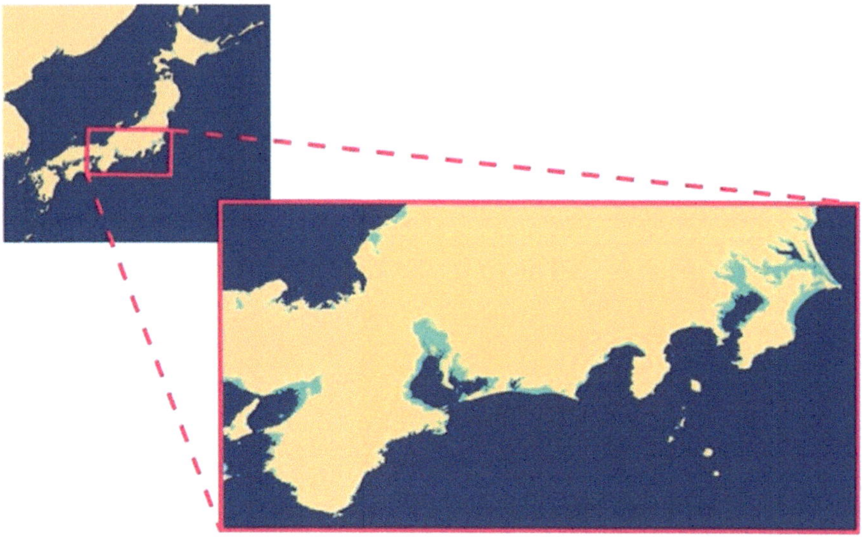

Fig. 8.2 Area lower than 6 m in Main part of Japan (By courtesy of Geospatial Information Authority of Japan)

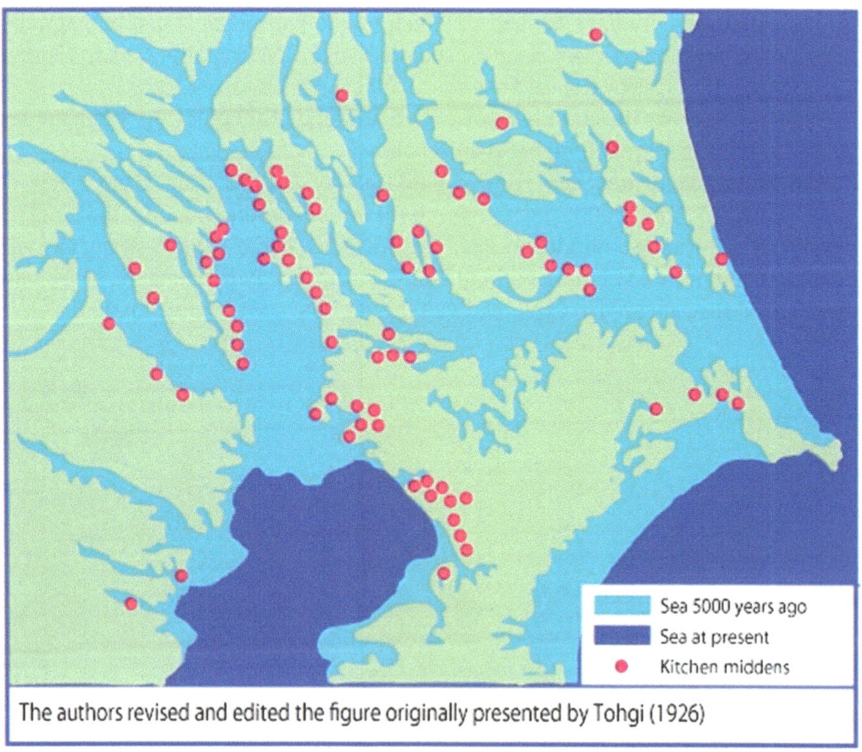

Fig. 8.3 Coastlines 5000 years ago and at present Tohgi (1926)

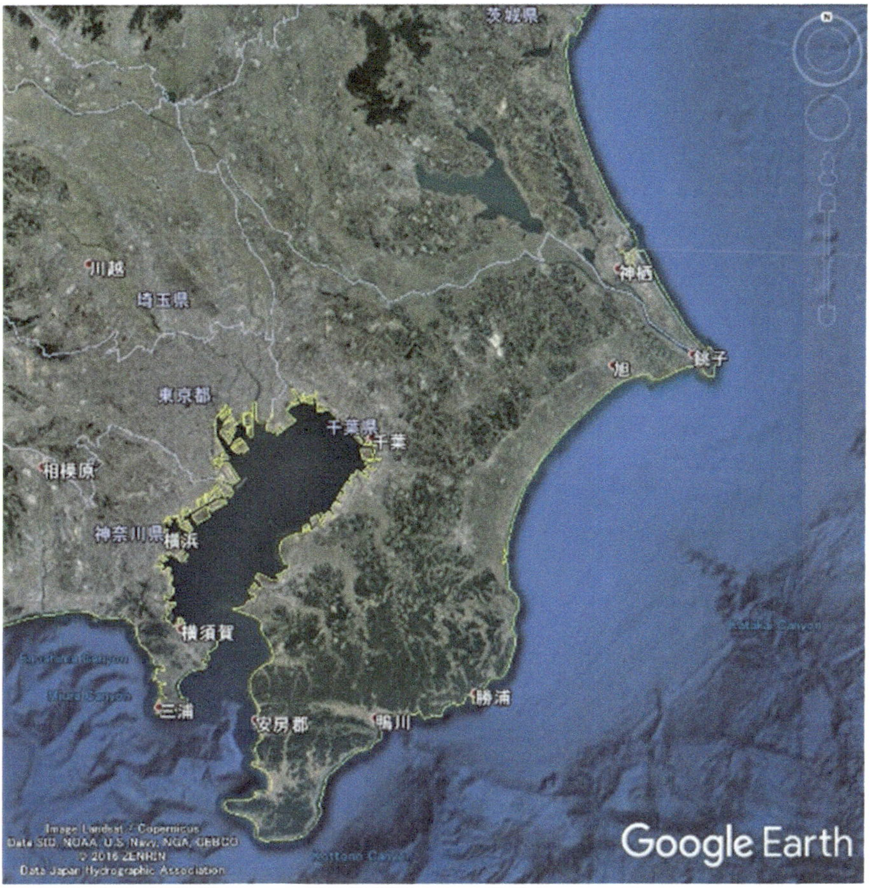

Fig. 8.4 Aerial view of Tokyo Metropolitan Area (Google Earth)

Bay (shown in Fig. 8.4) has effectively prevented the tsunami waves from intruding into the bay.

A heavy rainstorm accompanied by a typhoon hit Tokyo area in August, 1910 and caused Tone River and Arakawa River to flood Tokyo low-land area as seen in Fig. 8.5.

Since the damage was unexpectedly massive (toll of dead and missing: 1357 people, breach of riverbank: 7063 sites, number of affected people: 1.5 million), the government started constructing a large-scaled floodway of Arakawa River to drastically increase the capacity of discharging flash floods of that scale (Tsuchiya 2014). The construction work was supervised by Akira Aoyama who had been engaged in the Panama Project.

He employed then modern construction technology such as seen in Fig. 8.6. It took about 20 years to complete the Arakawa River Floodway project an aerial view of which is shown in Fig. 8.7.

Fig. 8.5 Flood of Tokyo low-land area (1910) (Website of Ministry of Land, Infrastructure, Transport and Tourism. Kanto Regional Development Bureau http://www.ktr.mlit.go.jp/ktr_content/content/000101340.pdf)

Fig. 8.6 Construction work of Arakawa River Floodway (Trucks/Excavator) (Website of Ministry of Land, Infrastructure, Transport and Tourism. Kanto Regional Development Bureau http://www.ktr.mlit.go.jp/ktr_content/content/000101340.pdf)

Demographic statistics of Japan in the past 130 years shown in Fig. 8.8 demonstrate that the population has gradually concentrated in Tokyo Metropolitan area after the Meiji Revolution that brought feudalism of shogunate system to an end. People who have not been allowed to change their places of living and work under the old shogunate system started to migrate to urban areas especially to Tokyo Metropolitan area.

Fig. 8.7 Aerial view of Arakawa River floodway (seen vertically at the centre) (Google Earth)

The migration made it possible to convert economic system under the old Shogunate which was dependent on agriculture and light industry into a westernized one based on heavy industries. Meiji regime, the new government, believed that westernization and industrialization of Japan were vitally needed in keeping the nation's independence by fending the colonial occupation by the allied western powers. Human resources to industrialize the country were supplied from rural agricultural areas, which resulted in the rapid change of land use in Tokyo Metropolitan area. The use of low-land area susceptible to flooding in its nature was subject to change. The area changed from agricultural fields to densely inhabited districts as clearly seen in Fig. 8.9a–c.

Facing Tokyo Bay, Tokyo low-land area is susceptible to storm tidal surges that are more likely to occur than tsunami waves. Storm surges are caused by the wind setup due to strong onshore winds over the sea surface and the inverse barometer effect associated with pressure drops in low-pressure systems (Hasegawa et al. 2012). A typhoon accompanied by a heavy rainstorm hit Tokyo area at the same time of the high tide in September, 1917. This unfortunate concurrence of two

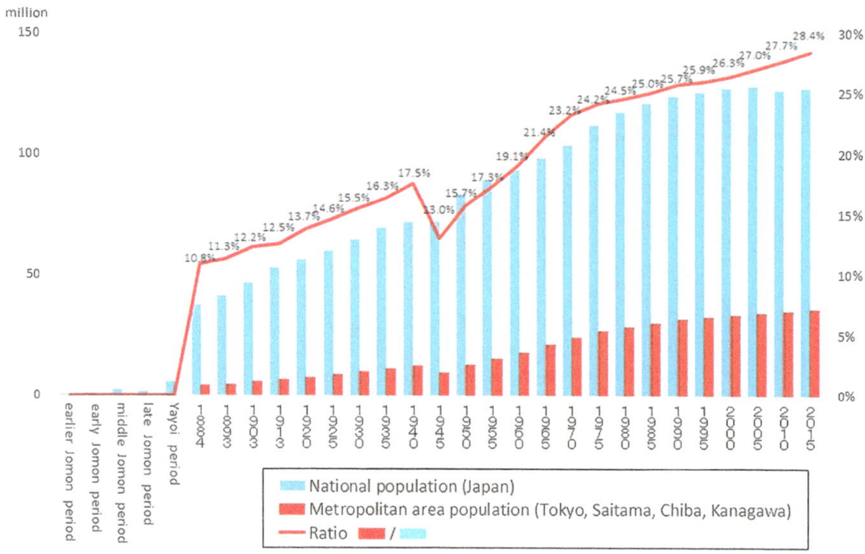

Fig. 8.8 Demographic statistics of Japan

Fig. 8.9 Change in land use (*green*: agricultural use) (**a**) 1910: before the Arakawa River floodway (**b**) 1928: after the Arakawa River floodway (**c**) 2015: current Arakawa River floodway

Fig. 8.10 Schematic map of Yamba Dam Project/photograph of dam works (Website of Ministry of Land, Infrastructure, Transport and Tourism. Kanto Regional Development Bureau http://www.ktr.mlit.go.jp/yanba/)

events resulted in breakage of some parts of the seawalls and the riverbanks of the floodway under construction. Resulting damage was serious and extensive (Tsuchiya 2014). The typhoon generated two storm surges which people mistook as tsunami waves. Kawata (2016) explained that the surge generated by the typhoon slowly advancing across Tokyo Bay happened to overlap with the wave periodically oscillating within the bay and formed two big waves.

Typhoon Kathleen that passed along the coast of the Pacific Ocean near Tokyo in September, 1947 accompanied a heavy rainstorm resulted in failure of some part of riverbank of Tone River, Arakawa River and Nakagawa River and flooded 114 km^2 of Tokyo low-land area (Nakajima 2015; Tsuchiya 2014). It was learnt that both discharge capacity of the rivers and storage capacity of upstream reservoirs need to be increased to cope with the flood of that scale. Four dams in the valley of Arakawa River, 14 flood control dams and retarding basins in the valley of Tone River have been built since then by Ministry of Land, Infrastructure, Transport and Tourism (MLIT) and Japan Water Agency. Yamba dam is currently under construction in the upstream valley of Tone river as shown in Fig. 8.10a and b.

Construction of the dam was started after a long period of preparation since it was planned in 1949. The construction was interrupted in 2010 to reappraise the project. Alternative measures for flood control were examined. However, the original project plan was recommissioned in 2011 as it was judged as the most suitable one.

Typhoon Kitty hit Tokyo area in August, 1949 when Tokyo Bay was in high tide. The storm surge flooded 92 km^2 of Tokyo low-land area (Nakajima 2015). High storm surges are caused by combination of the wind setup due to strong onshore winds over the sea surface and the inverse barometer effect associated with pressure drops in low-pressure systems (Hasegawa et al. 2012). They proved to be more dangerous for Tokyo low-land area when onslaught of low-pressure systems (i.e. typhoons) coincided with the high tide as in the cases of the typhoons in 1917 and 1949.

Coincidence of high tide and typhoon arrival in Tokyo Bay area seldom happens as seen in Table 8.1. However, probability of such coincidence is higher than that of

Table 8.1 Storm surges experienced in Japan (By courtesy of Ministry of Land, Infrastructure, Transport and Tourism)

Past main high tide disaster in Japan			
Date	Main cause	Main damage area	Highest tidal level (T.P m)
1917.10.01	Typhoon	Tokyo Bay	3.0
1927.09.13	Typhoon	Ariake sea	3.8
1934.09.21	Muroto typhoon	Osaka Bay	3.1
1942.08.27	Typhoon	Suonada	3.3
1945.09.17	Makurazaki typhoon	South Kyushu	2.6
1949.08.31	Kitty typhoon	Tokyo Bay	3.0
1950.09.03	Jane typhoon	Osaka Bay	2.7
1951.10.14	Ruth typhoon	South Kyushu	2.8
1953.09.25	Typhoon No.13	Ise Bay	2.4
1959.09.26	Isewan typhoon	Ise Bay	3.9
1961.09.16	The second Muroto typhoon	Osaka Bay	3.0
1970.08.21	Typhoon No.10	Tosa Bay	3.1
1985.08.30	Typhoon No.13	Ariake sea	3.3
1999.09.24	Typhoon No.18	Yatsushiro sea	4.2

mega-earthquakes in vicinity of Tokyo area. It should be noted that the likelihood is increasing with the gradual rise of both the sea water level and the sea water temperature that activates typhoons. Destructive power of a storm surge is far beyond our imagination. A storm surge is generated as a combination of (i) waves setup due to strong onshore winds over the sea surface, (ii) the inverse barometer effect associated with pressure drops in low-pressure system (typhoon) and in the worst case (iii) coincidence with high tide. This explanation may give us a false image of the phenomenon as if the sea water level gradually and gently gets higher especially when coincidence of high tide and arrival of typhoon is imagined. However, as seen in Fig. 8.11 taken after Typhoon Isewan (1959), a storm surge is violent enough to completely destroy concrete seawalls and make a large ship uncontrollably drifted and wrecked off the coast. It is as powerful as a tsunami.

8.3 Possible Flooding in Low-Land Area of Tokyo

As briefly mentioned in the introduction, Tokyo low-land area has experienced land subsidence of 4.5 m at maximum. This was, regrettably, the unrecoverable consequence of consolidation of soft clayey materials overlaying the permeable layers of aquifer from which the groundwater containing methane gas was excessively pumped up for about 100 years resulting in sharp drop of the groundwater level. Figure 8.12 shows that settlement of the ground surface has been accelerated in 1920s and continued until 1975 when regulation on the deep-well pumping of the groundwater for industrial use was introduced.

Fig. 8.11 Concrete seawall hit by the storm surge caused by Isewan Typhoon (Typhoon Vera) in 1959 and a wrecked large ship after uncontrollable drift (By courtesy of Professor Yoshiaki Kawata, Photo taken by the late Professor Yuichi Iwagaki)

31.5 km^2 of Tokyo low-land area is currently lower than the low tide level of Tokyo Bay. 92.8 km^2 of the area is lower than high tide level. As large area as 254.6 km^2 of the area is expected to be immersed by the sea water when storm surges hit Tokyo area at the worst timing (Fig. 8.13).

The comparison of ground surface levels measured in 1909 and 2016 in Fig. 8.14 shows extensive land subsidence in the area.

Figure 8.15a–c demonstrate the ground surface considerably lower than the water in the river and susceptible to flooding.

Land subsidence in Tokyo low-land area created situation in which numerous residences and industries will be under sea water level as if they stand on the bottom of Tokyo Bay. In order to avoid submersion of the settled areas, protective measures such as raising existing riverbanks, installation of seawalls and flood gates (and floodwall), construction of pumping stations, placement of emergency water-proof doors at the subway stations and underground shopping areas, and high-standard levees (super levees) have been progressively introduced. Anti-seismic strengthening of these facilities has been done in accelerated manners especially after experiencing Great East Japan Earthquake in 2011.

All of those protective measures which have been taken in and around Tokyo in the past 70 years worked well to prevent extremely serious flood damages from both rivers and the sea on the occasions of typhoons, heavy rainstorms (including

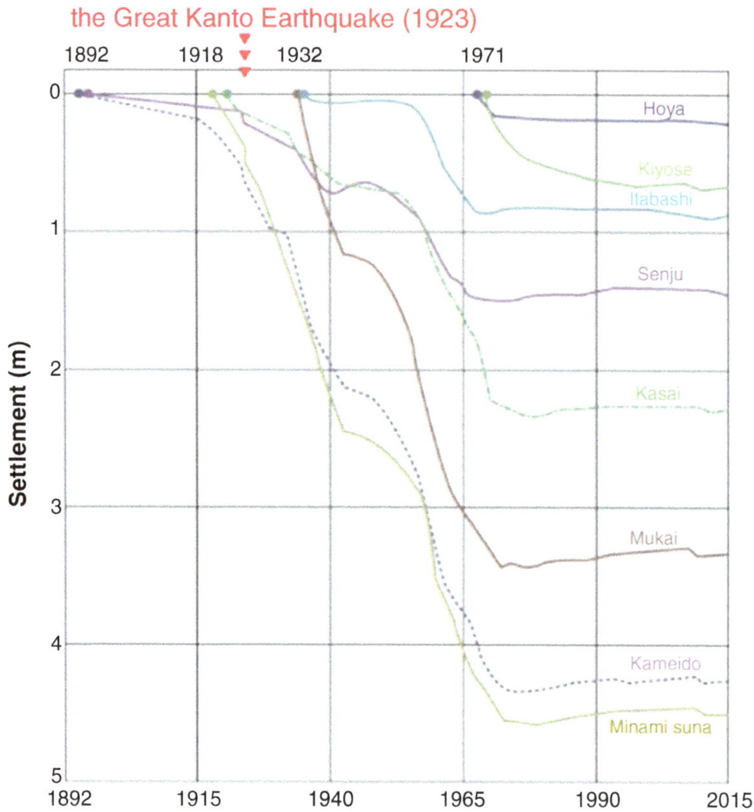

Fig. 8.12 Land subsidence (By courtesy of Tokyo Metropolitan Government)

so-called guerrilla rainstorms), and earthquakes of small and medium scales that may hit the area on regular base.

However, some of the specialists warned emerging risk by mega-floods especially in Tokyo low-land area despite the fact that the area has been successfully protected even after experiencing serious land subsidence that had made the situation much worse than before. The warning should not be overlooked. For instance, "Publication of Expected Flood Damage to Subways, etc." issued by the Expert Panel of Major Flood Control Measures of the Central Disaster Management Council of the Japan Cabinet Office (2010) predicted that 97 stations of 17 subway lines will be submerged in case the right bank of Arakawa River is breached at the 21 km point from the river mouth. Kawata (2016) argued that the damage will be more than what are presented in the report. Ikeuchi et al. (2012) demonstrated the possible scale of the damage to the subway system in such case. Kawata (2016) introduced lessons learnt from two hurricanes, Hurricane Katrina in 2005 and Hurricane Sandy in 2012. Disaster events and damages developed in unexpected, unexperienced ways in those two cases.

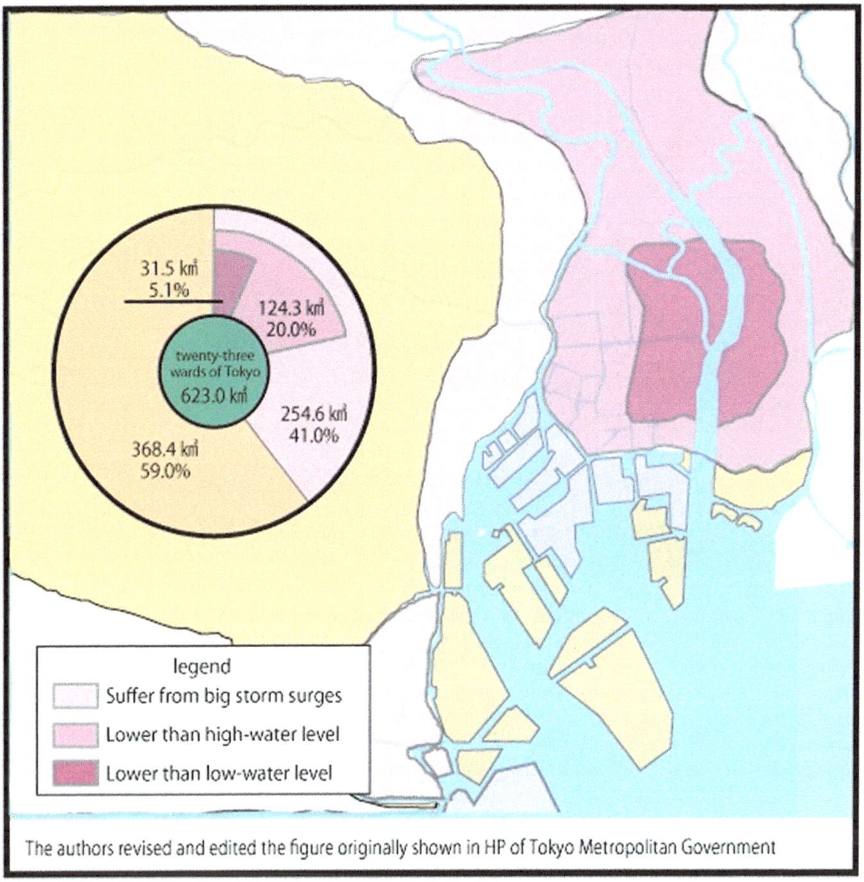

Fig. 8.13 The altitude of the ground surface of Tokyo (By courtesy of Tokyo Metropolitan Government)

Imamura et al. (2016) and Aoki et al. (2016) introduced the progress of installation of water-proof gates (Fig. 8.16a and b) and other facilities that prevent flood water from flowing into the subway system.

The Expert Panel of Major Flood Control Measures of the Central Disaster Management Council of the Japan Cabinet Office (2010) estimated damage caused by not only flood of Tone River and Arakawa River, but also sea water from Tokyo Bay. The Expert Panel studied several cases that could be much more serious than floods of rivers. Tsuchiya (2014) mentioned a possible situation in which the sea water overflows onto Tokyo low-land area. In such case, big storm surges attack the flood protective facilities such as seawalls, riverbanks, flood gates, floodwall gate and drainage pumping stations. He pointed out that some of these facilities, especially floodwall gate, may not be effective enough to protect the low-land from the invasion of the sea water if the storm surges are too big. Kawata (2016)

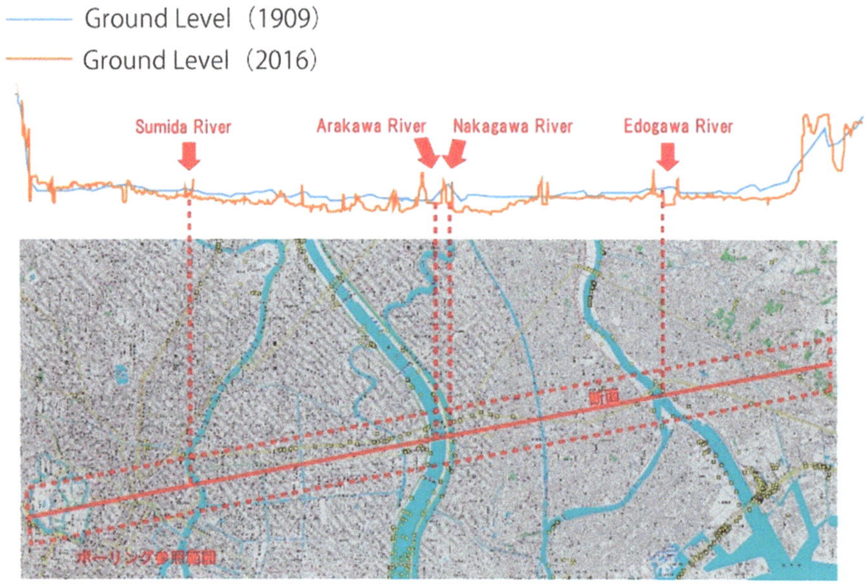

Fig. 8.14 Comparison of the attitude of the ground surface in 1909 and 2016

Fig. 8.15 (**a–c**) Ground surface lower than the water level of the river

Fig. 8.16 Water tight doors at the entrance of the subway (By courtesy of Tokyo Metro Co., Ltd.)

warned the possibility of uncontrollably drifting ships, such as seen in Fig. 8.11, hit and break a thin (about 0.3 m thick) concrete parapet walls at the time of big storm surges.

Even in the case that the sea water level is lower than the height of seawall, i.e. the sea water is successfully prevented from flowing onto the low-land over the top of the seawall, the water can burst out or seep from the ground surface when the water level is high enough compared with the ground level. Kawata (2016) reported that this happened at the time of the Second Muroto Typhoon (1961) in the districts called Chayamachi and Nakatsu in the vicinity of Yodogawa River running through the middle of Osaka City. Wide area of these districts was covered by the water erupting out from the ground surface. In his book, Kawata (2016) pointed out many different ways of possible flooding of Tokyo low-land area. For instance, he mentioned that flooding in the downstream area of Arakawa River would be very likely to occur when a big typhoon hits Tokyo Bay causing a storm surge (of the level of officially expected maximum scale) together with a heavy rainstorm (of the scale once in 200 years) in the upstream area. He also mentioned that the sea water in Tokyo Bay could flow onto Tokyo low-land area if any of the seawalls, riverbanks and/or flood gates settle instantaneously due to liquefaction of the ground caused by a possible mega-earthquake or due to a terrorist attack.

Nakabayashi (2015), Nakajima (2015), Asakawa (2015), and Kawata (2016) discussed about the possibility of having the situation in which a big storm surge is induced in Tokyo Bay by a big typhoon coming to Tokyo area sometime after Tokyo is hit by an earthquake big enough to seriously damage flood protective facilities (such as the flood gate shown in Fig. 8.17) through shaking and consequent liquefaction of the low-land area.

Figure 8.18 shows the area that Tokyo Metropolitan Government (2013) has announced as susceptible to the liquefaction at the time of big earthquakes. The entire Tokyo low-land area located inside the area is susceptible to liquefaction.

This implies that some of the flood protective facilities in Tokyo low-land area may be affected by the earthquake shaking and liquefaction of the ground to a degree that they need to be repaired or replaced after the earthquake taking some considerable time. Nakabayashi (2015), Nakajima (2015), Asakawa (2015), and Kawata (2016) are concerned about the situation of having a big storm surge before the repair and/or replacement works are finished. When this happens, Tokyo low-land area will possibly be inundated by water from Tokyo Bay with a maximum depth of 4.5 m.

It seems to the authors that flooding of Tokyo low-land area will most likely be triggered by a storm surge coming shortly after a severe enough earthquake that damages some of the flood protective facilities to an extent that they need repair works or replacement. Let us introduce two historical cases of such type of flooding.

At 16:13 on June 28th, 1948, a strong earthquake (magnitude of 7.1) hit the Fukui plain. The banks of Kuzuryu River located in the plain were seriously damaged by the earthquake as typically shown in Fig. 8.19a, b in which cracks generated in the slope are seen (rather horizontally in Fig. 8.19b) running along the bank. The maximum amount of settlement of the crest of a riverbank was 4.5 m.

Fig. 8.17 Flood gate

Countless number of cracks were found in the bank slopes longitudinally along the river. Failures of bank slope were also found in many places. About 1 month later, during a period of 23rd to 25th July, heavy rainstorm caused bank failures at Tohmyoji on the left bank of Kuzuryu River and downstream. An area of 19 km^2 in the Fukui plain was inundated by the water of maximum 2.4 m deep. The bank at the failure site settled about 2.5 m in average accompanied by cracks such as seen in Fig. 8.19. (Director General for Disaster Management, Japan Cabinet Office 2011).

At 5:46 on January 17th, 1995, Hyogo-ken Nanbu Earthquake (magnitude of 7.3 with an intensity of 7 on the Japanese scale) hit Kobe area and seriously damaged buildings, bridges, retaining walls, and harbour structures etc. The low-land areas near Kobe and Osaka seriously suffered from land liquefaction. The left bank of Yodogawa River at Torishima district settled maximum of about 3 m over a length of 1800 m (0.2–2.0 km from the river mouth) and laterally collapsed towards the river resulting in the overwhelming failure of riverbank and concrete parapet as seen in Fig. 8.20a, b.

In the nearby areas, convincing evidence for sand liquefaction, such as sand volcanoes, was widely observed. According to Watanabe (2010), the height of the bank after settlement was in a range of OP + 3.5–4.5 m (OP: Osaka Peil). Since the high tide at the mouth of Yodogawa River was OP + 2.20 m, the bank had almost completely lost its function as a riverbank after settlement. The emergency measures (double sheet-pile coffer dam) of riverbank rehabilitation was completed in June fortunately just before the typhoon season.

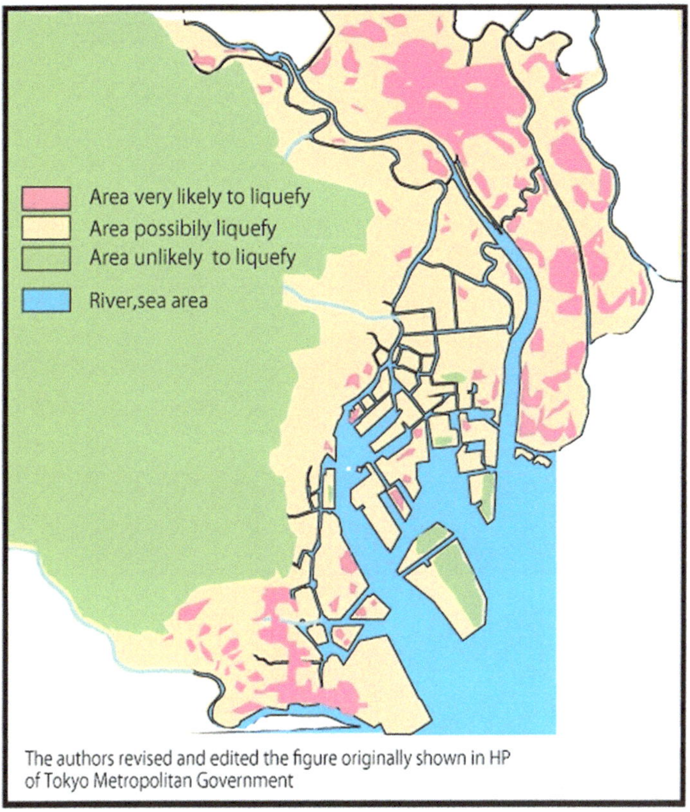

Fig. 8.18 Areas likely to suffer from liquefaction (By courtesy of Tokyo Metropolitan Government)

Oshiki and Sasaki (2001) introduced another four cases in addition to the above mentioned two cases of complex disaster of a rainstorm after an earthquake. They are (i) Great Kanto Earthquake (September 1st, 1923) followed by typhoon rainstorm on September 15th, 1923, (ii) Niigata Earthquake (June 16th, 1964) followed by rainstorm on July 7th flooded Niigata downtown area, (iii) Miyagiken-oki earthquake (June 12th, 1978) followed by four sequential rainstorms caused by seasonal rain fronts and (iv) Hokkaido-Nansei-oki Earthquake (July 12th, 1993) followed by a typhoon in the end of July. In all of those six cases, the rainstorms significantly damaged riverbanks that have seriously suffered from the preceding strong earthquake shaking. This type of complex disaster of a typhoon-induced rainstorm after strong earthquakes concerns the authors most.

It should be noted that flooding of some part of Tokyo low-land area lower than the low tide of Tokyo Bay will take place immediately after the earthquake when the earthquake is unfortunately powerful enough to severely destroy at least one of the flood protective facilities unavoidably letting the sea water flow onto Tokyo low-land. This is the worst scenario in which enormous number of people will inevitably suffer either on the ground or under the ground.

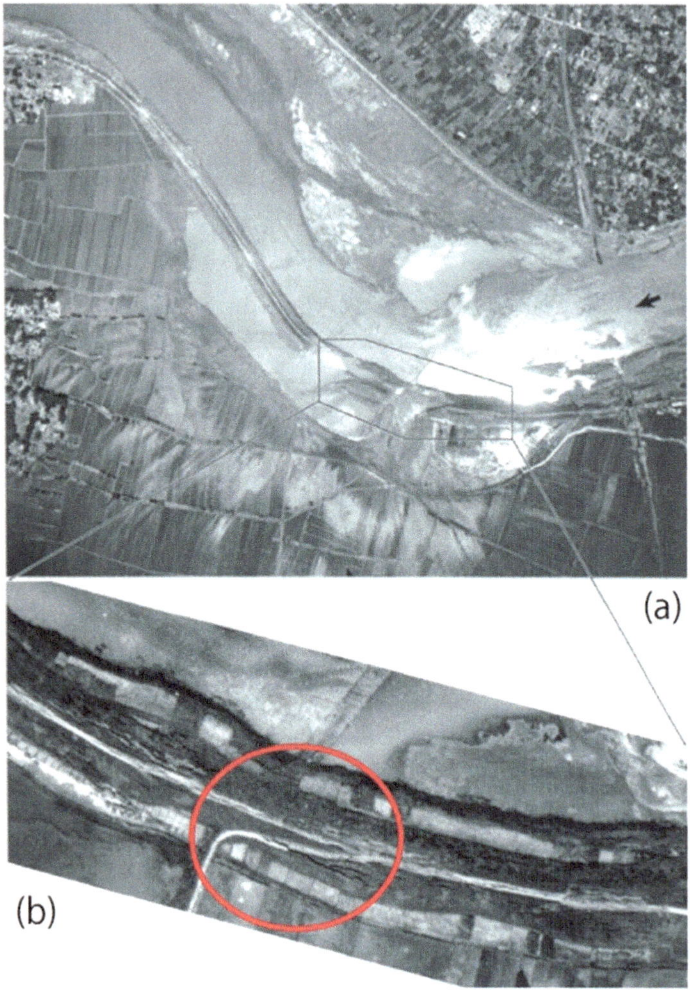

Fig. 8.19 Cracks in the slope of riverbank (By courtesy of Geospatial Information Authority of Japan)

8.4 Four Meter Deep Flooding in Tokyo Low-Land Area

Economic impacts of 4 m–deep flooding in Tokyo low-land area are examined in this section. The discussion focuses on the economic effect of possible submersion of the subway system in Tokyo. In case eastern part of Tokyo is inundated by water of 4 m deep, the subway system throughout the whole areas of Tokyo will be largely filled with water coming from some of the subway stations located in the submerged lowland areas. This was predicted by Publication of Expected Flood Damage to Subways, etc. issued by the Expert Panel of Major Flood Control

Fig. 8.20 Dyke suffered from the Hyogo-ken Nanbu Earthquake in 1995 (Website of Kinki Regional Development Bureau, /Yodogawa River Office http://www.yodogawa.kkr.mlit.go.jp/activity/maintenance/earthquake/)

Measures of the Central Disaster Management Council of the Japan Cabinet Office (2010) as introduced in the previous Sect. 8.3. Some preventive measures are being taken by Tokyo Metro that operates the subway system as introduced in Sect. 8.3, but these preventive facilities are not fully completed yet. The difficulty arises from many of the subway entrances that are complicatedly connected with the underground shopping areas and the basement of large-scaled buildings.

It is most likely that water from Tokyo Bay keeps coming until the seawalls, flood gates and dykes (damaged by the preceding big earthquake) are repaired hopefully in a month time. During this period of emergency repair works and water drainage, the subway system will remain submerged under muddy and salty water that irreparably damages mechanical and electrical facilities. All of such facilities have to be replaced by a set of new ones. This will take far more time than pumping the water out of the subway lines, may be several years. It should be noted that the pumping stations located in Tokyo low-land area currently standing on the ground surface will be under 4 m–deep sea water. They are not designed to keep working even when they are entirely under the water as mentioned by Tsuchiya (2014). This means that we need an alternative set of pumping stations designed to pump water out of the low-land area through properly designed another set of conduit networks.

Figure 8.21 shows the cross-section of Tokyo low-land area. The expected sea water level which is well above the current ground surface at the time of possible storm surges is indicated by a red dotted line. Subway tunnels shown in Fig. 8.21 are deep enough to be filled by water with high water pressure. Kawata (2016) discussed various scenarios on what will happen when such a mega-disaster hits Tokyo low-land area either in predictable ways or in unpredictable ways.

About ten million of people use the subway system of Tokyo Metropolitan area every day. When the subway system becomes unavailable due to flooding, the great majority of the ten million people have practically no alternative means of commuting. If this situation continues for a certain period of time, say, in the order of several months or years, Japanese workers living in Tokyo Metropolitan area will be unable to go to their offices and factories for work for a certain period of time,

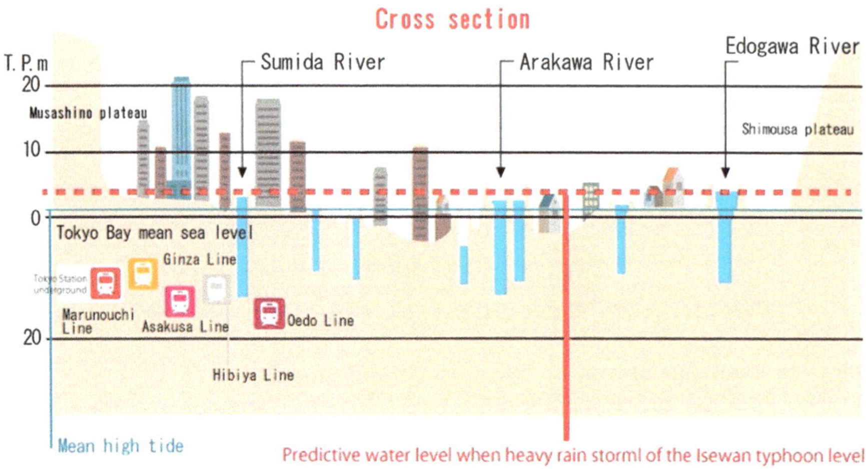

Fig. 8.21 Cross section of Tokyo low-land area

maybe half a year or several years. Many companies and businesses will suffer from a shortage of skilled staff members. If this happens, offices and factories in not only the low-land area but also the higher areas in Tokyo will not be able to produce their products because they have no workers and consequently nothing to sell. Most of them, companies and businesses, will get no money after the mega-disaster and eventually go bankrupt.

It is really a serious problem that the Japanese economy is too much dependent on the activities concentrated in the Tokyo Metropolitan area. In case that majority of the economic activities in Tokyo stops, almost all the economic activities in Japan will be most likely to stop too. Production activities of the Japanese companies highly depend on the supply chains consisting of great many pieces of parts produced by relatively small scaled companies. It is likely that some of the parts cannot be produced any more after the flooding of Tokyo low-land area. Deep flooding in the lowland area in Tokyo might bring serious difficulties to not only the people in Tokyo but also almost all of the people working in Japan. The authors assume that not many of companies and businesses in Tokyo do take such a situation into their Business Continuation Plan.

If companies and businesses are unable to pay to their employees, people (presumably of the order of several millions) will have practically no source of their income and have to move from Tokyo Metropolitan area to other regions in or outside Japan to seek their new jobs. Majority of the people in Japan used to purchase their flats and/or houses on housing loans. They will be unable to repay their loans and leave their flats or houses which are mortgaged. These houses and flats will hardly be sold because there will be practically no demand. Financial institutions will carry massive non-performing loans.

The extremely pessimistic scenario mentioned above is only a figment of the authors' imagination. But if this really happens, it will be unbelievably great loss to the shareholders of Japanese companies and financial institutions. The total amount of the loss may be more than that of Lehman shock. The Japanese economy will practically disappear and many of Japanese people will have no or minuscule way of earning money. There will be no practical need of publishing the 'After Action Review (AAR)'. Since this possible scenario is too striking and the likelihood is not clear, we need widely opened discussions from the various perspectives.

The authors believe that the only possible measure to avoid such a mega-disaster is to construct the high-standard riverbank (called the Super levee) along the waterfront of Tokyo Bay near the low-land area because the super levee can hardly be damaged to a fatal extent. In addition to preparing for such hardware with fail-safe system, complementary software systems are needed. It should be noted that the long-term after-effects of the subway submersion on the Japanese economy was first discussed by Aoyama (2016) at the meeting of Kyukokai.

8.5 Super Levee

The super levee is a wide (basement B = 30 × height h) riverbank with a gentle slope in the landside area as typically shown in Fig. 8.22. The extensively built-up landside areas are more spacious, enable effective land use, and provide with scenic beauty and comfortable residential environment as seen in Fig. 8.23 providing.

Conversion of an old-style riverbank into the super levee enables effective use of land. The scheme provides with more space for residents, better environment, and

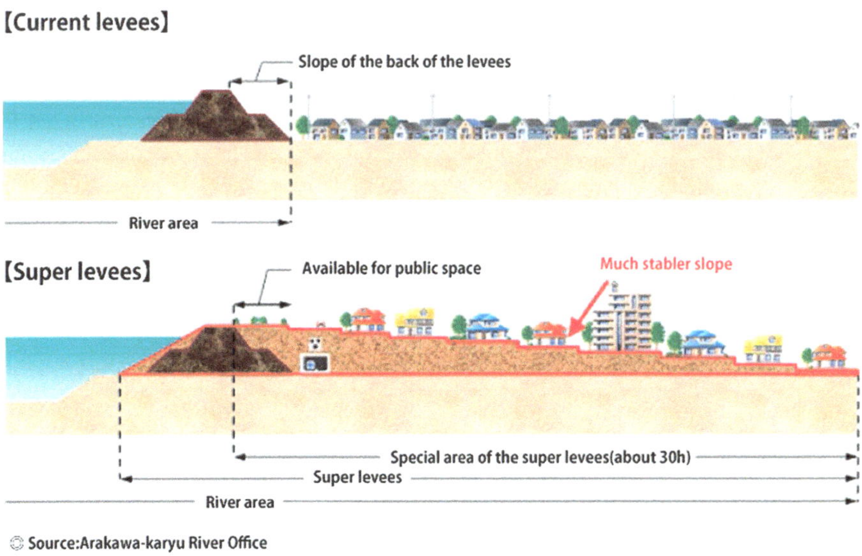

Fig. 8.22 Cross sectional image of the super levee (By courtesy of Kanto Regional Development Bureau)

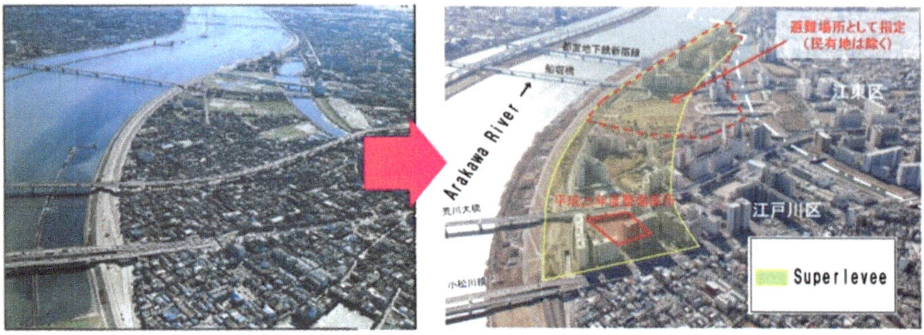

Fig. 8.23 Super levee before and after placement (By courtesy of Kanto Regional Development Bureau)

safe evacuation place in case of emergency such as flooding, earthquakes and fire. Since the evacuation site is in the immediate vicinity, people do not need cars in their evacuation trips and no traffic jams are expected at the time of emergency. On the other hand, much larger land is required for the super levee than for the normal riverbank. This has made the progress of the super levee projects extremely time-consuming because of Japan's very weak concept of eminent domain as pointed out by Brasor and Tsubuku (2015). The project was seriously criticized in the Japanese government's budget screening process by Government Revitalization Unit in 2009.

However, despite criticism over the slow progress of the project, the authors believe that the super levee is the only possible structure to prevent the mega-disaster caused by flooding of Tokyo low-land area. No promising alternatives are found. Some thoughtfully designed new procedures of planning and implementing the super levee project are needed. The super levee project (Fig. 8.24) of Yamato River in Osaka may be a good example.

In this project, lands owned by a local government were effectively utilized in the planning and implementation process. The authors believe that better and easier procedures can be created.

In order to reduce the impact of the possible mega-disaster triggered by the flooding of Tokyo low-land area, the authors emphasize the importance of the policy to encourage companies and businesses in Japan to disperse their productive activities over a wide area of the country.

8.6 Concluding Remarks

In case that a severe natural disaster such as sudden, large-scale tidal surging caused by a big typhoon soon after a strong earthquake or a human action such as terrorist attack results in Tokyo low-land area widely inundated by water of about 4 m deep, the floodwater will inevitably flow into the subway tunnels through entrances

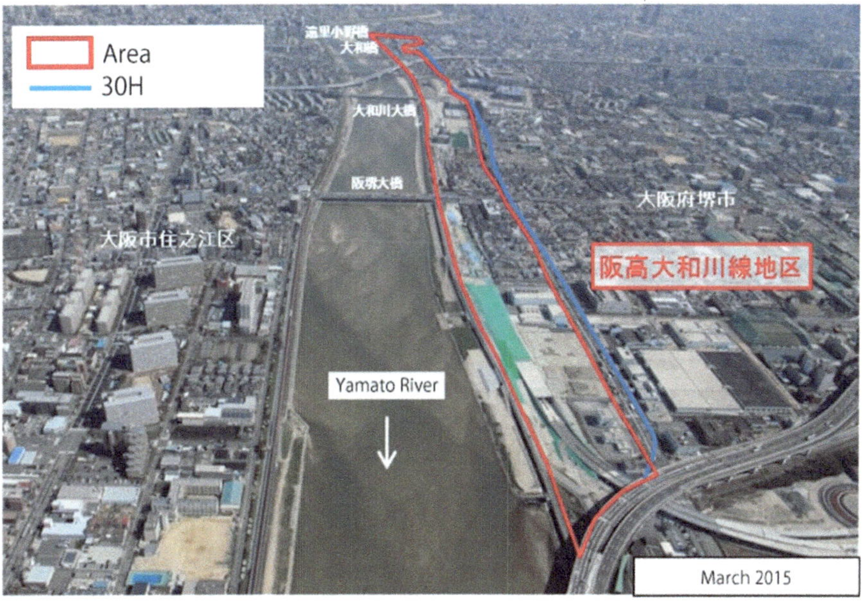

Fig. 8.24 Super levee project of the Yamato River (Website of Kinki Regional Development Bureau, https://www.kkr.mlit.go.jp/plan/zigyohyoka/pdf_2010_05/4.pdf)

located in lowlands and submerge the entire subway system in Tokyo. Enormous number of people will suffer from the floodwater either on the ground or under the ground. Submersion of the subway system will last over a considerably long period of time. This is likely to happen as Kawata (2016) warned in his book based on the detailed technical information obtained from an extremely wide range of sources.

About ten million/day of people use the subway system of Tokyo Metropolitan area. When the subway system becomes unavailable due to flooding, the great majority of the people have practically no alternative means of commuting. Many companies and businesses will suffer from a serious shortage of skilled staff. Offices and factories in Tokyo not only in the low-land area but also in the higher areas will be unable to produce their products and will eventually go bankrupt sometime after the mega-disaster. Production activities of the Japanese companies highly depend on the supply chains consisting of numerous pieces of parts produced by relatively small scaled companies. Some of the parts may not be produced any more after the flooding of Tokyo low-land area. Deep flooding in the lowland area in Tokyo might bring serious difficulties not only to the people in Tokyo but also to almost all of the people working in Japan.

When companies and businesses stop paying to their employees, people will have practically no source of their income and move from Tokyo Metropolitan area to other regions to seek their new jobs. They will have to stop repaying their loans and leave their flats or houses which are mortgaged. These houses and flats will hardly be sold because there is practically no demand. Financial institutions will carry massive non-performing loans. The Japanese economy will practically disappear and many of Japanese people will have no or minuscule way of earning money.

The authors believe that it is the only one possible measure to avoid such a mega-disaster to construct the high-standard riverbank (called the Super levee) along the waterfront of Tokyo Bay near the low-land area because the super levee can hardly be damaged to a fatal extent. In order to make rapid and steady progress of the super levee project, thoughtfully designed new procedures for planning and implementing the project are needed.

Acknowledgements The authors wish to express their gratitude to Mr. K. Irie (Metro Development), Dr. S. Konishi (Tokyo Metro), Mr. T. Sakamoto and Mr. N. Kato (Tokyo Metropolitan Government), Mr. H. Une (Geospatial Information Authority of Japan), Mr. S. Ochi (Foundation of River and Basin Integrated Communications), Dr. K. Hiroki (High-level Experts and Leaders Panel on Water and Disasters), Prof. S. Iai (Kyoto University) and Prof. M. J. Pender (University of Auckland) for their generous support.

This work was supported by JSPS Grant-in-Aid for Scientific Research (No. 26282103, 26420486).

References

Aoki Y, Yoshizawa A, Taminato T (2016) Anti-inundation measures for Underground Stations of Tokyo Metro. In: Proceedings of 15th world conference of Associated research Centers for the Urban Underground Space (ACUUS) 2016 -underground urbanization as a prerequisite for sustainable development- in Saint Petersburg Russia, pp 23–28

Aoyama T (2016) What should be prepared for creating a safer land of the nation, Invited lecture at the Autumn Meeting of Kyukokai (in Japanese)

Asakawa K (2015) Preparation in Edogawa for severe flood disasters, Emergency Evacuation in Flood Disasters -In preparation for big flash floods and storm surges in Capital Tokyo-, Japan Emergency Management Association and Oriental Consultants, Jihyo, 92–104 (in Japanese)

Brasor P, Tsubuku M (2015) Tokyo readies for the rise of the 'guerrilla rainstorm', The Guardian, Last modified on Tuesday 7 July 2015

Director General for Disaster Management, Japan Cabinet Office (2011) Report of 1948 Fukui Earthquake, pp 206–216 (in Japanese)

Expert Panel of Major Flood Control Measures of the Central Disaster Management Council of the Japan Cabinet Office (2010) Publication of expected flood damage to subways, etc. (in Japanese)

Hasegawa H, Kohno N, Hayashibara H (2012) JMA's storm surge prediction for the WMO Storm Surge Watch Scheme (SSWS), Technical Review No. 14, RSMC Tokyo-Typhoon Center, Japan Meteorological Agency. http://www.jma.go.jp/jma/jma-eng/jma-center/rsmc-hp-pub-eg/techrev.htm

Ikeuchi K, Ochi S, Yasuda G, Okamura J, Aono M (2012) Estimates of inundation of subways due to large-scale floods and an analysis of effectiveness of migration measures. J Jpn Soc Civ Eng Ser B1 68(3):136–147. (in Japanese)

Imamura T, Sasaki K, Kawakami K, Abe T, Konishi S (2016) New anti-inundation measures of subway tunnels. In: Proceedings of 11th World Conference of Railway Research (WCRR) 2016 in Milan Italy-Research and Innovation from Today Towards 2050-, OP60, No. 5, (ID Number 198488)

Japanese Government's budget screening process (WG-B-7) of Government Revitalization Unit in Cabinet Office (2009) (in Japanese)
Kawata Y (2016) Nihon Suibotsu (National Submergence in Japan), Asahi Shinsho N. 571, Asahi Shinbun Publication (ISBN 978-4-02-273671-0), p 109, p 116, p 119, pp 121–122, p 125, pp 131–132, pp 175–187, pp 228–245 (in Japanese)
Matsumoto K, Ohta H, Kobayashi A, Sandanbata I (1988) Analyses of deformation caused by excavating and filling on the reclaimed area at waterfront. In: Proceedings of 33rd symposium. Geotechnical Engineering, Tokyo, pp 117–122 (in Japanese)
Matsumoto K, Kobayashi A, Ohta H (2013) Chapter 8 Deformation prediction of a structure placed on soft clay in Tokyo Bay affected by heavy deep well pumping. In: Jian Chu, Sri P R Wardani, Atsushi IIzuka (eds) Geotechnical predictions and practice in dealing with geohazards. Springer
Nakabayashi I (2015) National resilience regional plan and needs of regional collaboration, Emergency Evacuation in Flood Disasters -In preparation for big flash floods and storm surges in Capital Tokyo-, Japan Emergency Management Association and Oriental Consultants, Jihyo, pp 36–54 (in Japanese)
Nakajima T (2015) Case histories of Typhoon Kathleen etc. in low-land area of eastern Tokyo, Emergency Evacuation in Flood Disasters -In preparation for big flash floods and storm surges in Capital Tokyo-, Japan Emergency Management Association and Oriental Consultants, Jihyo, pp 80–90 (in Japanese)
Oshiki H, Sasaki Y (2001) Restoration works of seismically damaged river dikes using remedial treatment of liquefiable layer. J Jpn Soc Civ Eng 686/VI-52:15–29. (in Japanese)
Tohgi R (1926) Ancient coastlines in Kanto lowland area estimated based on topography and distribution of kitchen middens (2). Geogr Rev Jpn 2(8):659–678. (in Japanese)
Tsuchiya N (2014) Shuto Suibotsu (Capital city submerged underwater), Bunshun Shinsho No. 980, Bungei Shunnju (ISBN 978-4-16-660980-2), pp 60–61, pp 96–98, p 104, p 142, p 172, pp 185–186 (in Japanese)
Watanabe T (2010) Survey on damage of rivers in Great Hanshin-Awaji Earthquake Disaster and damage rehabilitation works, 202–213, River Engineering Transmitted from Generation to Generation, River Bureau, Kinki Regional Development Bureau, Ministry of Land, Infrastructure and Transport (in Japanese)

Chapter 9
One Dimensional Shallow Foundation Macro-element

M.J. Pender, T.B. Algie, R. Salimath, and L.B. Storie

Abstract Recently a number of macro-element models have been formulated for assessing the performance of shallow foundations during earthquake loading. These provide a computational tool that represents the nonlinear dynamic behavior of the foundation in a manner much simpler than finite element modelling; consequently, they are useful for preliminary design. The basis of this chapter is the shallow foundation moment-rotation pushover curve, which is bracketed by the rotational stiffness at small deformations, determined by the small strain stiffness of the soil, and the moment capacity, which is a function of the soil shear strength and the vertical load carried by the foundation. Between these two limits there is a curved transition. The paper argues that when the vertical load carried by an embedded foundation is a small fraction of the vertical bearing strength, the moment-rotation behavior dominates the response. This means that the structure-foundation system can be reduced to a single degree of freedom (SDOF) model.

The form of the shallow foundation moment-rotation curve obtained from experimental and computational modelling is approximately hyperbolic; the nonlinear shape is due in part to the nonlinear deformation of the soil beneath the foundation but also to gradual loss of contact between the underside of the foundation and the soil below. The paper proposes a generalization of the pushover curve to give a shallow foundation cyclic moment-rotation relationship. The hysteretic damping properties of the model, as a function of the foundation rotation amplitude, are demonstrated as is the relation between secant stiffness and foundation rotation.

This chapter shows how the model can be applied in numerical simulation of structure-foundation systems subject to earthquake time histories. The significance

M.J. Pender (✉)
Department of Civil & Environmental Engineering, University of Auckland, Auckland 1142, New Zealand
e-mail: m.pender@auckland.ac.nz

T.B. Algie
Partners in Progress Ltd., Sydney, Australia

R. Salimath • L.B. Storie
Tonkin and Taylor Ltd., Auckland, New Zealand

S. Iai (ed.), *Developments in Earthquake Geotechnics*, Geotechnical, Geological and Earthquake Engineering 43, DOI 10.1007/978-3-319-62069-5_9

of the maximum displacement (foundation rotation) in relation to the damping and residual rotation at the end of the earthquake record are discussed.

9.1 Introduction

9.1.1 Background

A fundamental aspect of foundation behaviour is nonlinear interaction between the foundation and the surrounding soil. Often this interaction is idealised as elastic, but we must never overlook the fact that foundation-soil interaction is nonlinear. Having accepted this nonlinearity the problem is how to estimate the response of the foundation to imposed loading. There are well developed software systems that handle nonlinear behaviour which can also determine dynamic response. Thus our profession is well served with existing facilities. Unfortunately, there is an underlying problem in that the provision of numerical values for the various parameters required to make this numerical modelling possible is not a simple undertaking. Thus, from the point of view of foundation design, these highly sophisticated numerical tools may not be the best approach, at least for the initial stages of design. For most of the design process one needs simpler tools so that a number of scenarios can be evaluated quickly and the sensitivity of the response to variations in the input parameters assessed. At the final stage of the design process one might then want to use sophisticated numerical approaches to confirm the design decisions.

9.1.2 Macro-element Idealisation

Recently a number of so-called macro-elements have been developed as an aid to the design of foundations to resist earthquake loading (Cremer et al. 2001, 2002); Chatzigogos et al. 2007, 2009; Figini et al. 2012; Gajan and Kutter 2008; Toh 2008; Toh and Pender 2010). A macro-element is a computational tool that represents the behaviour of the foundation in a comparatively simple manner. If one was idealising foundation-soil interaction as linear, then an elastic macro-element would specify the vertical stiffness of the foundation, the horizontal stiffness and the rotational stiffness and any interaction between these stiffness parameters. In this way it is not necessary to model the distribution of stress and strain in the material beneath the foundation. This category of macro-element is well presented by Wolf and Deeks (2004). However, having said above that the soil behaviour is fundamentally nonlinear, alternative macro-elements, such as those cited above, are required. These are nevertheless quite sophisticated in that they employ the ideas of bounding surface plasticity using a bearing strength surface (the locus of all possible combinations of vertical load, horizontal shear, and moment – discussed further below) as the bounding surface. Nonlinear behaviour inside the surface is controlled by the distance between the current position on the action path and the bounding surface. Millen (2015) has implemented such a macro-element within the Ruaumoko, Carr (2016), nonlinear dynamic structural analysis software. Despite the success of these macro-elements, the behaviour of a very much simpler macro-element is explored herein.

9.1.3 Measured Shallow Foundation Response

Field experiments have been conducted a sites north of the Auckland CBD with shallow foundations supporting a simple single storey structure subject to gradual pull-back followed by snap-back release; more details are given by Algie et al. (2010, Algie (2011), and Salimath (2017). The dynamic response of the system was recorded after each snap release. An added bonus was the static load-deflection curve obtained during the pullback phase of the test. The pullback response is in effect a static pushover curve for the foundation. Elsewhere, the relations between the rocking period and the pull-back angular displacement and damping and pull-back angular displacement are presented (Algie 2011).

The site used for the Algie tests had a profile of stiff cohesive soil. The soil profile was investigated with CPT tests between the surface and depth of 8 m. In some of these the shear wave velocity of the soil was measured which indicated a reasonably consistent shear wave velocity for the materials at the site equivalent to a small strain shear modulus of about 40 MPa.

Figure 9.1 shows some of the pull-back data obtained by Algie. It is apparent that there is considerable nonlinearity in the moment-rotation curves but a simple hyperbolic relationship, which asymptotes to the moment capacity of the foundation, can be fitted through the data. It is evident that some degradation occurs from one snapback to the next, particularly for those tests following the snapback which applies the largest moment to the system.

Further testing of this type was completed late in 2016 at Silverdale, a site further north of Auckland, the results are given by Salimath (2017). In this case the shallow foundations were at the ground surface, whereas those for the Algie tests were embedded to a depth of 400 mm. The Salimath tests, with foundation static load bearing strength factors of safety as high 10, left barely perceptible marking on the ground surface once the test rig was removed. This means that for these tests the nonlinearity in the moment-rotation curve was mostly a consequence of gradual loss of contact between the underside of the foundation and the soil below.

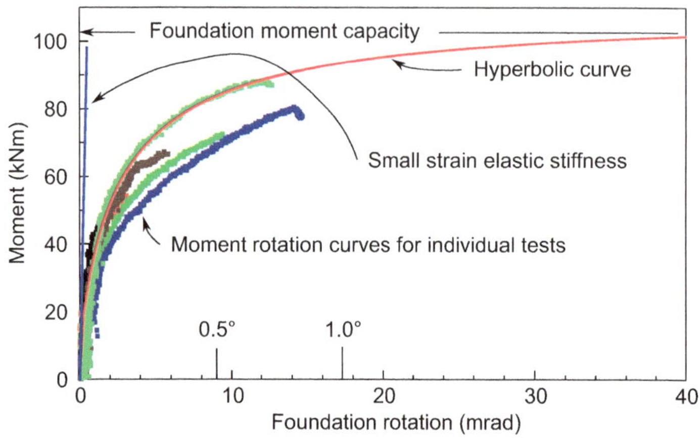

Fig. 9.1 Hyperbolic moment-rotation relation fitted to shallow foundation pull-back data (After Algie 2011)

9.1.4 *Simplification at Large Static Factors of Safety*

An important control on foundation response under earthquake loading is the static bearing strength factor of safety. A common suggestion is that mobilising about one third of the vertical bearing strength of the foundation will give satisfactory static vertical load only performance (satisfactory is usually determined in relation to the settlement under working loads). However, often the actual mobilisation of bearing strength will be less, even considerably less, because of conservative decisions regarding design values for the soil parameters. Additionally, if there is a basement beneath the building, the area of the foundation may occupy the full plan area of the structure, in such cases the bearing pressure generated by the vertical load may be a very small fraction of the vertical bearing strength of the foundation. When this occurs there is unlikely to be settlement of the foundation under earthquake actions. Likewise there will be little horizontal displacement as the lateral stiffness of the foundation will be enhanced by the lateral stiffness generated by the sidewalls of the basement. In this way the response of the foundation under earthquake will be dominated by the nonlinear rotational stiffness of the foundation.

9.1.5 *One Dimensional Macro-element Proposed in this Chapter*

Following from the above paragraph this chapter explores the behaviour of a one dimensional macro-element in which there is no settlement or horizontal deformation of the foundation; the only deformation is nonlinear rotation. Modelling shallow foundations in this way reflects our observation, following the February

22, 2011 earthquake in Christchurch, of the apparently satisfactory performance of foundations for multi-storey buildings founded on gravel (Storie et al. 2015; Storie 2017).

The foundation supports a multi-storey elastic viscously damped structure which is represented as a concentrated mass supported on an elastic viscously damped column. As there is only one mass in the system the structure-foundation system is a single degree of freedom (SDOF) model. It is intended that this will provide a simple method for foundation design sensitivity studies.

9.1.6 Substitute Structure

The shape of the shallow foundation moment-rotation curve in Fig. 9.1 is not unlike the moment-rotation curve for a reinforced concrete element. Figure 9.2 reproduces a diagram from the paper by Shibata and Sozen (1976). The procedure followed in this chapter is to make use of the ideas proposed by Shibata and Sozen, but to assume that the structural elements remain elastic whereas the shallow foundation is the source of nonlinearity in the structure-foundation system. Figure 9.3, taken from Priestley et al. (2007), shows how the approach of Shibata and Sozen can be used to reduce a multi-storey frame structure to an equivalent SDOF model; they refer to this as a substitute structure. Priestley et al. explain how the parameters for the substitute structure, h_e, m_e, and K_e, can be evaluated.

One additional aspect needs to be included in the substitute structure model: the compliance of the soil beneath and adjacent to the foundation. This is done by representing the structure-foundation system as a SDOF elastic structure supported on a nonlinear rotational spring, which is discussed further below and illustrated in Fig. 9.4a.

The essence of the Shibata and Sozen mechanism shown in Fig. 9.2 is the development of hysteretic damping through ductile deformation of structural members. Priestley et al. explain that, for a given displacement of the SDOF mass, elastic soil structure interaction reduces the ductility demand in the structure; in effect this results in a reduction of the hysteretic damping of the system. As will be explained below nonlinear foundation behaviour develops hysteretic damping and so satisfactory response the structure-foundation system to earthquake excitation may still be feasible even through the above ground structure remains elastic.

9.1.7 Performance Requirements

Prior to commencing the design of a foundation to resist earthquake loading performance criteria need to be specified. Since we are considering a one dimensional macro-element, this means specifying a permissible foundation rotation. After the Christchurch earthquake of February 2010 it was apparent that residual

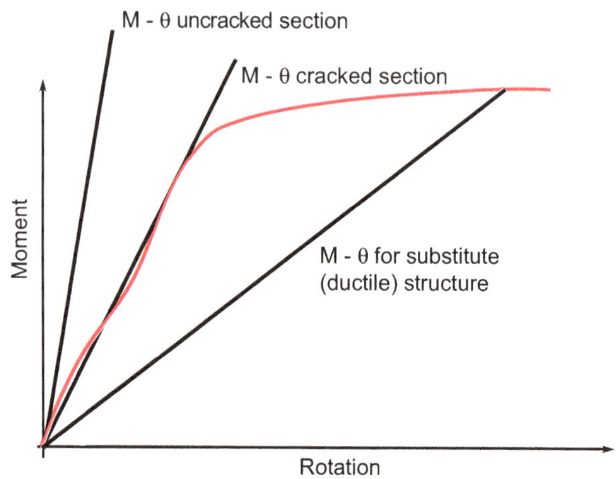

Fig. 9.2 The basis of Shibata and Sozen's substitute structure concept

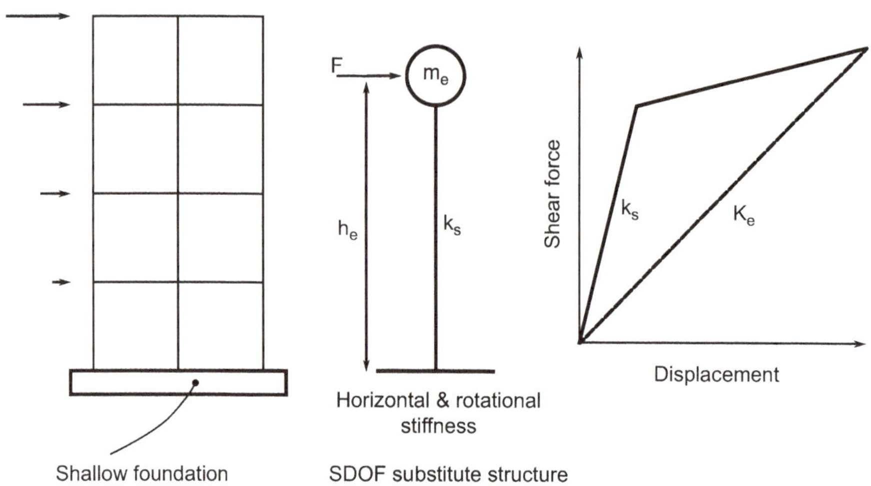

Fig. 9.3 Implementation of the substitute structure concept as presented by Priestley et al. (2007)

rotations of more than about 0.01 radian (≈0.5°) are easily perceived by the unaided eye. So if the design is based on a maximum rotation of 0.01 radian any residual rotation will not be readily visible. This is a conservative criterion as Deng et al. (2014) observed from centrifuge data that the residual rotation decreases as the pre-earthquake bearing strength static factor of safety increases. So for typical shallow foundation static design bearing strength factors of safety, the post-earthquake residual rotation can be expected to be significantly less than the peak rotation during the earthquake. As will be seen below the one dimensional macro-element model shows hysteretic damping even at small foundation rotations.

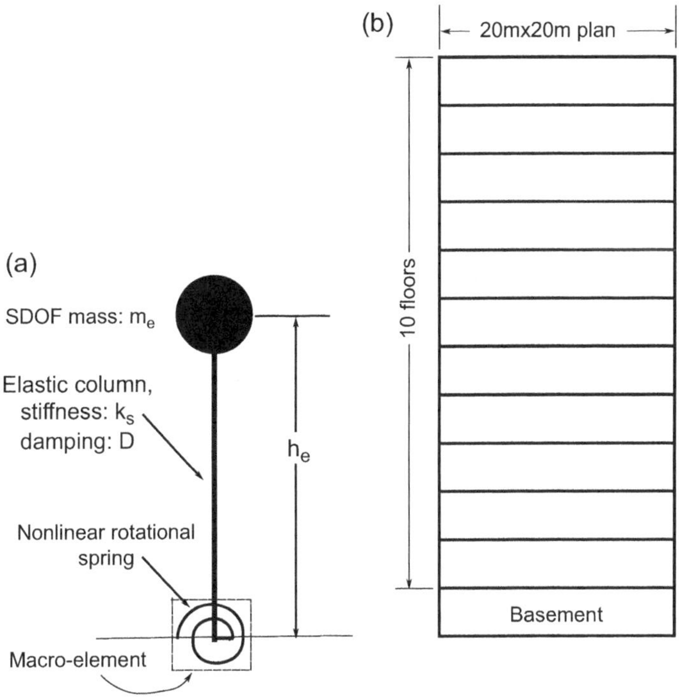

Fig. 9.4 Single degree of freedom model of the ten storey building with one basement level founded in dense gravel; (**a**) SDOF model, (**b**) actual structure

9.1.8 Foundation Bearing Failure

There have been many proposals for the design of shallow foundations pointing out that brief instances of bearing failure during the earthquake response are unlikely to generate significant permanent deformations but the transient reduction in stiffness will lead to a reduction in earthquake actions, for example: Anastasopoulos et al. (2010), Anastasopoulos and Kontoroupi (2014), Gajan and Kutter (2008), Gazetas (2015), Loli et al. (2015), Pecker and Teyssandier (1998), and Pender (2007). Although this is a convincing idea within the geotechnical community it is not readily accepted by the structural engineering fraternity who have a preference for restricting nonlinear deformations to above-ground structural members (a notable exception being Kelly (2009)). As will be seen below limiting the maximum foundation rotation to less than about 0.01 radians will provide benefits from nonlinear response but will still be well short of mobilising the moment capacity of the foundation, that is brief instances of bearing failure are unlikely at such modest rotations. Panagiotidou et al. (2012) found that for shallow foundations on stiff clay the onset of P-delta effects, which reduce the moment capacity of the system, occurs at rotations between 0.01 and 0.02 radians. Thus specifying a small value for the maximum design rotation has the added advantage that P-delta effects will not reduce the system capacity.

9.2 Equations of Motion

Wolf (1985), in his book *Dynamic Soil-Structure Interaction*, develops the equations of motion for a single degree of freedom model in which a lumped mass is positioned at some distance above a foundation resting on an elastic medium with radiation damping. There are three components to horizontal the displacement of the SDOF mass: the horizontal displacement of the foundation, the rotation of the foundation multiplied by the height of the mass, and the flexural displacement of the column supporting the mass, but since there is one mass in the system there is only one dynamic degree of freedom. Wolf shows how the different stiffness components contribute to the natural frequency of the structure-foundation system with the following equation:

$$\omega_e^2 = \frac{\omega_s^2}{1 + \frac{k_s}{k_h} + \frac{k_s h_e{}^2}{k_\varphi}} \tag{9.1}$$

where:

ω_e and ω_s are respectively the natural frequencies of the elastic structure-foundation system and the fixed base structure,
k_s is the flexural stiffness of the column supporting the SDOF mass,
k_h is the lateral stiffness of the foundation,
k_φ the rotational stiffness of the foundation,
and h_e is the height above foundation level of the SDOF lumped mass (m_e).

This equation shows that inclusion of foundation compliance reduces the natural frequency (lengthens the natural period) of the structure-foundation system. Equation 9.1 assumes that all stiffness components are elastic and gives the natural frequency of the system under steady state sinusoidal excitation, but in our case the foundation rotational stiffness is nonlinear so the natural frequency of the system under steady state excitation is variable being a function of the rotational amplitude of the foundation.

In this chapter we consider embedded foundations for which the lateral stiffness is large and does not have a significant effect on the system natural frequency given by the above equation. Consequently, the lateral stiffness of the foundation will not be included in the modelling below.

Under earthquake excitation the response is not steady state and so the rotational stiffness of the foundation, derived from the hyperbolic moment-rotation relationship, will vary from time step to time step. This means that the equation of motion for the SDOF mass needs to be integrated incrementally. The incremental equation of motion is:

$$\mathrm{m_e}\Delta\ddot{\mathrm{u}} + D\Delta\dot{\mathrm{u}}_\mathrm{m} + \mathrm{K_e}(\varphi)\Delta\mathrm{u_m} = -\mathrm{m_e}\Delta\ddot{\mathrm{u}}_\mathrm{g} \tag{9.2}$$

where:

$\mathrm{m_e}$ is the SDOF mass (tonnes),
D is the parameter for the equivalent viscous damping ratio of the column supporting the SDOF mass (tonnes/s),
φ is the foundation rotation (radians)
$\mathrm{K_e}(\varphi)$ is the equivalent system stiffness which includes the contribution from the foundation moment-rotation curve and the lateral stiffness of the column (kN/m),
$\Delta\ddot{u}, \Delta\dot{u}$, and Δu are respectively the horizontal acceleration increment, velocity increment, and displacement increment of the SDOF mass,
$\Delta\ddot{u}_g, \dot{u}$, and u is the increment in input acceleration (m/s^2)

Note that the velocity dependent damping term in Eq. 9.2 comes from the elastic column supporting the SDOF mass a distance $\mathrm{h_e}$ above the foundation level. All the foundation damping comes from hysteretic moment-deformation loops which are implicit in the foundation moment-rotation relationship discussed below. Frequency dependent elastic radiation damping is the usual energy dissipation mechanism used in elastic soil structure interaction. However, it is well known that elastic radiation damping for foundation rotation is very small so the hysteretic action is very important during foundation rocking.

The equivalent stiffness of the structure foundation system including the effect of the flexural stiffness of the column and the nonlinear rotational stiffness of the foundation is:

$$K_e(\varphi) = \frac{k_s K(\varphi)}{K(\varphi) + h_e^2 k_s} \tag{9.3}$$

where:

K(φ) is tangent rotational stiffness of the foundation at the current value of φ.

A possible extension of the above approach would be to include the mass and rotational inertia of the foundation. Doing this means that the structure-foundation system now has three dynamic degrees of freedom (lateral inertial displacement of the mass at a height $\mathrm{h_e}$ above foundation level, and lateral inertial displacement of the foundation, and inertial rotation of the foundation). Wolf (1985) states that pursuing this analysis does not produce results significantly different from those obtained with the SDOF model. Using modal analysis it is found that the two additional periods are well removed from that of the SDOF model and the first mode frequency is hardly different from that for the SDOF model. However, there is one significant factor in which the mass of the foundation block is important; the increased vertical load on the foundation enhances the moment capacity.

9.3 Foundation Moment Capacity

Definition of the nonlinear moment-rotation behaviour of the foundation is needed to calculate the response of the SDOF macro-element system to earthquake input. This is defined by two limiting conditions – the linear stiffness at very small rotations and the ultimate moment capacity of the foundation under the fixed vertical load. This section covers the estimation of the foundation moment capacity using the bearing strength surface.

9.3.1 Bearing Strength Surface

The following equation gives the standard estimation of the bearing strength of a shallow foundation under vertical load, horizontal shear, and moment:

$$q_u = c\lambda_{cs}\lambda_{cd}\lambda_{ci}N_c + q\lambda_{qs}\lambda_{qd}\lambda_{qi}N_q + \frac{1}{2}\gamma B\lambda_{\gamma s}\lambda_{\gamma d}\lambda_{\gamma i}N_\gamma \tag{9.4}$$

where:

λ_{cs}, λ_{qs}, and $\lambda_{\gamma s}$ are shape factors;
λ_{cd}, λ_{qd}, and $\lambda_{\gamma q}$ are depth factors;
λ_{ci}, λ_{qi} and $\lambda_{\gamma i}$ are inclined load factors;
and c, q, q_u, B, N_c, N_q and N_γ have their usual meanings.

A set of shape, depth, and inclined load factors for use in Eq. 9.4 are given by Pender (2017). Equation 9.4 provides a way of estimating what combinations of vertical load, horizontal shear, and moment mobilise all the available shear strength of the soil underlying and surrounding a shallow foundation. The locus of these combinations forms a three dimensional *Bearing Strength Surface* (BSS). There are two surfaces to consider, generated using appropriate rearrangements of Eq. 9.4, one for the undrained case and another for the drained case.

A convenient way of presenting the surfaces is to use axes defined in terms of dimensionless parameters, one for vertical load, another for horizontal shear, and a third for moment applied to the foundation. The normalising parameter is the ultimate bearing strength of the foundation subject to vertical load only; a possible suite of dimensionless parameters is then:

$$\mathrm{Vn} = \frac{\mathrm{V}}{\mathrm{V_{uo}}} \quad \mathrm{Hn} = \frac{\mathrm{H}}{\mathrm{V_{uo}}} \quad \mathrm{Mn} = \frac{\mathrm{M}}{\mathrm{BV_{uo}}} \tag{9.5}$$

where:

V, H and M are actions applied to the foundation,
Vn, Hn and Mn are normalized foundation actions,
B is the width of the foundation,
V_{uo} is the ultimate vertical load capacity of the foundation in the absence of shear and moment.

The Vn, Hn and Mn notation refers to general foundation actions. When referring to a bearing strength ultimate limit state we will use the notation Vnu, Hnu and Mnu.

When only vertical actions are to be applied to the foundation the usual terminology is to refer to the bearing capacity which has units of pressure (q_u in Eq. 9.4). When the vertical load is accompanied by shear and moment it is helpful to speak of bearing strength (q_u times the effective contact area) which has units of force rather than pressure; this terminology is used in the remainder of the chapter.

Views of the upper halves of the two versions of the surface are shown in Figs. 9.4 and 9.5. Algebraic equations for the bearing strength surfaces, developed from Eq. 9.1, are given by Pender (2017).

The bearing strength surfaces show that the capacity of a shallow foundation is not a single number but the combination of the vertical load, horizontal shear force and moment where the action path (a plot of the combinations of vertical load, horizontal shear and moment applied to the foundation) intersects the bearing strength surface (Fig. 9.6).

9.3.2 Moment Capacity at Constant Vertical Load with Moment and Shear

Estimation of the moment capacity of the foundation must consider the action path applied to the foundation, in our case the vertical load is constant so, in effect, we need to consider a Vn constant section through the BSS. In addition, to obtain the moment capacity of the foundation the solution process must allow for the dependence of eight of the nine λ factors ($\lambda_{\gamma d}$ is independent of moment as it has the value of unity) on the effective foundation width, B′, and this in turn is a function of the unknown ultimate moment. The inclined load terms, λ_{ci}, λ_{qi} and $\lambda_{\gamma i}$, are functions of the horizontal shear as well as B′, but the ultimate shear can usually be expressed as a multiple of the ultimate moment as the SDOF shear force is applied some distance above the foundation and so produces the foundation moment. Consequently we have an equation, albeit a rather complex one, for one unknown, the ultimate moment capacity under the fixed vertical load V:

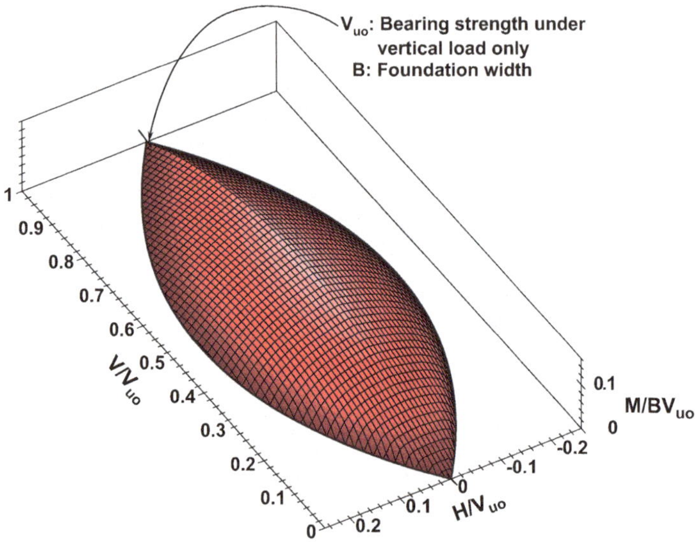

Fig. 9.5 Undrained bearing strength surface for a surface strip foundation

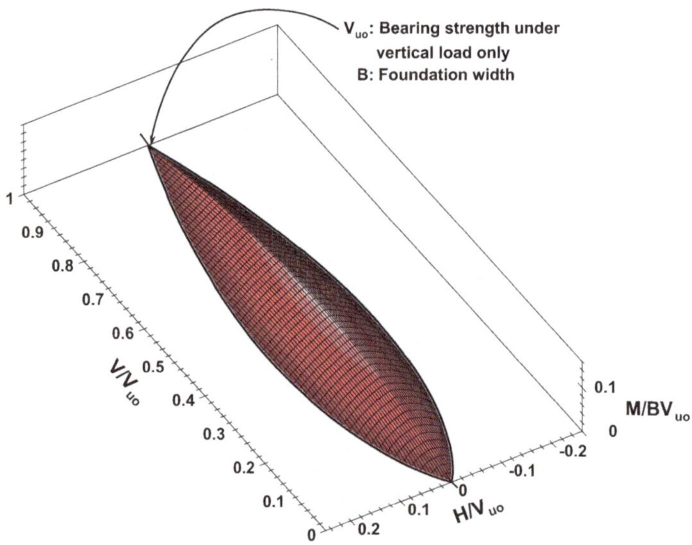

Fig. 9.6 Drained bearing strength surface for a surface strip foundation

$$\left\{\begin{array}{l} cN_c\lambda_{cs}(M_u)\lambda_{cd}(M_u)\lambda_{ci}(M_u) \\ +qN_q\lambda_{qs}(M_u)\lambda_{qd}(M_u)\lambda_{qi}(M_u) \\ +\frac{1}{2}\gamma B'(M_u)N_\gamma\lambda_{\gamma s}(M_u)\lambda_{\gamma i}(M_u)\lambda_{\gamma d} \end{array}\right\} B'(M_u)L - V = 0 \tag{9.6}$$

The solution to Eq. 9.6 gives the moment at which the effective width of the foundation is reduced to the extent that the value of q_u times the effective area of the foundation is equal to the applied vertical load.

9.4 Foundation Hyperbolic Moment-Rotation Relationship

Figure 9.1 presents the static moment-rotation curves obtained during the application of the pullback forces to one of the foundation sets at the Albany site. A hyperbolic curve is seen to fit through the data.

Hyperbolic curves have been used to approximate the stress-strain behavior of soil elements since the work of Kondner (1963). Duncan and Chang (1970) applied this model in finite element analysis. It is not surprising that a similar relation applies to shallow foundation behavior as the foundation response is the integration of the responses of a large number of soil elements each of which follows an approximately hyperbolic relationship. In this way it is possible to argue, heuristically, that the overall response of the foundation will be similar to that of the individual elements.

The form of the hyperbolic relationship shown in Fig. 9.2 is:

$$M = \frac{\varphi}{\frac{\Lambda}{K_{\varphi i}} + \frac{\Omega\varphi}{M_u}} \tag{9.7}$$

where:

M and φ are the current foundation moment and rotation respectively,
$K_{\varphi i}$ and M_u are the initial rotational stiffness and moment capacity of the foundation respectively,
and Λ and Ω are numerical parameters that may be used to refine the fit of the curve to the data (the curve fit in Fig. 9.1 has a value of unity for both of these parameters).

Salimath (2017) found that values for Λ and Ω depend on the shape of the foundation. The parameter Λ also has an effect on the amount of hysteretic damping in the cyclic hyperbolic relationship which is discussed below.

Equation 9.7 provides a pushover curve for a shallow foundation, which is bounded by the moment capacity of the shallow foundation at large values of the rotation and the elastic rotational stiffness at small rotations.

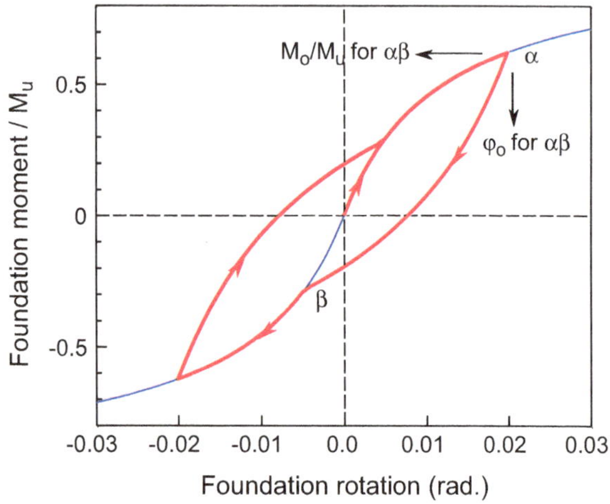

Fig. 9.7 Cyclic hyperbolic moment-rotation curve with the definition of the M_o and φ_o parameters

Equation 9.7 also acts as a backbone curve for the extension of hyperbolic relationship to cyclic loading given by the following equation:

$$M = \frac{K_{\varphi i}(AM_u - M_o)(\varphi - \varphi_o)}{\Lambda(AM_u - M_o) + \Omega K_{\varphi i}(\varphi - \varphi_o)} + M_o \tag{9.8}$$

where:

M_o and φ_o are the initial values for the current branch of the moment-rotation curve (defined in Fig. 9.7),

and A takes values of +1 and −1 indicating if the foundation is being loaded towards the moment capacity M_u or $-M_u$.

Whenever an unloading or reloading loop intersects the backbone curve the moment-rotation response then follows the backbone curve (this prevents the system developing unreasonably large foundation rotations). When the direction of loading reverses the stiffness reverts to the rotational stiffness at zero moment, $K_{\varphi i}$. The moment capacity can be reached when the rotation is to the left or the right, so that M_u limits the moment to the right and $-M_u$ limits it to the left. Figure 9.7 shows how Eq. 9.8 extends the pushover curve to give the cyclic response of a shallow foundation.

The rotational stiffness of the foundation is given by the slope of the current branch of the hysteretic moment-rotation loop and, as in apparent from Fig. 9.7, the slope of the M-φ curve changes continuously with moment. Because of this continuous variation in stiffness the nonlinear moment-rotation relationship is not associated with plastic yielding but rather is a type of hypoplasticity.

Equations 9.7 and 9.8 are relatively simple expressions representing complex behavior. Spring bed models indicate, for a shallow foundation under a fixed vertical load and gradually increasing moment, that there will be a gradual loss of contact at the edge of the foundation. The macro-element models of Cremer et al. and Chatzigogos et al. incorporate this effect specifically, but herein it is assumed that uplift effects are handled implicitly and part of the reason for the shape of the moment-rotation curve is a consequence of the reduction of contact area. Salimath (2017) reports on an extensive series of finite element analyses, using Plaxis 3D, (Plaxis 2012) which have a rigid rectangular foundation, under constant vertical load, on a clay soil with a no-tension interface between the foundation and the underlying clay. This modelling confirms that a hyperbolic curve is a reasonable representation of the shallow foundation moment-rotation push-over curve when uplift occurs.

9.5 Rotation Dependent Foundation Stiffness and Hysteretic Damping

Elastic soil-structure interaction modelling relies on radiation damping to take vibration energy away from the foundation. However, for rotation of the foundation damping available from elastic radiation is small. Consequently, herein we assume that all the foundation damping is derived from the area of the hysteresis loops. Figure 9.8 shows that the hysteretic damping increases as the rotation amplitude increases. The apparent foundation rotational stiffness and the equivalent viscous damping ratio associated with a given hysteretic moment-rotation loop are defined in Fig. 9.10a. Figure 9.10b gives the variation of the rotational stiffness with the logarithm of the rotation amplitude. Figure 9.10c gives the variation of the hysteretic damping against the logarithm of the rotation amplitude. When the rotation amplitude is 0.05 the hysteretic damping is 28 %. Ishihara (1996) states that the maximum

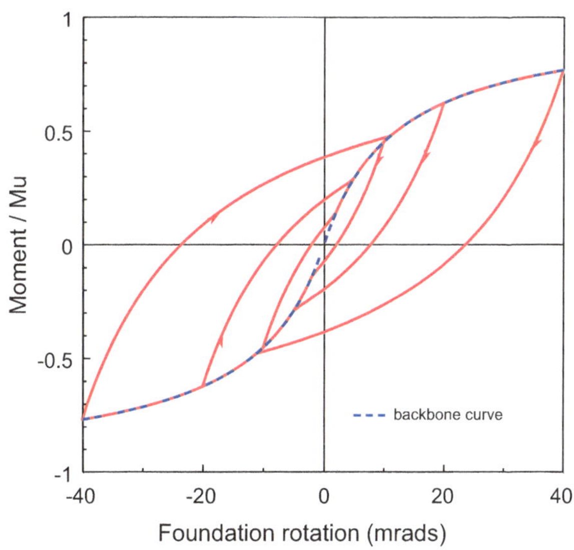

Fig. 9.8 Steady state hysteretic shallow foundation moment-rotation loops with the Λ parameter set to 1.0

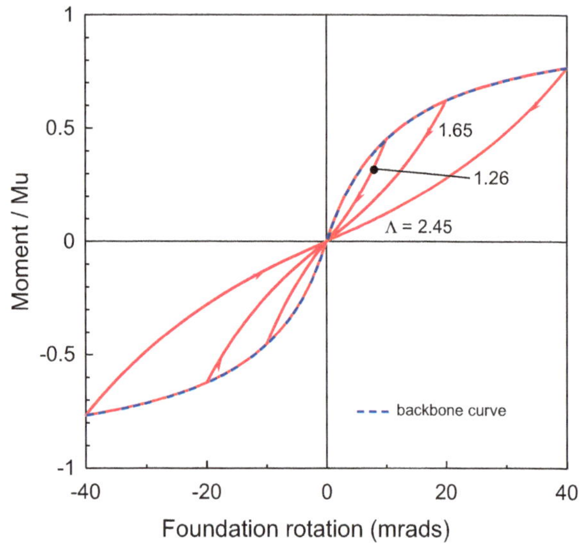

Fig. 9.9 Steady state hysteretic shallow foundation moment-rotation loops with the Λ parameter varied at each turning point so that the loops pass through the origin

damping ratio for the hyperbolic model is 64 %. Clearly the damping in the present model does not approach such a value at likely rotation amplitudes, a consequence of the hysteretic loops reattaching to the backbone curve rather than crossing it.

All the moment-rotation loops in Fig. 9.8 are generated with the Λ parameter set to a value of 1.0. The amount of damping can be reduced by decreasing the rotational stiffness of the foundation immediately after a change in direction of rotation which is achieved if Λ is given a value greater than unity. The effect of doing this is shown in Fig. 9.9 where the values of Λ have been adjusted so that the hysteresis loops pass through the origin.

As the foundation rotation increases the slope of the line between the end points of the hysteresis loops in Figs. 9.8 and 9.9 decreases; this gives the relationship, for the backbone curve, between the apparent rotational stiffness of the foundation which is plotted in Fig. 9.10b.

9.6 Example Calculations

In this section the response to earthquake excitation of a ten storey building with a single level basement and a foundation with a hyperbolic moment-rotation properties is presented. The input is the acceleration recorded in the 1940 El Centro earthquake (the N-S component) adjusted to a peak ground acceleration of 0.3 g. The nonlinear response of the structure-foundation system was calculated to the 0.3 g El Centro record scaled with factors of 0.5, 1.0, 2.0 and 3.0. The algorithm presented in Chap. 7 of Clough and Penzien (1993), for calculating the incremental response of a nonlinear single degree of freedom system, was used in these

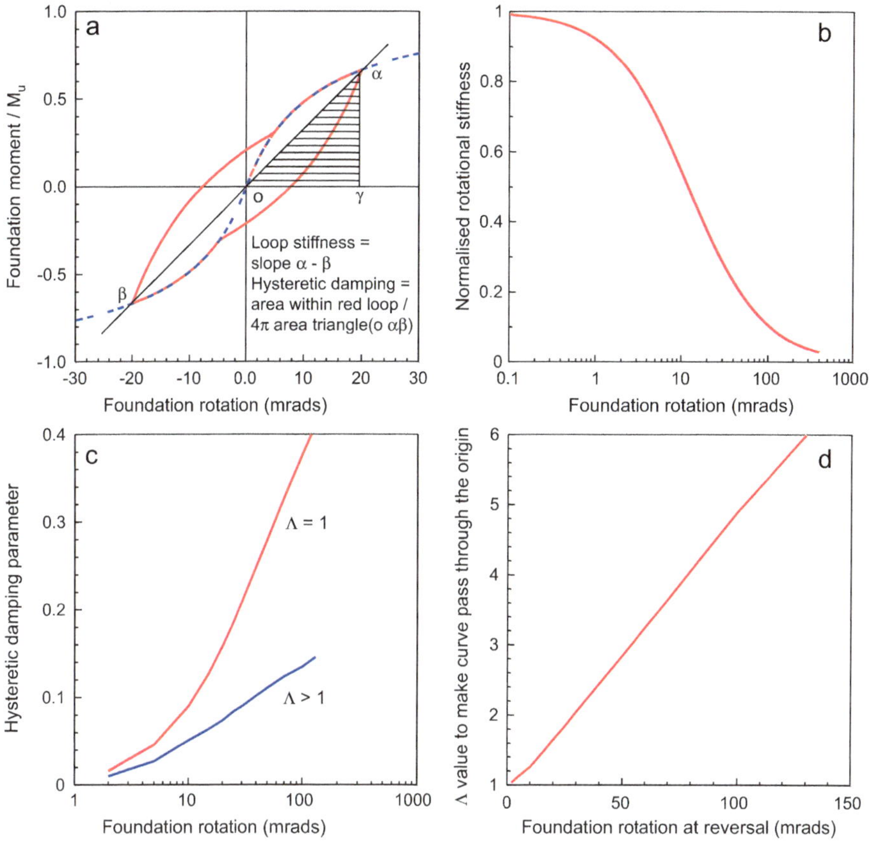

Fig. 9.10 Hysteretic equivalent viscous damping ratio and apparent rotational stiffness as functions of foundation rotation amplitude. (**a**) Parameter definitions; (**b**) foundation rotational stiffness as a function of rotation amplitude; (**c**) damping ratios against rotation amplitude; (**d**) value of the parameter Λ required to get the loops to pass through the origin as a function of rotation at the turning point

computations. In addition the response of the system when the foundation behaves elastically was evaluated. All the time history and response spectra calculations discussed in this chapter were done using Mathcad 15 (PTC 2015). The earthquake acceleration histories had a time step of 0.02 s.

The details of the ten storey building are shown in Fig. 9.4b, the plan dimensions are 20 × 20 m. The damping ratio for the elastic building structure was set at 3 % of the critical damping value. The building is founded on dense dry gravel with an angle of shearing resistance of 45° and a shear wave velocity of 280 m/s. The conversion from the actual structure, Fig. 9.4b, to the SDOF model, Fig. 9.4a, was done by assuming a uniform distribution of mass with height in the building and setting the SDOF mass to 70 % of the total mass acting at 70 % of the building height.

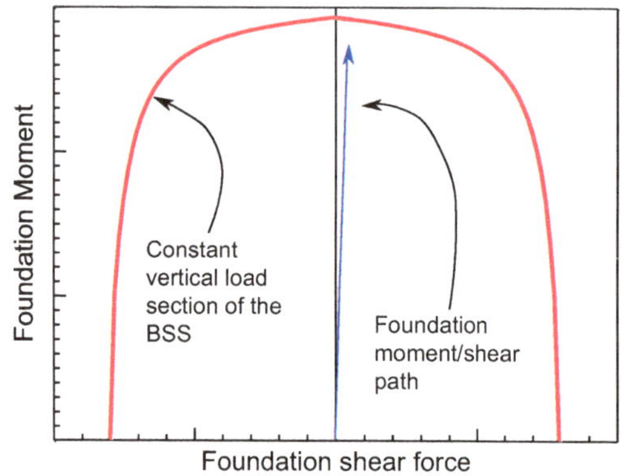

Fig. 9.11 Constant vertical load section of the bearing strength surface showing the path for foundation actions

Figure 9.11 gives a cross-section of the bearing strength surface at the vertical load imposed by the gravity loading (assumed constant during the earthquake). Also shown in Fig. 9.11 is the shear – moment action path for the foundation; because of the height of the SDOF mass the action path is close to vertical so the moment capacity of the foundation is close to the top of the section of the BSS.

The calculated foundation moment-rotation loops are plotted in Fig. 9.12. These show increasing hysteretic damping as the scale factor applied to the input motion is increased. Figure 9.13 shows the variation of the foundation tangent rotational stiffness, normalised with respect to the small rotation value ($K_{\varphi i}$), for the first 8 s of the response when the scale factor is 2.0 (the vertical lines indicate the abrupt change in stiffness when the direction of motion is reversed as then the rotational stiffness reverts to the small rotation value).

Figure 9.14 has the response spectra for the outputs from the scaled input motions, the spectra were obtained by processing the calculated horizontal acceleration histories of the SDOF mass. In addition there is a spectrum for the response when the foundation behaves elastically. All spectra were calculated for a damping ratio of 5 % and the plotted spectral accelerations are normalized with respect to the peak input acceleration for the scaled input, so the ordinates in Fig. 9.14 are dimensionless. It is apparent that the peak response in all cases occurs at a period close to 1 s, the natural period of the above ground part of the structure-foundation system. As the scaling of the input motion increases there is more hysteretic damping generated by the foundation as evidenced by the moment-rotation loops in Fig. 9.12. The effect of this is that the spectral peaks in the responses of the SDOF mass are reduced as the peak input acceleration increases. This is indicative of the increasing hysteretic damping in the foundation. The damping values plotted in Fig. 9.10c are for steady state cycling around moment-rotation loops of fixed rotation amplitude and so are not directly relevant to the damping obtained during an earthquake record. Pender et al. (2017), reporting an investigation of the performance of multi-storey buildings on shallow foundations having a hyperbolic moment-rotation curve, calculated numerically the SDOF response to snap-back release from various initial displacements. This revealed that there was

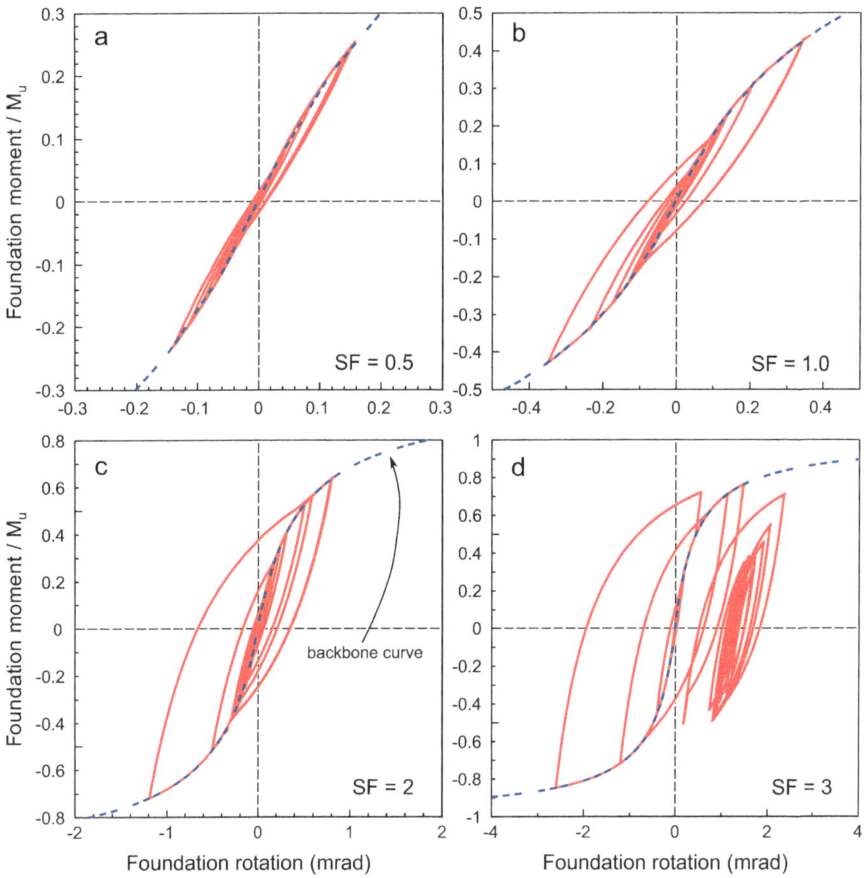

Fig. 9.12 Foundation moment-rotation loops calculated during the response to scaled El Centro ground acceleration input motions

considerable damping in the first half cycle of response and then a gradual reduction in damping.

The spectra in Fig. 9.14 are dominated by the response of the ten storey elastic structure supported on the nonlinear foundation. One might have expected, with the larger values of the scale factor, some period lengthening of the peak response because of nonlinearity in the foundation. From Fig. 9.12c we see that the peak foundation rotation during the motion when the scaling factor is 2 is about 1 milli-radian; if this is the peak rotation then the rotations for the remainder of the response will be smaller. Referring to Fig. 9.10b the equivalent stiffness for a rotation amplitude of 1 milli-radian is about 90 % of the elastic value. From Eq. 9.1 the period of the structure foundation system for elastic behavior 1.04 s, when 0.9 of the elastic foundation rotational stiffness is used the period is lengthened by 0.05 s. Given this, it is not surprising that the peaks of the response spectra

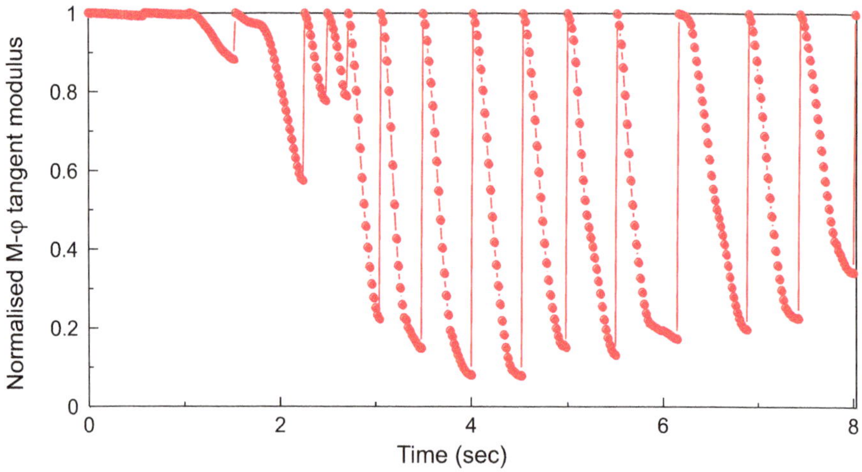

Fig. 9.13 Variation of the foundation tangent rotational stiffness during the first 8 s of the response to the El Centro input with a scale factor of 2.0. (The *vertical lines* indicate turning points when the foundation rotational stiffness reverts to the initial small rotation value)

Fig. 9.14 Response spectra, normalised with respect to the peak input acceleration, for the calculated horizontal acceleration of the SDOF mass to the scaled 1940 N-S El Centro record. The scale factors for the input ranged from 0.5 to 3.0, with the PGA for a scale factor of 1 being 0.3 g

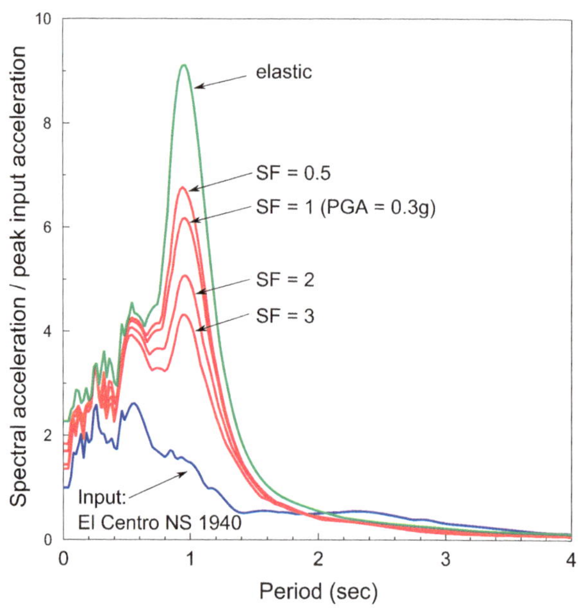

for the SDOF mass are close to the 1 s period of the above ground ten storey elastic structure. Figure 9.12 shows that, despite the small foundation rotations, there is still hysteretic damping and it is this damping that reduces the peaks in the response spectra as the scale factor is increased. In other words the hysteretic damping absorbs some of the incoming energy and prevents it being passed upwards into the elastic structure.

9.7 Conclusions

The motivation for the development of a one dimensional shallow foundation macro-element in this chapter is threefold: first, to use the structural engineering substitute structure concept but in a different context where the nonlinearity occurs in the foundation rather than the above-ground elastic structure (Figs. 9.2 and 9.3); second, to make things as simple as possible to ease sensitivity studies during the design process; third, to look specifically at situations dominated by rotational response of the foundation where settlement and sliding can be ignored. The following conclusions are reached:

- Small foundation rotations have a significant effect on the response not only because of reduced stiffness but also because of hysteretic damping, Fig. 9.12.
- The foundation moment-rotation response is controlled by the vertical load which means that the section of the bearing strength surface at the vertical load carried by the foundation must be considered in estimating M_u, Fig. 9.11 and Eq. 9.6.
- The continuous curvature of the hyperbolic curve means that plastic yielding of the foundation does not occur, rather the relationship is a type of hypoplasticity.
- Uplift is not considered specifically in the macro-element presented herein, but it is recognised that it is an important part of the explanation for the shape of the foundation moment-rotation curve.
- The rotation dependent equivalent stiffness and hysteretic damping of the model follow the expected trends, Fig. 9.10b, c.
- By altering the stiffness immediately after a change in direction of the rotation the amount of hysteretic damping can be controlled, Figs. 9.8 and 9.9.
- The hysteretic damping in the foundation shields the supported elastic structure from some of the incoming earthquake energy, Fig. 9.14.

References

Algie TB (2011). Nonlinear rotational behaviour of shallow foundations on cohesive soil. Ph.D. thesis, University of Auckland

Algie TB, Pender MJ, Orense RP (2010) Large scale field tests of rocking foundations on an Auckland residual soil. In: Orense R, Chouw N, Pender M (eds) Soil foundation structure interaction. CRC Press/Balkema, Leiden, pp 57–65

Anastasopoulos I, Kontoroupi T (2014) Simplified approximate method for analysis of rocking systems accounting for soil inelasticity and foundation uplifting. Soil Dyn Earthq Eng 56:28–43

Anastasopoulos I, Gazetas G, Loli M, Apostolou M, Gerolymos N (2010) Soil failure can be used for seismic protection of structures. Bull Earthq Eng 8(2):309–326

Carr A (2016) Ruaumoko. Carr Research Ltd, Christchurch

Chatzigogos CT, Pecker A, Salençon J (2007) A macro-element for dynamic soil-structure interaction analyses of shallow foundations. In: Proceedings of 4th international conference on earthquake geotechnical engineering, Thessaloniki, June 25–28

Chatzigogos, C. T., A. Pecker, and J. Salenc¸on (2009). Macroelement modeling of shallow foundations. Soil Dyn Earthq Eng 29, 765–781
Clough RW, Penzien J (1993) Dynamics of structures, 2nd edn. McGraw-Hill, New York
Cremer C, Pecker A, Davenne L (2001) Cyclic macro-element for soil-structure interaction: material and geometrical non-linearities. Int J Numer Anal Methods Geomech 25 (13):1257–1284
Cremer C, Pecker A, Davenne L (2002) Modelling of nonlinear dynamic behaviour of a shallow strip foundation with macro-element. J Earthq Eng:175–211
Deng L, Kutter B, Kunnath S (2014) Seismic design of rocking shallow foundations: displacement-based methodology. J Bridge Eng 19(11). doi:10.1061/(ASCE)BE.1943-5592.0000616
Duncan JM, Chang CY (1970) Nonlinear analysis of stress and strain in soils. Proc ASCE, J Soil Mech Found Div 96(SM5):1629–1653
Figini R, Paolucci R, Chatzigogos CT (2012) A macro-element model for non-linear soil-shallow foundation-structure interaction under seismic loads: theoretical development and experimental validation on large scale tests. Earthq Eng Struct Dyn 41(3):475–493
Gajan S, Kutter BL (2008) Capacity, settlement, and energy dissipation of shallow footings subjected to rocking. J Geotech Geoenviron Eng ASCE 134(8):1129–1141
Gazetas G (2015) Avoiding over-conservatism and conventional dogmas in seismic geotechnical design. Keynote address. 12th Australia-New Zealand conference on Geomechanics, Wellington, February
Ishihara K (1996) Soil behaviour in earthquake geotechnics. Oxford Science, Oxford
Kelly TE (2009) Tentative seismic design guidelines for rocking structures. Bull N Z Soc Earthq Eng 42(4):239–274
Kondner RL (1963) Hyperbolic stress-strain response: cohesive soils. Proc ASCE J Soil Mech Found Div 89(SM1):115–143
Loli M, Knappett JA, Brown MJ, Anastasopoulos I, Gazetas G (2015) Centrifuge testing of a bridge pier on a rocking isolated foundation supported on unconnected piles. In: Proceedings of 6th ICEGE, paper 362, Christchurch, November
Millen M. (2015). Integrated performance-based design of building-foundation systems. PhD thesis, University of Canterbury
Panagiotidou AI, Gazetas G, Gerolymos N (2012) Pushover and seismic response of foundations on stiff clay: analysis with P-delta effects. Earthquake Spectra 28(4):1589–1618
Pecker A, Teyssandier J-P (1998) Seismic design for the foundations of the Rion Antirion bridge. Proc Inst Civ Eng Geotech Eng 131(Jan.):4–11
Pender MJ (2007) Seismic design and performance of surface foundations. In: Pitalakis KD (ed) Chapter 10, Earthquake geotechnical engineering. Springer, pp 215–241
Pender MJ (2017) Bearing strength surfaces implied in conventional bearing capacity calculations. Geotechnique 67(4):313–424. doi:10.1680/geot./16-P-024
Pender MJ, Algie TB, Storie LB, Salimath R (2017) One dimensional moment-rotation macro-element for performance based design of shallow foundations. In: Proceedings performance based design III (PBDIII), Vancouver. Paper 332
Plaxis (2012) Plaxis 3D – version 2012. Plaxis b.v, The Netherlands
Priestley MJN, Calvi GM, Kowalsky MJ (2007) Displacement-based seismic design of structures. IUSS Press, Pavia
PTC (2015) Mathcad 15 (M045). PTC Corporation, Needham
Salimath R (2017) Experimental and finite element study of moment-rotation response of shallow foundations on clay. PhD thesis, University of Auckland
Shibata A, Sozen M (1976) Substitute structure method for the seismic design in reinforced concrete. J Struct Eng 102(1):1–18
Storie LB (2017) Soil-foundation-structure interaction in the earthquake performance of multistorey buildings on shallow foundations. PhD thesis, University of Auckland

Storie LB, Knappett JA, Pender MJ (2015) Centrifuge modelling of the energy dissipation characteristics of mid-rise buildings with raft foundations on dense cohesionless soil. In: Proceedings of 6th international conference on earthquake geotechnical engineering, Christchurch

Toh JCW (2008) Performance based aseismic design of shallow foundations. ME thesis, University of Auckland

Toh JCW, Pender MJ (2010) Design approaches and criteria for earthquake-resistant shallow foundation systems. In: Orense RP, Chouw N, Pender MJ (eds) Soil-foundation-structure-interaction. CRC Press/Balkema, Leiden, pp 173–180

Wolf JP (1985) Dynamic soil-structure interaction. Prentice Hall, Englewood Cliffs

Wolf JP, Deeks AJ (2004) Foundation vibration analysis: a strength of materials approach. Elsevier, Oxford

Chapter 10
Analysis of Liquefaction During 2011 East Japan Earthquake – Part 1: Seismic Ground Behavior in Tokyo Port at the 2011 Off Pacific Coast of Tohoku Earthquake – An Effective Stress Dynamic Analysis Focusing on the Impact of the Aftershock

Yukio Tamari, Junichi Hyodo, Koji Ichii, Takashi Nakama, and Atsushi Hosoo

Abstract Seismic ground motions and behavior of excess pore water pressures on a reclaimed land during the 2011 off Pacific coast of Tohoku earthquake are studied numerically by using a computer program, "FLIP ROSE" (Iai et al, Soils Found 32 (2):1–15, 1992; Int J Numer Anal Meth Geomech 35:360–392, 2011). Long duration time of 30 min which includes both the mainshock and the aftershock is taken into account in the dynamic analysis considering build-up and dissipation of excess pore pressures as well. The site studied is located in Tokyo port, Japan, where seismic ground motions are obtained with vertically arrayed seismographs by Tokyo metropolitan government. It is shown that observed peak ground accelerations at each depth of ground are reasonably simulated. The ratio of excess pore water pressure is calculated as maximum of about 0.4 during aftershock, agreeing with the fact of no significant liquefaction at the site. It is noted that significant rise of excess pore water pressure is calculated in the aftershock, suggesting significant effect on liquefaction.

Y. Tamari (✉) • J. Hyodo
TEPSCO (Tokyo Electric Power Services Co., Ltd.), Tokyo, Japan
e-mail: etamari@tepsco.co.jp

K. Ichii
Kansai University, Osaka, Japan

T. Nakama • A. Hosoo
Jishin Kogaku Kenkyusyo, Inc., Tokyo, Japan

S. Iai (ed.), *Developments in Earthquake Geotechnics*, Geotechnical, Geological and Earthquake Engineering 43, DOI 10.1007/978-3-319-62069-5_10

10.1 Introduction

During the the 2011 off Pacific coast of Tohoku earthquake in north eastern Japan ($M_w = 9.0$, hereafter, Earthquake 3.11), soil liquefaction occurred in a wide area of reclaimed land along Tokyo Bay. Despite of very large epicentral distance about 380 km to 400 km, and the relatively small peak ground accereration (PGA) of about 100–150 gal (Fukutake and Jang 2013), significant liquefaction occurred in Urayasu city resulting in sever damage of many houses, lifelines, civil structures. The security camera in a junior high school in Urayasu city captured the onset of sand boiling during the mainshock first, and it expanded in the aftershock ($M = 7.7$) which occurred about 30 min after the mainshock (Urayasu city government, 2011). In addition, it is reported that some inhabitants in Urayasu city testified sand boiling did not occurre during the mainshock but the aftershock (Yasuda et al. 2013). This suggests some amount of excess pore water pressure buildup by the cyclic loading due to mainshock and reach complete liquefaction by following aftershock. In this regard, some analytical studies have been conducted focusing on the effect of aftershock to liquefaction at Tokyo Bay area in the Earthquake 3.11.

Morikawa et al. (2011) undertook a numerical analysis on the ground in Urayasu city by effective stress method to clarify the effect of long duration time. It was shown that the characteristics of ground would have changed by aftereffect of stress by precedent earthquake, and turn easily into the state of liquefaction or re-liquefaction even by small intensity of aftershock. Fukutake and Jang (2013) studied the effect of aftershock on the settlement due to liquefaction by effective stress analysis for both liquefied and unliquefied ground in Tokyo Bay area. The mainshock and aftershock was considered in a row on the reponse analysis. It was mentioned that the excess pore water pressure continuously rise even after the cease of mainshock with a slower pase by the motion of low amplitude of acceleration. The aftershock would cause the large ammount of settlements after earthquake, estimating the 40% of all settlement could occurr in Urayasu city. Ueda et al. (2014) conducted the numerical study on the re-liquefaction phenomena targeting on the liquefied ground at Takasu district in Urayasu city. The excess pore water pressure buildup and dissipation during and after earthquake motion are considered in the study. It was shown that re-liquefaction would occur by after-shock in the condition of high excess pore water pressure due to mainshock.

It was thus clarified by past studies that there would be high possibility of re-liquefaction when the ground is in the process of dissipation of pore water pressure, supporting the existence of effect of aftershock on liquefaction. However, the past studies were focusing on mainly the liquefied ground where exhibition of liquefaction was observed. It will be meaningful in engineering point of view to study on non liquefied ground to add further case histories.

The authors have studied numerically the seismic ground behavior of non liquefied ground in Tokyo port at the earthquake 3.11. The reclaimed ground at Shin-Ariake in Tokyo Port is targeted where the seismic acceleration record is observed by vertically arrayed seismograph. First, ground behavior is assessed by reproducing stress strain hysteresis in the ground by using the recorded acceleration time histories. Next, the ground model is constituted at the site for dynamic effective stress response analysis. Finally, the effect of aftershock is studied using the developped model by considering reproduced mainshock and aftershock. The same constitutive model for soil as used in the past study by Ueda et al. (2014) is used which is utilised in a computer program "**F**inite element analysis program of **Li**quefaction **P**rocess/Response of Soil-structure systems during Earthquakes (FLIP ROSE, Iai et al. 1992, 2011).

10.2 Seismic Ground Behavior During the Earthquake 3.11 at Tokyo Port

10.2.1 Recording Station and Recorded Motions

The location of recording stations around Tokyo Port is illustrated in Fig. 10.1. The reported liquefied zones (Yasuda et al. 2013) at Earthquake 3.11 are shown in the figure as well. At the Shin-Ariake recording station, four set of seismograph are installed along the depth by Tokyo metropolitan government, illustrating soil profile and SPT N-values in Figs. 10.2 and 10.3. Typical observed acceleration time histories in Earthquake 3.11 of mainshock at GL-2.0 m, GL-16.0 m, GL-75.0 m are depicted in Fig. 10.2. PGAs are 1.38 m/s^2 (North-South direction, GL-2.0 m) and 0.67 m/s^2 (East-West direction, GL-75.0 m).

10.2.2 Reproduction of Stress Strain Hysteresis

To clarify the soil behavior during mainshock in Earthquake 3.11, stress strain hysteresis were reproduced using recorded accelerations. Shear stress in the ground was calculated by considering acceleration time history and mass density of soil. Shear strains were assessed by horizontal displacement time histories which were derived by time integration of acceleration, and vertical distance between seismographs (CDIT 1997). Typical stress strain hysteresis are illustrated in Fig. 10.4.

Figure 10.4a illustrates reproduced stress strain hysteresis at the depth between 2.0 to 16.0 m. Calculated shear modulus from hysteresis of about 130–150 m/s assuming the mass density of soil as 1.95 t/m^3 is less than the initial value of 140–175 m/s of bank as shown in following Table 10.1. Shear modulus seems to be degraded slightly by earthquake shaking, but it is not the level of shear modulus when the soil liquefy. Figure 10.4b illustrates stress strain hysteresis of the ground between the depth of 36.0 and 75.0 m. It is seen that the shear modulus calculated hysteresis is 335 m/s, being consistent with the existing data of 330 m/s at diluvial

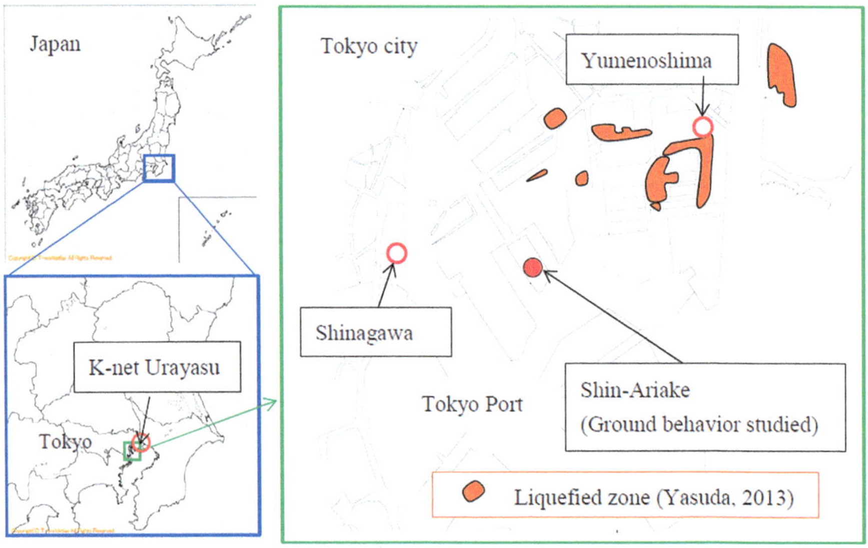

Fig. 10.1 Location of the recording station (Tokyo Port, Japan) (Maps: http://www.sekaichizu.jp/)

sandy layer of Esu (Ishihara et al. 1989) as shown in Table 10.1. It implies the reproduced procedure is valid. Thus, it is confirmed that the site we study did not liquefy at the earthquake 3.11.

10.3 Ground Model for Effective Stress Dynamic Analysis

10.3.1 Ground Model and Determination of Model Parameters

The effective stress model used in this study is composed of two models. One is the multiple shear mechanism for nonlinear stress strain relationship of soils (Towhata and Ishihara 1985). Basic idea of this mechanism is the simple shear mechanism in any direction of shear plane in soil. The effect of rotation of principal stress axes, which is known as characteristic nature of soils, is considered in this model. Also the behaviors of soil at the anisotropic consolidation can be realistically simulated (Iai et al. 1992). Principal model parameters are initial shear modulus, maximum damping coefficient, shear resistance angle and cohesion.

The other is the model concerning generation of excess pore water pressures. A stress-dilatancy relation (Iai et al. 2011) is incorporated in the multiple shear mechanism for cyclic behavior of sand. The contractive component of dilatancy is in proportion to the cumulative shear strain. This model has ten particular

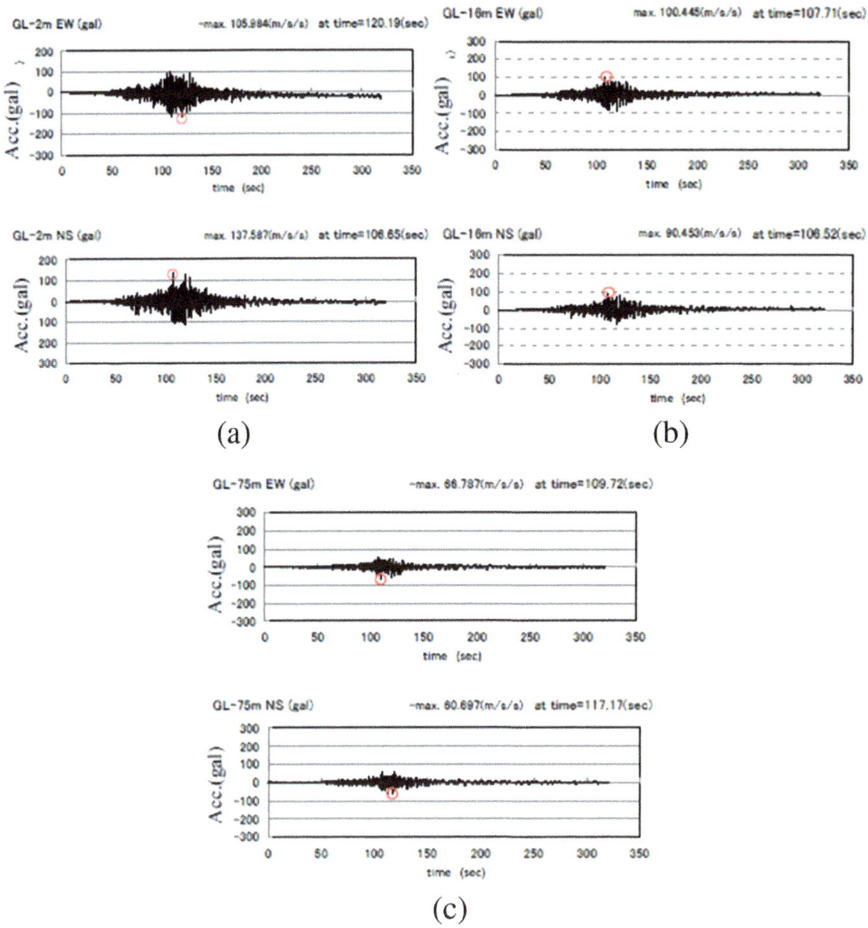

Fig. 10.2 Observed horizontal acceleration time histories at Shin-Ariake station in Tokyo Port. (**a**) GL-2.0 m. (**b**) GL-16.0 m. (**c**) GL-75.0 m

parameters, designating the control of contractive and dilative behavior, as well as volume compressibility after soil liquefaction. Solving the governing equations of the model with seepage flow equation in a coupled manner (Zienkiewicz and Bettess 1982), it becomes possible to calculate pore water pressures in soil. Volumetric strains of soil accompanied by dissipation of excess pore water pressures are simulated under the condition of given permeability of soil (e.g. Tamari et al. 2009).

The ground model at the recording station was developped by refering to the boring log near the site (Tokyo metropolitan government) and the geological cross section in Tokyo Port (Tokyo Geotechnical Consultant Association 2004). The ground model is illustrated in Table 10.1. The water table is assumed to be at

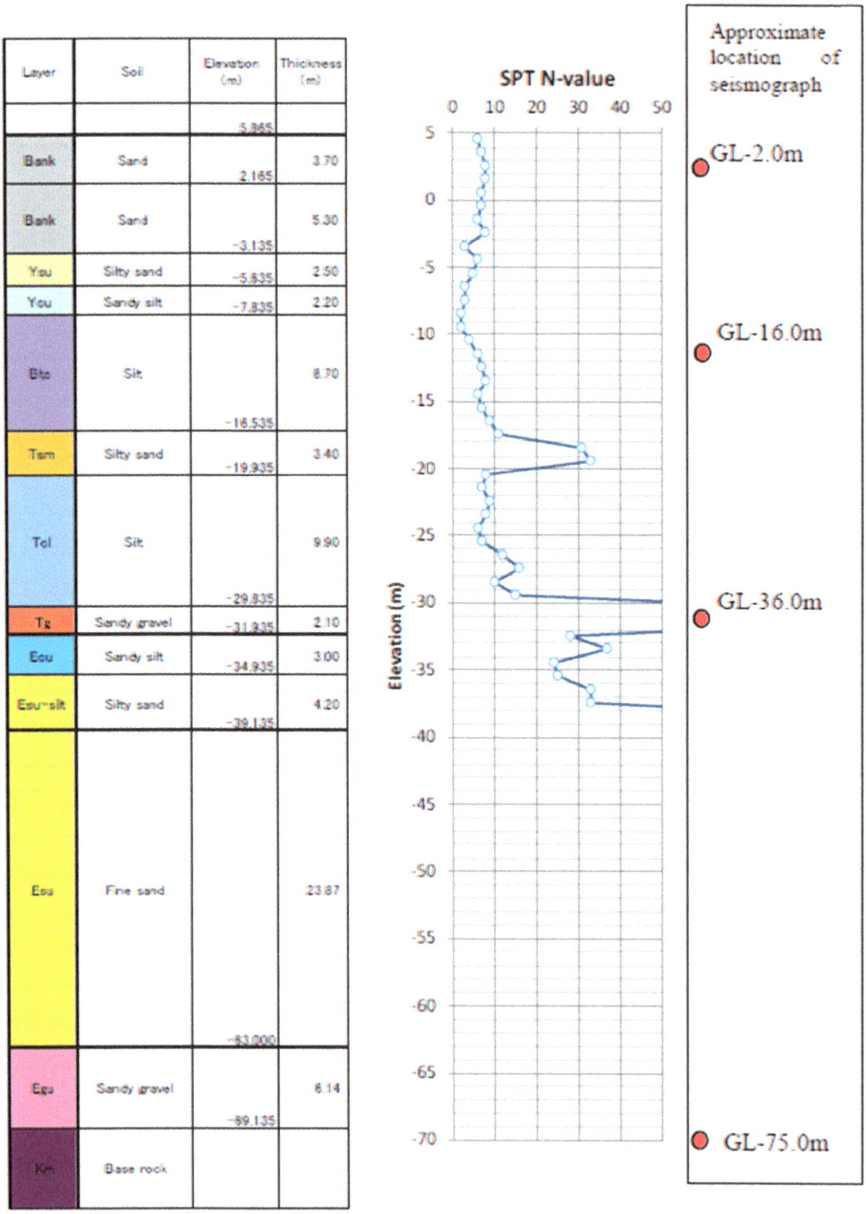

Fig. 10.3 Soil profile and SPT N-values near Shin-Ariake recording station

GL-3.7 m, liquefiable soil deposit to be the bank (Elevation of 2.165 m to −3.135 m) and alluvial sandy layer (Ysu, Elevation of −3.135 m to −5.635 m).

Model parameters for soil were assessed by refering to literal parameters and technical reports (Ishihara et al. 1989; TGCA 2004). Parameters of initial shear

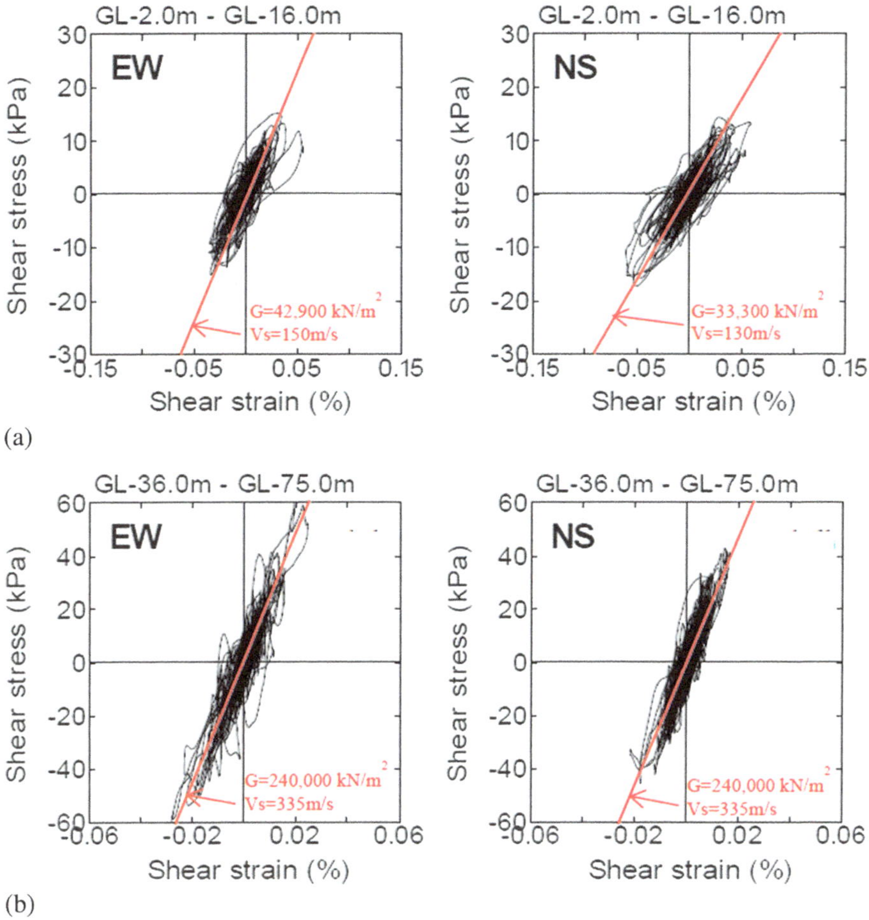

Fig. 10.4 Reproduced Stress strain hysteresis during mainshock by observed acceleration time histories at Shin-Ariake recording station. (**a**) GL-2.0–16.0 m. (**b**) GL-36.0–75.0 m

modulus, internal friction angle and cohesion for dynamic deformation characteristics are assessed by using simplified method to determine parameter of FLIP (Morita et al. 1997) refering to SPT N-values illustrated in Fig. 10.3 and fine contents of soil estimated based on description of soil classification in boring log. Parameters for phisical characteristics, dynamic deformation characteristics and permeability are summarized in Table 10.2. Dilatancy of soil and generation of excess pore water pressure was taken into account for bank material and alluvial sand (Ysu) below water table as liquefaction characteristics. Cyclic stress ratio is assessed based on the cyclic test using undisturbed sample at Sunamachi site in Tokyo port (Ishihara et al. 1989) which is near Yumenoshima recording station in Fig. 10.1. Volume compression characteristics are specified assuming relative density as around 60%, using relationship with maximum shear strain (Nagase and

Table 10.1 Summary of model parameters of dynamic deformation characteristics

Soil profile		Elevation (m)	Thickness of strata (m)	N value	γ_t (kN/m^3)	Vs (m/s)	Elatic shear module at a confining pressure of (σ_{ma}') G_{ma}(kPa)	Reference confining pressure (kpa)	Poisson's ratio	Maximum damping coefficient H_{max}	Shear resistance angle ϕ(°)	Cohesion C (kPa)	Coefficient of permeability k (m/s)	Approximate location of seismograph
		5.865												
Bank	Sand	2.165	3.70	7	18.0	140	71,402	98.0	0.33	0.240	39.1	0.0	1.0×10^{-5}	GL-2.0 m
Bank	Sand	−3.135	5.30	7	19.1	175	71,402	98.0	0.33	0.240	39.1	0.0	1.0×10^{-5}	
Ysu	Silty sand	−5.635	2.50	5	18.2	160	48,218	98.0	0.33	0.240	38.2	0.0	1.0×10^{-5}	
Ycu	Sandy silt	−7.835	2.20	3	15.0	138	27,876	98.0	0.33	0.150	0.0	40.0		
Btc	Silt	−16.535	8.70	6	15.0	172	39,383	98.0	0.33	0.150	0.0	56.7	1.0×10^{-7}	GL − 16.0 m
Tsm	Silty fine sand	−19.935	3.40	25	18.0	280	114,020	98.0	0.33	0.200	40.8	0.0	1.0×10^{-6}	
Tcl	Silt	−29.835	9.90	10	16.0	209	51,515	98.0	0.33	0.150	0.0	98.0	1.0×10^{-7}	
Tg	Gravel	−31.935	2.10	100	21.0	421	252,783	98.0	0.33	0.240	45.2	0.0	3.0×10^{-3}	GL − 36.0 m
Ecu	Sandy silt	−34.935	3.00	30	16.0	313	103,126	98.0	0.33	0.150	0.0	296.7	1.0×10^{-7}	
Esu − silt	Silty fine sand	−39.935	4.20	30	21.0	274	98,697	98.0	0.33	0.150	0.0	303.3	1.0×10^{-7}	
Esu	Fine sand	−63.000	23.87	113	21.0	330	233,357	98.0	0.478	0.240	44.7	0.0	1.0×10^{-5}	
Egu	Gravel	−69.135	6.14	–	21.0	560	672,000	505.2	0.478	0.240	44.7	0.0	3.0×10^{-3}	
Km	Rock	–	–	–	21.0	560	–	–	0.430	–	–	–	Impervious	GL-75.0 m

Table 10.2 Best fit liquefaction Parameters

Model parameters		Bank	Ysu	Remarks
Cyclic stress ratio (D.A. = 5% at number of cycles of 20)		0.21	0.21	Ishihara et al. (1989)
Volumetric strain due to consolidation following liquefaction (%)		2.7	2.7	Around 60% of relative density is assumed
Phase transformation angle $\phi_p(°)$		28.0	28.0	
Parameters for dilatancy	ε_d^{cm}	0.68	0.74	
	$r_{\varepsilon dc}$	0.958	0.928	
	$r_{\varepsilon d}$	0.68	0.74	
	q_1	1.0	1.0	
	q_2	0.5	0.6	
	q_4	1.0	1.0	
	q_{us}	–	–	
	r_γ	0.2	0.2	
	rrm_{tmp}	0.5	0.5	
	S_1	0.005	0.005	
Parameters for volume compressibility	l_k	2.0	2.0	
	r_k	0.147	0.135	
Parameter for limit of liquefaction	c_1	1.52	1.53	

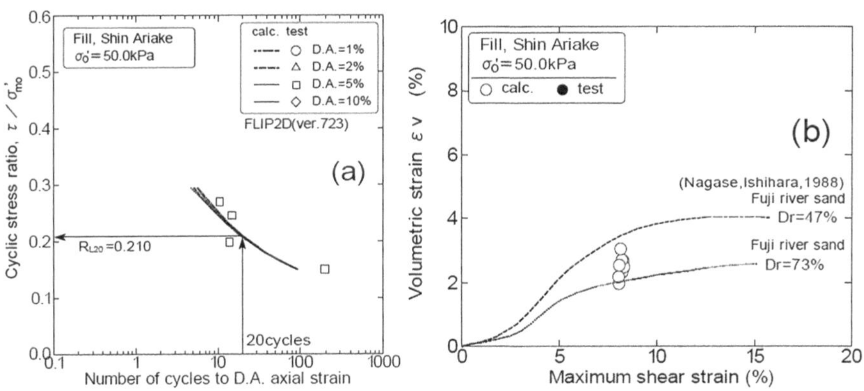

Fig. 10.5 Results of numerical simulation of cyclic test using best assessed parameter set. (**a**) Cyclic stress ratio versus number of cycles. (**b**) Relationships between reconsolidation volume change and shear stran

Ishihara 1988; Ishihara and Yoshimine 1992). Permeabilities are estimated based on the soil classification descrived in boring log. Parameters for liquefaction and volume compression are specified as best assessed parameter set by numerical simulation of undrained cyclic shear test. Result of simulation for cyclic strength and volume change following liquefaction is illustrated in Fig. 10.5. Those parameters are summarized in Table 10.2.

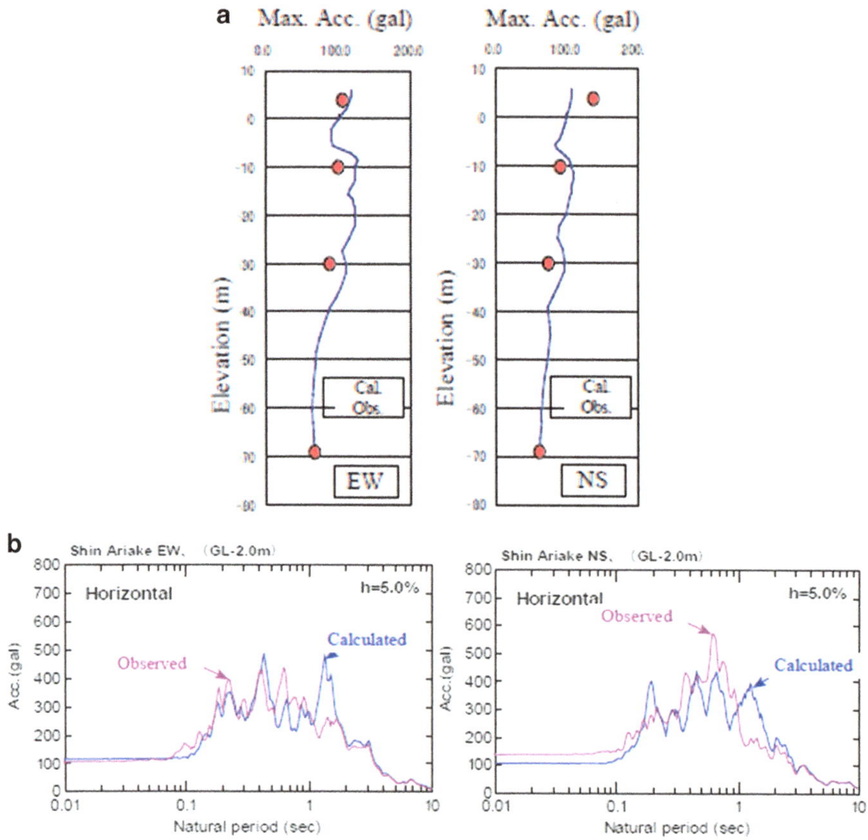

Fig. 10.6 Comparison of calculated and observed accelerations. (**a**) Distribution of maximum accelerations (**b**) Acceleration response spectra (GL-2.0 m)

10.3.2 Verification of the Numerical Ground Model

To verify the developed ground model, earthquake response analysis is conducted applying observed motion at GL-75.0 m to the bottom fixed boundary. Distribution of calculated maximum accelerations and their response spectrum are depicted in Fig. 10.6 with observed values. The calculated maximum accelerations show good agreement with observed values with difference of 10–30%. The maximum ratio of excess pore water pressure ($=1-\sigma'_m/\sigma'_{mo}$, where σ'_{mo}: effective mean stress) is calculated as about 0.1, being consistent with no liquefaction at the site. The calculated acceleration spectrum at GL-2.0 m is very similar to observed spectrum. These calculation results suggest the developed grand model is considered reasonable.

10.4 Reproduction of Ground Motion at Base for Mainshock and Aftershock

The earthquake motion both mainshock and aftershock at the base is reproduced for numerical study. The observed accelerations at the surface of recording station K-NET Urayasu (NIED), and at GL.-89.48 m of Yumenoshima (Tokyo metropolitan government) in Tokyo Port are used for calculation. Location of recording stations is shown in Fig. 10.1. The outline of ground models used is presented in Fig. 10.7. The model of Yumenoshima was developed by Ishihara et al. (1989), K-NET Urayasu by Urayasu municipal government (2011). The depth of the base are defined as GL-50.2 m at Yumenoshima, GL-42.8 m at K-NET Urayasu respectivery.

One dimensional response analysis program "SHAKE" (Schnabel et al. 1972) and "DYNEQ" (Sugito 1995; Yoshida and Suetomi 1996) are used for calculation, in which frequency dependency is taken into account in "DYNEQ". Caluculated base acceleration time historis and renponse spectra are depicted in Fig. 10.8 taking EW direction as a typical result. The acceleration time history by DYNEQ seems to be consistent with real phenomena in terms of maximum acceleration of mainshock (M = 9.0) and aftershock (M = 7.7). The estimated acceleration time histories at the base are shown in Fig. 10.8b.

Site Name		Yumenoshima		
Surface Elevation (T.P.m)		–		
Water Level (GL.-m)		–		
Soil	Depth of Lower End	Thickness	Unit Weight	Shear Wave Velocity
–	(m)	(m)	(kN/m3)	(m/s)
Bs1	5.4	5.4	18.0	230
Bs2	9.0	3.6	19.1	130
As1	13.7	4.7	18.2	170
As2	20.0	6.3	18.2	220
Ac1	26.0	6.0	17.0	150
Ac2	40.0	14.0	15.0	150
Ac3	44.2	4.2	15.0	170
Ac4	45.5	1.3	18.7	250
Ac5	50.2	4.7	16.9	250
Dg1	59.5	9.3	21.0	560
Dg2	73.8	14.3	18.7	330
Dg3	83.6	9.8	17.9	330
Gr	89.5	5.9	21.0	560
Gr(Base)	–	–	21.0	560

(a)

Site Name		K-net Urayasu		
Surface Elevation (T.P.m)		2.54		
Water Level (GL.-m)		2.5		
Soil	Depth of Lower End	Thickness	Unit Weight	Shear Wave Velocity
–	(m)	(m)	(kN/m3)	(m/s)
Bs	2.5	2.5	17.0	160
	5.2	2.7	17.0	110
As1	7.2	2.0	18.0	150
	10.8	3.6	18.0	160
	14.2	3.4	18.0	110
Ac1	23.7	9.5	16.0	110
	27.8	4.1	16.0	170
	36.8	9.0	16.0	140
Nac	39.6	2.8	17.0	150
	42.8	3.2	17.0	180
Kys(Base)	–	–	19.0	300

(b)

Fig. 10.7 Ground Model at recording stations. (**a**) Yumenoshima (for Mainshock). (**b**) K-net Urayasu (for Aftershock)

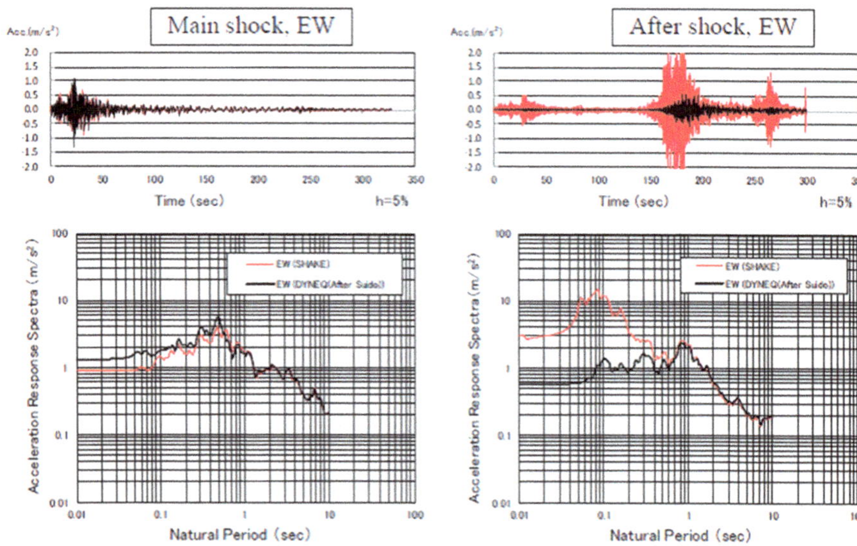

(a)

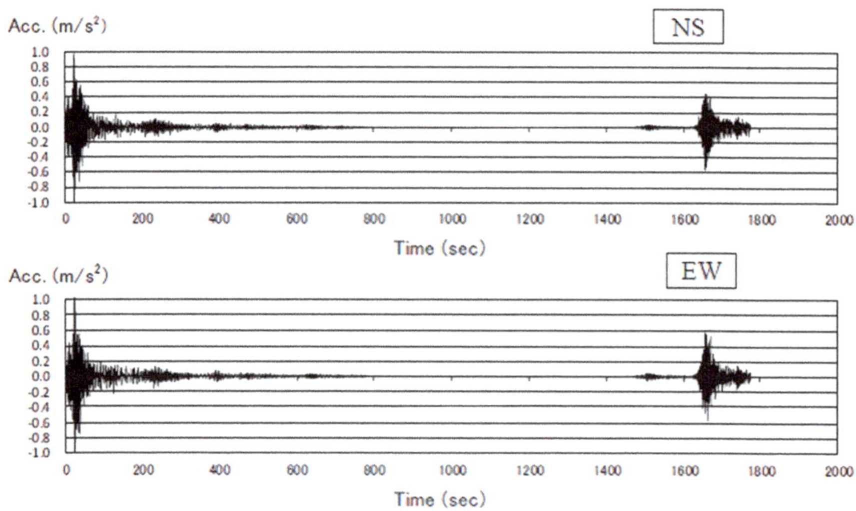

(b)

Fig. 10.8 Reproduced earthquake motion at the base. (**a**) Comparison of base earthquake motions. (**b**) Accerelation time histories (DYNEQ)

10.5 Dynamic Effective Stress Analysis Focusing on the Impact of Aftershock

A series of dynamic effective stress analysis were conducted considering buildup and dissipation of excess pore water pressure with total duration time of 180 min (10,800 s), including 30 min (1800 s) for mainshock and aftershock, and 150 min (9000 s) for pore water dissipation (case M+A). In addition, the analyses with input of only mainshock (case M) and only after shock (case A) are also carried out for comparison. Calculated maximum response values through the depth are illustrated in Fig. 10.9. First of all, calculated maximum horizontal accelerations are reasonably consistent with observed values through the depth in both EW and NS direction. It is seen that the maximum acceleration in case M+A and case M are completely same, indicating maximum acceleration occurs during mainshock. However, the maximum excess pore water pressure in case M+A of about 0.4 is greater than in case M of about 0.3. This implies that the maximum excess pore water pressure can appear during aftershock as presented in Fig. 10.11a.

The time histories of horizontal acceleration at the GL-2.0 m and settlement (vertical displacement, upward as positive) are depicted in Fig. 10.10. Settlement is calculated to be 0.005 m just before the aftershock and 0.018 m finally. It suggests that about half of all the settlement can occur due to aftershock, supporting observations by Fukutake and Jang (2013). To clear the initial state of soil at the beginning of aftershock, secant shear modulus is calculated from the stress strain hysteresis as shown in Fig. 10.11b. The secant shear stiffness are $G = 2.2 \times 10^4$ kPa in case M+A, and $G = 4.2 \times 10^4$ kPa in case A, implying about 50% of degradation of shear modulus. The stress path is illustrated in Fig. 10.11c. This indicates the effective mean stress before aftershock is reduced to about 80 kPa from the initial of 90 kPa. The degraded shear stiffness seems to cause large shear strain, leading to buildup of excess pore water pressure during aftershock.

10.6 Conclusions

The numerical study is conducted on the seismic ground behavior at the reclaimed ground at Shin-Ariake in Tokyo Port where no liquefied is reported at the Earthquake 3.11 to clarify the effect of aftershock on liquefaction. Following conclusions are derived:

(i) The stress strain hysteresis derived by recorded accelerations suggestd no liquefaction occurrence at the site during the mainshock in Earthquake 3.11.
(ii) By the dynamic effective stress analysis, it was found that the maximum pore water pressure and maximum shear strain appeared during aftershock not in main shock.

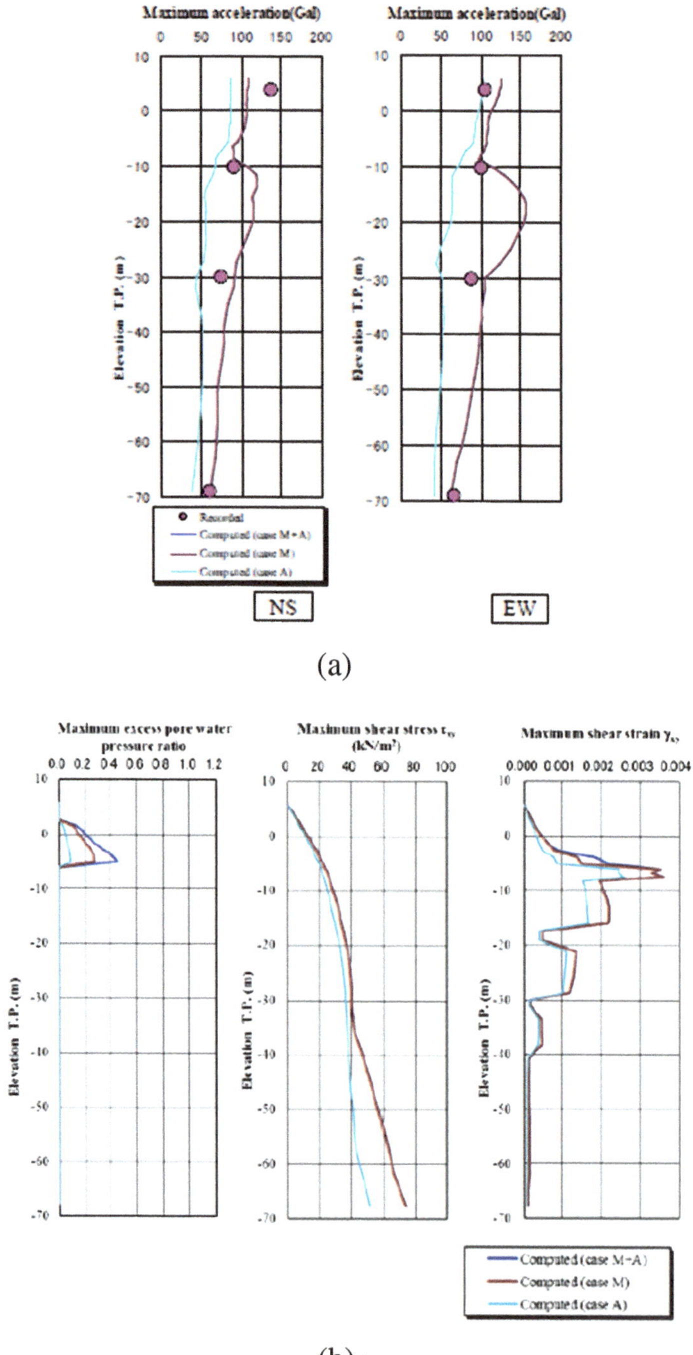

Fig. 10.9 Maximum response distributions. (**a**) Horizontal Acc. (**b**) ratio of E.P.W.Pr., Shear stress, Shear strain (EW direction)

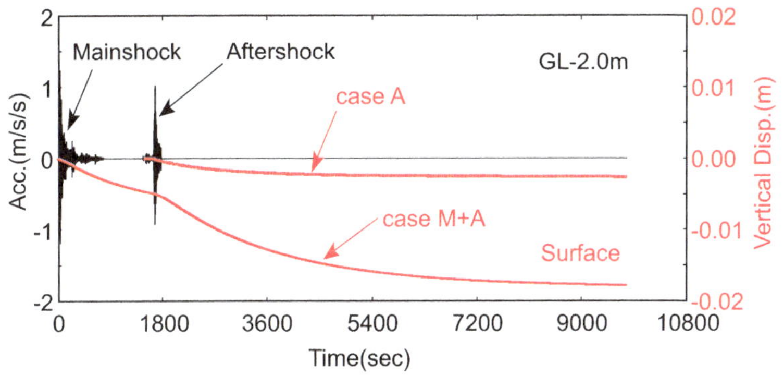

(a)

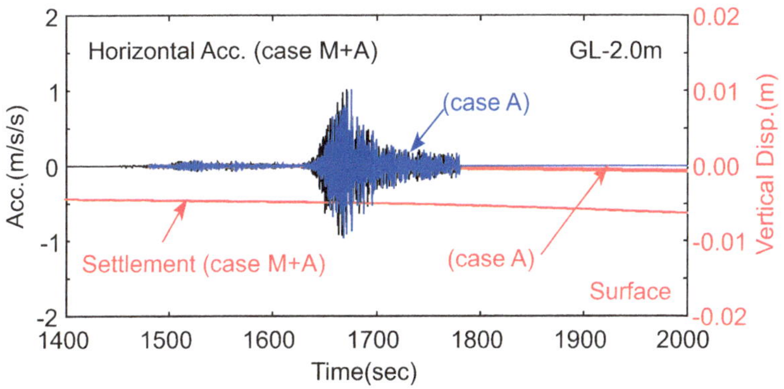

(b)

Fig. 10.10 Time histories of Acceleration and Settlement (EW component). (**a**) All duration of calculation. (**b**) After shock (enlarged time scale)

(iii) The ground which experienced the mainshock seems to become susceptive against liquefaction due to residual pore water pressure and degradetion of shear stiffness.

In engineering point of view, consideration of not only mainshock but also aftershock must be very important for assessment of excess pore water pressures in liquefiable soil layer.

Acknowledgments A series of study in this paper was carried out in the activity of workshop on "General Incorporated Association FLIP consortium", Japan in the period from 2012 to 2013. The observed acceleration time histories by seismography and boring log at Tokyo port was opened data through the web site by Tokyo metropolitan government, Japan. The authors deeply express their gratitude to all the members and related organizations.

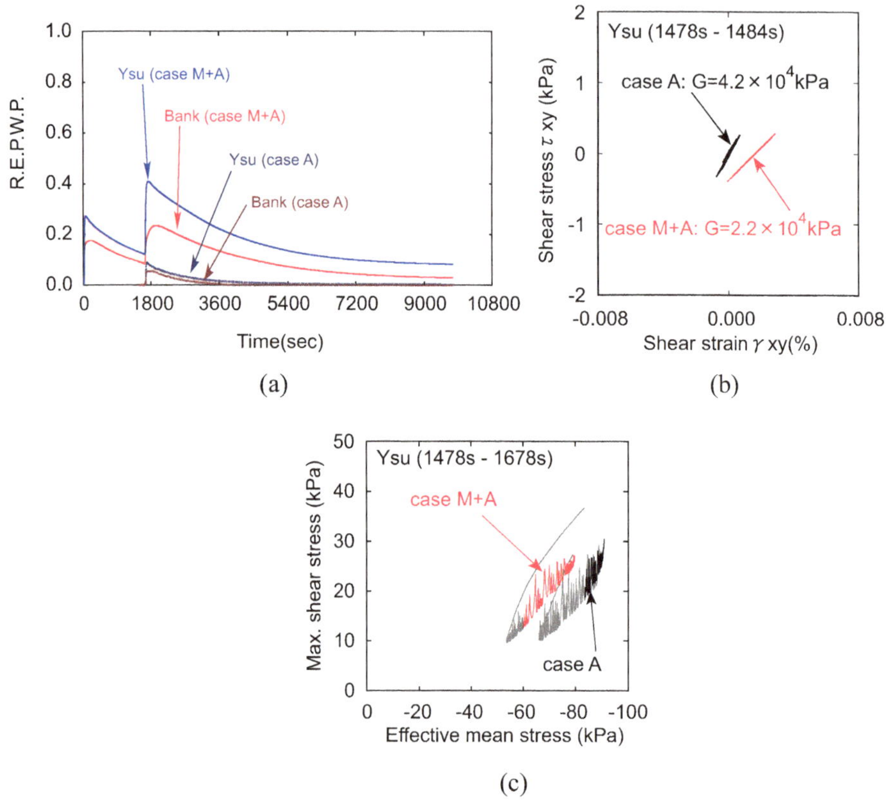

Fig. 10.11 Time histories and hysteresis. (**a**) Ratio of excess pore water pressure. (**b**) Stress Strain hysteresis (**c**) Stress path

References

Bureau of Port and Harbor, Tokyo Metropolitan Government. http://www.kouwan.metro.tokyo.jp/business/kisojoho/jishindou.html

Coastal Development Institute of Technology (CDIT) (1997) Handbook of liquefaction countermeasure for reclaimed ground (in Japanese)

Fukutake K, Jang J (2013) Studies on soil liquefaction and settlement in the Urayasu District using effective stress analyses for the 2011 off the Pacific Coast of Tohoku Earthquake. J JSCE 1:307–321

Iai S, Matsunaga Y, Kameoka T (1992) Strain space plasticity model for cyclic mobility. Soils Found 32(2):1–15

Iai S, Tobita T, Ozutsumi O, Ueda K (2011) Dilatancy of granular materials in a strain space multiple mechanism model. Int J Numer Anal Meth Geomech 35:360–392

Ishihara K, Muroi T, Towhata I (1989) In-situ pore pressures and ground motion during the 1987 Chiba touhou-oki earthquake. Soils Found 29(4):75–90

Ishihara K, Yoshimine M (1992) Evaluation of settlements in sand deposits following liquefaction during earthquakes. Soils Found 32(1):173–188

Morikawa Y, Xiaohua BAO, Maeda K, Imase T, Zhang F (2011) Importance of liquefaction analysis considering re-liquefaction due to aftershock of earthquake. J Geotech Eng 7 (2):389–397. (in Japanese)
Morita T, Iai S, Liu H, Ichii K, Sato Y (1997) Simplified method to determine parameter of FLIP, technical note of the port and Harbour Research Institute, Ministry of Transport, Japan, No.869, June 1997
Nagase H, Ishihara K (1988) Liquefaction-induced compaction and settlement of sand during earthquakes. Soils Found 28(1):65–76
National Research Institute for Earth Science and Disaster Prevention (NIED). K-NET WWW service, Japan (http://www.k-net.bosai.go.jp/)
Schnabel Per B, Lysmer John, Seed H Bolton (1972) SHAKE: a computer program for earthquake response analysis of horizontally layered sites, UCB/EERC-72/12, Earthquake Engineering Research Center, University of California, Berkeley, 1972–12
Sugito M (1995) Frequency-dependent equivalent strain for equi-linearized technique. In: Proceedings of IS-Tokyo '95, the first international Conference on earthquake geotechnical engineering, Tokyo, November 1995, pp 655–660
Tamari Y, Ozutsumi O, Iai S, Yokoyama N (2009) A numerical study on post liquefaction soil behavior by dynamic effective stress analysis, performance-based design in earthquake geotechnical engineering. Taylor & Francis Group, London, pp 1223–1231
Technical Committee on Measures Against Liquefaction in Urayasu City (2011) Report on ground feature and liquefaction factor analysis in Urayasu City. http://www.city.urayasu.lg.jp/_res/projects/default_project/_page_/001/002/934/04_siryo2-4-1jibanntokuseinohaaku.pdf (in Japanese)
Tokyo Geotechnical Consultant Association (2004) Technical note, No.37 – Tokyo Bay (in Japanese)
Towhata I, Ishihara K (1985) Modeling soil behavior under principal stress axes rotation. In: Proceedings of 5th international conference on numerical method in geomechanics, pp 523–530, Nagoya
Ueda K, Izawa J, Murono Y, Iai S (2014) Analytical study on the influence of aftershoks on the liquefaction behavior of ground. J JSCE, A1 70(4):I_578-I_585 (in Japanese)
Yasuda S, Towhata I, Ishii I, Sato S, Uchimura T (2013) Liquefaction-induced damage to structures during the 2011 great east Japan earthquake. J JSCE 1:181–193
Yoshida N, Suetomi I (1996) DYNEQ: Equivalent linear earthquake response analysis of ground, Technical report in Institute of Technology of Sato Kogyo, pp 61–70 (in Japanese)
Zienkiewicz OC, Bettess P (1982) Soils and other saturated media under transient, dynamic conditions; general formulation and the validity of various simplifying assumptions, soil mechanics – transient and cyclic loads, Wiley

Chapter 11
Analysis of Liquefaction During 2011 East Japan Earthquake – Part 2: Effective Stress Analysis Considering the Permeability of the Ground ~ Liquefaction at Reclaimed Land in the Main and After Shocks of the 2011 Great East Japan Earthquake

Shun-ichi Sawada, Kazuaki Uemura, Minoru Nobumoto, Makoto Yoshida, Shigeru Sato, Yoshiaki Kageji, Yukio Tamari, Junichi Hyodo, Takashi Nakama, Atsushi Hosoo, and Koji Ichii

Abstract A series of effective stress dynamic analyses considering permeability was performed in areas with and without liquefaction damage at reclaimed land in Urayasu city in the Tokyo Bay area due to the 2011 Tohoku Earthquake off the Pacific coast. Liquefaction damage was significant because the ground shook with earthquake motions for an extremely long duration (more than 2 min for the main shock) followed by an aftershock about 30 min later. In particular, the delayed sand boil, which occurred much later than the main shock, was monitored by a security camera at the site. In typical earthquake motions with a shorter duration, the permeability of the ground is ignored in the seismic response analyses. However, whether the effect of permeability can be ignored becomes questionable when the duration of earthquake motion is extremely long and additional effects occur due to

S.-i. Sawada (✉) • K. Uemura • M. Nobumoto
OYO Corporation, Tokyo, Japan
e-mail: sawada-shun@oyonet.oyo.co.jp; uemura-kazuaki@oyonet.oyo.co.jp

M. Yoshida
Penta-Ocean Construction Co., Ltd, Singapore, Singapore

S. Sato • Y. Kageji
Pacific Consultants Co., Ltd, Wellington, New Zealand

Y. Tamari • J. Hyodo
TEPSCO (Tokyo Electric Power Services Co., Ltd), Tokyo, Japan

T. Nakama • A. Hosoo
Jishin Kogaku Kenkyusyo, Inc, Tokyo, Japan

K. Ichii
Hiroshima University, Higashihiroshima, Japan

S. Iai (ed.), *Developments in Earthquake Geotechnics*, Geotechnical, Geological and Earthquake Engineering 43, DOI 10.1007/978-3-319-62069-5_11

an aftershock such as the case with the 2011 earthquake. The results of the effective stress analyses incorporating the effect of permeability are consistent with the behavior of the ground with and without liquefaction damage. In particular, the delayed sand boiling at the liquefied site is well simulated by the effective stress analysis for a specific combination of permeability assigned above and below the groundwater table, indicating that it is important to carefully evaluate the permeability in the analysis of liquefaction.

11.1 Introduction

Widespread liquefaction damage occurred in the Tokyo Bay area on March 11, 2011 due to the Tohoku Earthquake off the Pacific coast, which is also known as the Great East Japan Earthquake. The liquefaction damage was significant because the ground shook with earthquake motions for an extremely long duration (more than 2 min for the main shock) followed by an aftershock about 30 min later. In seismic response analyses of typical earthquake motions with shorter durations, the permeability of the ground is ignored, but this assumption may be questionable when the earthquake motion is extremely long and is accompanied by additional effects due to aftershocks as was the case with the 2011 earthquake.

In this study, a series of effective stress dynamic analyses considering the permeability of soils is performed for areas with and without liquefaction damage at reclaimed land in Urayasu city, Japan, using the computer program FLIP ROSE ver.7.2.3 (Iai 1992; Iai and Ozutsumi 2011) with the cocktail glass model (Iai et al. 2011).

11.2 Liquefaction Damage at the Urayasu Site

Widespread liquefaction damage occurred at reclaimed land in Urayasu city by the main shock ($M = 9.0$) and the after shock ($M = 7.7$) of the 2011 Great East Japan Earthquake. Figure 11.1 shows the areas in this study. The landfill in Area 1 is newer than Area 2. Area 1 sustained large liquefaction damage as shown in Picture 11.1, whereas Area 2 is the non-damaged area. The settlement after the earthquake in Area 1 was 20–30 cm, while the settlement was negligible in Area 2.

11.3 Analysis Conditions

The ground model for the effective stress analyses was developed as one-dimensional finite element models based on the soil profiles at liquefied (Area 1) and non-liquefied (Area 2) sites in Urayasu (Fig. 11.2). The model for

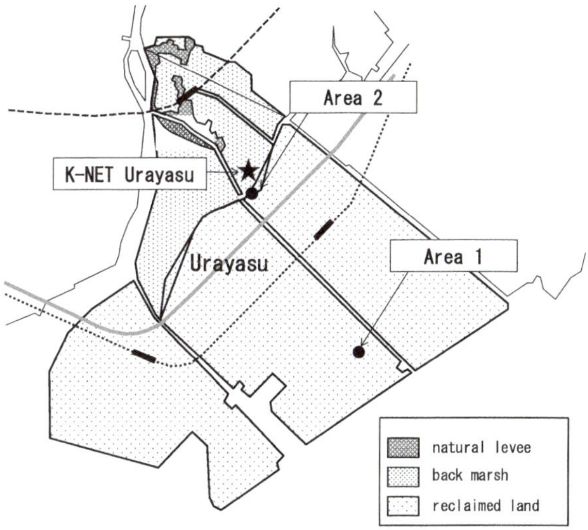

Fig. 11.1 Analysis area

Picture 11.1 Analysis area

Area 1 had liquefiable sandy layers Fs and As1 below the groundwater table with a total thickness of 10 m. In particular, the Fs layer was divided into four finer layers, which are designated as Fs_LGW, LGW1, LGW2, and LGW3, in order to investigate details of the special variation of excess pore water pressures in the vicinity of the groundwater table. The model at Area 2 had two liquefiable sand layers below the ground surface, designated as Bs_LGW and As1-2L and As1-3.

The pore water elements were also considered in the shallower ground above the groundwater table to simulate the rise in the water table due to the liquefaction of

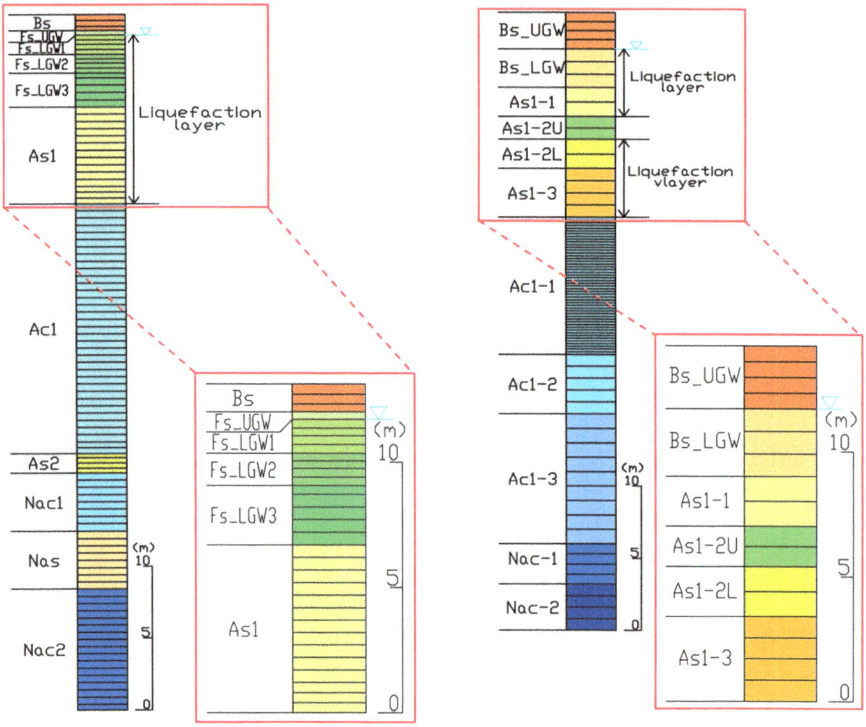

(a) Area 1 (b) Area 2

Fig. 11.2 Ground model

the sand layers beneath it. A negative pore water pressure was set at the ground surface to model the hydrostatic pressure in the ground.

The input earthquake motion used for the effective stress analysis is shown in Fig. 11.3. This input motion was obtained by combining the main shock and aftershock motions recorded at slightly different sites in the vicinity of or at the Urayasu site. The main shock portion was based on the recorded motion at the Yumenoshima site (5 km east of Urayasu) and the after shock portion was based on the recorded motion at the Urayasu site of K-NET. Deconvolution through an equivalent linear method (DYNEQ, Yoshida 1996) was applied to these recorded motions at the ground surface to estimate the input motion at the bedrock.

Table 11.1 lists the parameters of the ground for the effective stress analyses, which were determined from the soil type and the SPT-*N* value. The parameters for the liquefaction property were set based on a simulation of the soil element test. The liquefaction resistance curves computed by the element simulation are shown in Fig. 11.4. The target of the simulation was the laboratory liquefaction test result (liquefaction countermeasures and investigation committee in Urayasu city 2012). The volumetric strain after dissipation of the pore water pressure following liquefaction (Ishihara 1992) was also considered in the parameter determination.

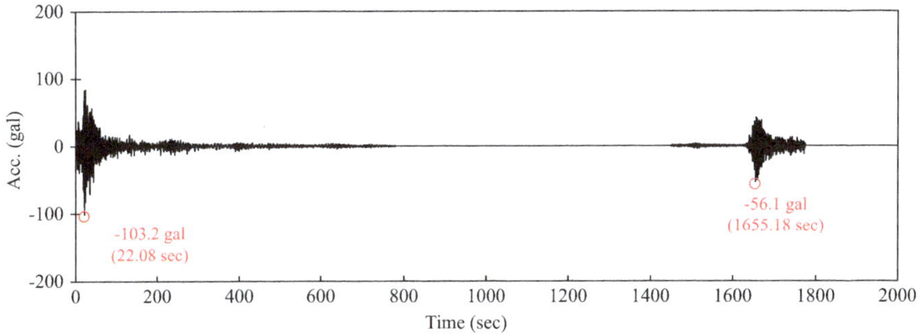

Fig. 11.3 Input motion

The permeability of the ground was determined based on an empirical equation (Creager 1944) using D_{20}, which is the particle size 20 % finer by weight. In addition, the parametric study of permeability was conducted because the permeability may change due to variations in the soil conditions (e.g., increase in the effective porosity during liquefaction, changes in the degree of saturation, and the occurrence of local crack of the ground). The permeability above the groundwater table varies, ranging from 1/100 to 100 times relative to the reference value as shown in Table 11.2. The permeability below the groundwater table in liquefiable sand also varies, ranging from the reference value to 100 times relative to it (Table 11.2). If the permeability above the groundwater table (non-liquefiable layer in Table 11.2) is specified to be lower than that below the groundwater table, the excess pore water pressure just below the groundwater table should accumulate to a higher pressure level. Thus, the combination of the permeability above and below the groundwater table is an important issue in the parametric study of the permeability of the ground.

A total of $3 \times 5 = 15$ cases of analyses were performed for each site. The boundary conditions were as follows. For the static analysis, the base was initially fixed in the vertical direction and the side was fixed in the horizontal direction. For the dynamic analysis, the viscous boundary condition in the base and the excess pore water pressure on the ground surface were zero.

11.4 Analysis Results

11.4.1 Area 1

Figure 11.5 shows the computed time histories of the excess pore water pressure ratios at the liquefied site (Area 1). The excess pore water pressure ratio at the top of layer Fs_LGW3 (green solid line) is the highest, reaching almost 1.0 over all the permeability conditions, indicating that liquefaction occurred in at least the Fs_LGW3 layer. This is consistent with the fact that Area1 liquefied.

Table 11.1 Ground parameters

(a) Area 1

Layer	Altitude[a] (lower end) T.P. (m)	Density ρ (g/cm^3)	Reference confining pressure σ'_{ma} (kPa)	Elastic shear modulus at a confining pressure G_{ma} (kPa)	Bulk modulus at a confining pressure $K_{LA}K_{UA}$ (kPa)	Cohesion c (kPa)	Shear resistance angle ϕ(°)	Poisson's ratio ν	Porosity n	Maximum damping coefficient h_{max}	Permeability estimated from D20 k(m/s)
Bs	2.72	1.7	98.0	57,200	149,200	0.0	41.3	0.33	0.45	0.24	5.0E-05
Fs_UGW	2.42	1.8	98.0	40,000	104,300	0.0	41.0	0.33	0.45	0.24	5.0E-05
Fs_LGW1	1.07	1.8	98.0	34,100	88,900	0.0	38.3	0.33	0.45	0.24	1.0E-06
Fs_LGW2	−0.23	1.8	98.0	32,500	87,800	0.0	35.6	0.33	0.45	0.24	1.0E-06
Fs_LGW3	−2.58	1.8	98.0	25,300	66,000	0.0	39.6	0.33	0.45	0.24	1.0E-06
Asl	−9.28	1.8	98.0	34,000	88,700	0.0	39.4	0.33	0.45	0.24	1.0E-05
Acl	−26.58	1.6	98.0	25,600	66,800	0.0	30.0	0.33	0.55	0.15	3.0E-08
As2	−27.93	1.8	98.0	34,900	91,000	0.0	40.0	0.33	0.45	0.24	9.0E-06
Nacl	−31.98	1.7	98.0	44,700	116,600	0.0	30.0	0.33	0.55	0.15	3.0E-08
Nas	−35.98	1.8	98.0	104,000	271,200	0.0	39.1	0.33	0.45	0.24	2.0E-05
Nac2	−44.38	1.7	98.0	39,100	102,000	0.0	30.0	0.33	0.55	0.15	3.0E-08

Layer	Phase transformation angle	Parameters for dilatancy											Parameters for volume compressibility		
	ϕ_p(°)	ε_{dcm}	$r\varepsilon_{dc}$	$r\varepsilon_d$	q_1	q_2	q_4	q_{us}	r_γ	$rmtmp$	S_1	C_1	l_k	r_k	r_k''
Fs_LGW1	28.0	0.500	0.958	0.500	1.0	0.5	1.0	0.0	0.2	0.5	0.005	1.52	2.0	0.200	0.200
Fs_LGW2	28.0	0.580	0.720	0.580	1.0	0.5	1.0	0.0	0.2	0.5	0.005	1.52	2.0	0.172	0.172
Fs_LGW3	28.0	0.400	0.958	0.400	1.0	0.5	1.0	0.0	0.2	0.5	0.005	1.52	2.0	0.250	0.250
Asl	28.0	1.000	0.500	1.000	6.0	2.1	1.0	0.0	0.2	0.5	0.005	1.60	2.0	0.100	0.100

(b) Area 2

Layer	Altitude[b] (lower end)	Density	Reference confining pressure	Elastic shear modulus at a confining pressure	Bulk modulus at a confining pressure	Cohesion	Shear resistance angle	Poisson's ratio	Porosity	Maximum damping coefficient	Permeability estimated from D20
	T.P.(m)	ρ (g/cm^3)	σ'_{ma}(kPa)	G_{ma}(kPa)	K_{LA}, K_{UA}(kPa)	c(kPa)	φ(°)	v	n	h_{max}	k(m/s)
Bs_UGW	0.04	1.7	15.9	44,408	115,810	0.0	41.3	0.33	0.45	0.24	4.0E-05
Bs_LGW	−2.66	1.7	39.0	20,990	54,738	0.0	41.3	0.33	0.45	0.24	4.0E-05
As 1-1	−4.66	1.8	52.1	41,327	107,773	0.0	39.4	0.33	0.45	0.24	9.0E-06
As l-2U	−8.26	1.8	62.9	47,020	122,622	0.0	39.4	0.33	0.45	0.24	9.0E-06
As l-2L	−6.26	1.8	73.7	47,020	122,622	0.0	39.4	0.33	0.45	0.24	9.0E-06
As 1-3	−11.66	1.8	89.9	22,224	57,958	0.0	39.4	0.33	0.45	0.24	9.0E-06
Ac 1-1	−21.16	1.6	121.4	19,755	51,518	0.0	30.0	0.33	0.55	0.15	3.0E-08
Ac 1-2	−25.26	1.6	152.0	47,184	123,048	0.0	30.0	0.33	0.55	0.15	3.0E-08
Ac 1-3	−34.26	1.6	181.5	32,000	83,451	0.0	30.0	0.33	0.55	0.15	3.0E-08
Nac-1	−37.06	1.7	209.1	39,031	101,786	0.0	30.0	0.33	0.55	0.15	3.0E-08
Nac-2	−40.26	1.7	224.9	56,204	146,571	0.0	30.0	0.33	0.55	0.15	3.0E-08

Layer	Phase transformation angle	Parameters for dilatancy											Parameters for volume compressibility		
	φ_p(°)	ε_{dcm}	$r\varepsilon_{dc}$	$r\varepsilon_d$	q_1	q_2	q_4	q_{us}	r_γ	$rmtmp$	S_1	C_1	l_k	r_k	r_k''
Bs_LGW	28.0	0.250	1.000	0.500	5.0	1.1	1.0	0.0	0.2	0.5	0.005	1.95	2.0	0.200	0.200
As 1-1	28.0	0.625	0.500	0.625	5.0	1.8	1.0	0.0	0.2	0.5	0.005	1.70	2.0	0.160	0.160
As l-2L	28.0	0.625	0.500	0.625	5.0	1.8	1.0	0.0	0.2	0.5	0.005	1.70	2.0	0.160	0.160
As 1-3	28.0	0.588	0.500	0.588	5.0	2.8	1.0	0.0	0.2	0.5	0.005	1.40	2.0	0.170	0.170

[a]Ground level = T. P. +3.82 m
[b]Ground level = T.P. + 2.54 m

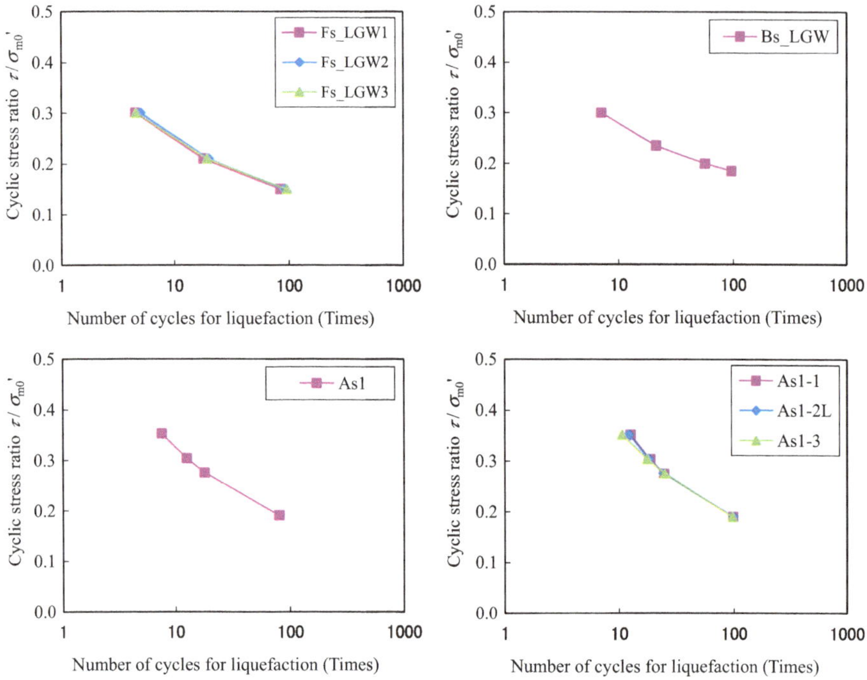

(a) Area 1 (b) Area 2

Fig. 11.4 Results of the element simulations

Table 11.2 Case of the parametric study (ground permeability)

		Permeability of liquelaction layer		
		Estimated value from D_{20}	×10	×100
Permeability of non-liquefaction layer	×1/100	Case 1–5	Case 2–5	Case 3–5
	×1/10	Case 1–4	Case 2–4	Case 3–4
	Estimated value from D_{20}	Case 1–1	Case 2–1	Case 3–1
	×10	Case 1–2	Case 2–2	Case 3–2
	×100	Case 1–3	Case 2–3	Case 3–3

Figure 11.5a shows the results for the cases where the permeability below the groundwater table (k_1) is constant, but the permeability above the groundwater table (k_2) is varied. When k_2 is smaller than k_1, the excess pore water pressure ratio of the top of the liquefiable layer (Fs_LGW1) (red solid line) becomes larger, indicating that the layer above the groundwater table with a lower permeability tends to

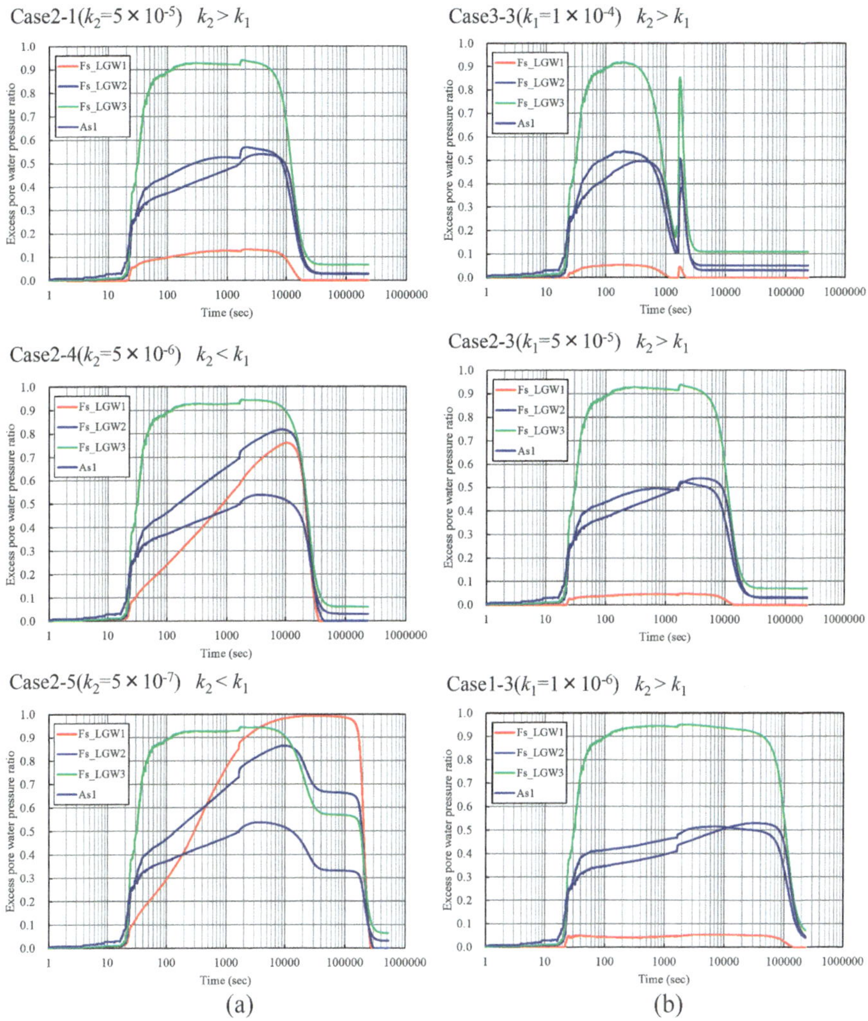

Fig. 11.5 Time history of the pore water pressure ratio (Area 1). (**a**) $k_1 = 1 \times 10^{-5}$ m/s. (**b**) $k_2 = 5 \times 10^{-3}$ m/s

prevent pore water from draining from the ground surface. Hence, the excess pore pressure tends to accumulate below the ground surface. In addition, the dissipation time for the water pressure tends to be longer when k_2 is smaller. This tendency is remarkable when the ratio of k_2 and k_1 becomes larger because the water pressure supplied from the bottom of the liquefiable layer (Fs_LGW3 and As1) is larger than the water pressure dissipated at the top of the liquefiable layer (Fs_LGW1).

Figure 11.5b shows the results for the cases where the permeability above the groundwater table (k_2) is constant, but the permeability below the groundwater

Table 11.3 Settlement after liquefaction (Area 1)

		Permeability of liquelaction layer		
		Estimated value from D_{20}	×10	×100
Permeability of non-liquefaction layer	×1/100	(Case 1–5) 18.4 cm	(Case 2–5) 19.8 cm	(Case 3–5) 22.4 cm
	×1/10	(Case 1–4) 18.0 cm	(Case 2–4) 19.1 cm	(Case 3–4) 22.4 cm
	Estimated value from D_{20}	(Case 1–1) 18.0 cm	(Case 2–1) 18.7 cm	(Case 3–1) 21.8 cm
	×10	(Case 1–2) 18.0 cm	(Case 2–2) 18.7 cm	(Case 3–2) 21.5 cm
	×100	(Case 1–3) 18.0 cm	(Case 2–3) 18.7 cm	(Case 3–3) 21.3 cm

table (k_1) is varied. When k_2 is larger than k_1, the difference in the maximum value of the excess pore water pressure ratio is small. The dissipation time for the water pressure tends to become longer when k_1 is small.

Table 11.3 shows the settlement after liquefaction. In each case, the settlement after liquefaction is around 20 cm, which agrees with the observed settlement in Area 1.

11.4.2 Area 2

Figure 11.6 shows the time history of the excess pore water pressure ratio. The maximum excess water pressure ratio is around 0.2 and liquefaction does not occur.

The settlement after liquefaction is shown in Table 11.4. In each case, the settlement after liquefaction is less than 2 cm, which is consistent with the observations.

11.5 Study for Delayed Sand Boiling at Area 1

The earthquake induced sand boiling at Area 1. According to the nearby security camera footage, boiling did not occur during the earthquake motion of the main shock, but occurred much later between the main shock and aftershock. Here, delayed sand boiling was investigated by focusing on the relationship between the overburden pressure and the pore water pressure during liquefaction at the groundwater level. In this study, sand boiling is judged to occur when the pore water pressure generated at the bottom of the layer above the groundwater table, W, becomes larger than the weight of that layer G. The ratio $W/G = 1.0$ was adopted as the criterion of the occurrence of sand boil.

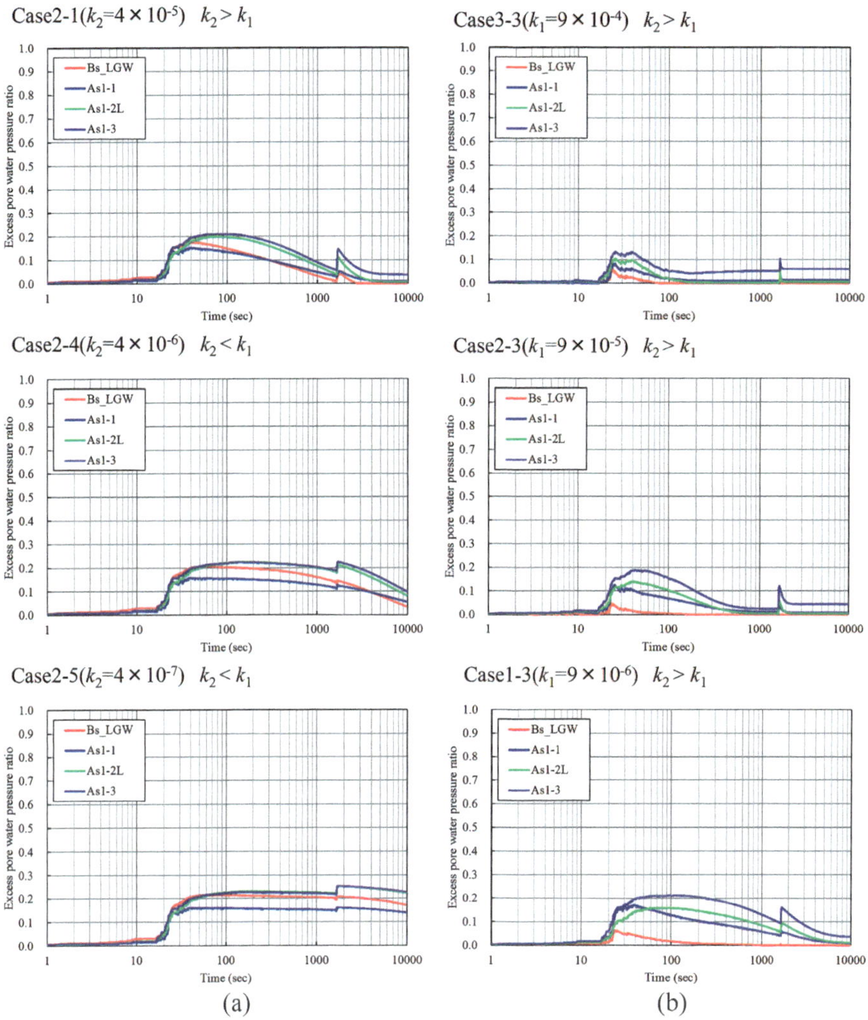

Fig. 11.6 Time history of the pore water pressure ratio (Area 2). (**a**) $k_1 = 9 \times 10^{-5}$ m/s. (**b**) $k_2 = 4 \times 10^{-3}$ m/s

Figure 11.7 shows the time histories of W/G in Case 1–5, Case 2–5, and Case 3–5. The permeability of the layer above the groundwater table k_2 is 1/100 of the estimated value from D_{20}. W/G in Case 2–5 (blue solid line) agrees with this observation since it also reaches 1.0 between the main shock and after shock. From this result, the permeability of the liquefaction layer may be larger than the estimated permeability from D_{20} through the empirical equation used in this study. In addition, sand boiling may be simulated by appropriately considering the permeability above the groundwater table relative to the permeability below the groundwater table.

Table 11.4 Settlement after liquefaction (Area 2)

		Permeability of liquefaction layer		
		Estimated value from D_{20}	×10	×100
Permeability of non-liquefaction layer	×1/100	(Case 1–5) 0.8 cm	(Case 2–5) 0.9 cm	(Case 3–5) 0.9 cm
	×1/10	(Case 1–4) 0.2 cm	(Case 2–4) 0.2 cm	(Case 3–4) 0.2 cm
	Estimated value from D_{20}	(Case 1–1) 1.3 cm	(Case 2–1) 1.5 cm	(Case 3–1) 1.5 cm
	×10	(Case 1–2) 1.3 cm	(Case 2–2) 1.6 cm	(Case 3–2) 1.7 cm
	×100	(Case 1–3) 1.4 cm	(Case 2–3) 1.6 cm	(Case 3–3) 2.1 cm

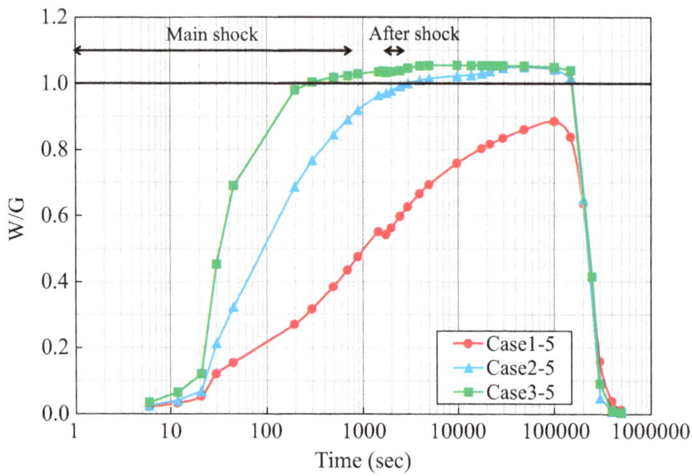

Fig. 11.7 Time history of *W*/*G*

11.6 Conclusions

This study examined a series of effective stress dynamic analyses considering the permeability for areas with and without liquefaction damage at reclaimed land in Urayasu city. The following conclusions are obtained:

1. The applicability of the cocktail glass model to simulate both areas with and without liquefaction damage is confirmed. In particular, the settlement of the ground after the dissipation of the pore water pressure in the liquefied area is consistent with the observations.
2. The results of the parameter study indicate that the layer above the groundwater table with a lower permeability tends to prevent drainage of pore water from the

ground surface. Consequently, the excess pore pressure tends to accumulate to a higher degree in the vicinity of the ground surface.

3. The delay of sand boiling at the Urayasu site, which occurs between the main shock and the after shock, can be well simulated by effective stress analysis through the cocktail glass model if the permeability below the groundwater table is set ten times larger than the estimated permeability from D_{20} based on Creager's empirical equation and the permeability above the groundwater table is set 1/100 of that value.

Acknowledgments A series of studies was carried out in the workshop on FLIP Consortium, Japan from 2012 to 2013. The observed acceleration time history by seismography was from open data through the website by the Tokyo Metropolitan Government and the National Research Institute for Earth Science and Disaster Prevention, Japan. The authors deeply express their gratitude to all the members and the related organizations.

References

Iai M (1992) Kameoka. Strain space plasticity model for cyclic mobility. Soils Found 32(2):1–15

Iai T, Ozutsumi U (2011) Dilatancy of granular materials in strain space multiple mechanism model. Int J Numel Meth Geomech 35:360–392

Ishihara Y (1992) Evaluation of settlements in sand deposits following liquefaction during earthquakes. Soils Found 32(1):173–188

Liquefaction countermeasures and investigation committee in Urayasu (2012) Report of liquefaction countermeasures and investigation committee in Urayasu (in Japanese)

National Research Institute for Earth Science and Disaster Prevention Kyoshin Net (K-net) (n.d.) http://www.kyoushin.bosai.go.jp/kyoshin/

Tokyo Metropolitan Government Bureau of Port and Harbor (n.d.) http://www.kouwan.metro.tokyo.jp/business/kisojoho/jishindou.html

Yoshida S (1996) DYNEQ:Earthquake response analysis program of the horizontal bedding ground based on the equivalent linear law. Report of research inst. of Sato Kogyo, pp 61–70

Chapter 12
Effective Stress Analysis of River Dikes During the 2011 East Japan Earthquake

Naoki Morishima

Abstract Effective stress analysis of the damage to river embankments during the 2011 Great East Japan earthquake ($M = 9.0$) is performed to study the applicability of the analysis method. The multi-spring model and the cocktail glass model are used for undrained and partially drained analysis, respectively. Both are defined in the framework of the strain space multiple mechanism model. The earthquake motion in this particular earthquake lasted more than 2 min, which is much longer than typical strong earthquake motions. Three damaged river embankments, 4–8 m high, constructed on liquefiable foundation soil and subjected to bedrock peak ground motions ranging from 0.15 to 0.6 g with and without the effect of an aftershock, are studied. The settlements of the embankment range from 0.8 to 4.3 m with and without differential settlements. The analysis results are mostly consistent with the damage to the river embankments both in terms of settlements and deformation modes. In particular, the partial drainage analysis using the cocktail glass model is applicable to the behavior of a river embankment subject to a seismic motion with a long duration, including the additional effect of the aftershock.

12.1 Introduction

The 2011 Great East Japan earthquake, which was off the Pacific coast of Tohoku, the northeastern portion of Honshu, caused widespread damage to river embankments. This damage was published in the "River embankment seismic measures emergency considerations committee document" (Japan Institute of Country-ology and Engineering (JICE) 2011). The type of damage is classified into three patterns (pattern I: liquefaction of the foundation ground, pattern II: partial liquefaction of the embankment, and pattern III: a mixture of foundation ground liquefaction and embankment liquefaction).

N. Morishima (✉)
Dia Consultants Co., Ltd., Tokyo, Japan
e-mail: n.morishima@diaconsult.co.jp

S. Iai (ed.), *Developments in Earthquake Geotechnics*, Geotechnical, Geological and Earthquake Engineering 43, DOI 10.1007/978-3-319-62069-5_12

Fig. 12.1 Locations of the selected river embankment sites relative to the epicenters of the main and after shocks of the 2011 Great East Japan earthquake

Location	
Site A	Tone River 26.5km
Site B	Yoshida River 16.0km
Site C	Tone River 39.5km

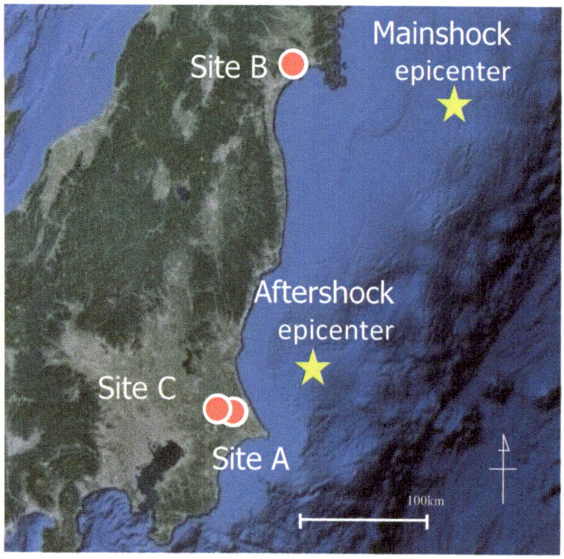

This study employs effective stress analysis using the FLIP code (Iai et al. 1992, 2011) to examine pattern I (liquefaction of foundation ground) of river embankment damage. Three damaged river embankments (4–8 m high) constructed on liquefiable foundation soil and subjected to bedrock peak ground motions ranging from 0.15 to 0.6 g with and without the effect of aftershock were selected. In particular, the river embankments at Sites A and C in Fig. 12.1 were subjected to bedrock peak ground accelerations of 0.15–0.2 g with additional intense effects from the aftershock. The embankment at Site B in Fig. 12.1 was subjected to a strong bedrock peak ground acceleration of 0.6 g.

The constitutive models used for the analysis were the multi-spring model (Iai et al. 1992) for the undrained analysis and the cocktail glass model (Iai et al. 2011) for the partially drained analysis. Both models are defined in the framework of the strain space multiple mechanism model. The undrained analysis was performed to capture the behavior of embankments primarily due to intense and strong shaking. The partially drained analysis using the cocktail glass model studied the effects of redistribution and dissipation of pore water on the behavior of a river embankment subjected to a long duration earthquake motion lasting over 2 min and the additional effects due to an aftershock.

A sufficiently high levee is needed on a river embankment to prevent secondary disasters following an earthquake such as a tsunami, flooding, or high tides. Consequently, the embankment subsidence following an earthquake must be precisely estimated. In addition, modeling of the deformation of the foot of the embankment is considered difficult using conventional analytical procedures. Examples of complex deformations include a combination of sliding along sloping ground and subsidence accompanying the dissipation of the excess pore water pressure, and shearing deformation of the embankment associated with lateral spreading of the foundation. This research focuses on the analysis of the levee crown settlement and the deformation mode of the embankment due to the main shock and an aftershock.

12.2 Damage to River Embankments and Stratigraphy of Foundation Ground

12.2.1 Site A (Tone River 26.5 km)

The river embankment at Site A suffered a 1.2-m settlement at the levee crown on the river side as a distinct differential settlement and a 4.32-m lateral displacement toward the river side at the river side foot of the embankment (Fig. 12.2). Near the crown at the differential settlement, a crack-like subsidence reaching a depth of 3 m from the crown was also observed.

In the vicinity of the damaged river embankment, the earthquake motions at the crown and the lower part of the levee were recorded at the Omigawa seismic observation station of the National Institute for Land and Infrastructure Management (NILIM), Ministry of Land, Infrastracture, Transport and Tourism (MLIT). These measurements were used as the input earthquake motions in the effective stress analyses.

The foundation ground consists of an approximately 2–3-m-thick alluvial sand layer (As and AsF layers) in the upper part (Fig. 12.3). This alluvial sand layer is susceptible to liquefaction. The geotechnical materials forming the embankment differ between the river side and the land side, as indicated by Bc and Bs. Distinct differential settlements occur at the boundary of different materials.

12.2.2 Site B (Yoshida River 16.0 km)

The river embankment at Site B suffered a 0.8-m settlement at the levee crown and a 2.6-m lateral displacement at the foot of the embankment (Fig. 12.4). A continuous crack occurred in the longitudinal direction at the small step on the land side.

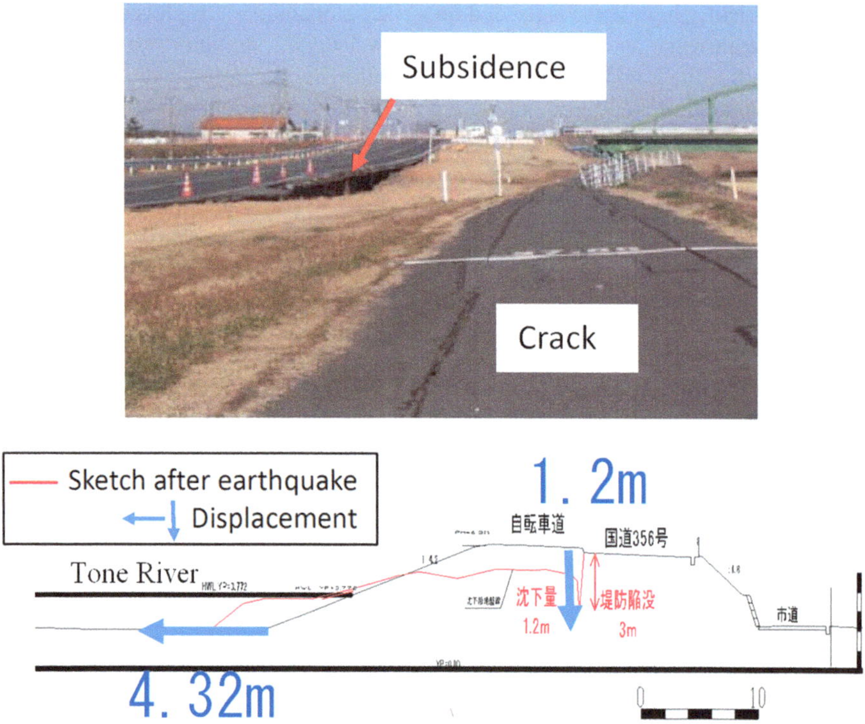

Fig. 12.2 Post-earthquake condition(Site A) (Japan Institute of Country-ology and Engineering (JICE) 2011)

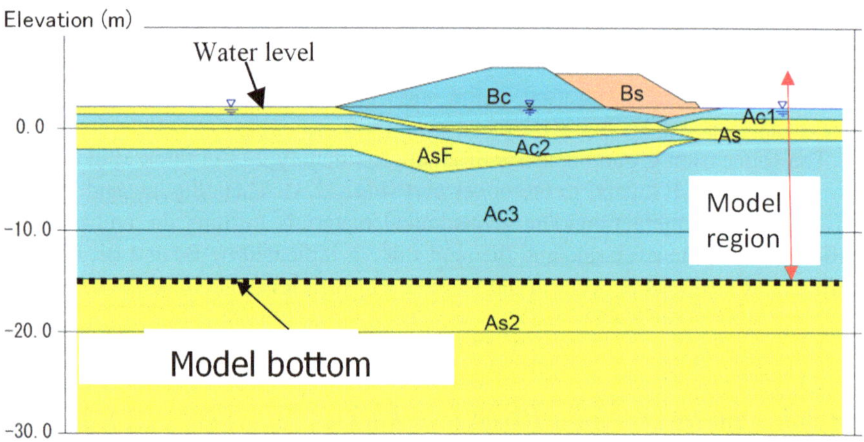

Fig. 12.3 Profile of the embankment and the foundation ground (Site A) (Japan Institute of Country-ology and Engineering (JICE) 2011)

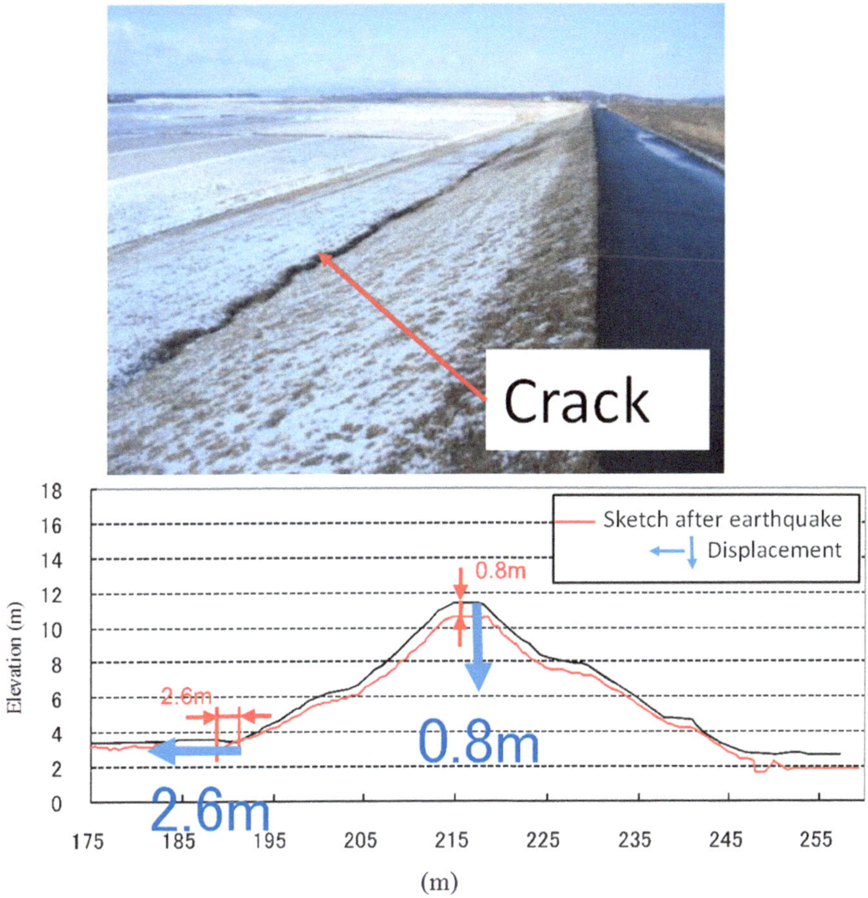

Fig. 12.4 Damage to the river embankment at Site B (Japan Institute of Country-ology and Engineering (JICE) 2011)

The earthquake motions at the levee crown and the Br layer (engineering bedrock) were recorded at the Yamazaki seismic observation station NILIM, MLIT, near the damaged river embankment. These recorded motions were used as the input earthquake motion for the effective stress analyses.

The foundation ground consists of a 1-m-thick Ac1 layer (upper section alluvial cohesive soil layer), which is underlain by a 3.5-m-thick As layer (alluvial sandy soil layer) (Fig. 12.5). The As layer is susceptible to liquefaction.

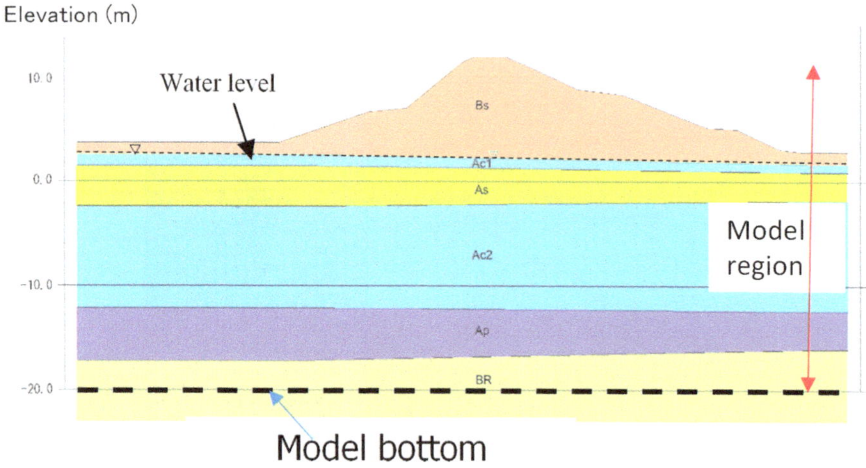

Fig. 12.5 Profile of the embankment and the foundation ground (Site B) (Japan Institute of Country-ology and Engineering (JICE) 2011)

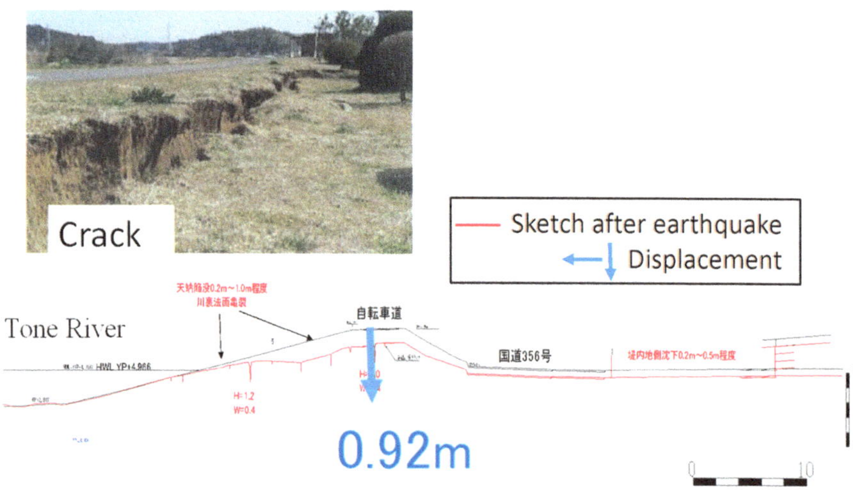

Fig. 12.6 Post-earthquake condition (Site C) (Japan Institute of Country-ology and Engineering (JICE) 2011)

12.2.3 Site C (Tone River 39.5 km)

The river embankment at Site C suffered a 0.92-m settlement at the levee crown and a 1.49-m lateral displacement at the foot of the embankment (Fig. 12.6). Differential settlements and cracks at the steps were confirmed in several places along the levee crown and the slope. The earthquake motions were recorded at the Sahara observation station on the strong-motion seismograph network K-NET (National

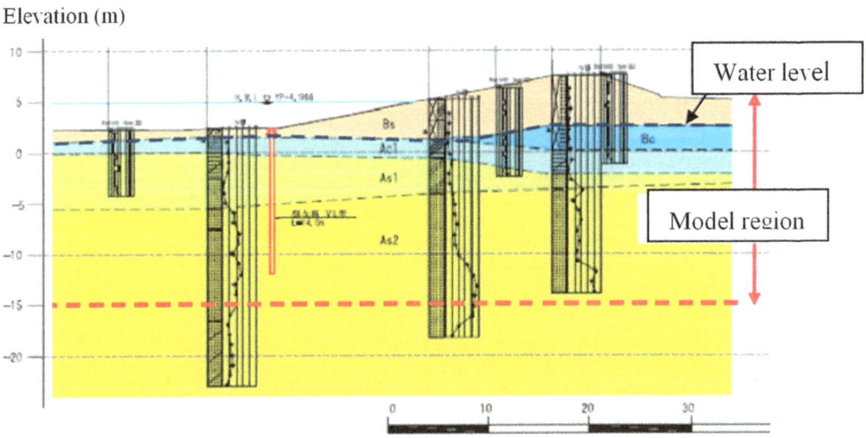

Fig. 12.7 Profile of the embankment and the foundation ground (Site C) (Japan Institute of Country-ology and Engineering (JICE) 2011)

Research Institute for Earth Science and Disaster Prevention (NIED)) near the damaged river embankment. These recorded earthquake motions were used as the input for the effective stress analyses.

The foundation ground consists of 1–5-m-thick Ac (alluvial cohesive soil layer) and Bc layers (embankment cohesive soil layer), which are underlain by 10-m thick As1 and As2 layers (alluvial sandy soil layer) (Fig. 12.7). The As1 and As2 layers are susceptible to liquefaction.

12.3 Finite Element Modeling of the River Embankments and the Model Parameters

12.3.1 Modeling of the River Embankment at Site A

Undrained effective stress analysis using the multi-spring model (Iai et al. 1992) was performed to capture the behavior of embankments primarily due to intense and strong shaking. Partially drained effective stress analysis using the cocktail glass model (Iai et al. 2011) was also performed to study the effects of redistribution and dissipation of pore water on the behavior of river embankment subjected to long duration (over 2 min) earthquake motion and the additional effects due to an aftershock.

The both constitutive models are defined within the framework of the strain space multiple mechanism model (Iai et al. 2011) and are consistent with each other in terms of the fundamental theoretical framework. However, these models differ in terms of modeling dilatancy. The dilatancy of the multi-spring model is defined in terms of the cumulative shear strain energy and the envelope called the

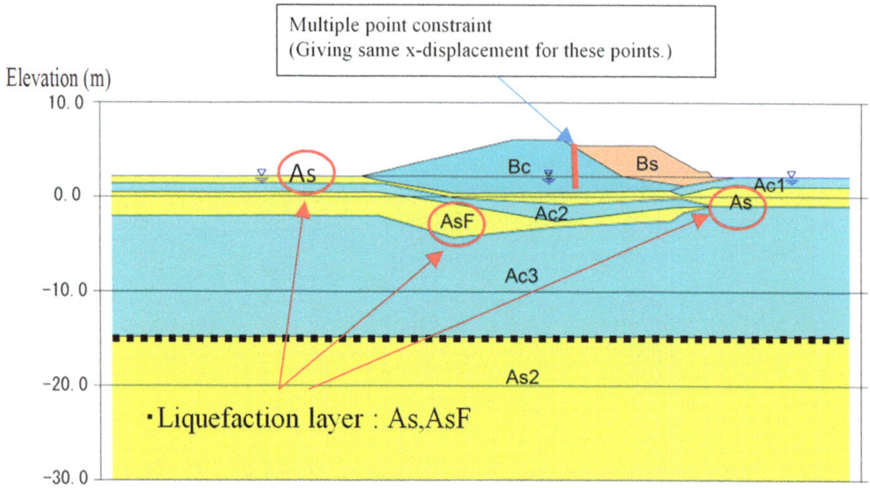

Fig. 12.8 Idealized cross-section of the river embankment for analysis (Site A)

liquefaction front, which is defined in the effective stress space. The dilatancy of the cocktail glass model is defined in terms of the contractive and the dilative components of dilatancy, where the contractive component is related to the multi directional virtual simple shear strain increment and the dilative component is defined as the energy-less strain component of the strain field in a granular material.

The cross-section of the river embankment at Site A was idealized for the analysis (Fig. 12.8). To idealize the discontinuity between the two different materials used to construct the embankment, an interface condition was imposed through a multiple point constraint condition (MPC) to allow frictionless vertical sliding at the boundary of Bs and Bc. This interface, which is defined in the embankment above the ground water table, should simulate the sliding along the opening of a crack reaching to a 3-m depth from the crown of the embankment.

The model parameters of the embankment materials and foundation soils were evaluated based on a simplified procedure (revised version) (Morita et al. 1997) from the SPT N-values and fines content (Table 12.1a, 12.1b, 12.1c). In particular, the materials, indicated by Bs, As, and AsF in Fig. 12.8, are susceptible to liquefaction. The liquefaction parameters were determined based on the cyclic resistance ratio at 20 load cycles, RL20 (Japan Institute of Country-ology and Engineering (JICE) 2011) (Fig. 12.9). The liquefaction parameters are shown in Table 12.1b.

The recorded earthquake motions at the Omigawa seismic observation station were converted into the transverse component to the embankment with PGA of about 0.13 g. The main shock (March 11, 2011, 14:47:00) and the aftershock that occurred approximately 29 min later (March 11, 2011, 15:15:53) were combined into a continuous time history with a duration of 3000 s as shown in Fig. 12.10, and used for the input at the bottom of the Ac3 layer for the analysis.

Table 12.1a Model parameters for analysis at Site A. Stiffness, strength, and coefficient of permeability

Soil layer	$\rho(t/m^3)$	Reference confining pressure $\sigma'ma$ (kN/m^2)	Elastic shear modulus at a confining pressure of (cr'ma) Gma (kN/m^2)	Poisson's ratio ν	Maximum damping coefficient Hmax	Cohesion C (kN/m^2)	Shear resistance angle $\varphi(°)$	Coefficient of permeability k(m/s)
Bs	1.90	57.75	46,204	0.33	0.24	0	35.1	1.0×10^{-5}
Be	1.80	63.00	23,800	0.33	0.24	0	30	1.0×10^{-8}
As	1.80	69.00	31,816	0.33	0.24	0	33.02	1.0×10^{-5}
AsF	1.80	84.75	31,816	0.33	0.24	0	32.77	1.0×10^{-5}
Ac 1	1.70	3.00	3,400	0.33	0.24	0	30	1.0×10^{-8}
Ac 2	1.70	75.75	3,400	0.33	0.24	0	30	1.0×10^{-8}
Ac 3	1.70	119.25	6,800	0.33	0.24	0	30	1.0×10^{-8}
As 2	2.00	–	–	–	–	–	–	–

Table 12.1b Model parameters for analysis at Site A. Parameters for the dilatancy of the multi-spring model

Soil layer		Bs	As	AsF
Phase transformation angle	ϕ_p(°)	28.0	28.0	28.0
Parameters for dilatancy	S_1	0.005	0.005	0.005
	w_1	6.50	3.06	3.16
	p_1	0.5	0.5	0.5
	p_2	0.486	0.419	0.407
	c_1	2.5	1.9	1.9
	S_{us} (kPa)	17.8	16.9	16.8

Table 12.1c Model parameters for analysis at Site A. Parameters for the dilatancy of the cocktail glass model

Soil layer		Bs	As	AsF
Phase transformation angle	ϕp (°)	28	28	28
Parameters for dilatancy	ε_d^{cm}	0.1	0.2	0.2
	$r\varepsilon_{dc}$	3.5	4	5
	$r\varepsilon_d$	0.15	0.2	0.2
	q_1	5.5	7	7
	q_2	1	1.5	1.5
	q_4	1	1	1
	q_{us}	–	–	–
	r_γ	0.2	0.2	0.2
	rrm_{tmp}	0.5	0.5	0.5
	S_1	0.005	0.005	0.005
Parameters for volume compressibility	I_k	2	2	2
	r_k	0.5	0.7	0.5
Parameter for limit of liquefaction	c1	2.6	1.98	1.98

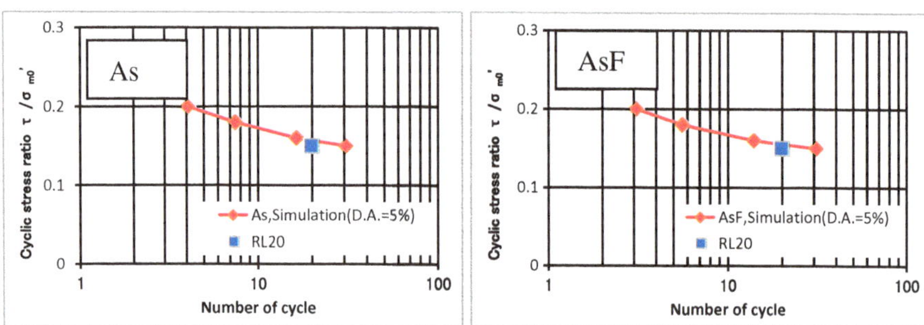

Fig. 12.9 Liquefaction resistance of the As and AsF layers (Site A) (*blue square*: measured, *red diamond*: computed by the cocktail glass model)

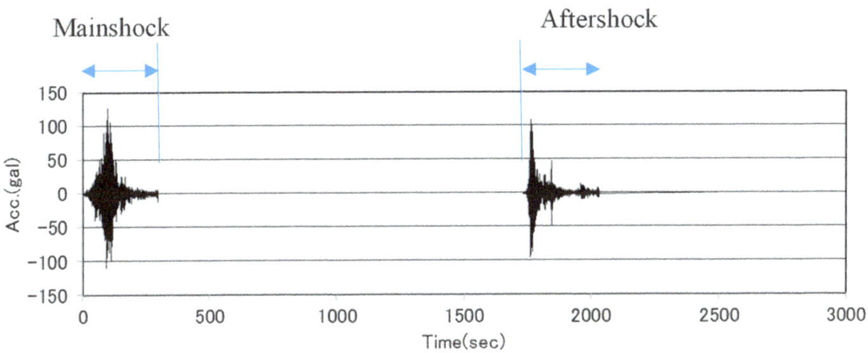

Fig. 12.10 Input seismic motions (Site A)

For the partially drained analysis, the motion with a duration of 3,000 s was used in the 10,000-s computation, allowing for full dissipation of the excess pore water pressure. For the undrained analysis, the aftershock motion was padded immediately after the main shock motion to form a combined earthquake motion with a 700-s duration.

12.3.2 *Modeling of the River Embankment at Site B*

Both the undrained effective stress analysis using the multi-spring model (Iai et al. 1992) and the partially drained effective stress analysis using the cocktail glass model (Iai et al. 2011) were performed to study the applicability of the models for the behavior of the river embankment. The cross-section of the river embankment at Site B was idealized in the analysis (Fig. 12.11). As mentioned earlier, the foundation ground at Site B consists of a 1-m thick Ac1 layer (upper section alluvial cohesive soil layer) underlain by a 3.5-m thick As layer (alluvial sandy soil layer) (Fig. 12.5). The As layer consists of loose sand with a SPT N-value of about 6, a fines content of 26 %, and is susceptible to liquefaction. These surface layers are underlain by a 10-m-thick Ac2 layer (lower section alluvial cohesive soil layer) and a 5-m-thick Ap layer (humus layer). The engineering bedrock at Site B is identified as a Br layer with a SPT N-value of 50 below the Ap layer. In the effective stress analysis, the ground water table was set to the elevation of the top of the Ac1 layer.

The model parameters of embankment material Bs and the foundation soils were evaluated based on a simplified procedure (revised version) (Morita et al. 1997) from the SPT N-values and the fines content (Table 12.2a). As previously mentioned, the materials of the As layer are susceptible to liquefaction, and the liquefaction parameters were determined based on the cyclic resistance ratio at 20 load cycles, RL20 (Japan Institute of Country-ology and Engineering (JICE) 2011) (Fig. 12.12). The liquefaction parameters are shown in Table 12.2b, 12.2c.

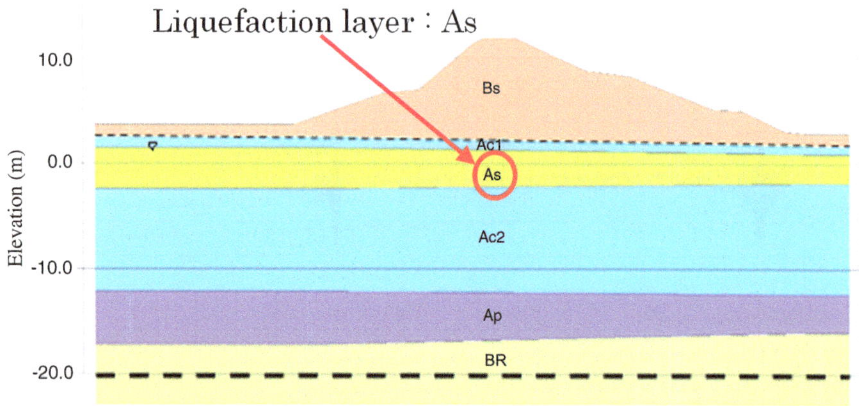

Fig. 12.11 Idealized cross-section of the river embankment for analysis (Site B)

Figure 12.13 shows the earthquake motions recorded at the Yamazaki seismic observation station near the river embankment at Site B, which registered a PGA of 0.6 g during the main shock. However, the earthquake motion of the aftershocks approximately 100 min after the main shock had a PGA of 0.04 g. Thus, the effect of the aftershock to the river embankment at Site B is considered small. Only the earthquake motion during the main shock was used for the analysis by converting the recorded components of the motions into the transverse component to the river embankment.

12.3.3 Modeling of the River Embankment at Site C

The effective stress analysis of the river embankment at Site C was performed through the partially drained effective stress analysis using the cocktail glass model (Iai et al. 2011) to study the applicability of the model on the behavior of the river embankment subjected to an earthquake motion of the main shock with a duration of about 2 min immediately followed by the earthquake motion from an aftershock.

The cross-section of the river embankment at Site C was idealized for the analysis (Fig. 12.14). As mentioned earlier, the foundation ground consists of 1–5-m-thick Ac (alluvial cohesive soil layer) and Bc layers (embankment cohesive soil layer) underlain by 10-m-thick As1 and As2 layers (alluvial sandy soil layer). The As1 and As2 layers are susceptible to liquefaction. Below a depth of 38 m, the Ac2 (alluvial cohesive soil layer) layer shows a SPT N-value of 50, and this elevation of layer is considered as the engineering bedrock.

To simulate cracks and local sliding differential settlements on the embankment slope, frictionless joint elements were used (Fig. 12.15). The joint elements were inserted at regular intervals to study which joint element is most mobilized,

Table 12.2a Model parameters for analysis at Site C. Stiffness, strength, and coefficient of permeability

Soil layer	ρ(t/m^3)	Reference confining pressure σ′ma (kN/m^2)	Elastic shear modulus at a confining pressure of (cr’ma) Gma (kN/m^2)	Poisson’s ratio ν	Maximum damping coefficient Hmax	Cohesion C (kN/m^2)	Shear resistance angle φ(°)	Coefficient of permeability k(m/s)
Bs	1.84	98	55,043	0.33	0.24	0	38.46	1.0×10^{-5}
Ac 1	1.63	98	31,367	0.33	0.2	0	30	1.0×10^{-8}
As	1.84	98	65,003	0.33	0.24	0	39	1.0×10^{-5}
Ac 2	1.63	98	18,785	0.33	0.2	0	30	1.0×10^{-8}
Ap	1.63	98	32,452	0.33	0.2	0	30	1.0×10^{-8}
Br	2.04	–	–	–	–	–	–	–

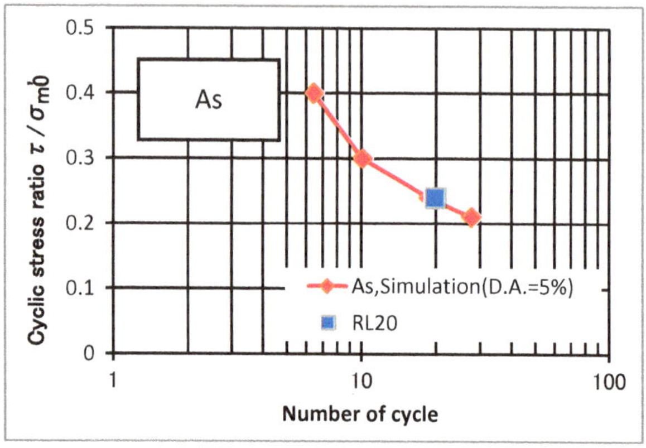

Fig. 12.12 Measured and computed liquefaction resistance of the As layer (Site B) (*blue square*: measured, *red diamond*: computed by the cocktail glass model)

Table 12.2b Model parameters for analysis at Site B. Parameters for the dilatancy of the multi-spring model

Soil layer		As
Phase transformation angle	ϕ_p(°)	28.0
Parameters for dilatancy	S_1	0.005
	w_1	8.7
	p_1	0.5
	p_2	0.93
	c_1	1.6
	S_{us}(kPa)	5.3

Table 12.2c Model parameters for analysis at Site B. Parameters for the dilatancy of the cocktail glass model

Soil layer		As
Phase transformation angle	φp (°)	28
Parameters for dilatancy	ε_d^{cm}	0.23
	$r\varepsilon_{dc}$	0.65
	$r\varepsilon_d$	0.2
	q_1	0.5
	q_2	1.38
	q_4	1
	q_{us}	5.3
	r_γ	0.2
	rrm_{tmp}	0.5
	S_1	0.005
Parameters for volume compressibility	I_k	2
	r_k	0.5
Parameter for limit of liquefaction	c1	0.95

allowing cracks as a form of separation and differential settlement as a form of sliding.

The model parameters of the embankment material Bs and the foundation soils were evaluated based on a simplified procedure (revised version) (Morita et al.

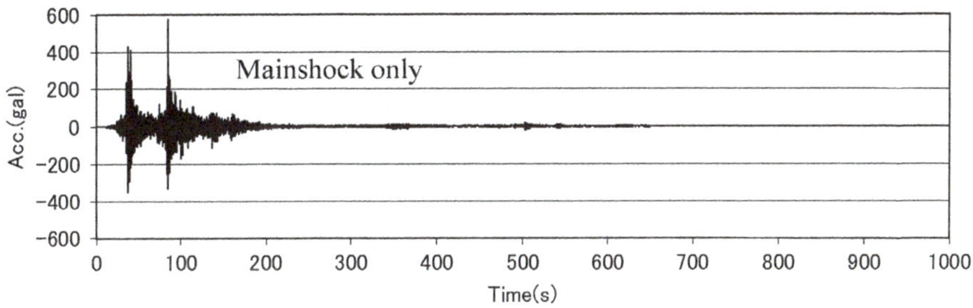

Fig. 12.13 Input seismic motions (Site B)

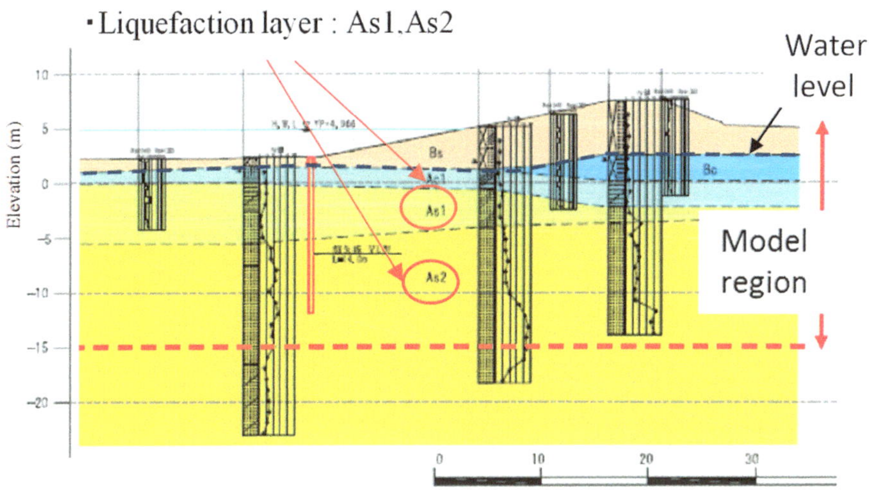

Fig. 12.14 Idealized cross-section of the river embankment for analysis at Site C

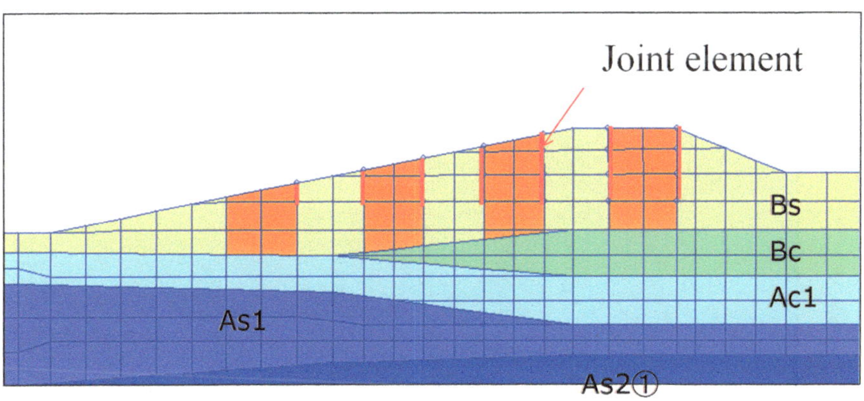

Fig. 12.15 Location of the joint elements, allowing the opening of a crack and local sliding differential settlement

Table 12.3a Model parameters for analysis at Site C. Stiffness, strength, and coefficient of permeability

Soil layer	ρ (t/m³)	Reference confining pressure σ′ma (kN/m²)	Elastic shear modulus at a confining pressure of (cr'ma) Gma (kN/m²)	Poisson's ratio ν	Maximum damping coefficient Hmax	Cohesion C (kN/m²)	Shear resistance angle φ(°)	Coefficient of permeability k (m/s)
Bs	1.73	98	47,704	0.33	0.24	0	38.13	
Bc	1.84	98	44,059	0.33	0.2	0	30	1.0×10^{-8}
Ac1	1.73	98	46,980	0.33	0.2	0	30	1.0×10^{-8}
As1	1.73	98	43,194	0.33	0.24	0	37.92	1.0×10^{-4}
As2	1.94	98	76,434	0.33	0.24	0	39.35	1.0×10^{-4}
Ac2	1.73	–	–	–	–	–	–	–

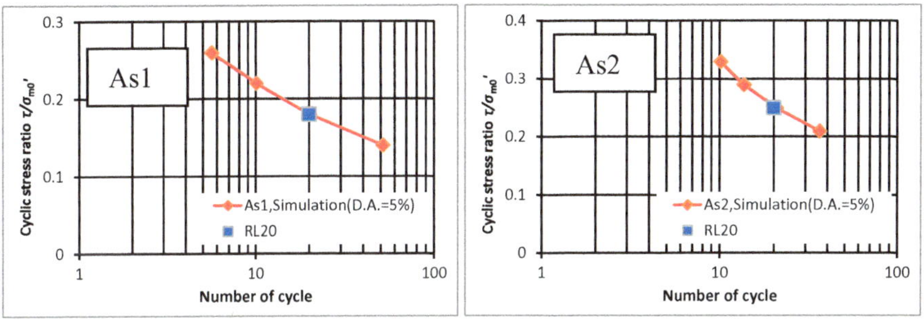

Fig. 12.16 Measured and computed liquefaction resistance of the As1 and As2 layers at Site C (*left*: As1, *right*: As2) (*blue square*: measured, *red diamond*: computed by the cocktail glass model)

Table 12.3b Model parameters for analysis at Site C. Parameters for the dilatancy of the cocktail glass model

Soil layer		As1	As2
Phase Transformation angle	φp (°)	28	28
Parameters for dilatancy	ε_d^{cm}	0.15	0.15
	$r\varepsilon_{dc}$	1	1
	$r\varepsilon_d$	0.2	0.2
	q_1	1	1
	q_2	0.2	1.85
	q_4	1	1
	q_{us}	–	–
	r_γ	0.2	0.2
	rm_{tmp}	0.5	0.5
	S_1	0.005	0.005
Parameters for volume compressibility	l_k	2	2
	r_k	0.5	0.5
Parameter for limit of liquefaction	c1	1.4	1.9

1997) from the SPT N-values and the fines content (Table 12.3a). In particular, the materials of the As1 and As2 layers are susceptible to liquefaction, and the liquefaction parameters were determined based on the cyclic resistance ratio at 20 load cycles, RL20 (Japan Institute of Country-ology and Engineering (JICE) 2011) (Fig. 12.16). The liquefaction parameters are shown in Table 12.3b.

The earthquake motions recorded at the Sahara observation station of the K-NET (National Research Institute for Earth Science and Disaster Prevention (NIED)) near the river embankment at Site C, registered a PGA of 0.25 g during the main shock followed by the earthquake motion of an aftershock with PGA of 0.18 g. The input earthquake motion used for the analysis was obtained by converting the recorded components of the earthquake motion into the transverse component to

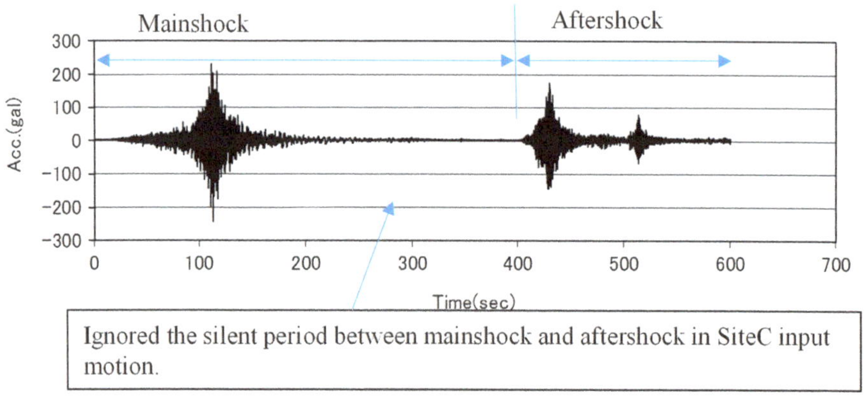

Fig. 12.17 Input seismic motion (Site C)

the river embankment (Fig. 12.17). For simplicity, the silent period between the main shock and aftershock was ignored in this analysis.

12.4 Analysis Results

12.4.1 Embankment at Site A (Morishima et al. 2013)

12.4.1.1 Results of the Undrained Analysis

To simplify the analysis, the effective stress analysis of the river embankment at Site A was performed by the undrained condition initially. The objective in the initial stage was to study the ability of the effective stress analysis to capture distinct failure modes of the river embankment with differential settlement at the crown.

As described earlier, the analysis allowed sliding at the opening of a crack to reach to a depth of 3 m from the crown by introducing an interface condition imposed between the two different materials used to construct the embankment. Using the model parameters shown in Table 12.1a, 12.1b, 12.1c, except for the shear resistance at the steady state Sus, which was altered from that shown in Table 12.1b to 1/2 the value, the earthquake motions of the main shock and the aftershock shown in Fig. 12.10 were inputted by ignoring the silent period between the main shock and the aftershock.

Figure 12.18 shows the computed results. Soil liquefaction at the As and AsF layers induces the deformation of the river embankment at Site A, resulting in a differential settlement at the crest between the Bc and Bs materials and the lateral movement toward the river side (left side). The computed residual deformation is consistent with that was measured after the earthquake (Fig. 12.19). This result

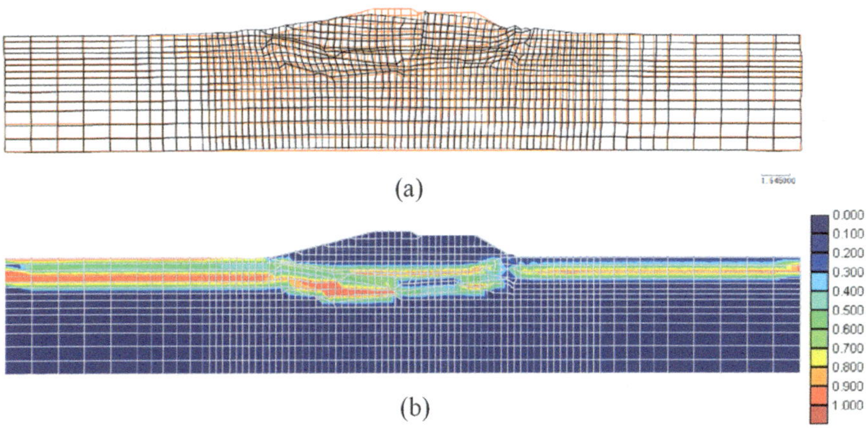

Fig. 12.18 Computed residual deformation and the excess pore water pressure ratio. (**a**) Residual deformation. (**b**) Excess pore water pressure ratio

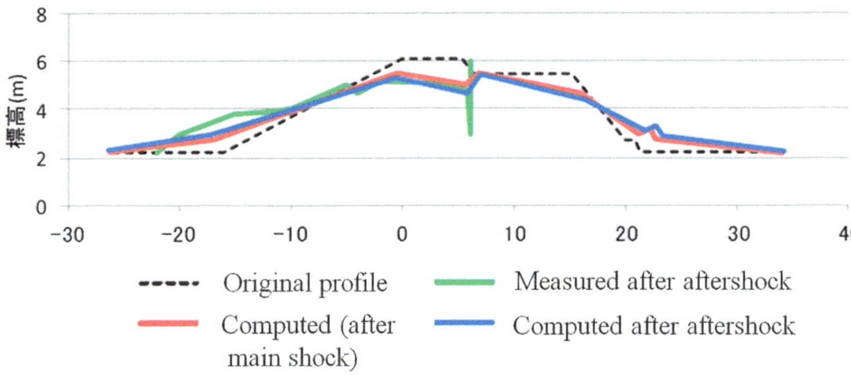

Fig. 12.19 Measured and computed residual deformations of the river embankment

indicates that the effective stress analysis in the undrained condition has the potential to capture distinct failure modes and the degree of deformation induced at the river embankment at Site A.

As a parameter study to evaluate the effect of sliding at the opening of a crack, a second set was analyzed when sliding is not allowed (Fig. 12.8, red vertical line). Only the earthquake response during the main shock was considered. A differential settlement is found around the crest (Fig. 12.20). Although a certain level of settlement at the crown is induced, the differential settlement between the two different materials Bc and Bs is unclear. The distinct failure mode of the river embankment at Site A was not computed.

To look into the details of the deformation mechanism, the shear strain distribution of this analysis case was also computed (Fig. 12.21). The computed localized shear strain zone, shown by the red and yellow counters, indicates that a

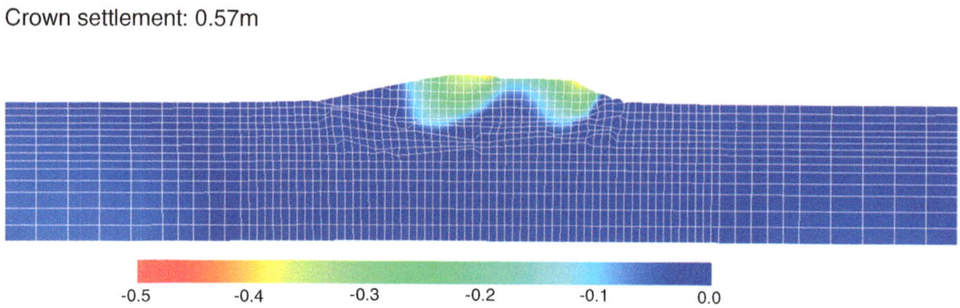

Fig. 12.20 Computed vertical displacement distribution after the main shock without sliding along the interface in the embankment

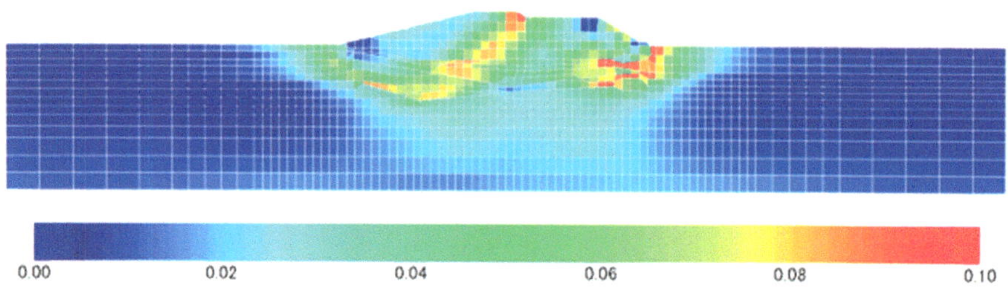

Fig. 12.21 Shear strain distribution after the main shock without sliding along the interface in the embankment

deformation toward the river side (left in figure) along a circular sliding plane going through the Bc embankment portion combined with a local failure at the foot of the other side of the embankment going through Bs. Although there is a tendency to settle and move toward the river side, the computed settlement at the crest does not fully capture the settlement observed during the earthquake. Consequently, the analysis has a limited ability to reproduce the failure and deformation sequence of the embankment. In particular, the initiation of the crack opening is not fully simulated despite the tendency of the lateral extension deformation of the embankment due to liquefaction of the foundation soil. Additional studies are necessary to resolve this problem.

12.4.1.2 Results of the Partially Drained Analysis

Similar to the undrained analysis described in the previous section, the partially undrained effective stress analysis was performed on the river embankment at Site A. The objective was to study the effects of redistribution and dissipation of pore water on the behavior of a river embankment subjected to a long duration (more than 2 min) earthquake motion and the additional effects due to an aftershock.

The analysis of a reference case (A-1) was performed using the coefficient of permeability shown in Table 12.1a, 12.1b, 12.1c. As a parameter study, the

Table 12.4 Coefficients of permeability used in the parameter study

	Coefficient of permeability k (cm/s)						
Case	Bc	Bs	As1	Ac1	AsF	Ac2	
A-1	1×10^{-6}	1×10^{-3}	1×10^{-3}	1×10^{-6}	1×10^{-3}	1×10^{-6}	Original
A-3			1×10^{-2}		1×10^{-2}		High permeability ($\times 10$) at liquefied layer
B-1			1×10^{-3}		1×10^{-3}	2×10^{-3}	High permeability at Ac2 (drainage for left-side ground surface)

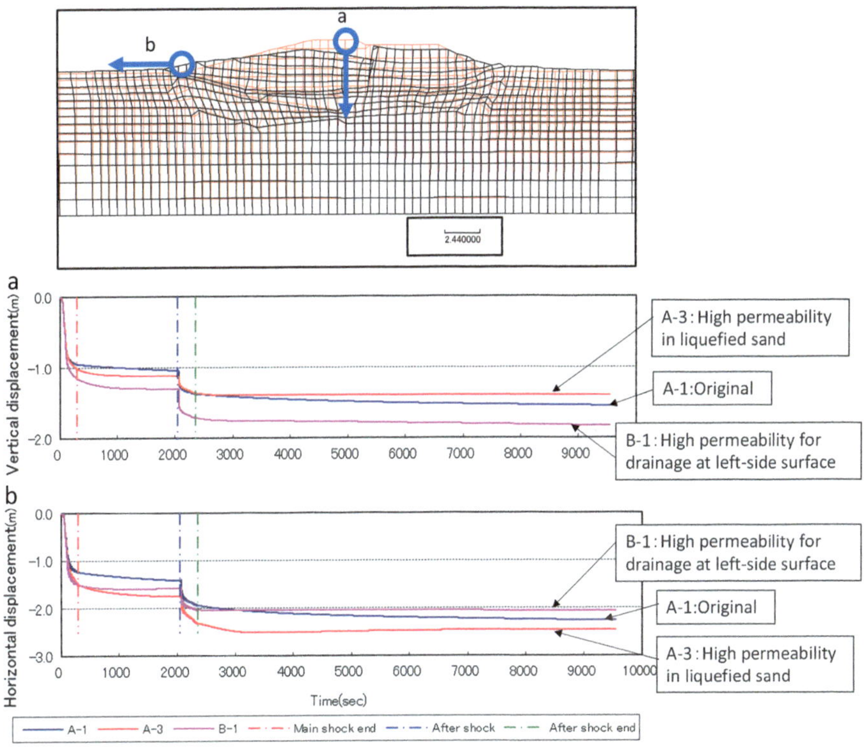

Fig. 12.22 Computed time histories of the vertical displacement at the crown (*point a*) and the lateral displacement at the foot of the river embankment (*point b*) at Site A

coefficients of permeability were altered from those of the reference case (Table 12.4). Case A-3 was performed using a higher permeability at the liquefiable layers As and AsF. Case B-1 used the higher permeability at the clay capping layer Ac2 at the ground surface on the river side (left side) of the embankment.

Figure 12.22 shows the computed time histories of the crown settlement (point a) and the lateral displacement at the foot of the embankment (point b) through the

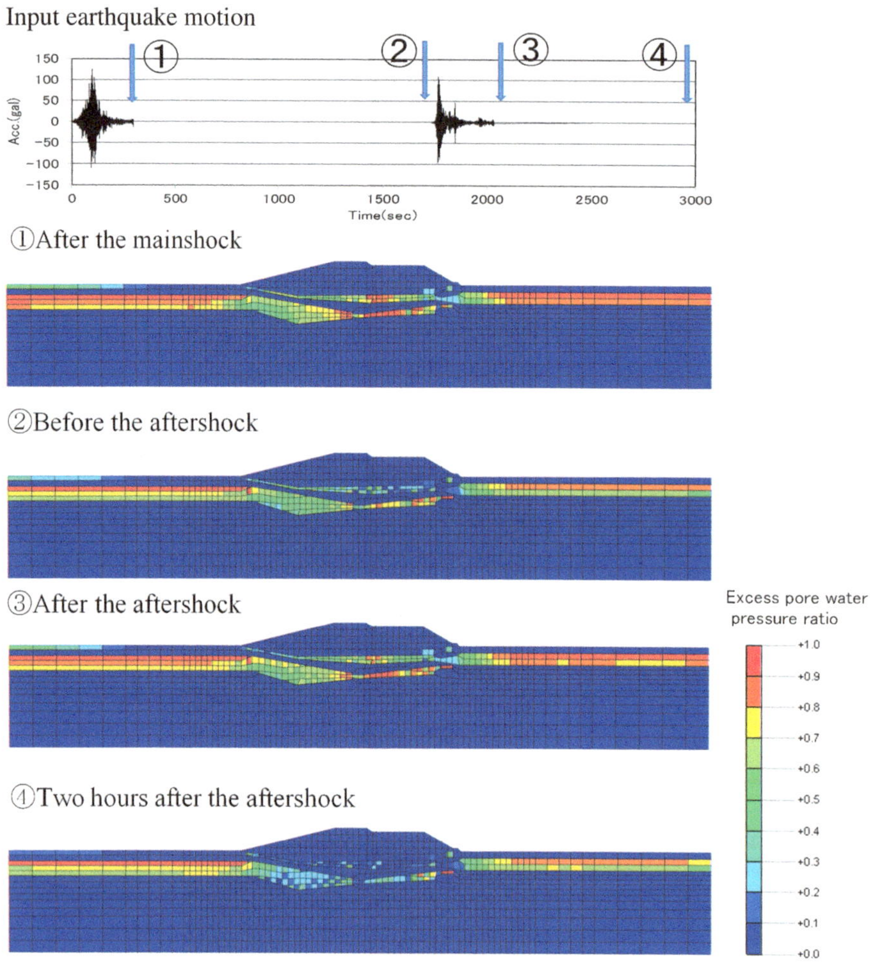

Fig. 12.23 Excess pore pressure ratio distribution (case A-3)

parameter study. All the cases show that (1) during the main shock, a large displacement on the order of 1 m is induced in both the crown settlement and the lateral displacement at the foot of the embankment, (2) after the main shock, moderate increases in the displacements are slowly induced, which are likely due to the redistribution or dissipation of the excess pore water pressure (to be confirmed later), (3) during the aftershock, further moderate displacement increases are induced, and (4) after the aftershock, moderate increases in the displacements are again induced slowly probably due to the redistribution or dissipation of excess pore water pressure (to be confirmed later).

In the parameter study, case A-3 with a higher permeability specified for the liquefiable layers As and AsF results in the smallest settlement at the crown and the

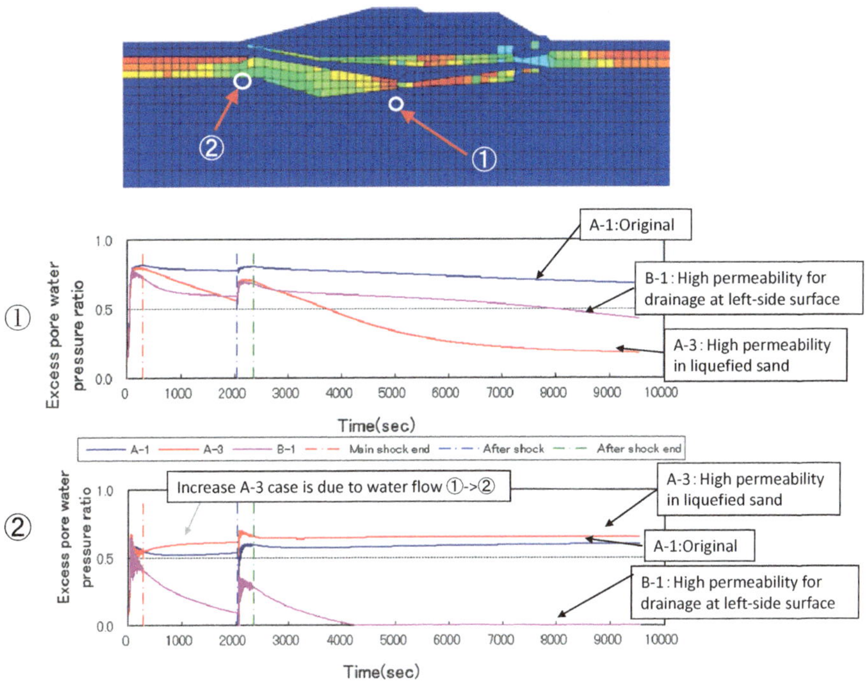

Fig. 12.24 Excess pore water pressure ratio time history (*Top*: at the foot of the embankment, *bottom*: below the levee center)

largest lateral displacement at the foot of the embankment, which are consistent with the degree of deformation induced at the site. The computed crown settlement for case A-3 is 1.39 m and the computed lateral displacement at the foot of the embankment is 2.45 m.

Figures 12.23 and 12.24 show the computed excess pore water pressure ratio distributions and the time histories at the bottom and the foot of the embankment, respectively. Liquefaction occurs in the As and AsF layers at the end of the main shock, inducing most of the deformation in the embankment (Fig. 12.23). Although some dissipation is allowed after the main shock, the excess pore water pressure does not fully dissipate before the aftershock. The excess pore water pressure increases again during the aftershock, but then should fully dissipate. The analysis over 10,000 s (about 3 h) is sufficient to allow full dissipation of the excess pore water pressure (Fig. 12.24).

The horizontal displacement at the foot of the embankment changes according to the dependence of the dissipation properties of the pore water pressure. In particular, the computed results in case A-1 relative to the case B-1 indicate that a less permeable layer such as the Ac layer, which exists above the liquefiable layer, slows the dissipation of the excess pore water pressure and increases the lateral

Table 12.5 Parameter study series performed for the embankment at Site B

	Shear resistance angle φ (°)		Cohesion C(kN/m^2)	
Case	Ac1	Ac2	Acl	Ac2
Basic case	30.0	30.0	0.0	0.0
Case. 1–1	15.0	30.0	0.0	0.0
Case. 1–2	20.0	30.0	0.0	0.0
Case. 1–3	25.0	30.0	0.0	0.0
Case. 2–1	15.0	15.0	0.0	0.0
Case. 2–2	20.0	20.0	0.0	0.0
Case. 2–3	25.0	25.0	0.0	0.0
Case. 3–1	0.0	30.0	5.0	0.0
Case. 3–2	0.0	30.0	10.0	0.0
Case. 3–3	0.0	30.0	20.0	0.0
Case. 3–4	0.0	30.0	50.0	0.0
Case. 4–1	0.0	0.0	5.0	5.0
Case. 4–2	0.0	0.0	10.0	10.0
Case. 4–3	0.0	0.0	20.0	20.0
Case. 4–4	0.0	0.0	50.0	50.0

(☆): qus = 0.0【Cocktail glass model】
(★): qus = 0.0【Multi sping model】

displacement at the foot of the embankment. The computed results in the A-3 case relative to the A-1 case indicate that more rapid pumping (redistribution) of the excess pore water pressure due to a higher permeability at the liquefiable layer induces a larger deformation of the embankment.

12.4.2 Embankment at Site B (Imai et al. 2013)

Both the undrained effective stress analysis using the multi-spring model (Iai et al. 1992) and the partially drained effective stress analysis using the cocktail glass model (Iai et al. 2011) were performed to study the models' abilities to reproduce the behavior of the river embankment at Site B.

The cross-section of the river embankment at Site B consists of interlayers of clay and sand; the sand layer As is between the Ac1 and Ac2 layers below the embankment (Fig. 12.5). Because these interlayers of clay and sand are considered the primary factors affecting the deformation of the embankment constructed above them, a series of parameter studies was performed by varying the shear strength parameters of the Ac1 and Ac2 layers (Table 12.5). Cases 1 and 2 are specified by varying the internal friction angle of the clay layers in terms of the effective stress analysis of clay. Cases 3 and 4 are specified by varying the cohesion of the clay in terms of the total stress analysis of clay.

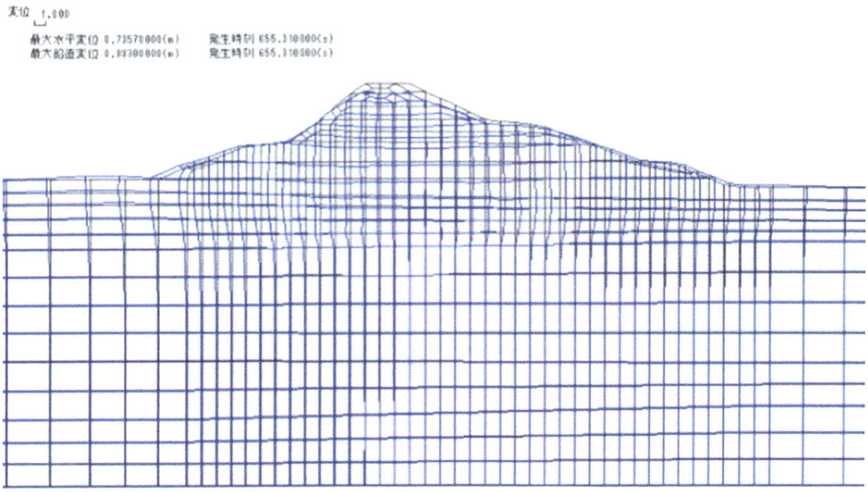

Fig. 12.25 Computed deformation of the embankment at Site B

Figure 12.25 shows a typical computed deformation of the embankment at Site B for the undrained effective stress analysis. In this analysis case, the shear resistance at the steady state Sus was ignored by setting the default value to 0 in the computer program of FLIP. The computed deformation mode exhibits settlement at the crest and lateral spreading at the foot of the embankment, which is generally consistent with the observations at the site. In particular, the computed and measured settlements were 0.87 m and 0.8 m, respectively. However, the computed lateral displacement is 0.67 m, which is much smaller than the measured displacement of 2.6 m.

The results of the parameter study are summarized in Fig. 12.26 in terms of the settlement at the crown (horizontal axis) and the lateral displacement at the foot of the embankment (vertical axis). The measured values of the settlement (0.8 m) and the lateral displacement (2.6 m), which are located just below the index, are plotted by open circles.

For the settlement at the crown, the computed values range from 0.8 to 2.4 times of the measured value. When the internal friction angle is reduced for the Ac1 and Ac2 layers (Cases 1 and 2), the computed settlement increases. However, when the cohesion is increased (Cases 3 and 4), the computed settlement increases.

As for the lateral displacement at the foot of the embankment (the river side), the computed values range from 0.1 to 0.5 times of the measured value. As the internal friction angle of the Ac1 and Ac2 layers (Cases 1 and 2) decreases, the computed lateral displacement increases becoming closer to the measured value. However, as the cohesion increases (Cases 3 and 4), the computed lateral displacement becomes larger.

The computed lateral displacement is consistently smaller than the measured values, which may be caused by a possible flow slide in the Ac1 layer due to the

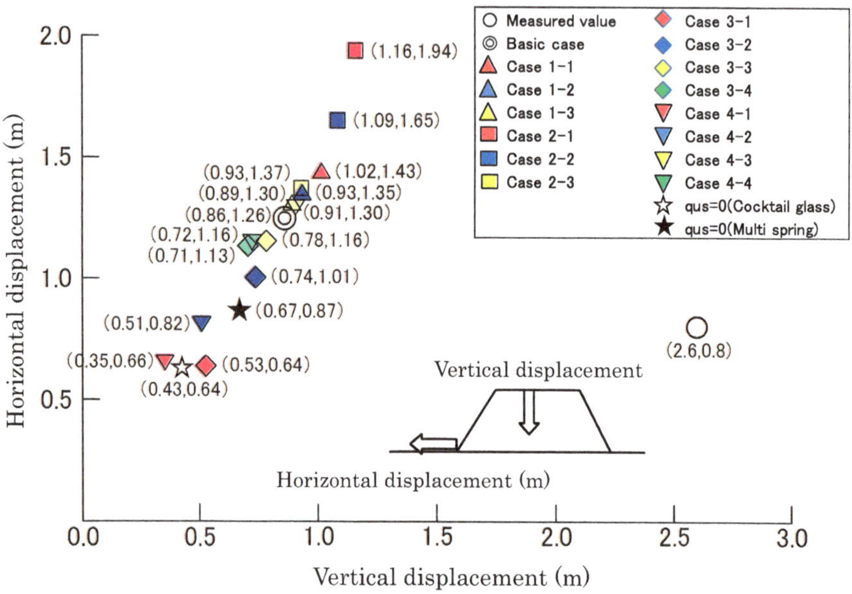

Fig. 12.26 Horizontal and vertical displacements for each case

pore water pressure redistribution in the As layer just below the Ac1 layer. Although the partial drainage analysis included this aspect, the effective stress analysis performed in this study does not completely simulate these phenomena.

In general, weaker foundation layers should induce a larger deformation in the embankment constructed above the foundation layer, which is consistent with the computed results by varying the internal friction angle of the foundation clay layers (Cases 1 and 2). However, the opposite trends are found by varying the cohesion of foundation clay layers (Cases 3 and 4). The change in the seismic response of the foundation embankment system due to the foundation layer strengths may explain this apparent contradiction. However, further studies are necessary to fully comprehend the mechanism behind the results of the parameter study.

12.4.3 Embankment at Site C

Partially drained effective stress analysis using the cocktail glass model (Iai et al. 2011) was performed to study the model's ability to reproduce the behavior of the river embankment at Site C. Analysis was performed with the input seismic motion lasting 600 s, including both the main shock and the aftershock followed by dissipation analysis over 59,400 s. The total analysis time was 60,000 s (16.7 h).

Figure 12.27 shows the residual deformation of the river embankment. The computed deformation mode shows settlement at the crown and the lateral

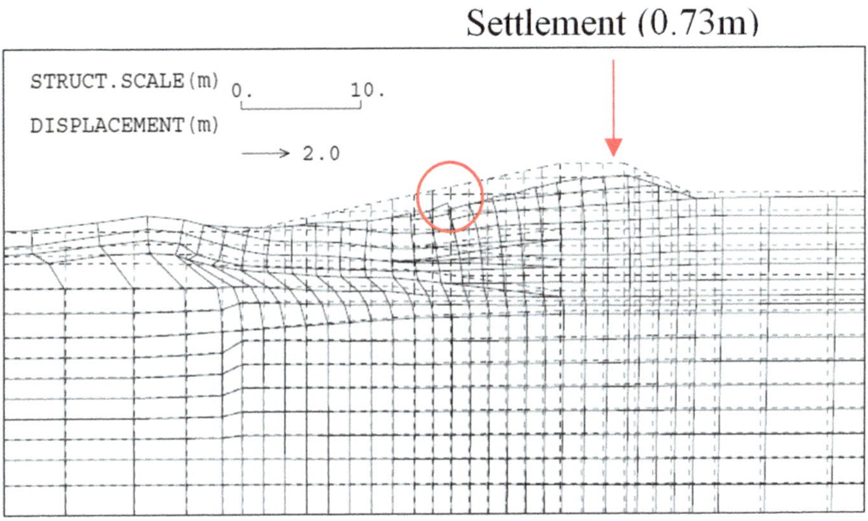

Fig. 12.27 Computed residual deformation of the river embankment at Site C

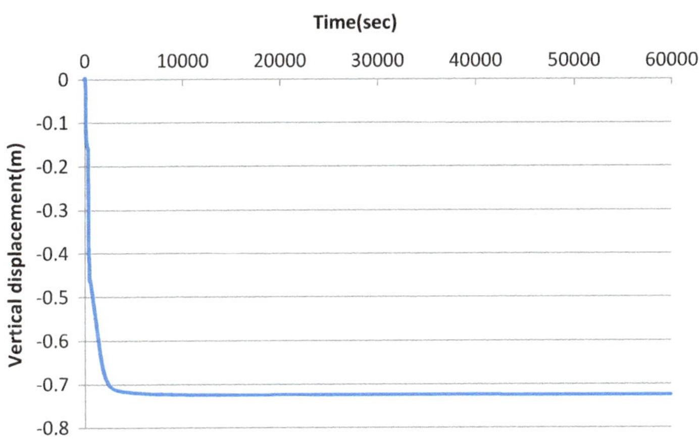

Fig. 12.28 Computed time history of the vertical displacement at the crown

displacement toward to the river side with some of the differential settlements induced along the slope (red circles) and is generally consistent with the observed deformation of the river embankment at Site C. The computed settlement at the crown is 0.73 m, which is close to or slightly smaller than the measured settlement of 0.92 m.

The computed deformation mode differs from that observed at the site as demonstrated by at the level ground portion of the river side ground surface. The computed deformation shows a heave, which is not present in the observed

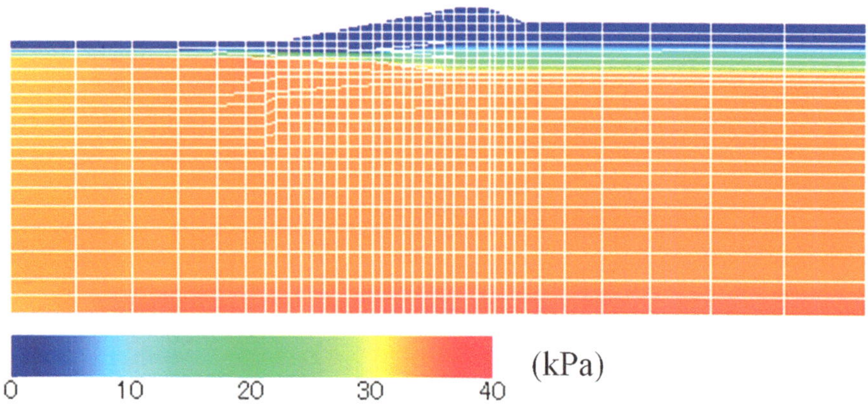

Fig. 12.29 Excess pore water pressure distribution at 60,000 s

deformation. The computed deformation might slightly exaggerate the local heave due to lateral spreading.

Figure 12.28 shows the computed time history of the crown settlement, and indicates that the primary portion of the settlement is induced during the main shock and the aftershock shaking for a duration of 600 s. Afterwards, a slight increase in the settlement associated with the dissipation of excess pore water pressure is recognized.

The excess pore water pressure remains 60,000 s after the earthquake (Fig. 12.29). A clay layer Ac1 cap over the liquefiable sand layer As1 may prevent a faster dissipation of the excess pore water pressure. If the computation is extended to allow complete dissipation of the pore water pressure, then the crown settlement may be slightly larger than the currently computed settlement and closer to the observed one.

12.5 Conclusions

In this study, damage to river embankments during the earthquake off the Pacific coast of Tohoku in 2011 was studied through effective stress analysis under undrained and partially drained conditions based on the strain space multiple mechanism models incorporated in the computer code FLIP. Three damaged river embankments, which were 4–8 m high, constructed on liquefiable foundation soil, and subjected to bedrock peak ground motions ranging from 0.15 to 0.6 g with and without the effect of an aftershock, were selected for the analysis. The analysis results are mostly consistent with the damage to the river embankments both in terms of the settlements and the deformation modes. In particular, the partial drainage analysis using the cocktail glass model is applicable to the behavior of a river embankment subjected to a seismic motion with a long duration, including the additional effect of aftershock.

Specifically, the following are obtained:

1. If cracks form in the embankment due to the overall lateral deformation associated with liquefaction of the foundation soil, differential settlement should occur due to sliding along the cracks. Realistic modeling of these effects of cracks through the use of multiple point constraint or joint elements adequately simulates these types of differential settlements.
2. Redistribution and dissipation of excess pore water pressures in the foundation soil significantly affect the degree of deformation of a river embankment. If the foundation ground is covered with an impermeable layer, then excess pore water pressures from the liquefied layer underneath tend to accumulate below the impermeable layer, increasing the deformation, especially at the foot of the embankment. Thus, appropriately evaluating the permeability and modeling of stratigraphy are crucial to evaluate the degree of damage to a river embankment.
3. Certain issues are for future studies. For example, the initial process of crack formation in an embankment is not fully simulated by the finite element modeling performed in this study. Although some of the model parameters such as the shear strength at the steady state of sand and the shear strength of clayey materials affect the computed degree of deformation of an embankment, the appropriate procedure to evaluate these parameters has yet to be fully developed.

Acknowledgements This investigation was implemented as a part of the activities of the general incorporated association FLIP Consortium WG, which was established to promote research into the efficient utilization and improvement of the FLIP program. We would like to express our gratitude to all those involved, the Ministry of Land, Infrastructure, Transport and Tourism (MLIT), the National Institute for Land and Infrastructure Management (NILIM), and the Risk Management Technology Research Center Earthquake Prevention Laboratory for providing very valuable seismic recorded numerical data from the Omigawa and the Yamazaki seismic observation stations.

References

Iai S, Matsunaga Y, Kameoka T (1992) Strain space plasticity model for cyclic mobility. Soils Found 32(2):1–15

Iai S, Tobita T, Ozutsumi O, Ueda K (2011) Dilatancy of granular materials in a strain space multiple mechanism model. Int J Numer Anal Meth Geomech 35:360–392

Imai Y, Suzuki A, Tobita T (2013) Example analysis of the right bank levee of Yoshidagawa River 16k point affected by the Great East Japan Earthquake in 2011. The 48th JGS geotechnical research conference, pp 1129–1130 (in Japanese)

Japan Institute of Country-ology and Engineering (JICE) (2011) River Embankment Earthquake Resistance Measures Emergency Review Committee, 8

Morishima N, Hayashi K, Kim K, Ichii K (2013) Example analysis of the right bank levee of Tonegawa River 26.5k point affected by the Great East Japan Earthquake in 2011. The 48th JGS geotechnical research conference, pp 1131–1132 (in Japanese)

Morita T, Iai S, Liu H, Ichii K, Sato Y (1997) Simplified method to determine parameter of FLIP, technical note of the port and Harbour research institute, Ministry of Transport. Japan, No.869, June 1997

National Research Institute for Earth Science and Disaster Prevention (NIED). "K-NET WWW service", Japan (http://www.k-net.bosai.go.jp/)

Chapter 13
Effective Stress Analysis of Quay Walls During the 2011 East Japan Earthquake

Soichi Tashiro

Abstract Numerous quay walls were damaged due to ground motions and subsequent tsunami during the 2011 Great East Japan Earthquake off the Pacific Coast of Tohoku, especially in the area south of the epicenter. Although it is assumed that liquefaction due to the ground motions occurring before the tsunami magnifies the tsunami damage, the severity of liquefaction and related damage to the quay walls before the tsunami is unknown. Various numerical simulations are performed for understanding the damage mechanisms caused by the ground motions using a strain space multiple mechanism model, called the cocktail glass model, as a constitutive model for soils. If a sheet pile type quay wall can retain structural stability and the excess pore water pressure dissipates before the arrival of a tsunami, the quay wall will not collapse due to a tsunami attack. However, a caisson-type quay wall, even if the quay wall retains its structural stability, may deform due to rubble foundation scouring caused by a tsunami attack. In addition, the cocktail glass model, which can consider the effect of pore water migration and dissipation, is applicable to evaluate the seismic performance such as deformation of quay walls against strong, long-duration ground motions.

13.1 Introduction

A number of quay walls were damaged over a wide area during the 2011 Great East Japan Earthquake off the Pacific Coast of Tohoku. In particular, the area south of the epicenter sustained more severe damage to the quay walls by both the ground motions and tsunami (Takahashi et al. 2011), which is attributed to the characteristics of the source rupture process of the earthquake and the influence of the site. As for the mechanism of damage to the quay walls, it is assumed that liquefaction due to the ground motions before the tsunami might magnify the damage during the tsunami. However, the tsunami attack destroyed evidence for liquefaction and damage to the quay walls due to the ground motions.

S. Tashiro (✉)
Toa Corporation, Shinjuku, Japan
e-mail: s_tashiro@toa-const.co.jp

S. Iai (ed.), *Developments in Earthquake Geotechnics*, Geotechnical, Geological and Earthquake Engineering 43, DOI 10.1007/978-3-319-62069-5_13

It is important to estimate the influence of soil liquefaction and the damage to quay walls due to the ground motions, which have characteristics such as site effects on the south side of the earthquake epicenter. To reveal the severity of soil liquefaction and to analyze the mechanism of damage to the quay walls due to ground motions on the south side of the earthquake epicenter, seismic response analyses are performed under various analytical conditions using a strain space multiple mechanism model, called the cocktail glass model, as a constitutive model for soils (Iai et al. 2011). Since the duration of the ground motions on the south side of the earthquake epicenter exceeded 100 s, the cocktail glass model is valuable because it considers the effects of pore water migration, dissipation, and dilatancy. In this study, a few sites with well-documented quay wall damage and known ground motions are selected to verify the applicability of the model.

13.2 Quay Walls for Analysis

13.2.1 Locations and Structural Types

Figure 13.1 shows the locations of the five quay walls used in the analysis. They are located in three ports: Sendai-Shiogama, Soma, and Onahama, which are located at about 162 km, 170 km, and 215 km from the earthquake epicenter, respectively. All three are south of the earthquake epicenter where a number of quay walls were severely damaged during the earthquake. Because none of the quay walls in this study were significantly damaged or collapsed, it is assumed that they were unaffected by the tsunami.

The quay wall structural types and water depths are:

- No. 1 quay wall at the Takasago wharf: Steel pipe sheet pile type (−12 m)
- No. 2 quay wall at the Takasago wharf: Steel pipe sheet pile type (−14 m)
- Quay wall at the No. 2 wharf at the Soma port: Steel pipe sheet pile type (−12 m)
- Quay wall at the No. 3 wharf at the Onahama port: Steel pipe sheet pile type (−10 m)
- Quay wall at the No. 5 wharf at the Onahama port: Caisson type (−12 m)

13.2.2 Ground Motions

In these ports, the observed peak ground acceleration (PGA) on the ground surface was 200–900 gal, and the duration of the ground motions exceeded 100 s. Nozu and Wakai (2011) estimated the acceleration time histories on the engineering bedrock from the strong motion records and the site characteristics in the ports. Figure 13.2 shows the acceleration time histories and acceleration response spectra with a dumping ratio of 5% in the direction perpendicular to the quay wall normal line.

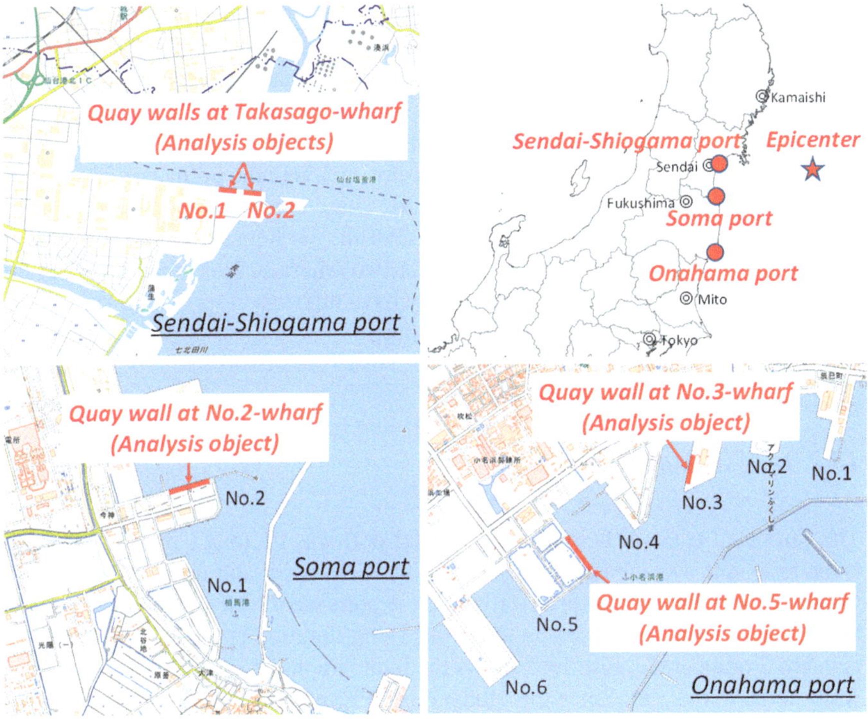

Fig. 13.1 Locations of the quay walls for simulation

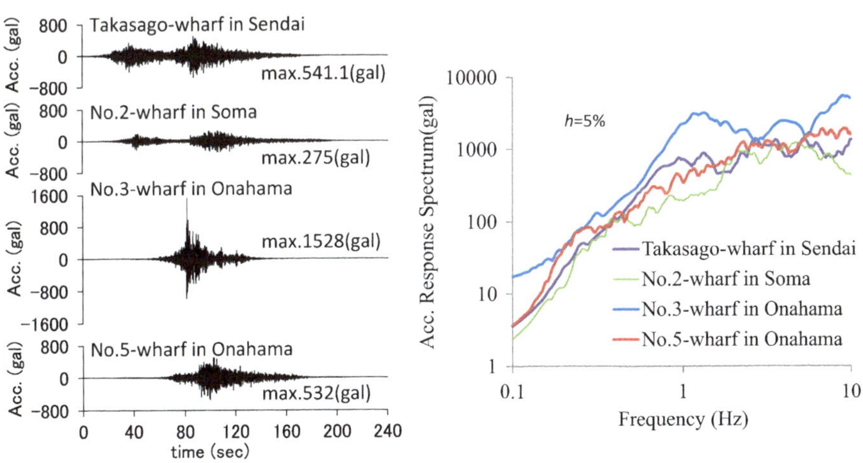

Fig. 13.2 Ground motions at the sites of the quay walls

The Takasago wharf in the Sendai-Shiogama port and the No. 2 wharf in the Soma port have double main shocks because they are near the earthquake epicenter. However, the No. 3 and No. 5 wharves at the Onahama port, which is farther from the earthquake epicenter, have a single main shock. The PGA at the No. 2 wharf in the Soma port is smaller than those for the two quay walls at the Onahama port as well as the response spectrum, although the Soma port is closer to the epicenter compared to the Onahama port. These observations indicate that the characteristics of the source rupture process of the earthquake and the site effects are influential. In addition, the PGAs at the two wharves, No. 3 and No. 5 at the Onahama port, differ from each other; they have different response spectra due to site effects, although they are in the same port.

13.2.3 Tsunami

The estimated tsunami height in the Sendai port was up to 7.2 m. The observed tsunami height was greater than 9.3 m in the Soma port and up to 3.33 m in the Onahama port. Numerous quay walls in these ports were seriously damaged due to the combined hazards caused by the strong, long-duration ground motions and the tsunami. For example, Fig. 13.3 shows the photographs of severely damaged quay walls at the Soma port. Since this study aims to reveal the damage mechanism due to only ground motions, such severely damaged quay walls are outside the scope of this paper.

13.2.4 Damage State and Cross Section for Simulations

13.2.4.1 Sendai-Shiogama Port

At the Takasago wharf in the Sendai-Shiogama port, there were two different damage states between adjacent quay walls, named No. 1 and No. 2. Figure 13.4 shows the horizontal displacement after the earthquake at the front sheet pile head and the ground surface settlement between the crane rails supported by steel pipe piles (Sendai Port and Airport Technology Survey Office). The No. 1 quay wall was only mildly damaged. A horizontal displacement of 10–19 cm at the pile head and a ground surface settlement of 8–16 cm between the crane rails were observed. On the other hand, the Part 2 area of the No. 2 quay wall was significantly damaged. This area underwent a horizontal displacement of 35–78 cm and a ground surface settlement of 58–88 cm between the crane rails.

Figure 13.5 shows representative cross sections of these quay walls before the earthquake. The red line indicates the deformed configuration after the earthquake. The gray hatches indicate liquefiable soil layers. However, it is unknown whether

Fig. 13.3 Example of severely damaged quay walls due to combined disaster in Soma port, after Takahashi et al. (2011)

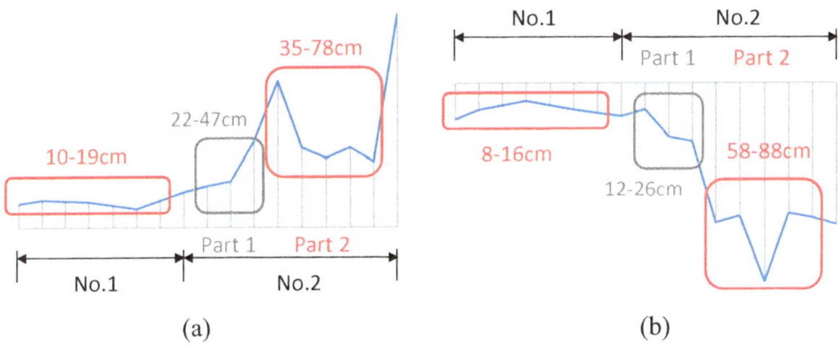

Fig. 13.4 Measured horizontal displacement and settlement distribution at the Takasago wharf. (**a**) Horizontal displacement. (**b**) Ground settlement

liquefaction actually occurred in the soil layers during the earthquake because the tsunami removed all traces of the liquefaction phenomenon.

Although these quay walls are both classified as steel pipe sheet pile–type quay walls, their construction procedures differ. The No. 1 quay wall was constructed through an excavation process of surrounding soils, but the No. 2 quay wall was constructed by backfilling methods due to the deep seabed. Crushed rock debris was used as the backfilling material to prevent liquefaction.

13.2.4.2 Soma Port

In the Soma port, some quay walls collapsed due to the combination of ground motions and the subsequent tsunami (Fig. 13.3). However, quay walls in the No. 2 wharf were only mildly damaged (Fig. 13.6). The normal line of the quay wall seems to be straight, but a slight difference in the level at the apron was observed. It is assumed that this quay wall was mainly damaged due to the ground motions and the tsunami attack had a negligible influence.

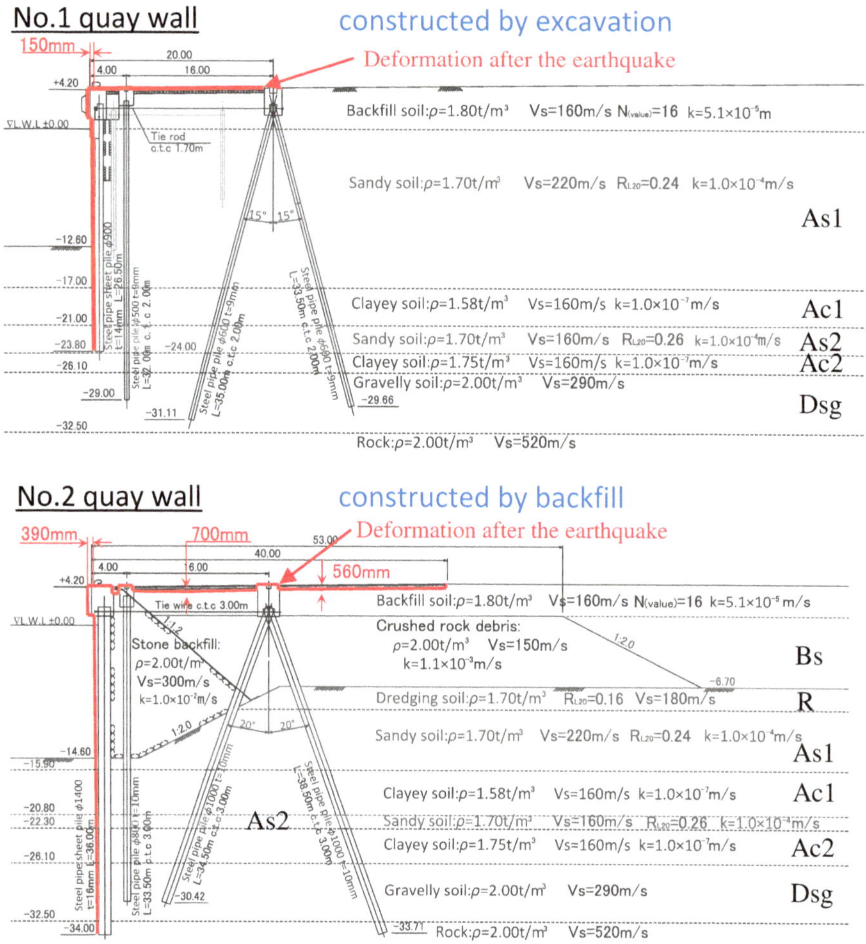

Fig. 13.5 Cross sections and deformations of the quay walls for simulations at Takasago-wharf

Fig. 13.6 Damage state photographs of the quay wall at No.2-wharf in Soma port

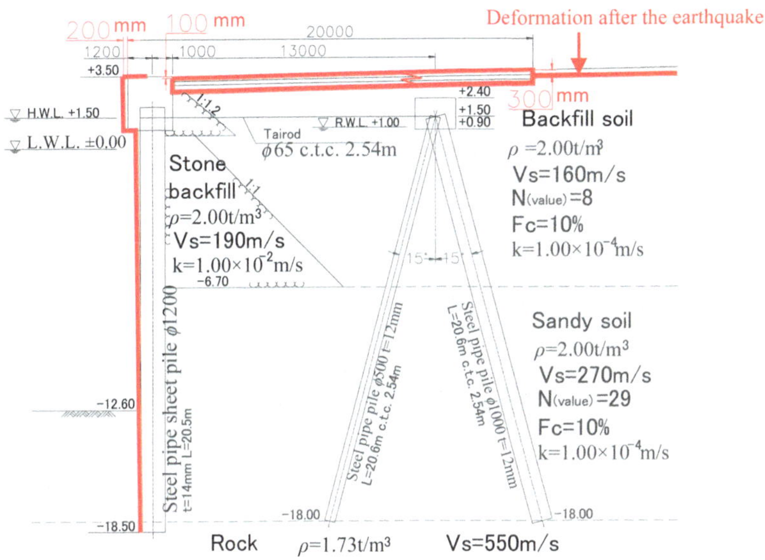

Fig. 13.7 Cross sections and deformations of the quay wall for simulations in Soma port

Figure 13.7 shows the cross section of the damaged quay wall at the No. 2 wharf where red indicates the deformed configuration. Although most parts of the soil layers are prone to liquefaction, it is unknown whether liquefaction actually occurred similar to the case for the Sendai-Shiogama port.

13.2.4.3 Onahama Port

In the Onahama port, a steel pipe sheet pile–type quay wall at the No. 3 wharf and a caisson-type quay wall at the No. 5 wharf with small amounts of damage similar to the Soma port were chosen for the simulations. Figure 13.8 shows a photograph of the damaged quay wall at the No. 3 wharf. At this wharf, the normal line of the quay wall seems to be straight, which is why the tsunami might not influence the quay wall damage. Figure 13.9 shows the damage to the quay wall at the No. 5 wharf. Although the normal line of the quay wall seems to be straight, the horizontal and vertical displacements as well as the incline angle of the caissons gradually increase as the distance from the wharf head decreases.

Figure 13.10 shows the cross section of the damaged quay wall at the No. 3 wharf, where red indicates the deformed configuration. Similarly, Fig. 13.11 shows that at the No. 5 wharf. Gray hatchings in both figures indicate the liquefiable soil layers. Similar to other ports, the tsunami might have erased signs of liquefaction. However, at the No. 3 wharf, there is evidence of significant liquefaction during the ground motions before the tsunami (Takahashi et al. 2011).

Fig. 13.8 Damage state photograph of the quay wall at No.3-wharf in Onahama port

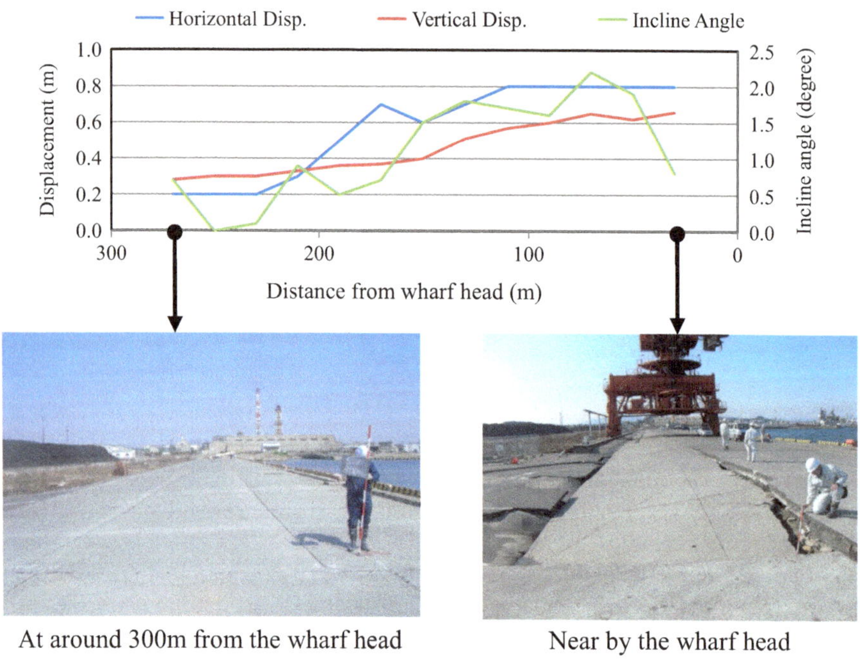

Fig. 13.9 Damage state photographs of the quay wall at No.5-wharf in Onahama port

13.3 Analysis Conditions

This study employed a two-dimensional effective stress analysis program, called FLIP ROSE, which is based on the finite element method. As a constitutive model for soils, two types of strain space multiple mechanism models, called the cocktail glass model and the multi-spring model, were implemented in the program (Iai et al.

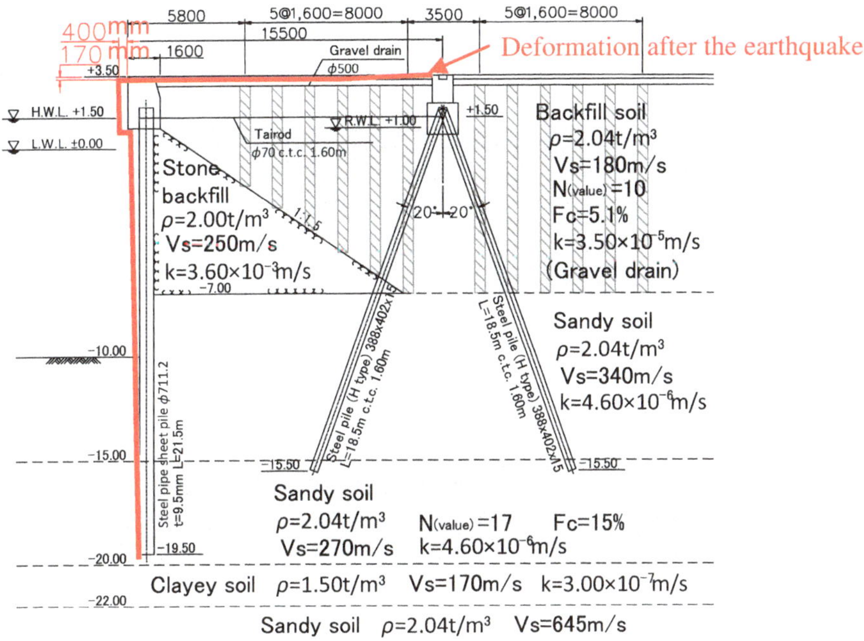

Fig. 13.10 Cross sections and deformations of the quay wall at No.3-wharf in Onahama port

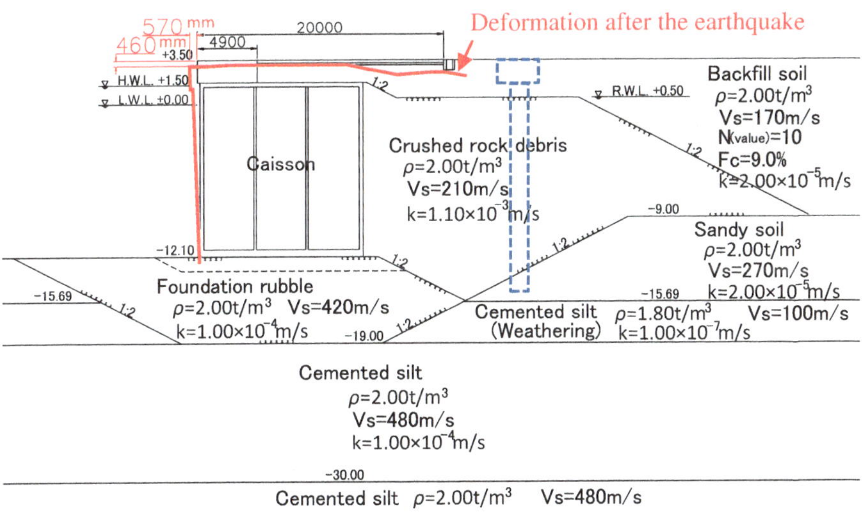

Fig. 13.11 Cross sections and deformations of the quay wall at No.5-wharf in Onahama port

1990). Although the multi-spring model was used only under undrained conditions, the model has been used in practice to evaluate the seismic performance of geotechnical works (Iai and Kameoka 1993, Iai et al. 1998). This is because the influences of pore water migration and dissipation are insignificant in the case of short-duration earthquakes. On the other hand, the cocktail glass model can consider the effect of pore water migration and dissipation.

In this study, numerical simulations were performed using both models in order to evaluate the effect of permeability. In addition to the above difference, the formulation of dilatancy also differs between these two models (Iai et al. 1992, 2011). Besides the soil elements, other types of elements (e.g., nonlinear beam elements, fluid elements, linear plane elements, and joint elements) have been implemented in the program in order to consider the effect of complex soil-structure-foundation–fluid interaction.

13.3.1 Soil Conditions

The physical properties of the soil layers at each quay wall are shown in the cross section (Figs. 13.5, 13.7, 13.10, and 13.11). Only at the Takasago wharf (Fig. 13.5), the shear velocities (Vs) and the liquefaction strength ratios (R_{L20}) were obtained from a PS logging test and undrained cyclic shear tests after the earthquake, respectively. Figure 13.12 shows the location of the borings for the PS logging test, the undrained cyclic shear tests, and the other standard penetration tests (SPTs). Figure 13.13 shows the liquefaction resistance curves obtained from laboratory tests, which were compared with the simulation results.

The PS logging tests and undrained cyclic shear tests were not conducted at the other quay walls shown in Figs. 13.7, 13.10, and 13.11. Thus, the shear velocities and the liquefaction characteristics were estimated in the following ways. The shear velocities were calculated at the center of each layer using the relationship between the shear velocity, density, and shear modulus. The shear moduli for all soil layers except the stone backfill and the foundation rubble were calculated from the SPT N-value using the method proposed by Morita et al. (1997). The shear velocities of the stone backfill and the foundation rubble were calculated at the center of each layer based on the shear modulus, which was obtained assuming that the shear

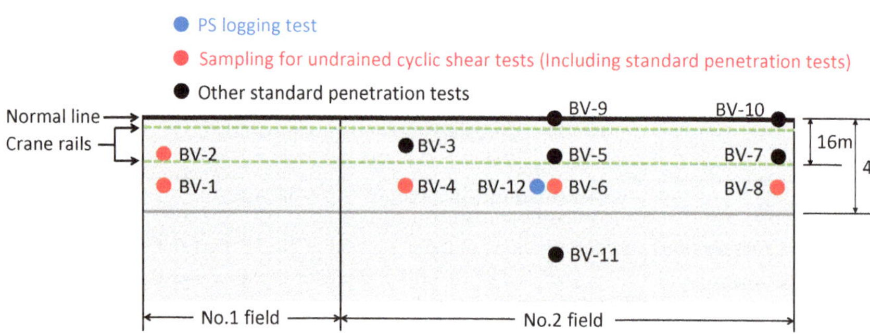

Fig. 13.12 Location of boring for PS logging test, undrained cyclic shear tests and other standard penetration tests at Takasago wharf

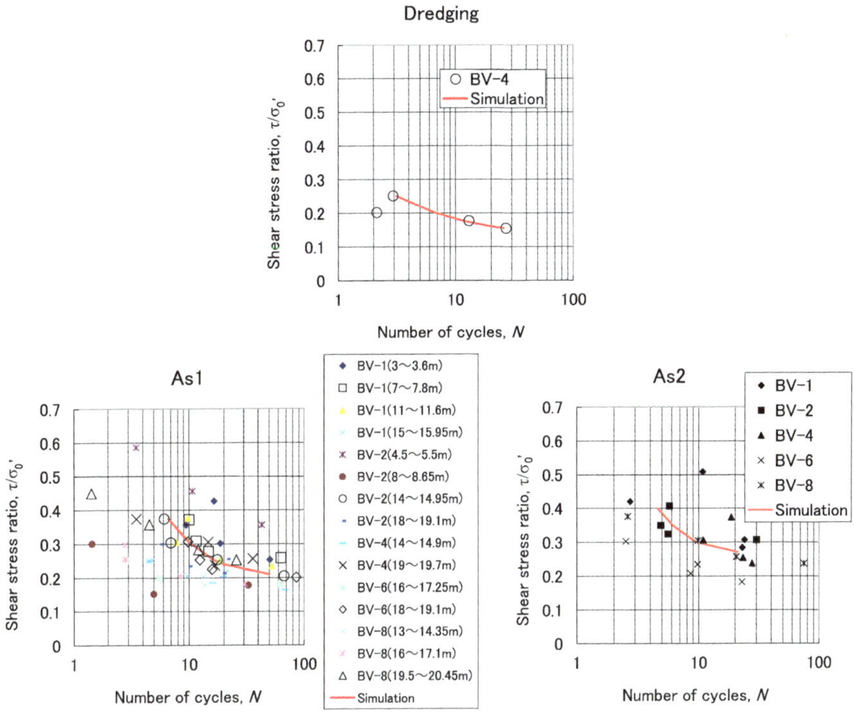

Fig. 13.13 Liquefaction resistance curve of saturated sandy soil layers at Takasago wharf

velocity is 300 m/s under an effective mean stress of 98 kPa and the shear modulus varies in proportion to the effective mean stress by a power index of 0.5.

In addition to the saturated backfilled soil layers, the sandy soil layer at the No. 2 wharf in the Soma port (Fig. 13.7) and that at an elevation between −15 and −20 m at the No. 3 wharf in the Onahama port were considered liquefiable. The model parameters of the multi-spring model, including the dilatancy parameters, for these soil layers were estimated from the SPT N-value and the fines content (Fc) using the simplified method proposed by Morita et al. (1997). The dilatancy parameters of the cocktail glass model were determined by referring to the liquefaction properties (e.g., liquefaction resistance curves, stress-strain hysteresis) that were obtained through the multi-spring model based on a simplified method. The fines content was obtained from laboratory tests for the Onahama port, but was unknown for the Soma port. The fines content at the No. 2 wharf in the Soma port (Fig. 13.7) was determined by trial numerical simulations using the multi-spring model so that the simulated deformation of the quay wall due to the liquefaction during the earthquake well explains the observed damage mechanism (Sakakibara et al. 2013).

At the No. 3 wharf in the Onahama port, part of the backfilled soil had been improved by the gravel drain method. To evaluate the drainage effects of the method, two simulations were performed (with and without considering the gravel drain). In the former case, an equivalent coefficient of the permeability in the horizontal direction was applied to the improved area by the gravel drain (Murakami et al. 2013).

The crushed rock debris was backfilled behind the sheet pile (Fig. 13.5) and the caisson (Fig. 13.11) in order to prevent liquefaction. Since the dilatancy effects of this debris are considered to affect the settlement on the ground surface, the dilatancy properties provided by Yamazaki et al. (2012) were used as the input parameters in some cases. The soil parameters are shown in Table 13.1a, 13.1b, 13.1c, 13.1d (Sumiya et al. 2013; Murakami et al. 2013; Shibata et al. 2014).

13.3.2 Structural Modeling

13.3.2.1 Steel Pipe Sheet Pile

Figure 13.14 shows how to model a steel pipe sheet pile–type quay wall, including structural members (i.e., a sheet pile, a tie rod, and anchored piles). The soil-structure interaction between the steel pipe–anchored piles and surrounding soils could be modeled in two different ways. In the conventional method (Fig. 13.14a), which is often used in finite element analyses for seismic design, the effect of friction between the anchored piles and soils is ignored in the direction parallel to the pile axis. In addition, the lower end of the anchored piles has the same degrees of freedom for both the lateral and vertical directions with the corresponding nodes of soils. In the other way, which has been proposed recently and is beginning to be used in design, the compression force at the lower end of the piles in the direction parallel to the pile axis is modeled with nonlinear spring elements that connect the pile end with the corresponding node of soils, but the effect of the tensile force is ignored (Fig. 13.14b). In addition, the friction between the anchored piles and soils is taken into account using joint elements. However, the friction resistance during earthquakes is difficult to accurately evaluate due to the lack of knowledge. Thus, two conditions about the friction were considered in the analyses of the quay wall at the No. 2 wharf in the Soma port: (1) a conventional value (15°) for wall friction and (2) an internal friction angle of the surrounding soils that are connected to the anchored piles for the friction angle.

The other contact conditions between soils and structures were as follows:

1. Between a sheet pile and soils on the seaward side:

The effect of both the friction and contact/non-contact in the parallel and normal direction, respectively, to the contact surface is taken into account using joint elements.

2. Between a sheet pile and soils on the landward side:

Only the contact/non-contact effect is taken into account using joint elements. The friction effect is ignored.

Table 13.1a Soil parameters for simulations. At Takasago-wharf in Sendai-Shiogama port

Common volumetric and shear parameters for cocktail glass and multi-spring model

Soil	ρ (t/m^3)	p_a (kPa)	K_{ma}, $K_{L/Ua}$ (kPa)	n_K	G_{ma} (kPa)	m_G	c (kPa)	ϕ_f (Deg.)	h_{max}	n	k (m/s)
Backfill	1.8	21	120,200	0.5	46,080	0.5	0	41	0.24	0.45	5.1×10^{-5}
Dredging	1.7	94	143,600	0.5	55,080	0.5	0	40	0.24	0.45	1.0×10^{-4}
As 1	1.7	124	214,600	0.5	82,280	0.5	0	40	0.24	0.45	1.0×10^{-4}
Ac 1	1.58	–	105,500	0.0	40,450	0.0	119	0	0.20	0.55	1.0×10^{-7}
As 2	1.7	166	113,500	0.5	43,520	0.5	0	39	0.24	0.45	1.0×10^{-4}
Ac 2	1.75	–	116,800	0.0	44,800	0.0	132	0	0.20	0.55	1.0×10^{-7}
Dsg	2	215	438,600	0.5	168,200	0.5	0	40	0.24	0.45	1.0×10^{-4}
Gc (CR)	2	66	117,400	0.5	45,000	0.5	0	40	0.24	0.45	1.0×10^{-4}
Stone B	2	98	469,400	0.5	180,000	0.5	20	35	0.24	0.45	1.0×10^{-2}

Dilatancy parameters

		For cocktail glass model											For multi-spring model				
Soil	Φ_p (Deg.)	ε_d^{cm}	$r_{\varepsilon_d^c}$	r_{ε_d}	r_K	l_K	q_1	q_2	s_1	c_1	r_r	r_{mtmp}	s_1	w_1	p_1	p_2	c_1
Backfill	28	0.10	0.5	0.75	0.5	2	7.0	2.0	0.005	1.50	0.10	0.5	0.005	17.86	0.5	0.71	6.75
Dredging	28	0.18	1.0	0.50	0.6	2	7.5	0.2	0.005	1.73	0.25	0.5	0.005	1.70	0.5	0.80	1.73
Asl	28	0.20	0.7	0.65	0.5	2	5.0	0.9	0.005	2.20	0.10	0.5	0.005	6.50	0.5	0.90	2.20
As2	28	0.20	1.3	0.50	0.5	2	7.0	2.0	0.005	2.60	0.30	0.5	0.005	4.00	0.5	0.51	3.30
Gc (CR)	39	0.20	1.0	1.82	0.1	2	1.0	0.0	0.005	1.00	0.50	0.5	–	–	–	–	–

Table 13.1b Soil parameters for simulations. At No.2-wharf in Soma port

Common volumetric and shear parameters for cocktail glass and multi-spring model

Soil	ρ (t/m^3)	p_a (kPa)	K_{ma}, $K_{L/Ua}$ (kPa)	n_K	G_{ma} (kPa)	m_G	c (kPa)	ϕ_f (Deg.)	h_{max}	n	k (m/s)
Backfill	1.8	98	196,230	0.5	75,246	0.5	0	39	0.24	0.45	1.0×10^{-4}
As	2	98	326,228	0.5	125,095	0.5	0	41	0.24	0.45	1.0×10^{-4}
Stone B	2	98	264,000	0.5	101,300	0.5	20	35	0.24	0.45	1.0×10^{-2}

Dilatancy parameters

		For cocktail glass model											For multi-spring model				
Soil	Φ_p (Deg.)	ε_d^{cm}	$r_{\varepsilon_d^c}$	r_{ε_d}	r_K	l_K	q_1	q_2	s_1	c_1	r_r	r_{mtmp}	s_1	w_1	p_1	p_2	c_1
Backfill	28	0.20	1.0	1.20	0.5	2	8.0	0.7	0.005	2.56	0.10	0.5	0.005	6.04	0.5	0.95	2.51
As	28	0.15	1.4	0.60	0.5	2	1.0	0.8	0.005	8.29	0.10	0.5	0.005	11.75	0.5	0.74	6.51

Table 13.1c Soil parameters for simulations. At No.3-wharf in Onahama port

Common volumetric and shear parameters for cocktail glass and multi-spring model

Soil	ρ (t/m^3)	p_a (kPa)	K_{ma}, $K_{L/Ua}$ (kPa)	n_K	G_{ma} (kPa)	m_G	c (kPa)	ϕ_f (Deg.)	h_{max}	n	k (m/s)
Backfill	1.8	98	221,000	0.5	84,700	0.5	0	40	0.24	0.45	3.5×10^{-4}
As 1	2	98	471,000	0.5	181,000	0.5	0	43	0.24	0.45	4.6×10^{-5}
As 2	2	98	247,000	0.5	94,900	0.5	0	40	0.24	0.45	4.6×10^{-5}
Ac	2	98	70,000	0.5	26,900	0.5	0	30	0.20	0.45	3.0×10^{-7}
Stone B	2	98	264,000	0.5	101,300	0.5	0	40	0.24	0.45	3.6×10^{-3}

Cofficient of permeability in gravel drain part: $k = 2.54 \times 10^{-3}$ (m/sec) in the case of considering gravel drain

Dilatancy parameters

		For cocktail glass model											For multi-spring model				
Soil	Φ_p (Deg.)	ε_d^{cm}	$r_{\varepsilon_d^c}$	r_{ε_d}	r_K	l_K	q_1	q_2	S_1	c_1	r_r	r_{mtmp}	S_1	w_1	p_1	p_2	c_1
Backfill	28	0.20	1.0	0.20	0.5	2	6.2	0.7	0.005	2.15	0.10	0.5	0.005	3.76	0.5	1.01	2.12
As2	28	0.20	0.9	0.20	0.5	2	10.0	1.6	0.005	3.80	0.10	0.5	0.005	7.50	0.5	0.85	4.20

Table 13.1d Soil parameters for simulations. At No.5-wharf in Onahama port

Common volumetric and shear parameters for cocktail glass and multi-spring model

Soil	ρ (t/m^3)	p_a (kPa)	K_{ma}, $K_{L/Ua}$ (kPa)	n_K	G_{ma} (kPa)	m_G	c (kPa)	ϕ_f (Deg.)	h_{max}	n	k (m/s)
Backfill	1.8	98	196,600	0.5	75,400	0.5	0	39	0.24	0.45	2.0×10^{-5}
As	2	98	366,700	0.5	140,600	0.5	0	41	0.24	0.45	2.0×10^{-5}
C.Silt	2	172	26,600	0.5	10,200	0.5	30	0	0.20	0.55	1.0×10^{-7}
Mound	2	98	479,800	0.5	184,000	0.5	20	35	0.24	0.45	1.0×10^{-3}

Parameters of Cruched Reck are reffered from the crushed reck of Takasago-wharf

Dilatancy parameters

		For cocktail glass model											For multi-spring model				
Soil	Φ_p (Deg.)	ε_d^{cm}	$r_{\varepsilon_d^c}$	r_{ε_d}	r_K	l_K	q_1	q_2	s_1	c_1	r_r	r_{mtmp}	s_1	w_1	p_1	p_2	c_1
Backfill	28	0.20	1.5	0.20	0.5	2	10.0	1.3	0.005	2.50	0.10	0.5	0.005	5.41	0.50	0.97	2.40

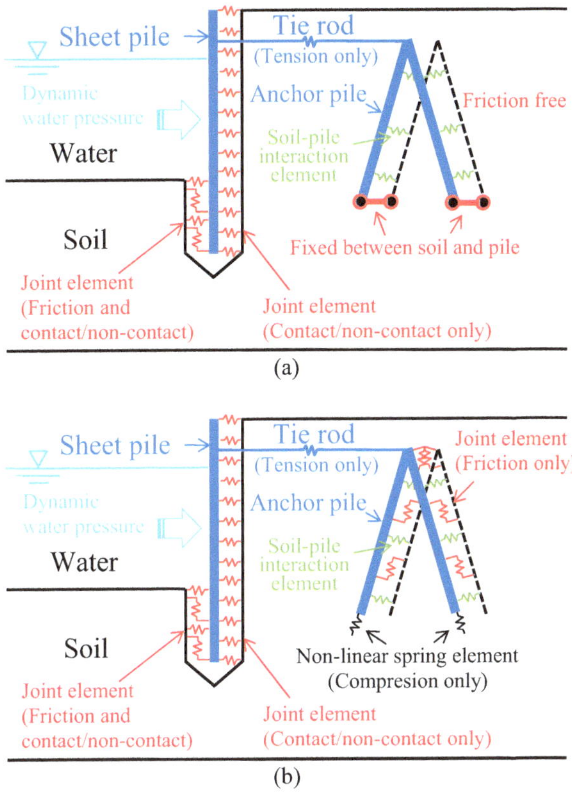

Fig. 13.14 Structural modelling concept of steel pipe sheet pile type quay wall. (**a**) Conventional contact condition. (**b**) Recent standard contact condition

3. Between anchor piles and soils in the normal direction to the pile axis:

The effect of the three-dimensional interaction between the piles and soils in the direction is taken into account using the pile–soil interaction spring elements.

The sheet pile and the anchored piles were modeled by nonlinear beam elements in order to appropriately evaluate the seismic performance of the members even when the stress condition exceeds the yield point. In the modeling of the tie rod connecting the sheet pile and the anchored pile head, only tension was considered using a nonlinear spring element. To take the dynamic water pressure acting on the sheet pile during shaking into account, fluid-structure interface elements and fluid elements were used.

13.3.2.2 Caisson

Figure 13.15 shows how to model a caisson-type quay wall. Joint elements were used to consider the effects of friction and contact/non-contact between the caisson and surrounding soils (i.e., the backfill and the foundation). The caisson was modeled by a linear elastic material. As described in the previous section, fluid–

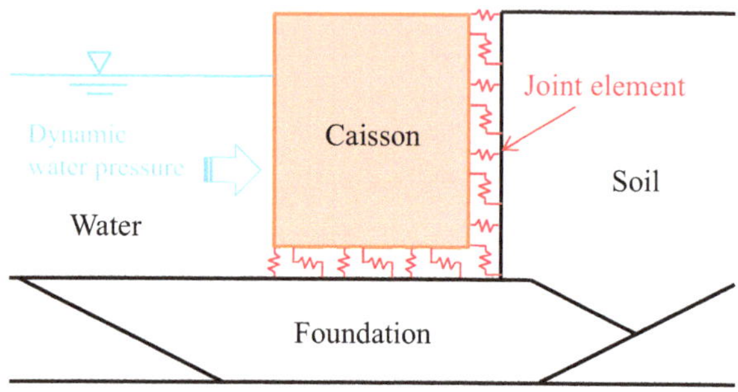

Fig. 13.15 Structural modeling concept of a caisson-type quay wall

structure interface elements and fluid elements were used to consider the dynamic water pressure.

13.3.3 Simulation

13.3.3.1 No. 1 and No. 2 Quay Walls at the Takasago Wharf in the Sendai-Shiogama Port

Table 13.2 shows the simulation cases in the analysis of the No. 1 and the No. 2 quay walls at the Takasago wharf in the Sendai-Shiogama port. The conventional way in Fig. 13.14a was adopted for Cases 1–1 through 1–3 and Cases 2–1 through 2–3. The proposed way in Fig. 13.14b was used in Cases 1–4 through 1–6 and Cases 2–4 through 2–6 in order to determine the influence of the modeling difference in the contact conditions.

The liquefaction resistance of the alluvium soil layers (i.e., As1, As2) varies greatly as shown in Fig. 13.13. Because the layers may have a higher resistance in situ than the average one, they were modeled as non-liquefiable in some cases for comparison. In addition, the effect of dilatancy was considered in the modeling of the crushed rock debris for Cases 2–3 and 2–6 at the No. 2 quay wall because it is assumed that the volumetric strain due to the dilatancy induces a large settlement on the ground surface, as mentioned in Sect. 13.3.1.

13.3.3.2 Quay Wall at the No. 2 Wharf in the Soma Port

Table 13.3 shows the simulation cases for the analysis of the quay wall at the No. 2 wharf in the Soma port. Comparing Cases 1 and 3 elucidated the influence of the difference in the formulation of dilatancy between the multi-spring and cocktail glass models on the analytical results. In addition, the effect of the difference in the modeling of friction between the anchored piles and soils should be determined by comparing Cases 2 and 3.

Table 13.2 Simulation cases for No.1 and No.2 quay walls at Takasago wharf

Quay wall	Case name	Soil model	Contact condition on anchor lower end	Friction between anchor pile and soil	Considering liquefaction of As layers	Considering dilatancy of crushed rock debris
No.1	Case1–1	Multi	Fix	Free	Yes	–
	Case1–2	Multi	Fix	Free	No	–
	Case1–3	Ctgl	Fix	Free	Yes	–
	Case1–4	Multi	NLS	IFAS	Yes	–
	Case1–5	Multi	NLS	IFAS	No	–
	Case1–6	Ctgl	NLS	IFAS	Yes	–
No.2	Case2–1	Multi	Fix	Free	Yes	No
	Case2–2	Ctgl	Fix	Free	Yes	No
	Case2–3	Ctgl	Fix	Free	Yes	Yes
	Case2–4	Multi	NLS	IFAS	Yes	No
	Case2–5	Multi	NLS	IFAS	No	No
	Case2–6	Ctgl	NLS	IFAS	No	Yes

Multi Multi-spring model, *Ctgl* Cocktail Glass model, *NLS* Non-Linear Spring by considering only compression, *IFAS* Internal Frictional Angle of contact Soil on pile

Table 13.3 Simulation cases for quay wall at No.2 wharf in Soma port

Case name	Soil model	Contact condition on anchor lower end	Friction between anchor pile and soil
Case-1	Multi	NLS	IFAS
Case-2	Ctgl	NLS	15°
Case-3	Ctgl	NLS	IFAS

Multi Multi-spring model, *Ctgl* Cocktail Glass model, *NLS* Non-Linear Spring by considering only compression, *IFAS* Internal Frictional Angle of contact Soil on pile

13.3.3.3 Quay Wall at the No. 3 Wharf in the Onahama Port

Table 13.4 shows the simulation for the analysis of the quay wall at the No. 3 wharf in the Onahama port. Comparing Cases 1 and 2 clearly demonstrated the difference between the multi-spring and cocktail glass models. In addition, Case 3 was carried out in order to investigate the drainage effect of the gravel drain. In contrast to the cases in Table 13.3, the friction angle was set to 15° for all three cases. This is because the difference in the angle does not significantly affect the analytical results in the case of the quay wall at the No. 3 wharf in the Onahama port.

13.3.3.4 Quay Wall at the No. 5 Wharf in the Onahama Port

Table 13.5 shows the simulation for the analysis of the quay wall at the No. 5 wharf in the Onahama port. In Case 3, the effect of dilatancy was considered for the crushed rock debris as mentioned in Sect. 13.3.3.1.

Table 13.4 Simulation cases for quay wall at No.3 wharf in Onahama port

Case name	Soil model	Contact condition on anchor lower end	Friction between anchor pile and soil	Considering gravel drain
Case-1	Multi	NLS	15°	No
Case-2	Ctgl	NLS	15°	No
Case-3	Ctgl	NLS	15°	Yes

Multi Multi-spring model, *Ctgl* Cocktail Glass model, *NLS* Non-Linear Spring by considering only compression

Table 13.5 Simulation cases for quay wall at No.5 wharf in Onahama port

Case name	Soil model	Considering dilatancy of crushed rock debris
Case-1	Multi-Spring model	No
Case-2	Cocktail Glass model	No
Case-3	Cocktail Glass model	Yes

13.4 Analysis Results

13.4.1 Takasago Wharf in the Sendai Shiogama Port

Figures 13.16, 13.17, 13.18 and 13.19 show the simulated results at the Takasago wharf (the No. 1 and No. 2 quay walls). In addition to the simulated deformation, the distributions of the excess pore pressure ratios at 100 and 240 s after the end of the second main shock are given in Figs. 13.18 and 13.19. Figure 13.16 shows the lateral displacements at the quay wall head at a time of 240 s, while Fig. 13.17 shows the vertical displacements on the ground surface of the backfill at 240 s. The stress states of the steel members at the end of the analysis are given in Table 13.6.

Figure 13.16 reveals that the computed lateral displacements at the quay wall head fall into the observed range or very close to the upper limit for both Cases 1–1 through 1–3 and Cases 2–1 through 2–3, in which the fixed condition was used between the lower end of the anchored piles and the surrounding soils for both No. 1 and No. 2 quay walls. In addition to the settlement, the computed damage condition of the sheet pile in Case 1–1 for the No. 1 quay wall differs from the observation and that of the tie rod in Cases 2–1 through 2–3 for the No. 2 quay wall (Table 13.6). Comparing Cases 2–2 and 2–3 for the No. 2 quay wall indicated whether considering the dilatancy of the crushed rock debris has a significant effect on the vertical displacement. Although the observed lateral displacement is well simulated in both Cases 1–1 through 1–3 and Cases 2–1 through 2–3, the residual settlements in Cases 1–1 and 1–3 for the No. 1 quay wall overestimate the observation as shown in Fig. 13.17a. Because the As layers were considered liquefiable in these cases, it is highly possible that no significant liquefaction actually occurred in the layers.

Unlike Cases 1–1 through 1–3 and Cases 2–1 through 2–3, nonlinear spring elements were used to model the compression force at the lower end of the piles in

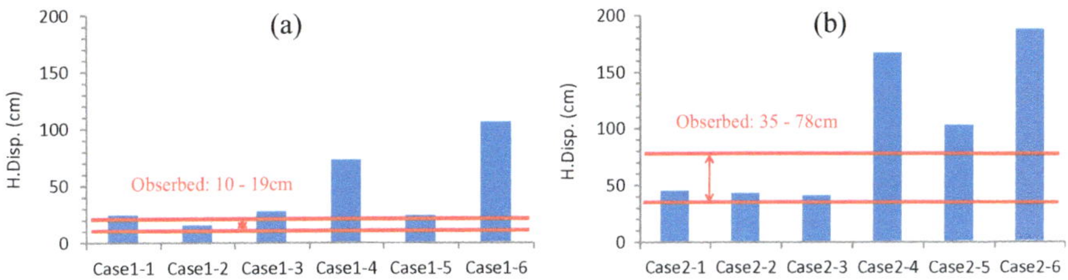

Fig. 13.16 Simulated horizontal displacements at sheet pile head at Takasago-wharf in Sendai-Shiogama port. (**a**) No.1 quay wall. (**b**) No.2 quay wall

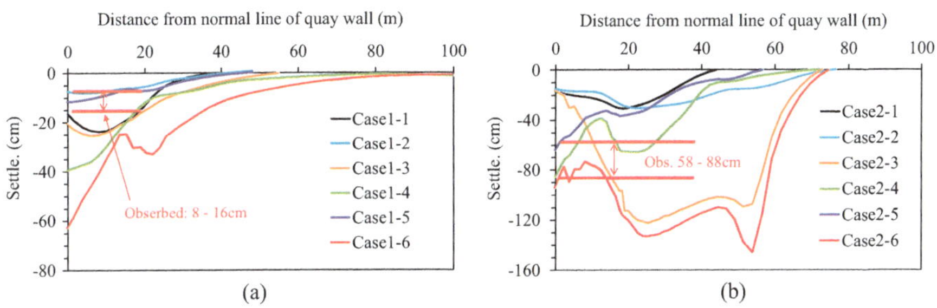

Fig. 13.17 Simulated vertical displacements on ground surface at Takasago-wharf in Sendai-Shiogama port. (**a**) No.1 quay wall. (**b**) No.2 quay wall

the direction parallel to the pile axis in Cases 1–4 through 1–6 and Cases 2–4 through 2–6. In addition, the friction between the anchored piles and soils was considered using joint elements. Among these cases, the simulation well reproduced the observed residual lateral and vertical displacements in Cases 1–5 and 2–5, where the As layers are modeled as non-liquefiable as well as the damage condition of the steel members (Table 13.6) for the No. 1 and the No. 2 quay walls. Comparing the lateral displacements between Cases 1–4 and 1–5 for the No. 1 quay wall indicated that the computed displacement in Case 1–4 is more than twice the upper limit of the observation, suggesting that the As layers might not have been liquefied as described above. The same tendency is recognized in a comparison between Cases 2–4 and 2–5 for the No. 2 quay wall.

Figure 13.17b shows the effect of dilatancy of the crushed rock debris on the residual settlement. In Cases 2–3 and 2–6, the maximum value of the settlement is two or three times larger than that in the case without considering the dilatancy effect. Although whether the induced volumetric strain due to dilatancy is significant at the No. 2 quay wall remains controversial, the dilatancy effect of the crushed rock debris should be appropriately considered in the simulation in order to estimate the ground settlement with a high degree of accuracy.

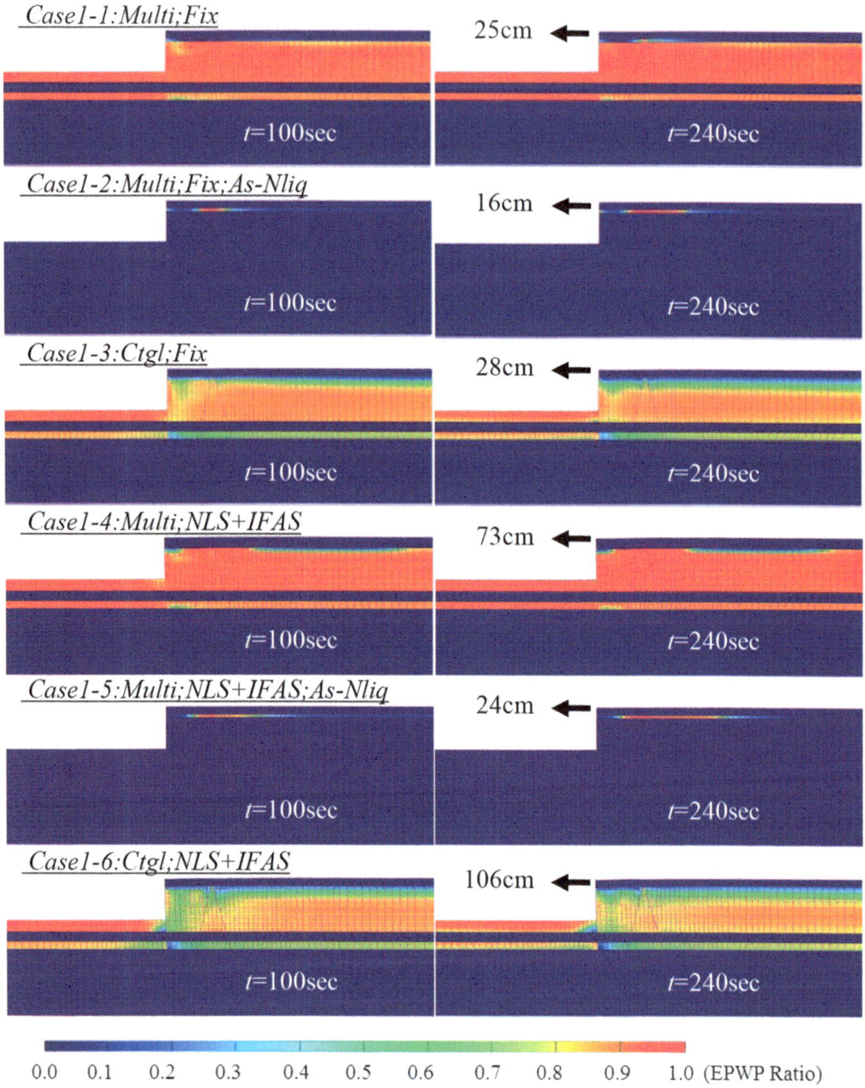

Fig. 13.18 Simulated deformations and excess pore water pressure ratios at No.1 quay wall at Takasago-wharf in Sendai-Shiogama port

13.4.2 Quay Wall at the No. 2 Wharf in the Soma Port

Figures 13.20, 13.21 and Table 13.7 show the simulated results of the quay wall at the No. 2 wharf in the Soma port. The simulation well reproduced the observed lateral displacement in Case 1. Additionally, the computed damage condition of the steel members is coincident with the observation. This indicates that the multi-

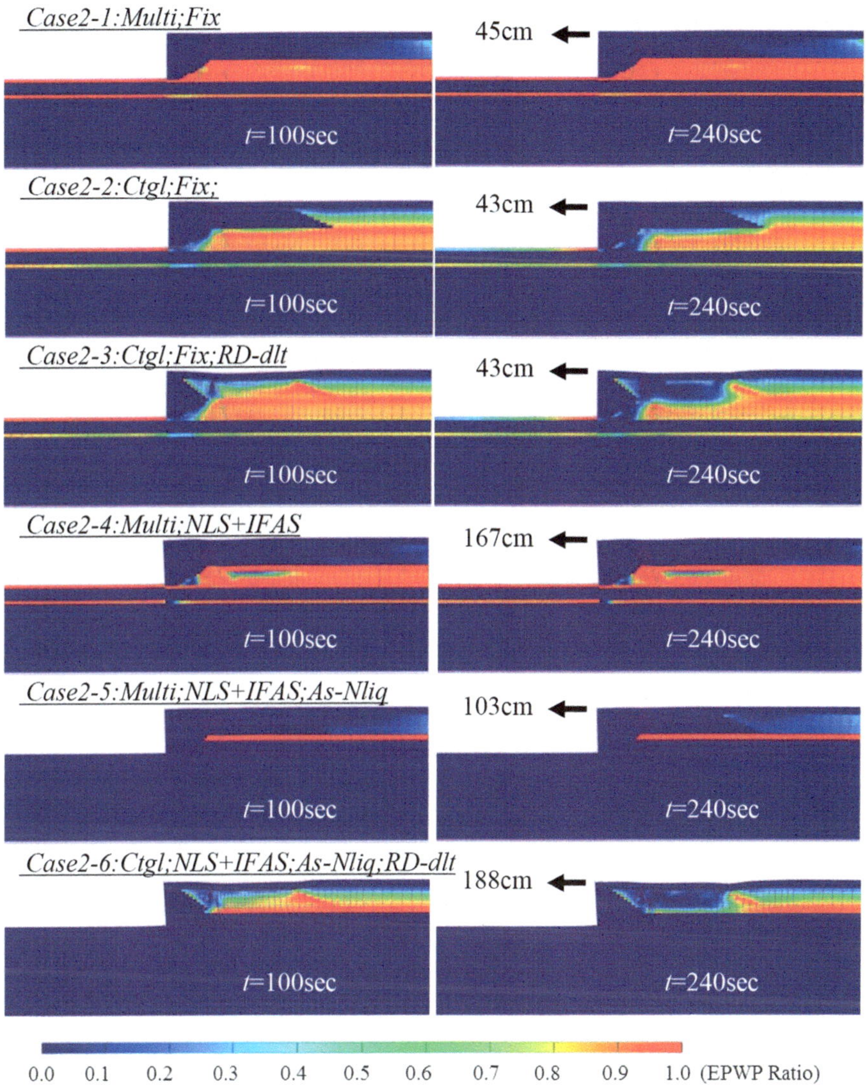

Fig. 13.19 Simulated deformations and excess pore water pressure ratios at No.2 quay wall at Takasago-wharf in Sendai-Shiogama port

spring model may be applicable to evaluate the seismic performance of quay walls by appropriately modeling the interaction between structures (e.g., anchored piles) and soils even if the duration of ground motions exceeds 100 s. However, the computed residual settlement is less than the observation because the model cannot take the effect of pore water dissipation into account.

Comparing Cases 1 and 3 indicated that the observed settlement, the lateral displacement, and the damage condition of steel members are well simulated by

Table 13.6 Stress damage states of steel members at Takasago-wharf in Sendai-Shiogama port

No.1 quay wall				No.2 quay wall			
Name	Sheet pile	Anchor	Tie rod	Name	Sheet pile	Anchor	Tie rod
Case 1–1	Plastic	No	No	Case 2–1	No	No	Break
Case 1–2	No	No	No	Case 2–2	No	No	Break
Case 1–3	No	No	No	Case 2–3	No	No	Break
Case 1–4	No	No	No	Case 2–4	No	Plastic	No
Case 1–5	No	No	No	Case 2–5	No	No	No
Case 1–6	No	Plastic	No	Case 2–6	No	No	No
Observed	No	Unknown	No	Observed	No	Unknown	No

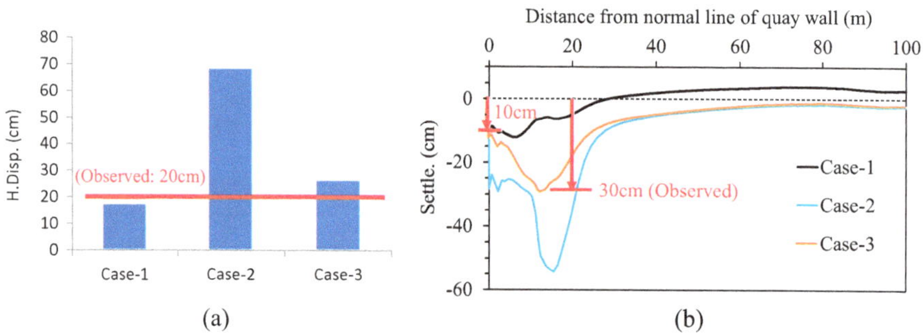

Fig. 13.20 Simulated horizontal displacements at sheet pile head and vertical displacements on ground surface at No.2-wharf in Soma port. (**a**) Horizontal displacements. (**b**) Vertical displacements

considering the drainage effect with the cocktail glass model. These results suggest that the cocktail glass model provides a higher accuracy than the multi-spring model when the duration of ground motions is long and consequently, the effect of pore water migration and dissipation cannot be ignored, especially when evaluating settlements. However, the computed lateral and vertical displacements in Case 2 are much larger than the observation even when the cocktail glass model is used. In addition, the damage condition of the anchored pile differs between the simulation and the observation. This may be because the friction (15°) between the anchored pile and surrounding soils in Case 2 does not represent the actual condition in situ. These results imply that both how the interaction between structures and soils is modeled and the selection of constitutive models of soils are important to precisely evaluate the seismic performance of quay walls.

The excess pore pressure ratio builds up to almost 1.0 at 120 s for all cases, but decreases to about 0.5, on average, at 240 s in the backfilled soil due to the dissipation of the excess pore pressure in Cases 2 and 3 (Fig. 13.21). This suggests that the tsunami attack may not have severely damaged the quay wall because the excess pore pressure dissipated to some extent before the tsunami, allowing the backfilled soils and the quay wall to regain sufficient strengths to withstand the tsunami force.

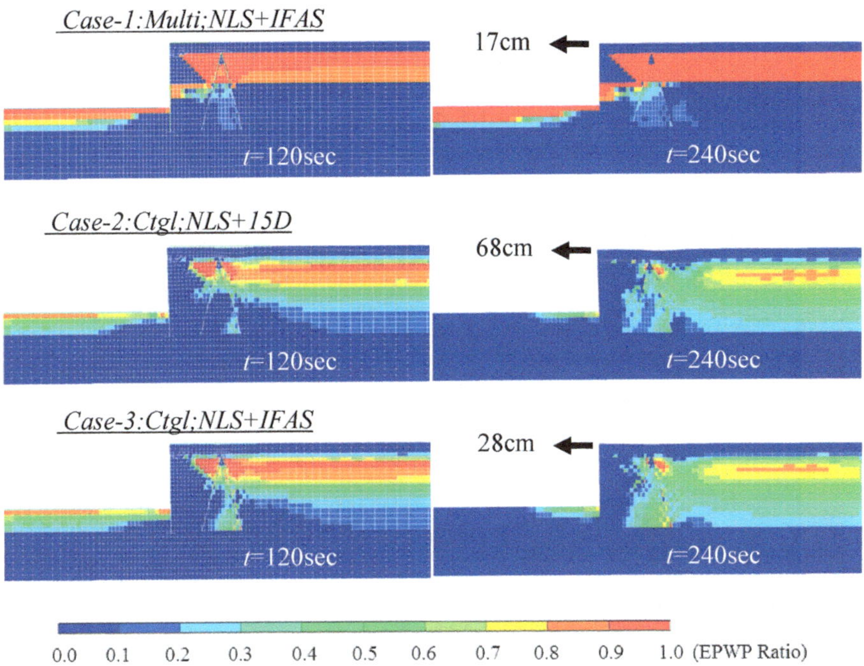

Fig. 13.21 Simulated deformations and excess pore water pressure ratios at No.2-wharf in Soma port

Table 13.7 Stress damage states of steel members at No.2-wharf in Soma port

Name	Sheet pile	Anchor	Tie rod
Case-1	No	No	No
Case-2	No	Plastic	No
Case-3	No	No	No
Observed	No	Unknown	No

13.4.3 Quay Wall at the No. 3 Wharf in the Onahama Port

Figures 13.22, 13.23 and Table 13.8 show the simulated results for the Onahama No. 3 wharf. Although all cases overestimate the observed lateral displacement (Fig. 13.22a), the computed result in Case 3 becomes closer to the observation upon considering the high permeability of the gravel drain with the cocktail glass model. The differences in the lateral displacements of Cases 1 and 2 are attributed mainly to the discrepancy in the distribution of the excess pore pressure ratio at 100 s (i.e., the ratio is higher in the multi-spring model) in Fig. 13.23 rather than the effect of the pore water dissipation. Comparing Cases 2 and 3 in Fig. 13.22b indicated that the computed settlement approaches the observation due to the effect of the gravel drain in the backfilled soil, and the drain effect may reduce the damage of quay walls under strong, long-duration ground motions.

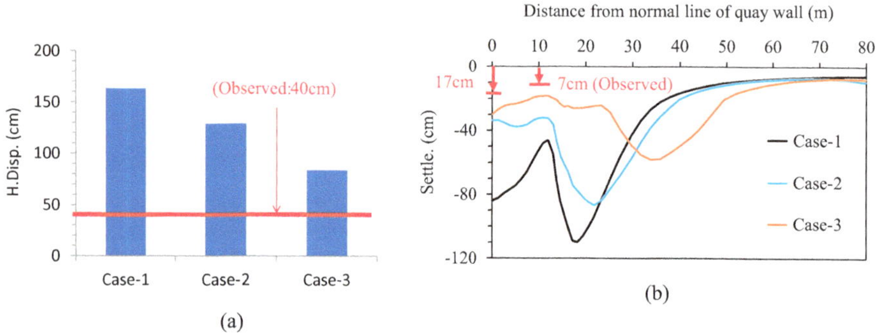

Fig. 13.22 Simulated horizontal displacements at sheet pile head and vertical displacements on ground surface at No.3-wharf in Onahama port. (**a**) Horizontal displacements. (**b**) Vertical displacements

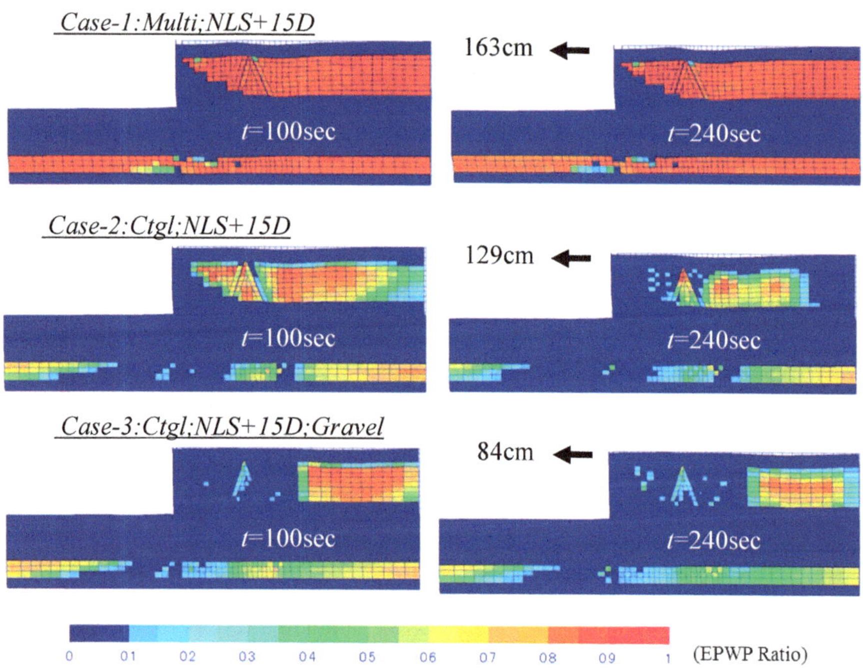

Fig. 13.23 Simulated deformations and excess pore water pressure ratios at the quay wall in No.3-wharf in Onahama port

Table 13.8 Stress damage states of steel members at No.3-wharf in Onahama port

Name	Sheet pile	Anchor	Tie rod
Case-1	Plastic	No	No
Case-2	Plastic	Plastic	No
Case-3	No	Plastic	No
Observed	No	Unknown	No

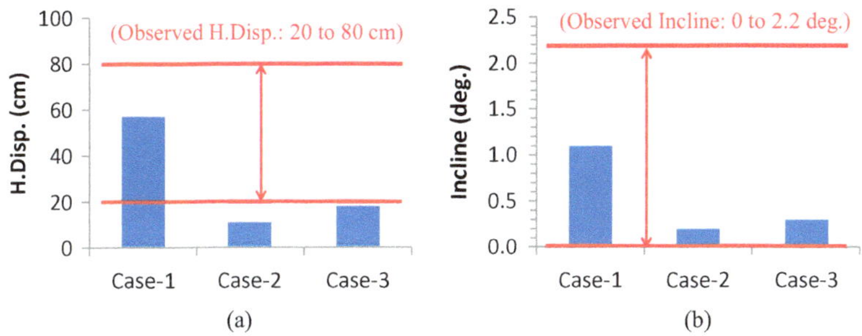

Fig. 13.24 Simulated horizontal displacements at caisson head and inclinations of caisson at No.5-wharf in Onahama port. (**a**) Horizontal displacements. (**b**) Caisson inclinations

Fig. 13.25 Simulated vertical displacements on ground surface at No.5-wharf in Onahama port

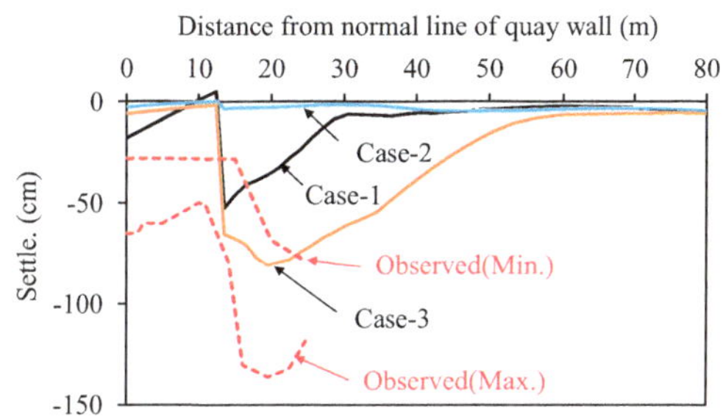

13.4.4 *Quay Wall at the No. 5 Wharf in the Onahama Port*

Figures 13.24, 13.25 and 13.26 show the simulated results of the quay wall at the No. 5 wharf in the Onahama port. In addition to the lateral and vertical displacements, the inclinations of the caisson are given. While the computed lateral displacement with the multi-spring model (Case 1) falls into the observed range, the displacements with the cocktail glass model (Cases 2 and 3) are slightly below the lower limit of the observation. Comparing Cases 2 and 3 in Figs. 13.24a and 13.25 indicated that the computed displacements become closer to the observations by considering the dilatancy of crushed rock debris.

Although the observed lateral and vertical displacements are not perfectly simulated in Case 3, the cocktail glass model can be applied to evaluate seismic performance of caisson-type quay walls under strong, long-duration ground motions. This is because the observed displacements also include the influence of scouring of the foundation underneath the caisson due to the tsunami attack as well as the effect of liquefaction due to shaking.

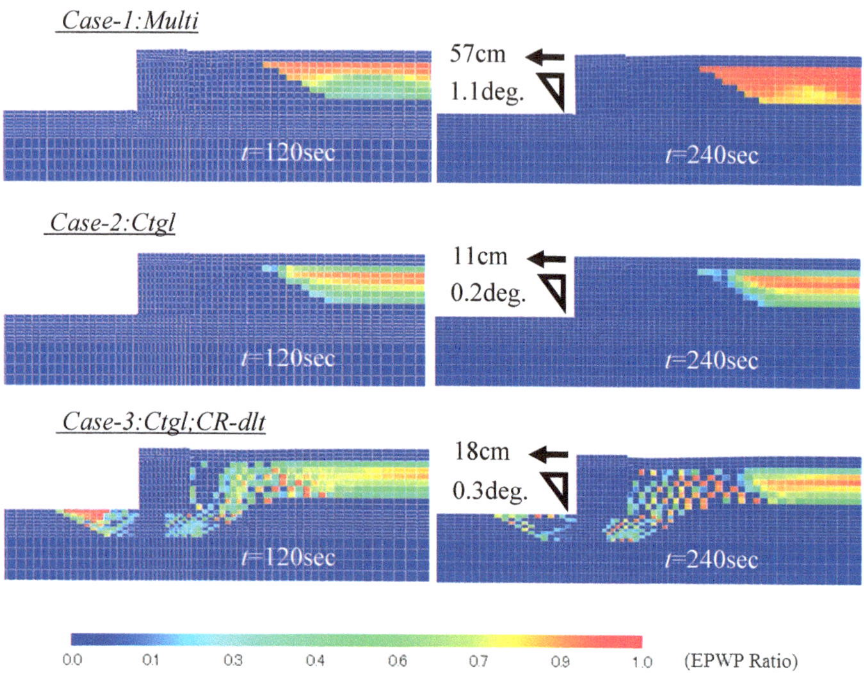

Fig. 13.26 Simulated deformations and excess pore water pressure ratios at the quay wall in No.5-wharf in Onahama port

13.5 Conclusions

The results obtained by the numerical simulations of the damaged quay walls due to the strong, long-duration ground motions using the strain space multiple mechanism models (i.e., the multi-spring model and the cocktail glass model) are summarized as follows:

1. Some quay walls might be damaged mainly due to liquefaction during shaking. The impact of the subsequent tsunami attack is insignificant. This is because the excess pore pressure in the backfilled soils dissipated to some extent before the tsunami's arrival, allowing the soils to regain a sufficient strength to withstand the tsunami's force.
2. The multi-spring model cannot consider the effect of pore water migration and dissipation. Strictly speaking, it should be used under undrained conditions. However, the model may be applicable to evaluate the seismic performance of port structures (e.g., sheet pile–type and caisson-type quay walls), even against strong, long-duration ground motions, if the effects of pore water migration and dissipation are insignificant.

3. The cocktail glass model, which can take the effect of pore water migration and dissipation during and after shaking into account, is effective to evaluate the seismic performance of port structures against strong, long-duration ground motions, particularly when the effect of pore water migration and dissipation cannot be ignored (e.g., in the case of gravel drain).
4. To precisely evaluate the seismic performance of port structures, it is necessary to appropriately model the interaction between structures (e.g., anchored piles) and surrounding soils in the simulations as well as to select an appropriate constitutive model of soils. In addition, the dilatancy effect of crushed rock debris should be appropriately considered in simulations to estimate the ground settlement with a high degree of accuracy.

Acknowledgements We recognize the Sendai Port and Airport Technology survey office for providing valuable data about the damaged quay walls after the earthquake.

This paper is a compilation of the results of the collective efforts of the working group of FLIP consortium for the analysis of damaged quay walls during the 2011 Great East Japan Earthquake off the Pacific Coast of Tohoku. The working group consists of the following members: Susumu Iai, Tetsuo Tobita, Kyohei Ueda, Koji Ichii, Eiji Kohama, Yousuke Ohya, Makoto Yoshida, Keiichi Sumiya, Tsutomu Sakakibara, Dengun Kyoku, Seiji Nishiyama, Takumi Miyazawa, Taro Isayama, Seiichi Sato, Atsushi Mori, Hiroshi Kaneko, Daisuke Shibata, Koichi Sugihara, Koichi Murakami, Tomonari Imono, and Hironobu Murakami. The author, as the coordinator of the working group, hereby gratefully acknowledges the contributions of the working group.

References

Iai S, Kameoka T (1993) Finite element analysis of earth-quake induced damage to anchored sheet pile quay walls. Soils Found 33(1):71–91

Iai S, Matsunaga Y, Kameoka T (1992) Strain space plasticity model for cyclic mobility. Soils Found 32(2):1–15

Iai S, Ichii K, Liu H, Morita T (1998) Effective stress analyses of port structures. Soils and foundations, special issue on geotechnical aspects of the 17 January 1995 Hyogoken-Nambu. Earth 2:97–114

Iai S, Tobita T, Ozutsumi O, Ueda K (2011) Dilatancy of granular materials in a strain space multiple mechanism model. Int J Numer Anal Methods Geomech 35(3):360–392

Morita T, Iai S, Liu H, Ichii K, Sato Y (1997) Simplified method to determine parameter of FLIP. Technical Note of the Port and Airport Research Institute; No.869 (in Japanese)

Murakami K, Sugihara K, Murakami H, Tashiro S, Kohama E, Tobita T (2013) The replicative simulation analysis on sheet pile quay wall with the backfill ground improved by gravel drain in the Onahama harbor No.3 wharf. 33th JSCE annual conference on earthquake engineering; CD-ROM (in Japanese)

Nozu A, Wakai A (2011) Characteristics of ground motions at damaged ports during the 2011 Great East Japan Earthquake Disaster. Technical Note of the Port and Airport Research Institute; No.1244 (in Japanese)

Sakakibara T, Tashiro S, Kyoku D, Ohya Y, Tobita T (2013) Effective stress analysis of quay wall damaged due to liquefaction in Soma port during the 2011 off the Pacific Coast of Tohoku earthquake. 68th JSCE annual meeting; 503–504 (in Japanese)

Shibata D, Kaneko H, Mori A, Sato S, Tashiro S, Ohya Y, Iai S (2014) Simulation analysis of high earthquake resistance Onahama berth No.5(-12m) damaged by the 2011 off the Pacific Coast of Tohoku earthquake. J JSCE 70(4):I_80–I_88. (in Japanese)

Sumiya K, Sakakibara T, Tashiro S, Kyoku D, Kohama E, Ichii K (2013) Numerical simulations for maritime structures damaged due to ground motion during the 2011 off the Pacific Coast of Tohoku earthquake – with quay walls anchored steel tube piles as a target–. 33th JSCE annual conference on earthquake engineering; CD-ROM (in Japanese)

Takahashi S, Toda K, Kikuchi Y, Sugano T, Kuriyama Y, Yamazaki H, Nagao T, Shimosako K, Negi T, Ugeno J, Tomita T, Kawai H, Nakagawa Y, Nozu A, Okamoto O, Suzuki K, Morikawa Y, Arikawa T, Iwanami M, Mizutani T, Kohama E, Yamaji T, Kumagai K, Tatsumi D, Washizaki M, Izumiyama T, Seki K, Yeom G, Takenobu M, Kashima H, Banno M, Fukunaga Y, Sakunaka J, Watanabe Y (2011) Urgent Survey for 2011 Great East Japan Earthquake and Tsunami Disaster in Ports and Coasts. Technical Note of the Port and Airport Research Institute 2011; No.1231 (in Japanese)

Yamazaki H, Gotoh Y, Kohama E, Ohya Y (2012) Experiment of crushed rock debris specimens on volumetric strains due to cyclic loads. 47th JGS annual conference; 411–412 (in Japanese)

Chapter 14
Modeling the Axial Behavior of Piles

Junichi Hyodo, Yoshio Shiozaki, Akito Sone, Osamu Ozutsumi, and Koji Ichii

Abstract The end bearing capacity of a pile is modeled for two-dimensional effective stress analysis by initially conducting a vertical loading test in a centrifugal field and then simulating the vertical loading test by three-dimensional analysis to validate the three-dimensional analysis. Finally, a model, which uses nonlinear spring elements for two-dimensional analysis, is proposed and validated via case studies. In the proposed model, the nonlinear spring elements are characterized by the hyperbolic relationship determined in the three-dimensional analysis.

14.1 Introduction

Effective stress analysis is commonly used to verify the seismic performance in practical designs of port structures such as caisson-type quay walls, sheet pile quay walls, and piled piers (Coastal Development Institute of Technology 2007, 2011). To properly consider the pile–soil interaction, the piles in the structure must be appropriately modeled in dynamic analyses of sheet pile quay walls and piled piers. FLIP ROSE (Iai et al. 1992) (Finite element analysis program of LIquefation Process/Response Of Soil-structure during Earthquakes) is used in most seismic design processes for port structures in Japan, specializing in "pile–soil interaction spring elements" (Ozutsumi et al. 2003) and is available in FLIP to accurately consider pile–soil interactions. With this element, dynamic pile–soil interactions in complex situations (e.g., piles in liquefied ground) can be considered.

J. Hyodo (✉)
TEPSCO (Tokyo Electric Power Services Co., Ltd), Tokyo, Japan
e-mail: hyoudou@tepsco.co.jp

Y. Shiozaki
JFE Steel Corporation, Kawasaki, Japan

A. Sone
NEWJEC.Inc, Osaka, Japan

O. Ozutsumi
Meisosha Corporation, Tokyo, Japan

K. Ichii
Kansai University, Osaka, Japan

S. Iai (ed.), *Developments in Earthquake Geotechnics*, Geotechnical, Geological and Earthquake Engineering 43, DOI 10.1007/978-3-319-62069-5_14

Most studies have focused on the equilibrium forces in the horizontal direction because dynamic pile–soil interactions during an earthquake are due to the horizontal movement of piles and ground deformation. A dynamic pile–soil interaction in the vertical direction is complicated and is not well considered in numerical modeling. In current seismic design scheme applications of FLIP ROSE, a simple model of the pile bottom (e.g., beam element) shares a node with the ground (soil) element at a given position. Consequently, various issues may arise, including the mesh size effect, potential pore water pressure buildup at the ground (soil) element under undrained conditions, and the scale settings of the anti-plane direction in two-dimensional analysis (Ozutsumi et al. 2013). Regardless of these obstacles, practitioners working on seismic designs of port structures expect the pile–soil interaction to be appropriately modeled in the vertical (along the pile) direction.

Some studies have investigated these issues. Hirayama (1990) proposed a hyperbolic approximation of the load-settlement relationship, and the hyperbolic expression agrees well with the results of the in situ pile loading test. Wakita (1998) sought to find the most appropriate formula to express the numerous in situ pile loading test results. The parameters for his formula were set by pile type and pile condition. The hyperbolic approximation was adapted for a stage with loads below a certain level. Yasufuku et al. (2001, 2000) proposed the end bearing capacity of a pile from the spherical cavity expansion theory at the pile end. They reported that a combination of the ultimate bearing capacity from the theory and the hyperbolic approximation of the load–settlement relationship agrees well with the results of the in situ pile loading test.

The pile group effect on the end bearing capacity of a pile has also been examined. AIJ Standards for Design of Building Foundations (2001) defined the pile group efficiency factor η from the vertical bearing capacity at the pile head, which includes the pile group effect on pile friction. In the case of sand, the pile group efficiency factor exceeds 1.0. Vesic (1985, 1969) reported that the group pile effect does not influence the pile tip bearing capacity. (The group pile effect was not evaluated for pile friction). He also reported that the pile group efficiency factor for the pile head bearing capacity can be maximized at 1.3 with a pile interval 3 to 4 times the pile diameter. Aoyama et al. (2013) carried out experiments on pile groups with a closed end pile model in sand. The pile end resistance mainly induced ground deformation; deformation in the case of a pile group can be expressed as the superposition of the deformation for a single pile. Thus, he concluded that pile group effect is insignificant.

This study focuses on the most appropriate modeling of piles with the hyperbolic approximation of the load settlement relationship. Specifically, the group pile effect in the anti-plane direction is examined when modeling a pile via two-dimensional analysis.

14.2 Modeling of Piles

Two methods (Method A and Method B) are used to research pile–soil interactions in the horizontal direction. Method A is used as the pile–soil interaction with a Winkler-type spring between the pile and a fixed point (Architectural Institute of

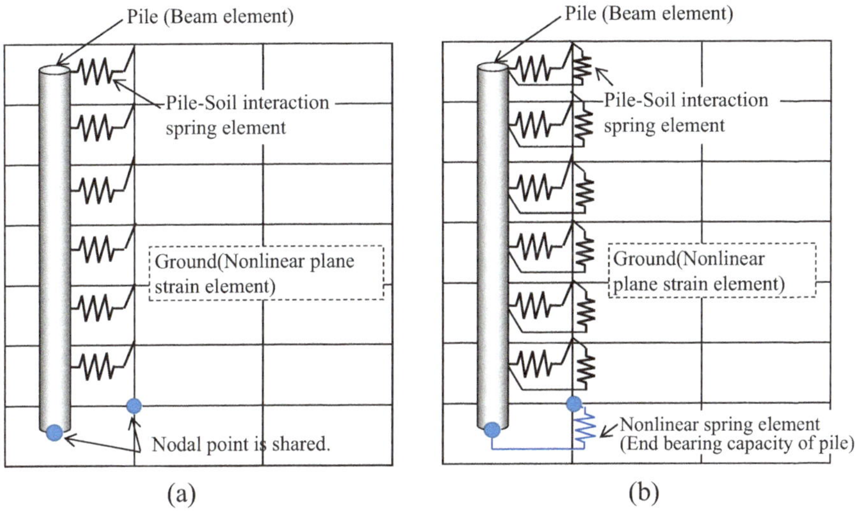

Fig. 14.1 Modeling the pile–soil interaction. (**a**) Method A. (**b**) Method B

Japan 2001). In Method B, we modeled the subgrade reaction in the vertical (along the pile) direction and considered the end bearing capacity of the pile (nonlinear spring element) as well as the function of the shaft friction of the pile.

Method A (Fig. 14.1a) models the subgrade reaction in the horizontal direction by the pile–soil interaction element (Ozutsumi et al. 2003) and the subgrade reaction in the vertical (along the pile) direction by the same displacement between the pile tip and the ground (Coastal Development Institute of Technology 2011). A potential drawback of Method A is that it may not appropriately model the subgrade reaction in the vertical direction of a pile. In particular, the end bearing capacity of a pile in two-dimensional analysis (Method A) is influenced by the pile tip ground, the mesh size, and the width of the anti-plane direction (Ozutsumi et al. 2013).

In Method B (Fig. 14.1b), the end bearing capacity of a pile is modeled by the nonlinear spring element, and the influences of the mesh size for two-dimensional analysis and the anti-plane width are removed from the estimated settlement of the piles. It is possible to double count ground deformation because Method B connects a nonlinear spring element and pile tip ground in a series. Because Method B considers both the deformation of the nonlinear spring element and the deformation of the ground element, it may overestimate the displacement. However, the possible double counting of the ground deformation can be prevented by sufficiently constraining the nodes of the ground to prevent localized deformation.

This paper makes the following contributions:

1. A centrifugal model test of a pile is developed using Method B.
2. Three-dimensional analysis of the centrifugal model test of a pile is implemented.
3. A nonlinear spring element is proposed to simulate the end bearing capacity of a pile.
4. Case study of the nonlinear spring element to simulate the end bearing capacity of a pile is conducted.

14.3 Centrifugal Model Test of a Pile

14.3.1 Experiment Outline

An experiment (Sone et al. 2014) in a 40-G centrifugal field was conducted at the Disaster Prevention Research Institute, Kyoto University. An object was pushed into the sand layer with a brass pile (12-mm diameter and 0.8-mm thick). Figure 14.2 shows the brass pile where the tip of the pile is blocked and the four positions where strain gauge values are obtained. Toyoura standard sand was used for the experiment. The relative density (D_r) in the upper layer was 60% and D_r of the lower layer was 60 or 90%. The coefficient of friction between the pile and the sand was approximately 0.3. Table 14.1 shows the experimental case where the model container had dimensions of 450-mm long, 150-mm wide, and 300-mm deep. The lower layer was formed by an aerial fall or an underwater fall. The penetration length to the lower layer of the pile was set at 1.0 times the diameter of the pile and the upper layer was formed by the aerial fall or water fall.

Figure 14.3 shows the experimental model. We performed the experiment by attaching an electrical actuator to the model container and inserting a 12-mm diameter pile at a speed of 0.02 mm/s. The pile head displacement was measured using a laser displacement meter. In addition, the pile head load was measured using a load cell.

14.3.2 Experimental Results

The pile end resistance was calculated using the value of the strain gauge at point d, which was located 15 mm from the pile end. Table 14.2 shows the parameters for the brass pile. The shaft friction, which was considered to be the range from the pile

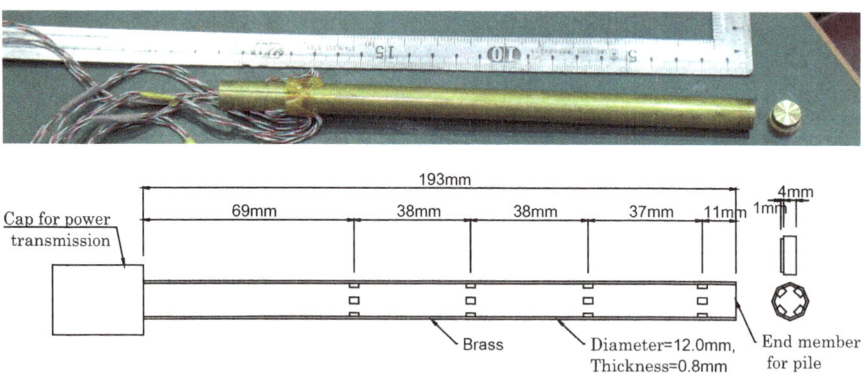

Fig. 14.2 Brass pile

Table 14.1 Experimental case

Case	Drainage condition	Relative density (%)	
		Upper layer	Lower layer
1	Dry	60.0	90.0
2	Saturated	60.0	More than 90
3	Saturated	60.0	60.0

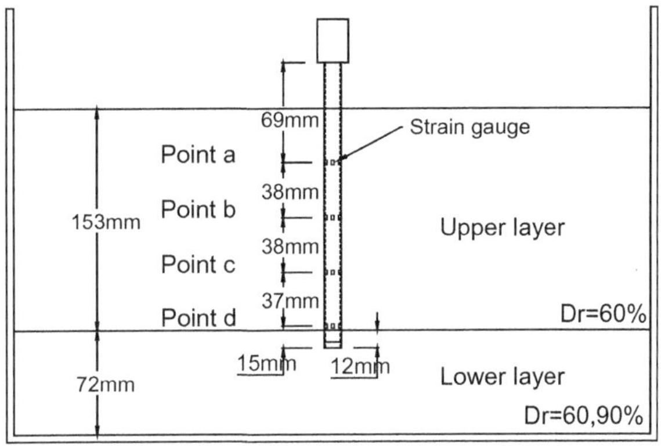

Fig. 14.3 Model in the experiment

Table 14.2 Parameters for the brass pile

Young's modulus E (kN/m^2)	Diameter D (mm)	Thickness t (mm)	Cross-sectional area A (mm^2)	EA (kN)
1.05×10^9	12.000	0.800	28.149	2.956×10^4

head to the 15-mm upper part of the pile tip, was calculated from the difference between the pile end resistance and the pile head load measured by the load cell. The vertical displacement and the load indicated the values on the actual scale. Figure 14.4 shows the relationship between the shaft friction and the vertical displacement in each case. The shaft friction increases rapidly in the initial load stage, but gradually increases in the other stages.

The coefficient of friction was 0.3. The shaft friction was 96 kN in Case 1, but was 58 kN in Case 2 and Case 3. When the vertical displacement is large, the difference between the shaft friction under dry conditions (Case 1) and submerged conditions (Case 2 and Case 3) is small. Figure 14.5 shows the relationship between the pile head load and the vertical displacement in each case.

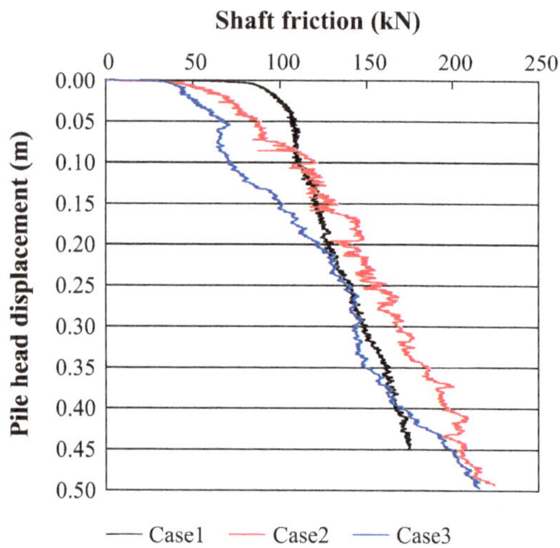

Fig. 14.4 Relationship between the shaft friction and the vertical displacement

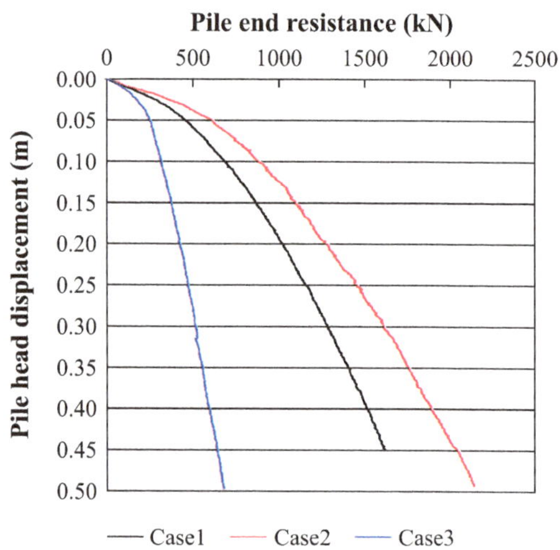

Fig. 14.5 Relationship between the pile head load and the vertical displacement

14.4 Three-Dimensional Analysis of the Centrifugal Model Test

14.4.1 Target and Analysis Condition

A three-dimensional analysis was performed on the centrifugal model test mentioned in the previous section. A single hollow brass pile (prototype scale with a 0.48-m diameter) pressed into sandy ground in a 40-G centrifugal field was simulated. The experimental model [12-m high, 18-m long, and 6-m wide

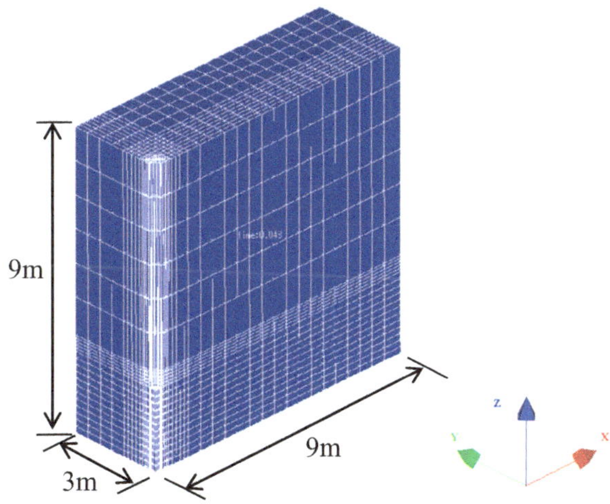

Fig. 14.6 Three-dimensional mesh model for the analysis

(to prototype scale)] was modeled. Case 2 of the test consisted of saturated Toyoura sand in two layers (upper layer: Dr = 60%, lower layer: Dr = 90%).

The three-dimensional analysis program in the FLIP series (FLIP three-dimensional Iai 1993 Ver.1.4.0) was used. Both the fully drained conditions and the fully undrained conditions were analyzed. A quarter (1/4) of all targets were modeled to save computational time (Fig. 14.6). The pile penetration was given by enforced displacement at the pile head nodes. The x-, y-, and z- axes were the length (horizontal), the width (horizontal), and vertical axis, respectively.

The relative densities were used to determine the parameters for the soil types. The corresponding equivalent SPT N values were applied. The N values for an effective overburden stress of 65 kPa following the design standards for Japanese ports (The Ports and Harbours Association of Japan 2007) were determined by the relative density from the Mayerhof (1957) formula and the simplified parameter setting procedures for FLIP (revised version) (Coastal Development Institute of Technology 2007, 2011) was applied. Tables 14.3 and 14.4 summarize the parameters for the soil and the piles, respectively. Three-dimensional multi-spring model elements with the Mohr-Coulomb's failure criteria were used for the soil. The plate bending elements (fixed movement in the x- and y-directions) were used for the pile. Joint elements were inserted between the soil and the piles, and the entire node at the pile tip and the ground to contact the pile tip were combined to have the same movement in the z-direction. Roller boundary conditions were applied to the side boundary (xz-plane and yz-plane), while the bottom boundary was fixed. On the symmetrical plane (y = 0), the movement in the y-direction and the rotational movement around the x-axis and z-axis were fixed. On another symmetric plane (x = 0), the movement in the x-direction and the rotational movement around the y-axis and z-axis were fixed. To focus on the end bearing capacity of the pile, the pile–soil friction at the joint elements was set as zero. Thus, the bearing capacity at the pile head is the same as that at the tip.

Table 14.3 Soil parameters

Layer	Wet density ρ (t/m^3)	Shear modulus G_{ma} (kPa)	Bulk modulus K_{ma} (kPa)	Reference confining effective stress σ'_{ma} (kPa)	Reference parameter m_G, m_K	Cohesion C (kPa)	Internal friction angle ϕ (degrees)
Lower layer	1.93	8.97×10^4	2.34×10^5	98.0	0.5	0.0	39.86
Bearing stratum	1.99	1.51×10^5	3.94×10^5	98.0	0.5	0.0	42.05

Table 14.4 Pile parameters

Diameter D (m)	Thickness t (m)	Young's modulus E (kPa)	Poisson's ratio n	Area A (m^2)	Second moment of area I (m^4)
0.48	0.03	1.05×10^9	0.35	0.045	0.00114

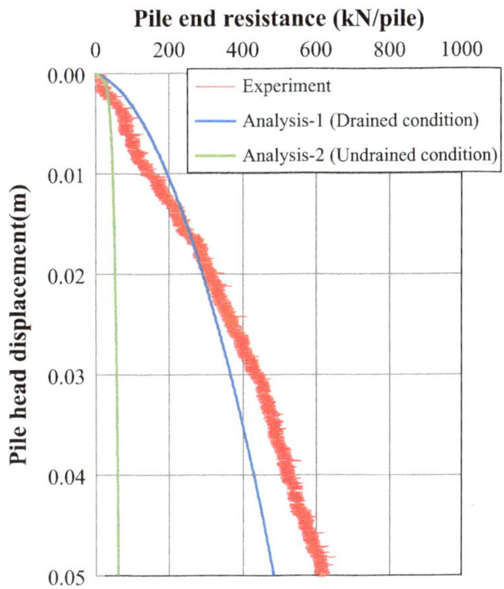

Fig. 14.7 Three-dimensional mesh model for the analysis. Shaft friction is eliminated in both the test and analysis results

14.4.2 Analysis Results

Figure 14.7 shows the load-settlement relationship. In this study, the end bearing capacity of a pile is defined as the load at settlement of 10 % of the pile diameter. In the analysis, the bearing capacity in fully drained conditions is larger than that in fully undrained conditions, which agrees well with the test results. In the analysis, the mechanism for the increased bearing capacity is the increase in the strength of the soil element accompanying the increase in the effective confining stress in the undrained conditions. In the undrained conditions, the load acting on the pile head is supported by both the soil element and the pore water element in the analysis, and the increase in the effective confining stress in the soil element is insignificant. Hence, consideration of the loading velocity and the soil permeability is important.

Figures 14.8 and 14.9 show the deformation of the ground around the pile tip (at the moment of settlement for 10 % of the pile diameter) with the distribution of the effective vertical stress σ_z' and the deformation of the ground with the maximum shear strain γ_{max}, respectively. The shape of the pressure bulb can be observed as the region of high vertical stress and large shear strain. In this case, the amount of highly concentrated stress and large strain is approximately two to three times the pile diameter in both the vertical and horizontal directions.

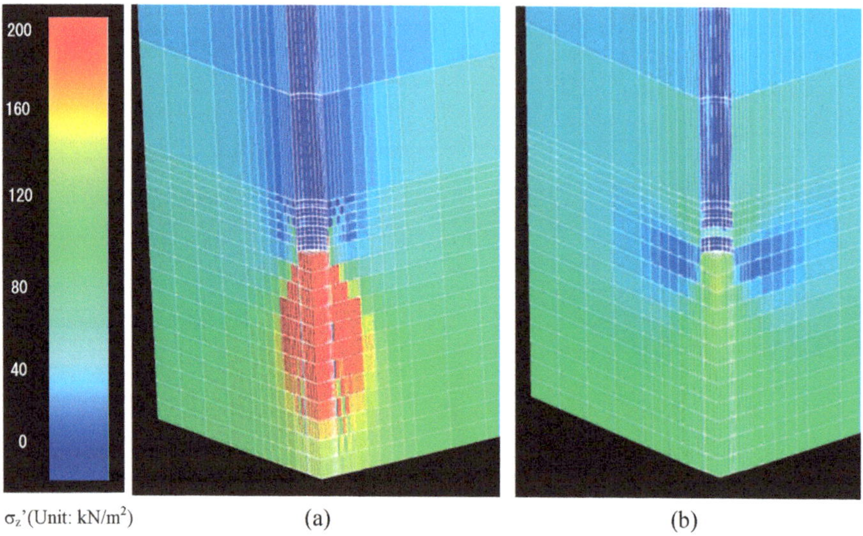

Fig. 14.8 Deformation (at settlement of 10 % of the pile diameter) and the vertical effective stress σ'_v. (**a**) Fully drained condition. (**b**) Fully undrained condition

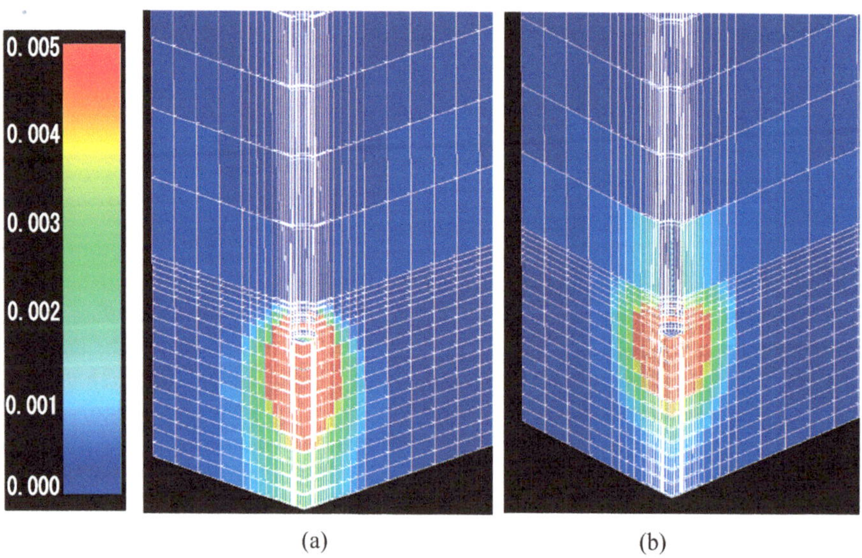

Fig. 14.9 Deformation (at settlement of 10 % of the pile diameter) and the maximum shear strain $\gamma_{\max}$. (**a**) Fully drained condition. (**b**) Fully undrained condition

14.4.3 Conclusion of Three-Dimensional Analysis

The load-settlement relationship observed in the centrifugal model test is successfully simulated by three-dimensional analysis in fully drained conditions. Since the applicability of the three-dimensional analysis is confirmed, we decided to examine the proposed nonlinear spring element for the end bearing capacity of a pile in practical two-dimensional analysis.

14.5 Nonlinear Spring Element to Simulate the End Bearing Capacity of a Pile

14.5.1 Proposal to Consider the Differences Between the Two-Dimensional and Three-Dimensional Analyses and Our Target

FEM simulations of the pile end resistance indicate that three-dimensional analysis generally produces better results than two-dimensional analysis. In practice, two-dimensional analysis is preferable to three-dimensional analysis due to its simplicity.

The soil behavior around the pile tip depends on the pile penetration. Because the affected area spreads three-dimensionally around the pile tip, it is difficult for two-dimensional analysis to appropriately consider the influence area as it inherently difficult to accurately simulate the behavior of soil around the pile tip in two-dimensional analysis. Furthermore, since the pile end resistance is related to the influence area in spreading soil, the out-plane direction cannot be considered in the analysis, and the settlement of the pile may be overestimated in two-dimensional analysis due to an underestimation of the pile end resistance.

Although two-dimensional analysis has the aforementioned difficulties, it can be applied if the target is clarified. In our proposal, the simulation of the local soil behavior around the pile tip is skipped, and we focus on settlement of the pile tip and load transfer to the ground. To appropriately simulate pile settlement, we propose a nonlinear spring element between the pile tip and the ground. For the load transfer from the pile to the ground, we propose constraints for the nodal points in the FEM mesh.

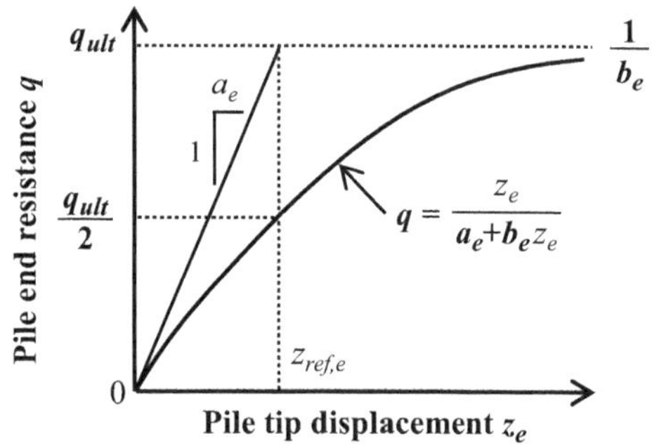

Fig. 14.10 Relationship between the pile tip displacement and the end resistance (Hirayama 1990)

14.5.2 Nonlinear Spring Element Between the Pile Tip and the Ground

Hirayama (1990) proposed a hyperbolic approximation between the pile end resistance q and the settlement (vertical displacement) of a pile, as shown in Eq. 14.1 and Fig. 14.10. Parameters a_e and b_e are given by Eq. 14.2. The hyperbolic approximation is consistent with the observed results of the in situ test (Hirayama 1990). Furthermore, only two parameters are used in the hyperbolic correlation, improving the practicality. Thus, we used the hyperbolic approximation as the nonlinear spring element between the pile tip and the ground.

Since there are only two parameters, a wide range of settlements must be considered because the observations and model behavior may not be consistent. To avoid this difficulty, the level of settlement should be carefully considered in advance, and the appropriate parameters for the estimated level of settlement should be given.

$$q = \frac{z_e}{a_e + b_e z_e} \tag{14.1}$$

$$a_e = \frac{z_{ref,e}}{q_{ult}} = \frac{0.25D_e}{q_{ult}}, b_e = \frac{1}{q_{ult}} \tag{14.2}$$

Reference displacement $z_{ref,e} = 0.25D_\mathrm{e}$ (m) in Eq. 14.2 is given for sandy soil. D_e is the pile diameter and q_{ult} is the ultimate value of the pile end resistance. The reference displacement is provided by Hirayama (1990) from the loading test results for sandy soil (i.e., BCP Committee 1971). In this paper, Eq. 14.2 is used as the basis to determine the parameters.

If any in situ loading test results are given, the hyperbolic approximation can be adjusted to the test data by changing the value of the reference displacement $z_{ref,e}$ and the maximum value of the tip resistance q_{ult}. Hirayama (Yasufuku et al. 2001) proposed a reference displacement of $z_{ref,e} = (0.16–0.31)D_e$(m).

Generally, applying a load corresponding to the maximum value in the tip resistance q_{ult} is difficult in an in situ loading test. Thus, for the hyperbolic approximation based on in situ loading test results, the load corresponding to a small settlement such as $0.1D_e$ is preferred over the maximum value of the pile end resistance q_{ult}. In addition, some design standards regard the load corresponding to the pile tip settlement to be approximated by $0.1D_e$ as the end bearing capacity of pile (Architectural Institute of Japan 2001).

The hyperbolic approximation of the in situ loading test data using the pile tip resistance $q_{0.1}$ (kN/m^2) at a pile tip settlement of $z_e = 0.1D_e$ (m) is determined via the following procedure:

Equation 14.1 for the pile tip settlement of $z_e = 0.1D_e$ (m) gives

$$q_{0.1} = \frac{0.1D_e}{a_e + b_e \cdot 0.1D_e}. \tag{14.3}$$

Equations 14.3 and 14.2 give

$$q_{0.1} = 0.286q_{ult}. \tag{14.4}$$

Then Eqs. 14.4 and 14.2 give the following equation for the parameters a_e and b_e.

$$a_e = \frac{0.0715D_e}{q_{0.1}}, b_e = \frac{0.286}{q_{0.1}} \tag{14.5}$$

The hyperbolic correlation between the pile tip settlement and the pile end resistance was examined using the results of three-dimensional FEM analysis described in Sect. 14.4. Table 14.5 summarizes the parameters in the hyperbolic approximation. Cases with Dr = 60 and 75% have the same conditions as the three-dimensional analysis of Case 2 in Sect. 14.3, except that the relative densities for the lower layer differ. The parameters used for the pile in the three-dimensional analysis are the same as those shown in Table 14.4.

Figure 14.11 shows the relationship between the pile tip settlement (m) and the pile end resistance (kN/m^2). As a reference, the hyperbolic approximation based on the test data of the pile tip resistance $q_{0.05}$ (kN/m^2) with a pile tip settlement of $z_e = 0.05D_e$ (m) is shown in Fig. 14.12. For the cases in Fig. 14.12, Eqs. 14.3, 14.4 and 14.5 are similarly modified. As an index of the fitting error, the differences between the three-dimensional analysis results and the hyperbolic approximation are evaluated by the average of the squared residual errors. For the lines in Fig. 14.11, the averages of the squared residual errors are 4.81×10^4 (kN/m^2)^2 (for the settlement range of 0–0.096 m) and 6.53×10^4 (kN/m^2)^2 (for a settlement range of 0–0.048 m). Similarly the averages of the squared residual errors in Fig. 14.12 are 3.04×10^5 (kN/m^2)^2 (for the settlement range of 0–0.096 m) and 5.55×10^4 (kN/m^2)^2 (for the settlement range of 0–0.048 m). Thus, for a settlement range of 0–0.096 m, employing the test data at $z_e = 0.1D_e$ in the hyperbolic approximation provides better results than the test data at $z_e = 0.05D_e$. On the

Table 14.5 Parameters for Hirayama's model

	Case 1	Case 2	Case 3	
D_r of lower layer	60%	75%	90%	Unit
$q_{0.1}$	1763	2183	2628	(kN/m^2)
a_e	1.947×10^{-5}	1.572×10^{-5}	1.306×10^{-5}	(m/(kN/m^2))
b_e	1.622×10^{-4}	1.310×10^{-4}	1.088×10^{-4}	(1/(kN/m^2))
q_{ult}	6164	7631	9188	(kN/m^2)
$A \cdot q_{ult}$	1115	1381	1663	(kN/pile)

Fig. 14.11 Relationship between the pile end resistance and the pile head displacement

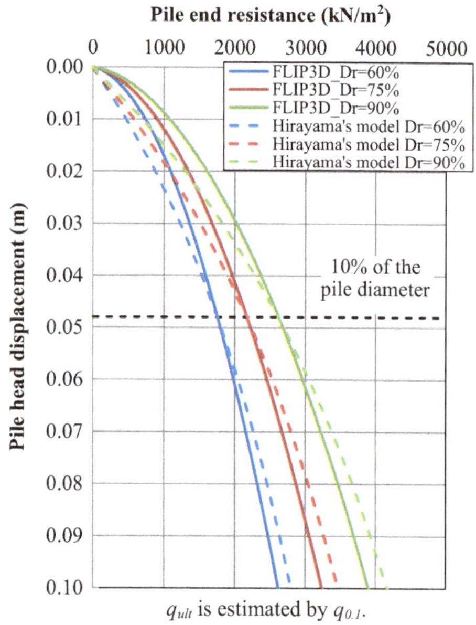

other hand, if the settlement is limited to a range 0–0.048, using the test data at $z_e = 0.05D_e$ in the hyperbolic approximation provides better results.

Consequently, these results demonstrate the difficulty in obtaining agreement between the observation and the model behavior when considering a wide range of possible settlements. However, these results also indicate that parameters corresponding to the possible settlement range can be provided and that the observed settlement is close to test data when determining parameters ($z_e = 0.1D_e$ in Fig. 14.11 and $z_e = 0.05D_e$ in Fig. 14.12). The approximation agrees well with the three-dimensional analysis results.

The possible degree of settlement and the required level of agreement in the approximation may be case dependent (e.g., structure type and expected performance of the structure). If a possible settlement range of 0–0.096 m (20% of the pile diameter) is assumed, the hyperbolic approximation proposed by Hirayama (Ozutsumi et al. 2013) in our proposed parameter setting from the test data at $z_e = 0.1D_e$ may be applied as the skeleton curve (the relationship for the initial

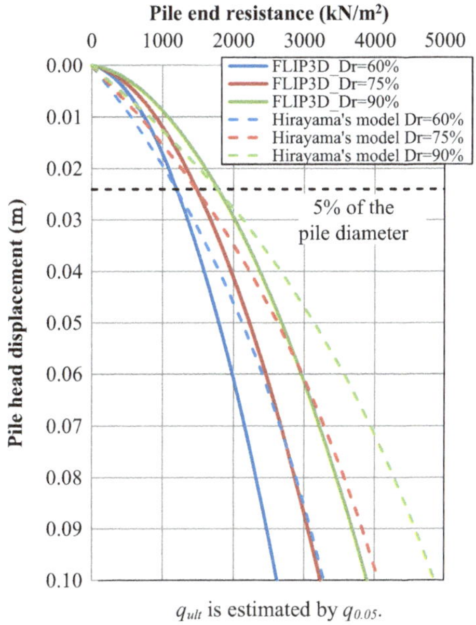

Fig. 14.12 Relationship between the pile end resistance and the pile head displacement

monotonic loading case between the pile tip settlement and the pile end resistance) because this agrees well with the three-dimensional analyses results shown in Fig. 14.11. There are only three parameters for the spring: the pile diameter D_e(m), the pile end resistance at the pile tip settlement of $0.1D_e$: $q_{0.1}$ (kN/m^2), and the reference displacement $z_{ref,e}(=0.25D_e)$. Note that the proposal of Hirayama was for an open-end pile, whereas we assume that it is applicable to a close-end pile. Thus, the applicability of the hyperbolic relationship and the appropriate value for reference displacement $z_{ref,e}$ $(= 0.25D_e)$ remain for a future study.

14.5.3 Cyclic Behavior of a Nonlinear Spring Element Between the Pile Tip and the Ground

We propose installing a nonlinear spring element between the pile tip and the ground. The spring can demonstrate the relationship between the pile loading in the axial direction and the pile settlement using the hyperbolic approximation proposed by Hirayama (1990) and Yasufuku et al. (2001). Hysteresis of the spring is modeled as follows. For the unloading stage, the displacement rate of the pile is assumed to be same as that of the pile settlement in the initial (virgin) loading stage. However, the pullout resistance is not considered. In other words, the pile end resistance is zero when the pile tip settlement value becomes negative. Thus, the pile end resistance is always positive or zero.

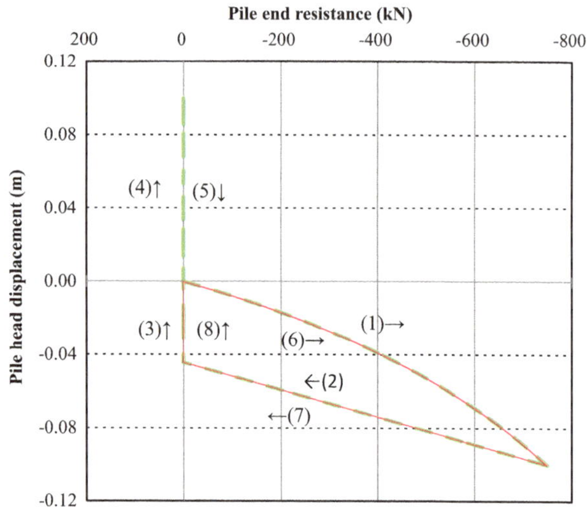

Fig. 14.13 Hysteresis loop of the proposed nonlinear spring element

In this modeling, the movement of the ground-side nodal point of the spring is an important issue. If the nodal point moves, settlement of the pile tip may be double-counted due to the behavior of the spring (the relative displacement of the pile- side nodal point to the ground-side nodal point of the spring) and the displacement of the ground-side nodal point of the spring from the original position. Thus, displacement of the ground-side nodal point of the spring is a constraint. However, if the nodal point is fixed, the load from the pile is not transferred to the ground around the pile tip.

Figure 14.13 illustrates the hysteresis loop of the proposed nonlinear spring between the pile tip and the ground. The ground-side nodal point of the spring is fixed, and element behavior is modeled as loading (1), unloading (2 and 3), pullout (4), pushback (5), reloading (6), or re-unloading (7 and 8).

14.5.4 Analysis of the Vertical Loading Test of the Pile and Nodal Point Constraints Around the Pile Tip

Analysis of the vertical loading test of a pile in the reference (Case 2 in Sect. 14.3) was attempted using the proposed nonlinear spring element between the pile tip and the ground. The enforced displacement was applied at the pile tip, and shaft friction of the pile was not considered.

The mesh size may influence the results. In this analysis, each mesh was designed to be approximately 1 m^2. The anti-plane width (the scale of mesh size in the out-plane direction) was 6 m, which is the width of the soil tank used in the test. Therefore, the unit of the pile-end resistance was “kN/pile”. Analysis was carried out under the plane strain conditions, corresponding to the case where the

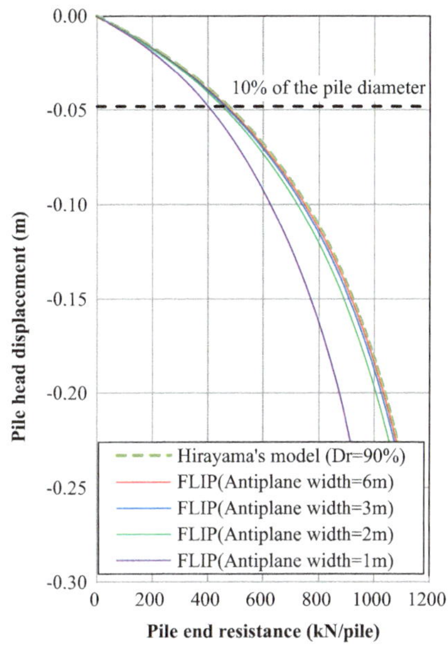

Fig. 14.14 Relationship between the pile end resistance and the pile head displacement under fully drained conditions

piles are located at 6 m intervals. In the calculation, the element stiffness matrix and element mass matrix for the ground were six times that of the matrixes at an anti-plane width of 1 m. (The mesh size scale in the out-plane direction was 1 m.) This suggests that considering the appropriate anti-plane width can suppress the nodal point of the ground around the pile tip. Although it is quite difficult to accurately simulate the soil behavior around the pile tip in two-dimensional analysis, this pseudo-three-dimensional modeling of the ground increases the tip resistance related to the influence area of soil spreading in the anti-plane direction, which may lead to an improved estimation of the pile settlement.

Figure 14.14 shows the effect of the anti-plane width (scale of mesh size in the out-plane direction) under the drained condition. At an anti-plane width of 6 m, the ground stiffness (constraint effect of the ground-side nodal point) is sufficiently large and the analysis results agree well with the hyperbolic approximation. However, when the anti-plane width is 1 m, pile settlement is overestimated and the computed bearing capacity is insufficient. This can be considered as the group pile effect in the anti-plane direction.

Figure 14.15 shows the results of vertical loading under undrained conditions. Here, the pile load increases the pore water pressure of the soil, but an increase in the effective stress of the ground is not observed. Consequently, the ground stiffness or strength does not increase. Thus, the ground-side nodal point tends to settle, and the bearing capacity of pile is underestimated in the analysis.

In summary, the condition of the soil around the pile tip is very important because the pile end resistance is obtained by the reaction force in a series of nonlinear spring and FEM elements of the soil around the pile tip. When the soil

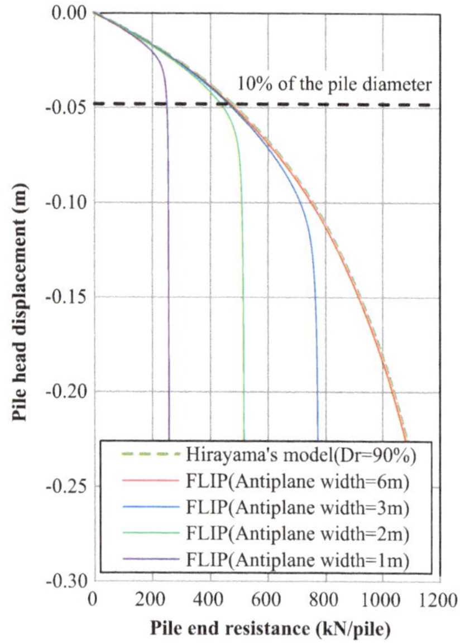

Fig. 14.15 Relationship between the pile end resistance and the pile head displacement under undrained conditions

around the pile tip is softened by an excessive increase of pore water pressure, the effect of the nonlinear spring between the pile tip and the ground is negligible, and the end bearing capacity of the pile is significantly reduced. Qualitatively, these results may agree with the observation, but due to the lack of sufficient test data results, a future qualitative consideration is necessary. In an actual application of the proposed method, the deformation characteristics and the pore water pressure level of the soil elements around the pile tip should be verified to confirm the appropriateness of the analysis.

14.6 Applicability of the Proposed Nonlinear Spring Element: Case Studies

14.6.1 Case Study Target

We studied the applicability of the nonlinear spring elements in order to simulate the end bearing capacity of a pile using a case from the 1993 Kushiro-oki Earthquake. In the Fishery Pier, X-type sheet piles were used for the sheet pile quay wall with coupled piles and steel piles for the anchor piles (Fig. 14.16).

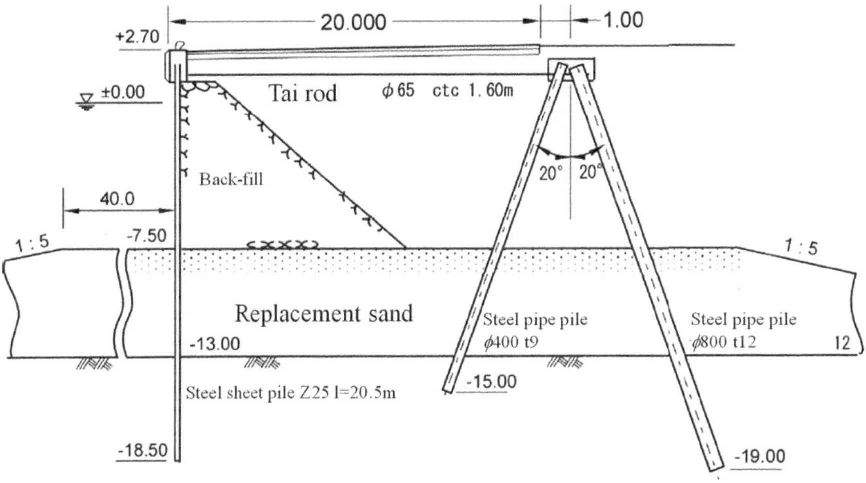

Fig. 14.16 Fishing port facilities before the earthquake (Ueda et al. 1993; Shiozaki et al. 2004)

14.6.2 Analysis Conditions

We modeled the ground using multi-spring model elements. The lateral boundary condition was a viscous boundary. The soil parameters were determined by the relative density (simplified parameter setting procedures for FLIP ROSE (revised version) [1]). Nonlinear beam elements were used for the steel sheet pile and steel pipe pile, while nonlinear spring elements were used for the tie rod. Table 14.6 shows the analysis cases. To model the interaction between the anchor piles and the ground, Method A (Case A) and Method B (Case C) were used (Fig. 14.1).

In addition, the shaft friction force of the pile in Method B was that of a 15-degree shaft friction angle. Nonlinear spring elements were installed between the pile tip and the ground to simulate the end bearing capacity of the pile. The upper limit of the nonlinear spring elements was determined by the N-value. The anti-plane width (the interval between anchor piles) was 1.6 m. The Rayleigh damping β (β = 0.0005) was decided by seismic response analysis (non-liquefied analysis). The input motion was 28.5 s (17.5–46.0 s) for the principal shock (Fig. 14.17). For effective stress analysis in the FLIP series, FLIP ver7.1.9-6-2-5 test was used.

14.6.3 Analysis Results

Figure 14.18 shows the residual deformation. Various sheet pile quay walls have almost identical displacements. The displacement of the anchor pile head is as follows:

Table 14.6 Analysis cases

	End bearing capacity	Shaft friction
Case A	Pile side nodal point shares with ground side nodal point at same position	–
Case B	Pile side nodal point shares with ground side nodal point at same position	Pile-soil interaction spring element $\phi_j = 15°$
Case C	Non-linear spring element	Pile-soil interaction spring element $\phi_j = 15°$

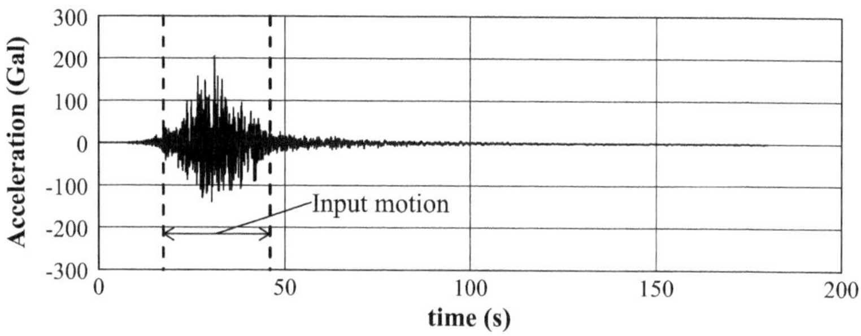

Fig. 14.17 Input motion

Method A (Case A, B) < Method B (Case C)

This occurs because we estimated the difference in connecting the condition between the pile and the ground in the anchor pile tip. (In Method A, the horizontal and vertical displacements are almost same. In Method B, the nonlinear spring elements of the pile tip are used). Comparison with the actual occurrence is difficult due to accuracy issues in the case history data. Thus, an appropriate model of the end bearing capacity or the pile is needed.

Figures 14.19, 14.20 and 14.21 summarize the maximum section force. Both Method A and Method B have almost the same section force of the steel sheet pile. The axial force of the pulling pile of anchor piles is small from the pile head to the pile tip due to the shaft friction of the pile.

14.6.4 Summary

To investigate the applicability of nonlinear spring element modeling for the end bearing capacity of a pile, we simulated the sheet pile quay walls damaged by the 1993 Kushiro-oki Earthquake. The analysis results of Case A (Method A) and Case C (Method B) are similar to the data from the actual occurrence. Additionally, both methods provide almost identical sheet pile behaviors. In contrast, the steel pipe

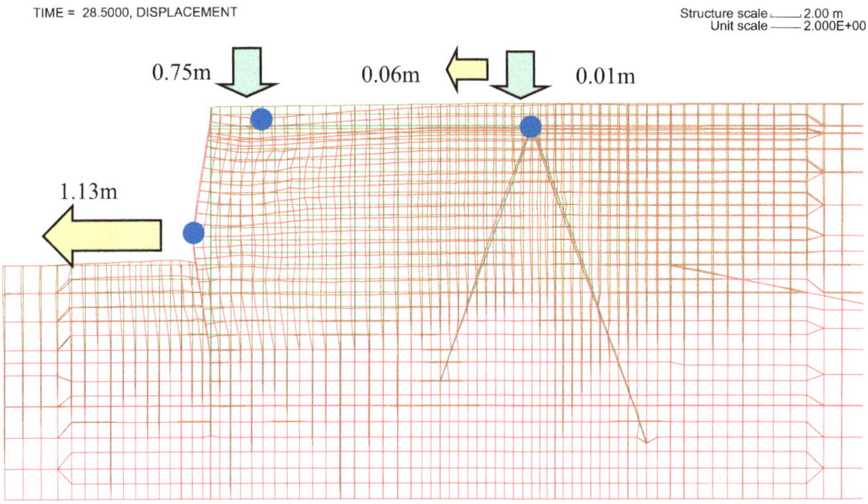

(a)

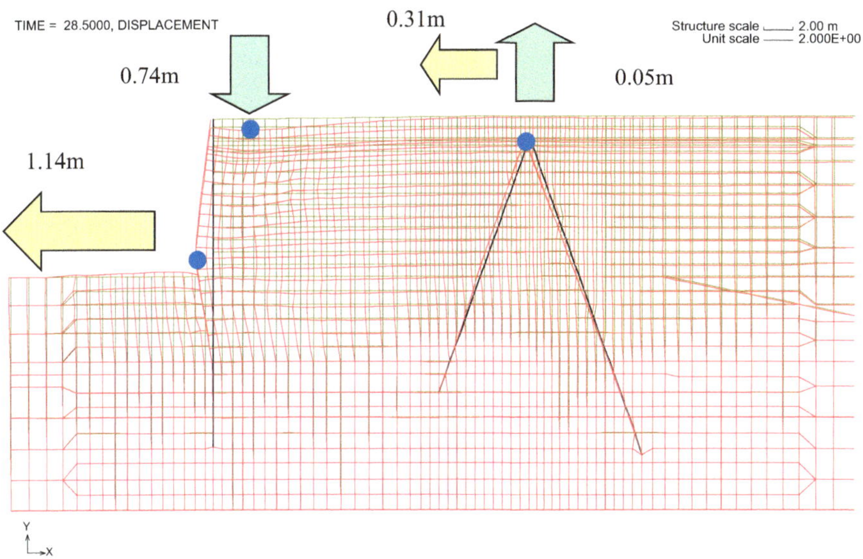

(b)

Fig. 14.18 Residual deformation. (**a**) Case A. (**b**) Case C

pile behavior depends on the method. Method A evaluated the deformation of the anchor piles to a lesser degree than the actual occurrence, but the model appropriately predicted the end bearing capacity of the pile. Using Method B, we proposed a potential model that can predict the end bearing capacity of a pile.

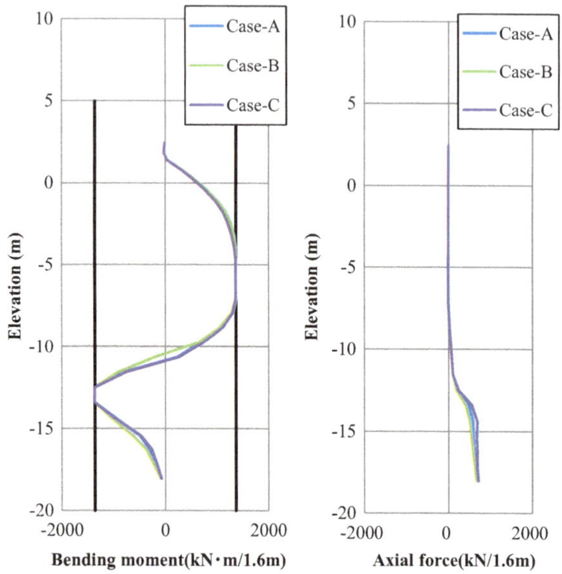

Fig. 14.19 Maximum section force of the sheet pile

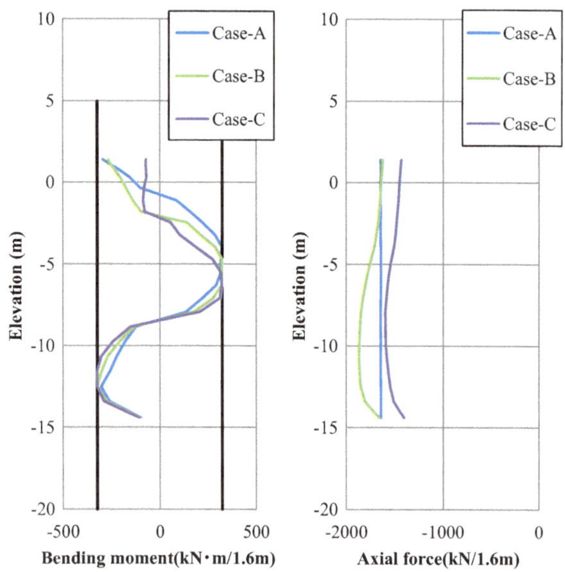

Fig. 14.20 Maximum section force of the counter pile (pushing pile)

14.7 Conclusions

The centrifugal model test using three-dimensional analysis is appropriately simulated. The relationship between pile tip displacement and the end bearing capacity of a pile is confirmed using the hyperbolic relationship proposed by Hirayama.

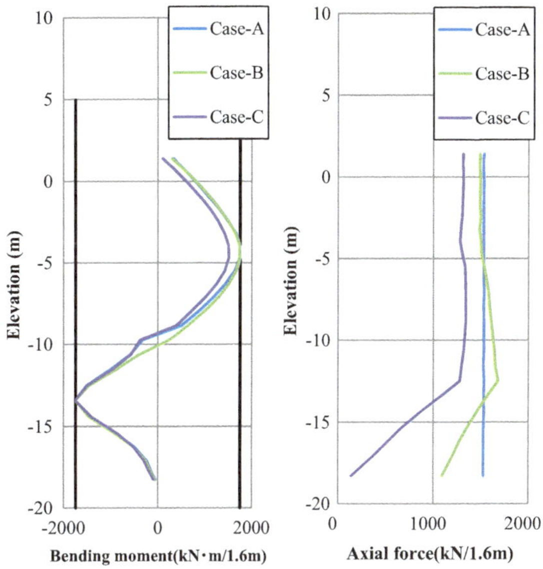

Fig. 14.21 Maximum section force of the counter pile (pulling pile)

We proposed modeling the end bearing capacity of a pile using Hirayama's proposed hyperbolic relationship. Then we simulated case studies. Compared to Method A, Method B provides the following improvements:

1. The end bearing capacity of a pile is clarified.
2. Overvaluation of the pulling resistance is reduced.

Currently we are analyzing the characteristics of the proposed method.

References

Aoyama S, Luki D, Goto S, Towhata I (2013) Study of ground deformation during group pile loading by PIV analysis. In: The 48th Japan national conference on geotechnical engineering, pp 1273–1274 (in Japanese)

Architectural Institute of Japan (2001) Recommendations for design of building foundations (in Japanese)

Committee BCP (1971) Field tests on piles in sand. Soils Found 11(2):29–49

Coastal development institute of technology (2007) Case studies on seismic design of port structures (in Japanese)

Coastal development institute of technology (2011) Dynamic analysis practices of FLIP (in Japanese)

Hirayama H (1990) Load-settlement analysis for bored piles using hyperbolic transfer functions. Soils Found 30(1):55–64

Iai S (1993) Three dimensional formulation and objectivity of a strain space multiple mechanism model for sand. Soils Found 33(1):192–199

Iai S, Matsunaga Y, Kameoka T (1992) Strain space plasticity model for cyclic mobility. Soils Found 32(2):1–15

Japan Geotechnical Society (1985) Design method and explanation of pile foundations (in Japanese)

Meyerhof GG (1957) Discussion. In: Proceedings of 4th international conference on soil mechanics and foundation engineering, vol 3, p 110

Ozutsumi O, Tamari Y, Oka Y, Ichii K, Iai S, Umeki Y (2003) Modeling of soil-pile interaction subjected to soil liquefaction in plane strain analysis. In: The 28th Japan national conference on geotechnical engineering, pp 1899–1900 (in Japanese)

Ozutsumi O, Ueno D, Hyodo J, Tobita T (2013) Analytical consideration of the end bearing capacity of the piles. In: The 48th Japan national conference on geotechnical engineering, pp 1293–1294 (in Japanese)

Shiozaki Y, Sugano T, Kohama E (2004) Experiment and analysis of earthquake performance of sheet pile quaywalls. Ann J Civ Eng Ocean JSCE 20:131–136. (in Japanese)

Sone A, Kuwabara N, Tobita T (2014) Loading test of pile in a centrifugal model experiment, The 49th Japan national conference on geotechnical engineering, pp 1191–1192 (in Japanese)

The Ports and harbours association of Japan (2007) Technical standards and comment arise for port and harbor facilities in Japan (in Japanese)

Ueda S, Inatomi T, Uwabe T, Iai S, Kazama M, Matsunaga Y, Hujimoto T, Kikuchi Y, Miyai S, Sekiguchi S, Hujimoto Y (1993) Damage to port structures by the 1993 Kusiro-oki earthquake, technical note of the port and harbor research institute ministry of transport, No. 766 (in Japanese)

Vesic AS (1969) Experiments with instrumented pile groups in sand. ASTM STP 444:177–222

Wakita E (1998) Study on standard settlement characteristic of pile. Proceedings of JSCE, No. 603/III-44, pp 45–52

Yasufuku N, Ochiai H, Ohno S (2000) Geotechnical analysis for predicting pile bearing capacity. The 45th Japan national symposium on geotechnical engineering, pp 163–168

Yasufuku N, Ochiai H, Ohno S (2001) Pile end bearing capacity of sandy ground related to soil compressibility and it's application, Tsuchi-to-Kiso, vol 49, No 3

Chapter 15
Modelling of Cohesive Soils: Soil Element Behaviors

Tomohiro Nakahara, Kyohei Ueda, and Susumu Iai

Abstract The strain space multiple mechanism model, which was originally developed for the cyclic behavior of granular materials such as sand, is adapted to idealize the stress–strain behavior of clay under monotonic and cyclic loads. Compared to the conventional elasto-plastic models of the Cam-clay type, advantages of the proposed model include (1) the arbitrary initial K_0 state can be analyzed by static gravity analysis, (2) the stress-induced anisotropy (i.e., the effect of initial shear) in the steady (critical) state can be analyzed based on Shibata's dilatancy model (Ann Disaster Prev Res Inst Kyoto Univ 6:128–134, 1963), (3) over-consolidated clay can be analyzed by defining the dilatancy at the steady state based on the over-consolidation ratio, and (4) the strain-rate effects for monotonic and cyclic shears can be analyzed based on the Isotach/TESRA model proposed by Tatsuoka et al. (Soils Found 42(2):103–129, 2002) in a strain rate ranging from zero to infinity as well as by the conventional strain-rate effects of the secondary consolidation (creep) type. Simulations of the drained/undrained behaviors of clay under monotonic and cyclic loadings are used to demonstrate the performance of the proposed model.

15.1 Introduction

Combining geotechnical hazards provides a forum on new and challenging issues, including combined failure due to a tsunami after an earthquake, the effects of aftershocks on liquefaction after the mainshock, and the effects of deformation due to consolidation of clay before an earthquake. Although these issues have not been studied in the past, recent case histories during earthquakes have exposed them as new engineering problems to be solved by advanced and sophisticated approaches.

T. Nakahara (✉)
Penta-Ocean Constraction Co. Ltd., Tokyo, Japan
e-mail: tomohiro.nakahara@mail.penta-ocean.co.jp

K. Ueda • S. Iai
Disaster Prevention Research Institute, Kyoto University, Kyoto, Japan

S. Iai (ed.), *Developments in Earthquake Geotechnics*, Geotechnical, Geological and Earthquake Engineering 43, DOI 10.1007/978-3-319-62069-5_15

This paper presents the formulation and fundamental performance of a new constitutive model for clay based on the framework of the strain space multiple mechanism model of granular materials. Although numerous elasto-plastic models for the conventional Cam-clay type have been proposed for clay over the years, practical engineering applications of these models to the seismic response of clayey ground or clay and sand mixtures are non-existent or extremely limited to the knowledge of the authors. Major drawbacks of conventional elasto-plastic (or visco-elasto-plastic) models for clay are: (1) the initial K_0 state, which has significant effects on the deformation due to seismic load, is specified as soil parameters instead of evaluating via static gravity analysis, which represents the realistic initial stress field in soil-structure systems, and (2) the viscous effects of clay are typically formulated to simulate the secondary consolidation behavior and not earthquake loading, which may include an extremely high strain rate in the shear strain. This study aims to solve these problems through the framework of the strain space multiple mechanism model.

This paper is a compilation of the collective efforts of the working group of the FLIP consortium (see Acknowledgements), primarily based on Iai et al. (2015) and Iai and Ueda (2016), with additional examples on the performance of the proposed model.

15.2 Modelling of Cohesive Soils in the Strain Space Multiple Mechanism Model

15.2.1 Basic Form of the Strain Space Multiple Mechanism Model

The basic form of the strain space multiple mechanism model for sand is given in two dimensions as the relation between the macroscopic stress tensor and the strain tensor as (Iai et al. 2011)

$$\boldsymbol{\sigma} = -p\mathbf{I} + \int_0^{\pi} q\langle \mathbf{t} \otimes \mathbf{n} \rangle \mathrm{d}\omega \langle \mathbf{t} \otimes \mathbf{n} \rangle = \mathbf{t} \otimes \mathbf{n} + \mathbf{n} \otimes \mathbf{t} \tag{15.1}$$

where $\mathbf{I}$ denotes the identity tensor. $\mathbf{n}$ and $\mathbf{t}$ are the direction vectors normal and tangential to the particle contacts within a granular material, respectively. Scalars p and q are functions with history of the volumetric strain ε, volumetric strain due to dilatancy ε_{d}, and virtual simple shear strain γ, and are defined as

$$p = p(\varepsilon'), q = q(\gamma), \varepsilon' = \varepsilon - \varepsilon_{\mathrm{d}}, \varepsilon = \mathbf{I} : \boldsymbol{\varepsilon}, \gamma = \langle \mathbf{t} \otimes \mathbf{n} \rangle : \boldsymbol{\varepsilon} \tag{15.2}$$

In particular, q is a hysteretic function that approximates the hysteretic hyperbolic curve of the macroscopic shear stress and the shear strain in a cyclic simple shear.

The tangential bulk modulus and the elastic shear modulus are given by power functions of the confining pressure p from the values at a reference state $p = p_{\mathrm{a}}$ as

$$K = K_{\mathrm{La}}(p/p_a)^{n_{\mathrm{K}}}, G = G_{\mathrm{ma}}(p/p_{\mathrm{a}})^{m_{\mathrm{G}}} \tag{15.3}$$

Under the condition of K_0 consolidation where the lateral normal strain is fixed, the vertical normal strain governs both the volumetric and deviator strains. Using the same power index $n_{\mathrm{K}} = m_{\mathrm{G}} = m$, this K_0 consolidation leads to the following equation, which relates the coefficient of earth pressure at rest K_0 with the internal angle ϕ_{f}, as

$$K_{\mathrm{La}} = \frac{1}{1-m}\frac{(1+\sin\phi_{\mathrm{f}})K_0 - (1-\sin\phi_{\mathrm{f}})}{(1-K_0)\sin\phi_{\mathrm{f}}}G_{\mathrm{ma}} \tag{15.4}$$

For analysis, the power index can be set to $m = 0.5$ (or any number except for 1) for static gravity analysis to achieve a realistic K_0 state level ground.

15.2.2 Steady State and Stress-Induced Anisotropy

In the strain space multiple mechanism model, the steady-state line in the p–ε plane is defined by a vertical shift in the normal consolidation curve (Fig. 15.1). The normal consolidation curve to analyze clay is given with the reduction factor r_{K} and the power index $l_{\mathrm{K}} = l_{\mathrm{G}} = 1$ as

$$K_{\mathrm{L/U}} = r_{\mathrm{K}}K_{\mathrm{U0}}(p/p_0)^{l_{\mathrm{K}}}, G = G_{\mathrm{m0}}(p/p_0)^{l_{\mathrm{G}}} \tag{15.5}$$

where the subscript 0 implies the initial state for consolidation or dynamic analysis after static gravity analysis representing the initial stress–strain field.

The integrated form of the volumetric mechanism in Eq. 15.5 with $l_{\mathrm{K}} = 1$ is given as a typical straight line in the e–ln$p$ plane. The vertical shift of the curve is measured by the contractive component of dilatancy at the steady state where the subscript US implies the steady (ultimate) state. This component is defined based on Shibata's dilatancy model (Shibata 1963) as

$$\varepsilon_{\mathrm{d}}^{\mathrm{c}} = -D\eta^*, \eta^* = \left(1/\sqrt{2}\right)|\overline{\boldsymbol{\sigma}}/p - \overline{\boldsymbol{\sigma}}_{\mathbf{0}}/p_0|, \overline{\boldsymbol{\sigma}} = \boldsymbol{\sigma} - p\mathbf{I} \tag{15.6}$$

where the initial stress ratio term $\overline{\boldsymbol{\sigma}}_{\mathbf{0}}/p_0$ represents the stress-induced anisotropy (Sekiguchi and Ohta 1977).

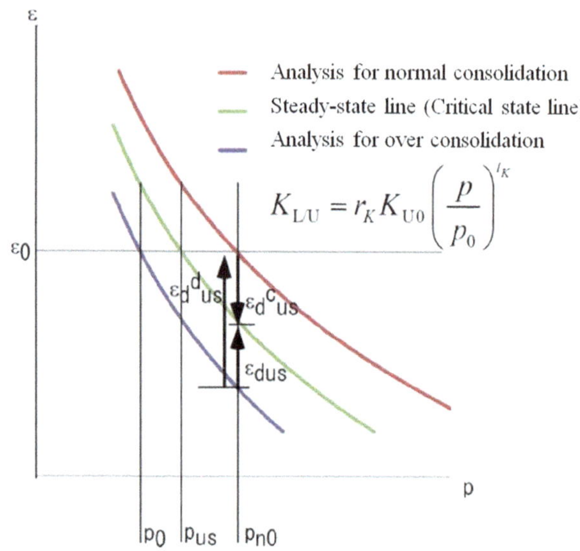

Fig. 15.1 Steady-state lines

The parameter D is related to the parameters for the Cam-clay model M, λ, κ, and the initial void ratio e_0 as (Iizuka and Ohta 1987)

$$DM = (1 - \kappa/\lambda)\varepsilon_{m0}, \quad \varepsilon_{m0} = \lambda/(1 + e_0) = p_0/(r_K K_{U0}) \tag{15.7}$$

Consequently, the contractive component of dilatancy in the steady state is given by the following equation and is substituted for the upper limit of the contractive component of dilatancy ε_d^{cm} in the strain space multiple mechanism model.

$$\varepsilon_{dus}^{c} = -\Lambda\left(\eta_{us}^{*}/M\right)\varepsilon_{m0}, \quad \Lambda = (1 - \kappa/\lambda) \tag{15.8}$$

The stress ratio at the steady state η_{us}^{*}, which is indicated by the length of the arrow in Fig. 15.2, represents the stress-induced anisotropy of clay.

To analyze over-consolidated clay, the curve governing the volumetric mechanism is defined by the vertical shift of the normal consolidation curve. The initial confining pressure p_0 at the over-consolidated state and the corresponding confining pressure at the normal consolidation line p_{n0} are defined as $p_{n0} = r_{p_{n0}} p_0$, where $r_{p_{n0}}$ is related with the over-consolidation ratio OCR, λ, and κ.

$$p_{n0} = r_{p_{n0}} p_0, \quad r_{p_{n0}} = \mathrm{OCR}^{(1-\kappa/\lambda)} \tag{15.9}$$

Then the dilative component of the dilatancy in a steady state is given by the following equation.

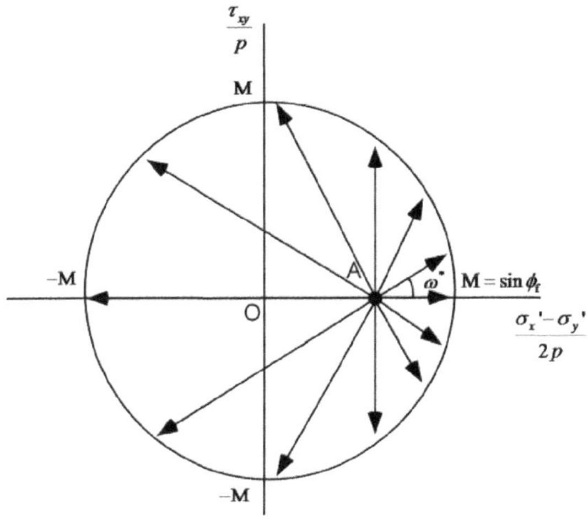

Fig. 15.2 Stress ratio at the steady state η^*_{us}

$$\varepsilon^{d}_{dus} = \varepsilon_{m0} \ln(p_{no}/p_0) \tag{15.10}$$

Over-consolidated clay can be analyzed by defining the dilatancy at the steady state based on the over-consolidation ratio. Consequently, the dilatancy at the steady state is given by

$$\varepsilon_{dus} = \varepsilon^{c}_{dus} + \varepsilon^{d}_{dus} \tag{15.11}$$

15.2.3 Shear Strength of over-Consolidated Clay

For an undrained shear initiated from the over-consolidated state, the stress path crosses the shear failure lines and reaches the steady state on the shear failure lines. To model this behavior, the shear strength τ_m is given by the cohesion c and the internal friction angle ϕ_f. Further cohesion is given by a function of the dilative component of dilatancy ε^{d}_{d}.

$$\begin{aligned} \tau_m &= c^* \cos\phi_f + p \sin\phi_f, c^* = 4\varepsilon^*(1-\varepsilon^*)c, \\ c &= (1 - p_0/p_a)c_a(p_{n0}/p_a), \varepsilon^* = \varepsilon^{d}_{d}/\varepsilon^{d}_{dus} \end{aligned} \tag{15.12}$$

15.2.4 Strain-Rate Effects

The strain-rate effects in the proposed model are formulated in the strain space multiple mechanism model with the Isotach/TESRA mechanism representing the viscosity during shear (Tatsuoka et al. 2002) in addition to the conventional creep mechanism for the secondary consolidation. The constitutive equation is written by adding the viscous terms $\boldsymbol{\sigma}^*$ to Eq. 15.1 as

$$\boldsymbol{\sigma} = -p\mathbf{I} + \int_0^{\pi} q\langle \mathbf{t} \otimes \mathbf{n} \rangle \mathrm{d}\omega + \boldsymbol{\sigma}^* \quad \boldsymbol{\sigma}^* = \int_0^{\pi} q^* \langle \mathbf{t} \otimes \mathbf{n} \rangle \mathrm{d}\omega \tag{15.13}$$

The term of the viscous effect q^* is formulated in the strain space multiple mechanism model by considering the viscosity function of the Isotach model as the upper limit of linear viscous damping at a given strain rate regime. This formulation has an advantage because numerical analysis of clayey ground during an earthquake is stable compared to the original formulation of the Isotach/TESRA model by Tatsuoka et al. (2002). This formulation is written as

$$q^* = r_{\mathrm{ISO}} r_{\mathrm{q}} \frac{\dot{\gamma}^*}{\max|\dot{\gamma}^*|} \left(q_{\mathrm{v}} \frac{\max|\xi|}{1 + \max|\xi|} \right) \quad r_{\mathrm{q}} = 1 - \exp\left(1 - (1 + \max|\dot{\gamma}^*|)^{q_{\mathrm{ISO}}}\right) \tag{15.14}$$

where $\xi = \gamma/\gamma_{\mathrm{v}}$ and $\dot{\gamma}^* = r_{\dot{\gamma}} \dot{\gamma}$. The Isotach/TESRA parameters used by Tatsuoka et al. (2002) are replaced by $r_{\mathrm{ISO}} = \alpha, r_{\dot{\gamma}} = 1/\dot{\varepsilon}_{\mathrm{r}}^{\mathrm{ir}}$, and $q_{\mathrm{ISO}} = m$. The shear strength q_{v} and the reference strain γ_{v} in the virtual simple shear mechanism are given by the macroscopic shear strength and the reference strain as $q_{\mathrm{v}} = \tau_{\mathrm{m}}/2$ and $\gamma_{\mathrm{v}} = (\pi/4)\gamma_{\mathrm{m}}$.

The specific characteristics of the Isotach/TESRA model relative to the simple linear viscous damping are attributed to the parameter q_{ISO}, which is typically so small that the viscous term is insensitive to the strain rate level. With the parameter $q_{\mathrm{ISO}} = 0.04$ for Kaolin, a shear strain rate ranging from $\dot{\gamma}^* = 10^{-5}$ to 10^{+7} corresponds to the magnitude of the viscous function r_{q}, which ranges from 0.1 to 0.9.

15.2.5 Model Parameters

Table 15.1 summarizes the model parameters described above. Many of the parameters used to model the behavior of sand are not required to model that of clay, and are set to default values.

The specific parameters for clay are indicated below. The power index, which describes the confining pressure dependency of the tangent stiffness, is $l_{\mathrm{K}} = l_{\mathrm{G}} = 1.0$ (0.5 in the case of sand). The phase transfer angle is equal to the internal friction angle $\phi_{\mathrm{p}} = \phi_{\mathrm{f}}$. A parameter describing the contractive and dilative dilatancy

Table 15.1 Model parameters

Symbol	Mechanism	Parameter designation
Model parameters for representing shear and dilatancy		
$K_{L/Ua}$	Volumetric	Bulk modulus at reference state
n_K, m_G	Volumetric/shear	Power index for bulk/shear moduli at static gravity analysis
r_K	Volumetric	Reduction factor of bulk modulus for consolidation analysis
r_{pn0}	Volumetric	Over-consolidation state
G_{ma}	Shear	Shear modulus at reference state
ϕ_f	Shear	Internal friction angle ($M = \sin\phi_f$)
c_a	Shear	Cohesion at reference state (for over-consolidation regime)
h_{max}	Shear	Upper bound for hystcretic damping factor
$r_{\varepsilon_d^c}$	Dilatancy	Parameter controlling contractive component
q_2	Dilatancy	Parameter controlling later phase of contractive component
Λ	Dilatancy	Parameter controlling dilatancy at steady state ($\Lambda = (1 - \kappa/\lambda)$)
Model parameters tor representing strain rate effects		
α_c	Volumetric	Secondary consolidation coefficient
$\dot{v}_0$	Volumetric	Initial volumetric creep strain rate
r_{ISO}	Shear	Isotach parameter representing max amplitude $r_{ISO} = \alpha$
$r_{\dot{\gamma}}$	Shear	Isotach parameter for normalizing strain rate $r_{\dot{\gamma}} = 1/\dot{\varepsilon}_r^{ir}$
q_{ISO}	Shear	Isotach parameter for representing strain rate dependency $q_{ISO} = m$

is $r_{\varepsilon_d} = 1$ (about 0.2 in case of sand). A parameter for the initial phase of dilatancy is $q_1 = 1$. The upper limit for the contractive dilatancy is $\varepsilon_d^{cm} = \varepsilon_{duc}^{c}$ (by substitution as part of an algorithm). A small positive number for numerical stability at liquefaction is given as $S_1 = 0.005$. The parameter defining elastic range for liquefaction is $c_1 = 1.0$. The steady-state shear strength is replaced by the parameter $\Lambda = (1 - \kappa/\lambda)$ to control the dilatancy in the steady state.

15.3 Model Performance Under Drained/Undrained Monotonic Loadings

15.3.1 Performance Under an Drained Shear Compounded with Consolidation and Shear

The performance of the proposed model to simulate the strain behavior under drained monotonic loading was examined using the parameters listed in Table 15.2. A clay element was initially consolidated $K_0 = 0.6$, and then strain controlled tests were performed under various strain paths with the volumetric strain dominant regime (Fig. 15.3) and the deviator strain dominant regime

Table 15.2 Model parameters for the simulation (p_a = 135.5 kPa)

$K_{L/Ua}$ (kN/m^2)	n_K, m_G	r_K	r_{pn0}	G_{ma} (kN/m^2)	ϕ_f (°)	c_a (kN/m^2)	h_{max}	$r_{\varepsilon_d^c}$	q_2	Λ
2155	0.5	0.348	1.0	352	33.3	20	0.24	0.85	1.0	0.544

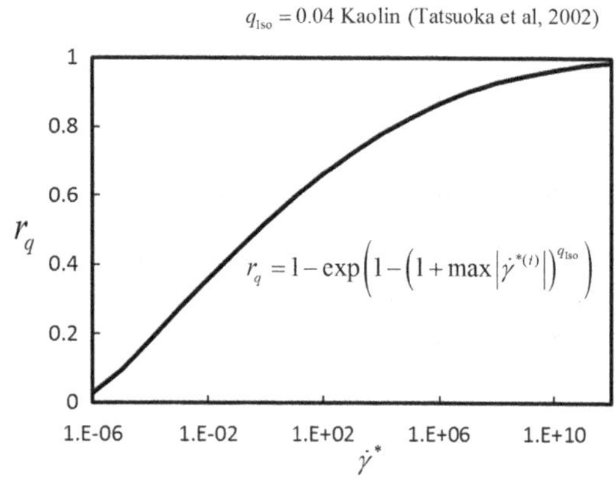

Fig. 15.3 Strain-rate effects with the Isotach mechanism representing the viscosity during a shear

(Fig. 15.4a). The computed results in the stress paths and the consolidation curves are shown in Figs. 15.4b, c and 15.5b, c.

The primary findings for the proposed model are:

1. After K_0 consolidation, the stress path following the initial K_0 line (Figs. 15.4b, c and 15.5b, c, purple line) is given by the strain path with a volumetric strain increase and constant deviator strain (Figs. 15.4a and 15.5a, purple line).
2. After K_0 consolidation, the stress path given by the strain paths with a fixed lateral normal strain (i.e. $\varepsilon_x - \varepsilon_y = -(\varepsilon_x + \varepsilon_y)$) (Fig. 15.4a, green and yellow) are directed closer to the shear failure line (Fig. 15.4b, green and yellow).
3. After K_0 consolidation, the consolidation curves approach the steady-state lines affected by the stress-induced anisotropy (Fig. 15.5b, c).
4. The proposed model does not exhibit the "metastability" phenomenon (Takeyama et al. 2013) that traps a certain range of the stress paths into a single stress path on the same K_0 line (Fig. 15.4c).

15.3.2 Performance Under the Undrained Shear

As previously mentioned, the proposed model allows the stress-induced anisotropy (i.e., the effect of the initial shear) at the steady state to be analyzed based on Shibata's dilatancy model (1963). Additionally, the proposed model allows over-consolidated clay to be analyzed by defining the dilatancy in the steady state based on the over-consolidation ratio. Actually, the contractive dilatancy in the steady state can be designated by an analysis option switch ICSW, as described below.

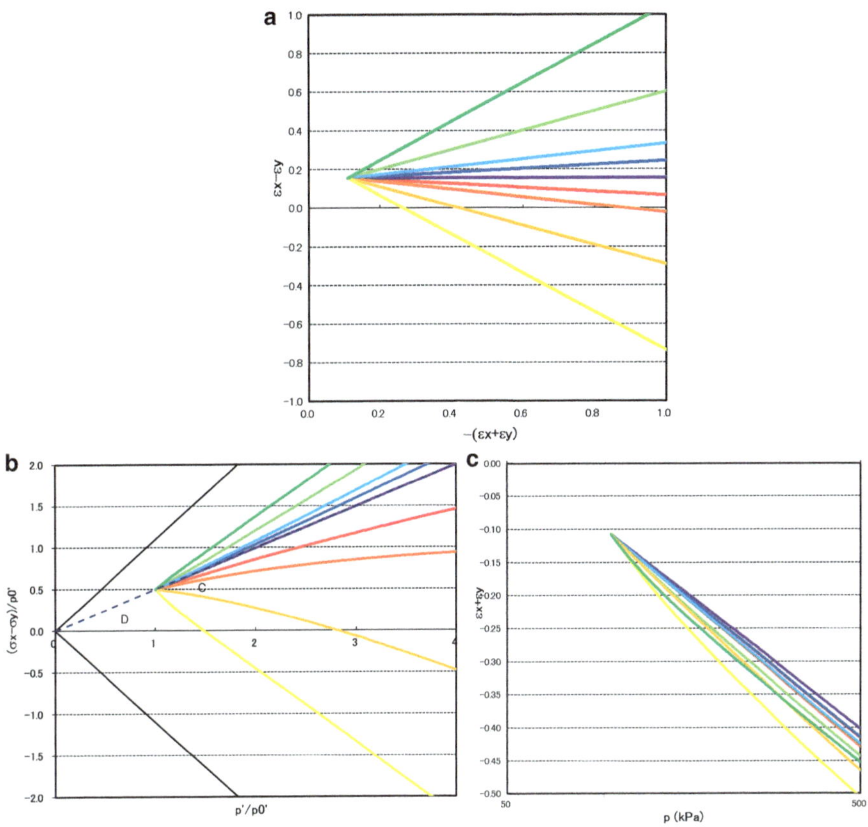

Fig. 15.4 Performance under drained monotonic loading (for the volumetric strain dominant regime). (**a**) Strain paths. (**b**) Computed stress paths. (**c**) Computed consolidation curves

$$\begin{aligned}
\varepsilon_{\mathrm{dus}}^{\mathrm{c}} &= -\Lambda\left(\eta_{\mathrm{us}}^{*}/\mathrm{M}\right)\varepsilon_{\mathrm{m0}} && \mathrm{ICSW}=1:\text{Sekiguchi and Ohta model type}\\
\varepsilon_{\mathrm{dus}}^{\mathrm{c}} &= -\Lambda\varepsilon_{\mathrm{m0}} && \mathrm{ICSW}=2:\text{Cam-clay model type}\\
\varepsilon_{\mathrm{dus}}^{\mathrm{c}} &= -\ln\left(p_{\mathrm{us}}/p_{\mathrm{n0}}\right)\varepsilon_{\mathrm{m0}} && \mathrm{ICSW}=3:\text{Steady state}\mathrm{q}_{\mathrm{us}}\text{ is designated directly}
\end{aligned} \tag{15.15}$$

Contractive dilatancy until it reaches the steady state is given by the following equation. The behavior approaching the upper limit of the contractive dilatancy $\varepsilon_{\mathrm{d}}^{\mathrm{cm}}$ can be controlled flexibly and precisely using parameters to control the increment of contractive dilatancy $r_{\varepsilon_{\mathrm{d}}}$, the contribution of contractive dilatancy near the ultimate limit state r_{tmp}, etc.

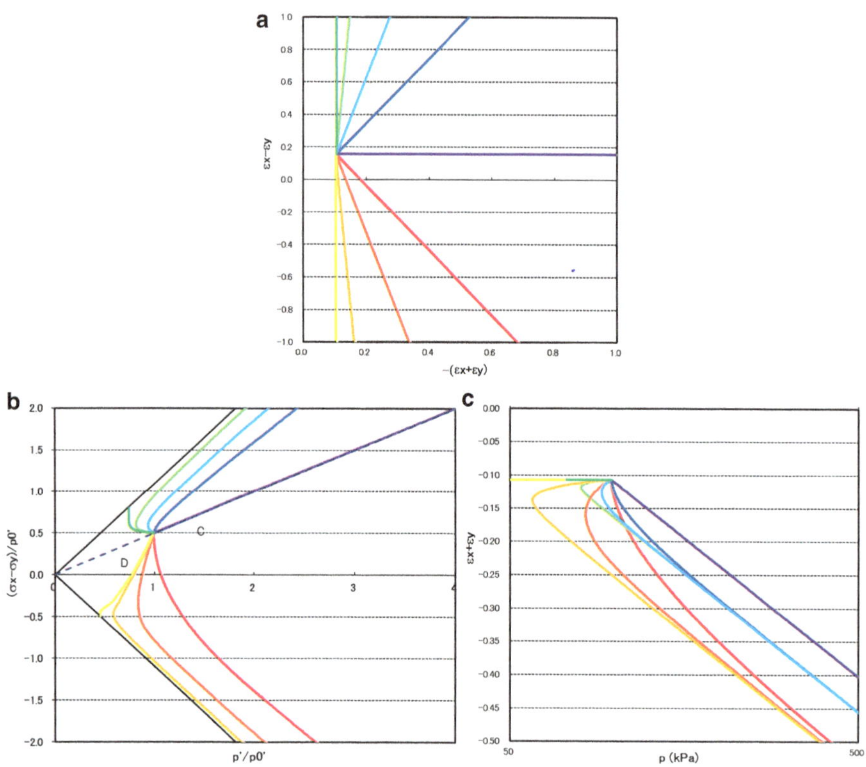

Fig. 15.5 Performance under drained monotonic loading (for the deviator dominant regime). (**a**) Strain paths. (**b**) Computed stress paths. (**c**) Computed consolidation curves

$$\varepsilon_{\rm d}^{\rm c} = -r_{\varepsilon_{\rm d}} r_{\varepsilon_{\rm d}^{\rm c}} r_{S_0} r_{\rm tmp} \sum_{i=1}^{I} \left(1 - \frac{\left(-\varepsilon_{\rm d}^{\rm c}\right)}{\varepsilon_{\rm d}^{\rm cm}} \right) {\rm M}_{\rm v0} \left| {\rm d}\gamma_{\rm p}^{(i)} \right| \Delta\omega^{(i)}$$
$${\rm M}_{\rm v0} = \frac{{\rm M}_{\rm p}}{\sum_{i=1}^{\rm I} |\sin\omega_i| \Delta\omega_i} \qquad {\rm M}_{\rm p} = \sin\phi_{\rm p} \tag{15.16}$$

The ability of the proposed model to simulate the stress–strain behavior under undrained monotonic loading was evaluated using the parameters shown in Table 15.3. A clay element was initially consolidated at $K_0 = 0.6$ with over-consolidation ratios of OCR = 1, 2, 5, and 20 to achieve the same initial void ratio of $e_0 = 0.84$, and then sheared in compression and extension. ICSW = 1, the contractive dilatancy at the steady state, is designated as the Sekiguchi and Ohta model type. Figure 15.6a illustrates the element simulation.

The initial confining pressure p_0 for the initial volumetric strain ε_0 can be defined based on formulation (15.9) $p_{\rm n0} = r_{p_{\rm n0}} p_0$. The initial confining pressure p_0 is related with the corresponding confining pressure at the normal consolidation line $p_{\rm n0}$, the

Table 15.3 Model parameters for the undrained monotonic loading simulation

p_a (kN/m²)	G_{ma} (kN/m²)	n_K, m_G	K_{La} (kN/m²)	K_{Ua} (kN/m²)	ϕ_f, ϕ_p (°)	c_a (kN/m²)	$r_{\varepsilon_d^{cm}}$	$r_{\varepsilon_d^c}$	h_{max}	n
135.5	352	0.5	2155	4310	33.3	20	0.5	0.85	0.24	0.67
r_{ε_d}	q_1	q_2	l_K	r_K	q_{us} (kN/m²)	q_4	ICSW	Λ	$r_{p_{n0}}$	$r_{m_{tmp3}}$
1.0	1,0	0.1	1.0	0.174	30	0.001	1	0.544	1.0	1.0

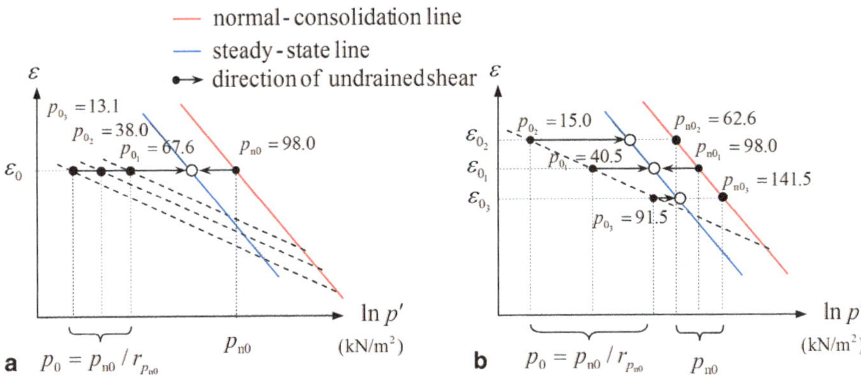

Fig. 15.6 Images of the element simulation. (**a**) Initial void ratio e_0 is fixed. (**b**) Initial void ratio e_0 is variable

over-consolidation ratio OCR, λ, and κ. By this formulation, the over-consolidation state can be evaluated without using the relation to the stress–strain path in consolidation and unloading analyses. The parameters are shown in Table 15.4a.

Where the subscript 0 implies initial state for consolidation or dynamic analysis after static gravity analysis for representing the initial stress-strain field.

Furthermore, to confirm the void-ratio dependence of the steady state and the stress–strain path, the clay element was initially consolidated at $K_0 = 0.6$ with various initial void ratios of $e_0 = 0.75$, 0.84, and 0.95 (OCR = 2.21, 5, and 13.5), and then sheared in compression and extension. Figure 15.6b illustrates the element simulation. The initial confining pressure and p_0 for the initial volumetric strain ε_0 are defined in Table 15.4b.

The computed results shown in Figs. 15.7 and 15.8 indicate that the proposed model reasonably simulates the steady-state strengths affected by the stress-induced anisotropy and reproduces the contractive and dilative behaviors in the over-consolidated stress–strain field.

The primary findings for the proposed model are:

1. Even if shearing begins from a different over-consolidation ratio under the condition with the same initial void ratio, the stress–strain curve reaches the same steady state (Fig. 15.7a).

Table 15.4 Parameters for the confining pressure p0 and the over-consolidation state

	Over consolidation ratio OCR	Initial void ratio e_0	Consolidation pressure p_0(kN/m^2)	Pressure on the normal consolidation line p_{n0}(kN/m^2)	$r_{p_{n0}} = p_{n0}/p_0$
(a) Initial void ratio e_0 is fixed					
Normal consolidation	1	0.84	98.0	98.0	1.0
Over consolidation	2	0.84	67.6	98.0	1.45
	5	0.84	38.0	98.0	2.58
	20	0.84	13.1	98.0	7.48
(b) Initial void ratio e_0 is variable					
Normal-consolidation	1.0	0.84	98.0	98.0	1.0
Over-consolidation	5.0	0.84	40.5	98.0	2.42
	13.5	0.95	15.0	62.6	4.17
	2.21	0.75	91.5	141.5	1.55

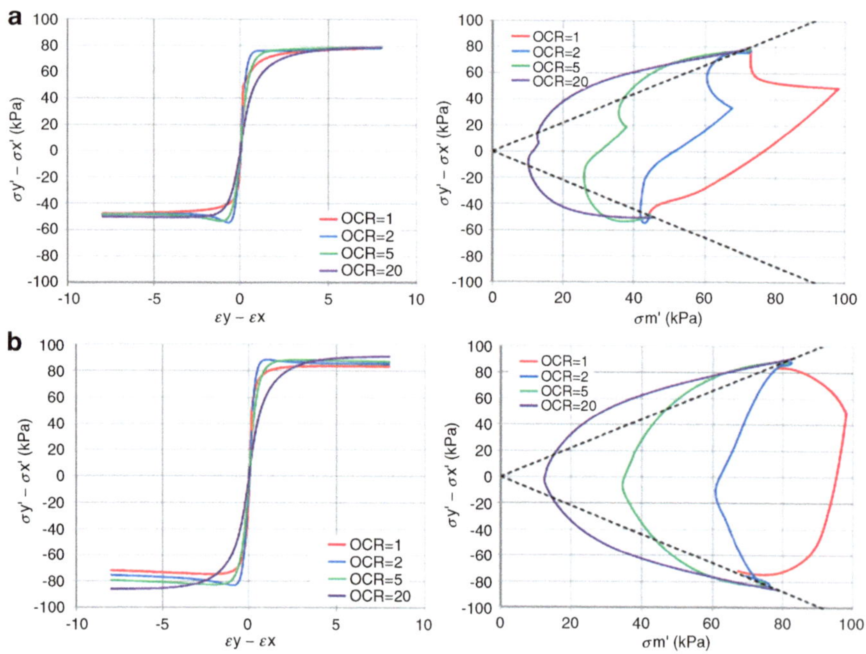

Fig. 15.7 Performance under undrained monotonic loading (initial void ratio e_0 is fixed. OCR = 1, 2, 5, and 20, $\Lambda = 0.549$, and $c_a = 20\text{kN/m}^2$). (**a**) $r_{\varepsilon dc} = 0.85$. (**b**) $r_{\varepsilon dc} = 0.085$

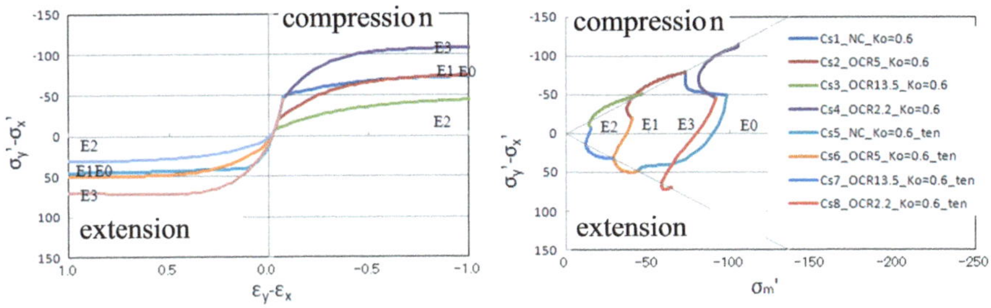

Fig. 15.8 Performance under undrained monotonic loading (initial void ratio e_0 is variable, $r_{\varepsilon dc} = 1.48$, $\Lambda = 0.549$, and $c_a = 10\text{kN/m}^2$)

2. On the other hand, if shearing begins under the condition with different initial void ratios, the stress–strain curve reaches a different steady state (Fig. 15.8).
3. The behavior of the compression side differs from that of the extension side, confirming stress-induced anisotropy in the steady state (Fig. 15.7a).
4. In an over-consolidated field, the effective stress path exhibits a contractive behavior in the initial shearing stage, and the effective stress path consists of a soft curved line like a cookie (Figs. 15.7a and 15.8).
5. A contractive and dilative behavior is observed in the stress path until it reaches the steady state, and can be controlled precisely by a parameter (Fig. 15.7a, b).
6. In the undrained shear beginning from the over-consolidated state, the stress path crosses the shear failure lines and reaches the steady state on the shear failure lines. The behavior can be controlled by a parameter.

To confirm the application and the characteristics of the proposed model, a comparative analysis was performed using computer code DACSAR for a simulation based on Sekiguchi-Ohta model (1977), which is recognized to design code for clay ground. The primary findings are:

1. The clay element is initially consolidated at $K_0 = 0.6$ with the over-consolidation ratios, and then sheared due to compression and extension under undrained monotonic loading. The steady state of the proposed model reaches the same state as DACSAR (Fig. 15.9). Even if the clay element is sheared from conditions with different initial void ratios, the steady state of the proposed model reaches the same steady state as DACSAR (Fig. 15.9). Consequently, the proposed model can express the stress-induced anisotropy of the Sekiguchi-Ohta-type model.
2. In the over-consolidated region, the effective stress path of the proposed model exhibits a contractive behavior in the initial shearing stage. Because the effective stress path consists of a soft curved line like a cookie, the proposed model is called the "Cookie Model". On the other hand, the effective stress path of DACSAR indicates an elastic behavior not due to dilatancy, which reaches and follows the yield surface, and eventually achieves a state. Thus, it was confirmed that both the Cookie Model and DACSAR indicate a characteristic behavior in the over-consolidated region.

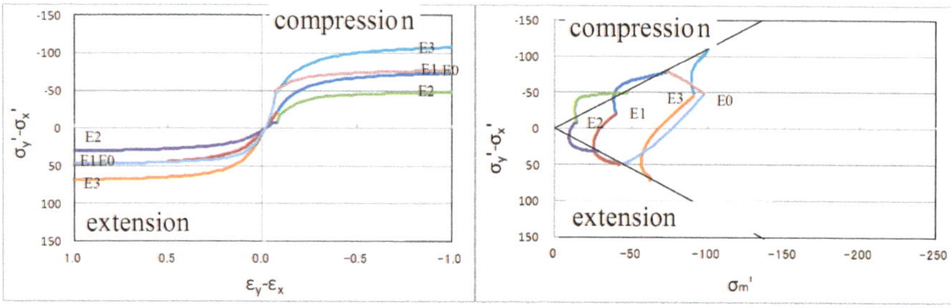

(a)

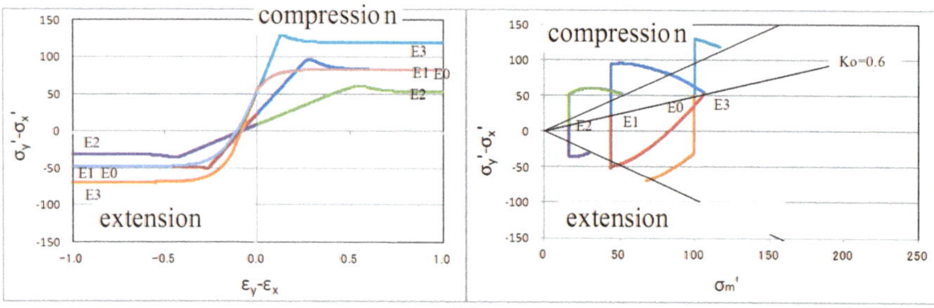

(b)

Fig. 15.9 Performance of the proposed model (cookie model) with reference to DACSAR (initial void ratio e_0 is variable and is 0.75, 0.84, or 0.95) (**a**) Cookie model (**b**) DACSAR

15.4 Model Performance Under Undrained Cyclic Loadings

The ability of the proposed model to simulate the strain-rate effects under undrained monotonic and cyclic loading (volumetric creep effect is excluded in this analysis) was investigated. The Isotach parameters to represent Kaolin were $r_{TSR} = 0.5$, $r_{\dot{\gamma}} = 10^6 s$, and $q_{ISO} = 0.04$. The parameter controlling contractive dilatancy was $r_{\varepsilon_d^c} = 0.85$, while the parameter for the later phase of contractive dilatancy was $q_2 = 0.01$, and the parameter for controlling dilatancy in the steady state was $\Lambda = (1 - r_{K_{us}}) = 0.544$. Table 15.5 lists the other parameters.

First, undrained monotonic loading simulations were performed. A clay element was initially consolidated at an isotropic stress $K_0 = 1.0$, and then strain-controlled tests were performed under various conditions (Table 15.6). Case 1 is the ordinary case with a non-viscous model, Case 2 is an Isotach/Rayleigh viscous model, Case 3 is a non-viscous model with a strain rate 1000 times higher than Case 1, and Case 4 is an Isotach/Rayleigh viscous model with a strain rate 1000 times higher than Case 2.

Table 15.5 Model parameters for the undrained monotonic loading simulation

p_a (kN/m^2)	G_{ma} (kN/m^2)	n_K, m_G	K_{La}, K_{Ua} (kN/m^2)	ϕ_f, ϕ_p (°)	c_a (kN/m^2)	$r_{\varepsilon_d^{cm}}$	$r_{\varepsilon_d^c}$	h_{max}
135.5	352	0.5	2155	33.3	0	0.5	0.85	0.24

r_{ε_d}	q_1	q_2	l_K	r_K	q_{us} (kN/m^2)	q_4	ICSW	Λ	$r_{p_{n0}}$	$r_{m_{tmp3}}$
1.0	1.0	0.01	1.0	0.348	30	0.001	1	0.544	1.0	1.0

Table 15.6 Case of undrained monotonic loading

Case	Viscous model	Strain rate of monotonic loading
Case 1	Ordinary case with non-viscous model	1 time
Case 2	Isotach/Rayleigh viscous model $r_{ISO} = 0.5$	1 time
Case 3	Ordinary case with non-viscous model	1000 times
Case 4	Isotach/Rayleigh viscous model $r_{ISO} = 0.5$	1000 times

Table 15.7 Case of undrained cyclic loading

Case	Viscous model	Strain rate of cyclic loading
Case 5	Ordinary case with non-viscous model	1 Hz
Case 6	Isotach/Rayleigh viscous model $r_{ISO} = 0.5$	1 Hz

Continuously, undrained cyclic loading simulations were performed. A clay element was initially consolidated at an isotropic stress $K_0 = 1.0$, and then stress-controlled simulations were performed under the conditions with a sinusoidal motion of 1 Hz, cyclic number of 60 times, and a shear stress amplitude of $\tau/\sigma'_m = 0.15$ (Table 15.7). Case 5 is for the non-viscous model and Case 6 is for the Isotach/Rayleigh viscous model.

Figures 15.10, 15.11, and 15.12 show the computed results. The primary findings for the proposed model are:

1. The proposed model can analyze the strain-rate effects of clay.
2. The strain-rate effects of the Isotach/Rayleigh viscous model under undrained step loading can be analyzed based on the Isotach model proposed by Tatsuoka et al. (2002) (Fig. 15.10).
3. When the strain rate effect of the Isotach/Rayleigh viscous model type is taken into consideration, the extension of the shear strain and excess pore water pressure are reduced. These reductions become remarkable in proportion to the shear-strain rate (Figs. 15.11 and 15.12).
4. The computed results of the proposed model reproduce the test results of clay.

Fig. 15.10 Strain-rate effects of the Isotach viscous model under undrained step loading

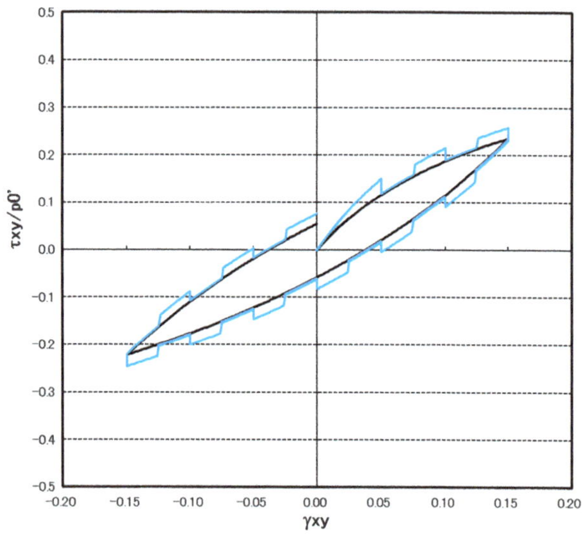

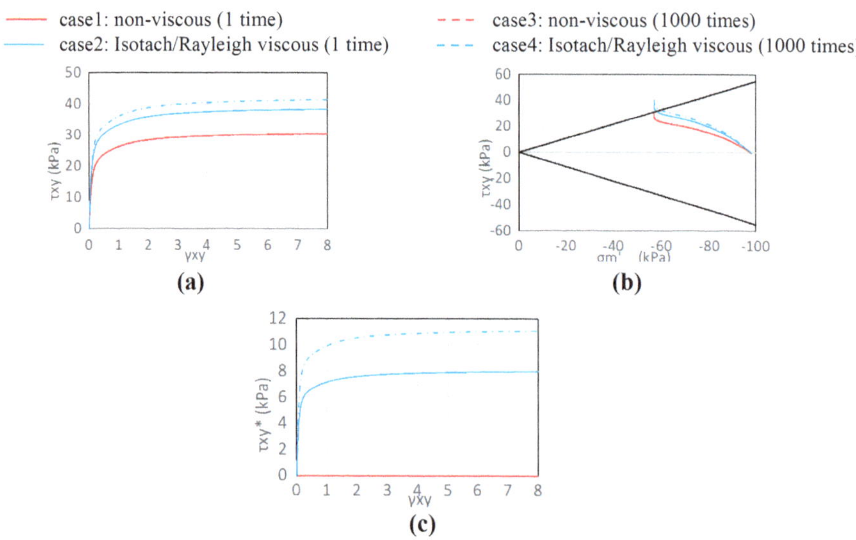

Fig. 15.11 Strain-rate effects of the Isotach viscous model under undrained monotonic loading. (**a**) Shear stress τ_{xy}– shear strain γ_{xy}. (**b**) Effective stress path. (**c**) Isotach viscous damping

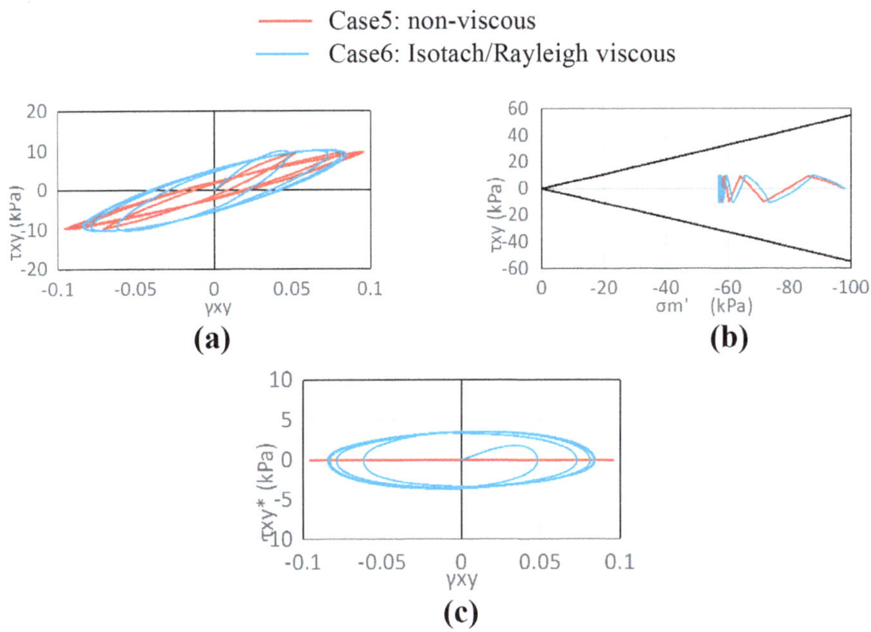

Fig. 15.12 Strain-rate effects of the Isotach viscous model under undrained cyclic loading. (**a**) Shear stress τ_{xy}– shear strain γ_{xy}. (**b**) Effective stress path. (**c**) Isotach viscous damping

15.5 Model Performance for Seismic Response Analysis After Consolidation Settlement Analysis

In the previous section, the performance of the proposed model was demonstrated through simulations of the drained/undrained behavior of clay under monotonic and cyclic loading. However, to ensure that the proposed model is applicable to more complex soil-structure systems, a one-dimensional seismic response analysis after a consolidation settlement analysis was also investigated. It is confirmed that the stress, strain, and dilatancy after consolidation settlement analysis can be used as approximations of the initial values in seismic response analysis.

Figure 15.13 shows the one-dimensional analysis model used for this study, which consists of a uniform clay ground model (10-m thick) and the three elements from the upper end subdivided by a small 0.1-m mesh. Initially the $K_0 = 0.66$ state was set through static gravity analysis, and then a 1000-kPa consolidation load was applied at upper end, and the analysis continued until the excess pore water pressure disappeared. Continuously, the seismic response analysis was performed under level-1 earthquake motion (Fig. 15.14) and continued consolidation settlement analysis for about 100 days. Table 15.8 shows the model parameters used for this study. The coefficient of permeability was 1.91×10^{-7} (m/s).

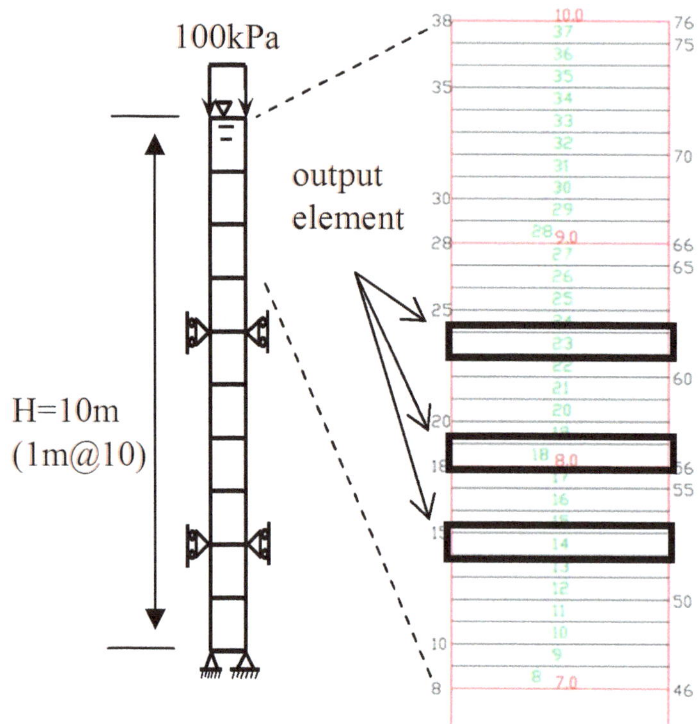

Fig. 15.13 One-dimensional analysis model

Table 15.8 Model parameters

ρ (t/m³)	p_a (kN/m²)	G_{ma} (kN/m²)	n_K, m_G	K_{La}, K_{Ua} (kN/m²)	ϕ_f, ϕ_p (°)	c_a (kN/m²)	$r_{\varepsilon_d^{cm}}$	$r_{\varepsilon_d^c}$	h_{max}
1.48	135.5	352	0.5	2155	33.3	0	0.062 (con.) 0.5 (seis.)	0.85	0.24

r_{ε_d}	q_1	q_2	l_K	r_K	q_{us} (kN/m²)	q_4	ICSW	Λ	$r_{p_{n0}}$	$r_{m_{tmp3}}$	$r_{m_{tmp3}}$
1.0	1.0	0.01	1.0	0.174	30	0.001(con.) 1.0 (seis.)	1	0.544	1.0	1.0	

r_{ISO}	$r_{\dot{\gamma}}$	q_{ISO}	$r_{S_o}^{dmin}$	q_5	$r_{\varepsilon_{dus}^c}$	$r''_{\varepsilon_{dus}^c}$
0.5	1.0×10^6	0.04	0.001	5.0	1.0	1.0

Figure 15.15 shows the analysis results. The primary findings for the proposed model are:

1. The arbitrary initial K_0 state can be analyzed by static gravity analysis (Fig. 15.15a).

Fig. 15.14 Earthquake motion

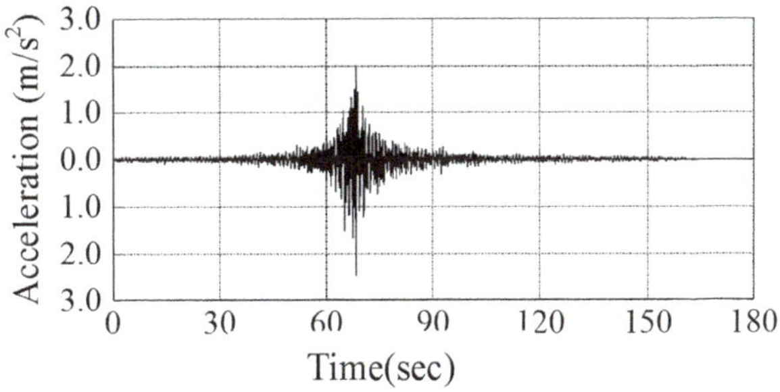

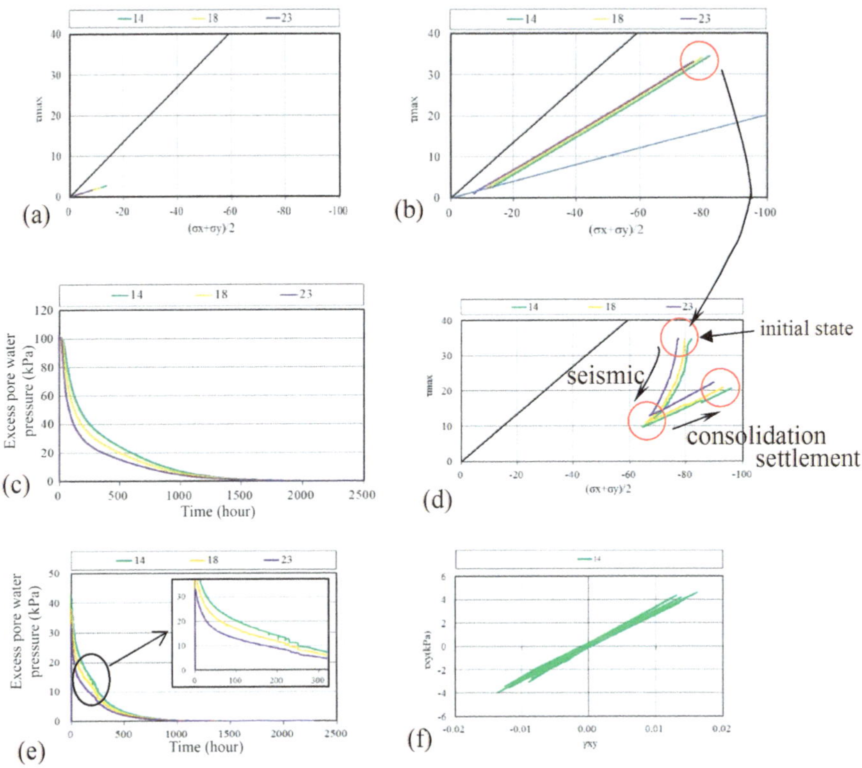

Fig. 15.15 Performance under seismic response analysis after consolidation settlement analysis. (**a**) Effective stress path (static gravity analysis, $K_0 = 0.66$). (**b**) Effective stress path (consolidation settlement analysis). (**c**) Excess pore pressure (consolidation settlement analysis). (**d**) Effective stress path (seismic response and consolidation analysis)

2. After K_0 consolidation, the stress path given by the consolidation settlement analysis does not follow the K_0 line, but is closer to the shear failure line. It depends on the dilatancy because the lateral normal strain is restricted (Fig. 15.15b).
3. The stress–strain state, including dilatancy at the end of consolidation settlement analysis, can be used as the initial state in seismic response analysis (Fig. 15.15b, d).
4. During seismic response analysis, the shear stress is reduced in proportion to the increase in the excess pore water pressure. Through consolidation settlement analysis after an earthquake, the shear stress increases in proportion to the disappearance of the excess pore pressure (Fig. 15.15d, e)
5. The proposed model can simulate seismic response analysis after consolidation settlement analysis (Fig. 15.15f).

15.6 Model Performance for the Isotach/Rayleigh Viscous Effects Under Earthquake Loading

The performance of the proposed model for viscous effects was confirmed through simulations of the undrained behavior of clay under monotonic and cyclic loadings. Additionally, to ensure its applicability to the Isotach/Rayleigh viscous effects through earthquake loading, one-dimensional seismic response analysis after consolidation settlement analysis was investigated. The analysis model and model parameters used for this study were same as those of previous chapter shown in Fig. 15.13 and Table 15.8. A sine wave with an input acceleration of 3.0 m was given as the earthquake loading. The frequency was changed to confirm the influence of strain-rate effects; Case 1 was set to 0.1 Hz and Case 2 was set to 0.01 Hz.

The function controls the upper limit of the shear strain amplitude in the steady state under cyclic loading (Case 3). The degree of the steady state was expressed by the liquefied front parameter S_0. The contractive dilatancy, which estimates S_0, was designated as the virtual contractive dilatancy as ε''^{c}_{d}. ε''^{c}_{d} is controlled by the parameter $r''_{\varepsilon^c_d}$. The upper limit of the shear strain amplitude can be controlled precisely by the parameter $r''_{\varepsilon^c_d}$.

Table 15.9 shows these cases, conditions, and model parameters.

Figures 15.16, 15.17, 15.18, and 15.19 show the analysis results. The primary findings are:

1. The shear strain–rate effects of clay under earthquake loading can be analyzed by the Isotach/Rayleigh viscous–type model.

Table 15.9 Case, condition, and model parameters used in the analysis

Case	Contents	Parameter			
		$r_{S_o}^{\mathrm{dmin}}$	q_5	$r_{\varepsilon^{\mathrm{c}}_{\mathrm{dus}}}$	$r''_{\varepsilon^{\mathrm{c}}_{\mathrm{dus}}}$
Case 1	Isotach viscous effects. Input motion 0.1 Hz	0.001	5.0	1.0	1.0
Case 2	Isotach viscous effects. Input motion 0.01 Hz	0.001	5.0	1.0	1.0
Case 3	Isotach viscous effects. Input motion 0.1 Hz, controls the shear strain amplitude at the steady state	1.000	1.0	1.0	5.0

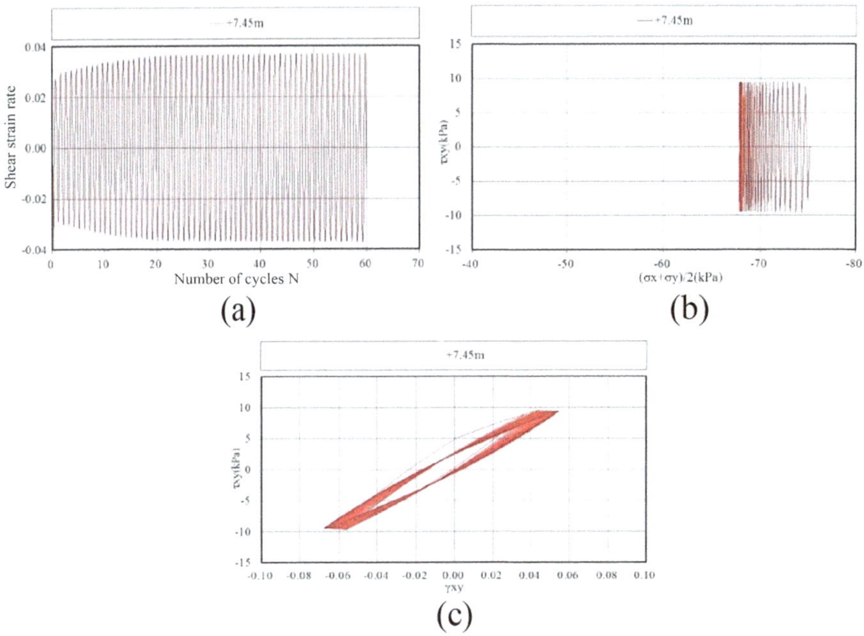

Fig. 15.16 Isotach/Rayleigh viscous effects for Case 1 (loading frequency 0.1 Hz and $r''_{\varepsilon^{\mathrm{c}}_{\mathrm{d}}} = 1.0$) (**a**) Velocity of shear strain rate. (**b**) Effective stress path. (**c**) Shear-stress strain relations

2. The number of cyclic loads to reach the steady state tends to decrease while the extension of shear strain tends to increase in proportion to the shear strain rate (Figs. 15.16 and 15.17).
3. The steady-state behavior of the shear stress–strain relations and the upper limit of the shear strain amplitude can be controlled by the parameter $r''_{\varepsilon^{\mathrm{c}}_{\mathrm{d}}}\left(r_{\mathrm{eduspp}}\right)$.

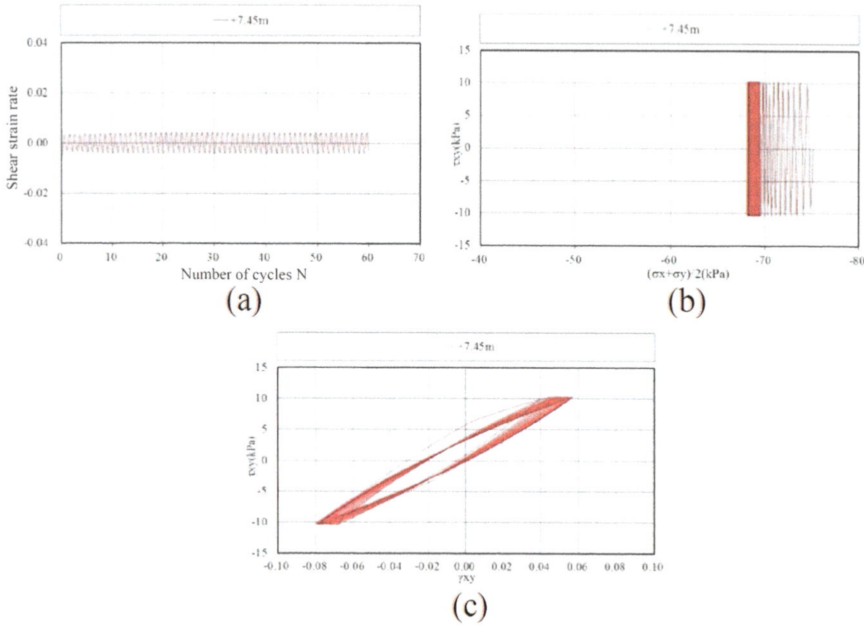

Fig. 15.17 Isotach/Rayleigh viscous effects for Case 2 (loading frequency 0.01 Hz and $r''_{\varepsilon_d^c} = 1.0$) (**a**) Velocity of shear strain rate. (**b**) Effective stress path. (**c**) Shear-stress strain relations

15.7 Conclusions

The strain space multiple mechanism model, originally developed for the cyclic behavior of granular materials such as sand is adapted to idealize the stress–strain behavior of clay. Compared to the conventional elasto-plastic (or visco-elasto-plastic) models for clay, the proposed model has many advantages. The performance of the proposed model is demonstrated through simulations of the drained/undrained behavior of clay under monotonic and cyclic loading.

The primary findings for the proposed model are:

1. The arbitrary initial K_0 state can be analyzed by static gravity analysis.
2. Stress-induced anisotropy (i.e., the effect of the initial shear) in the steady (critical) state can be analyzed based on Shibata's dilatancy model (1963).
3. Over-consolidated clay can be analyzed by defining the dilatancy at the steady state based on the over-consolidation ratio.
4. Strain-rate effects for monotonic and cyclic shears can be analyzed based on the Isotach/TESRA model proposed by Tatsuoka et al. (2002) in the strain rate ranging from zero to infinity.
5. The proposed model can simulate seismic response analysis after consolidation settlement analysis. The stress–strain state, including dilatancy at the end of

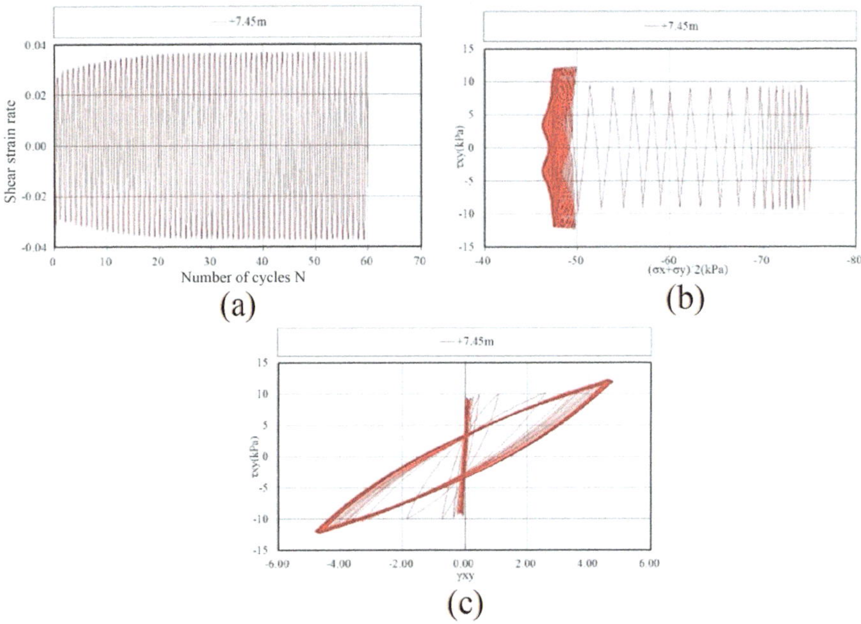

Fig. 15.18 Isotach/Rayleigh viscous effects for Case 3 (loading frequency 0.1 Hz and $r''_{\varepsilon_d^c} = 1.0$) (**a**) Velocity of shear strain rate. (**b**) Effective stress path. (**c**) Shear-stress strain relations

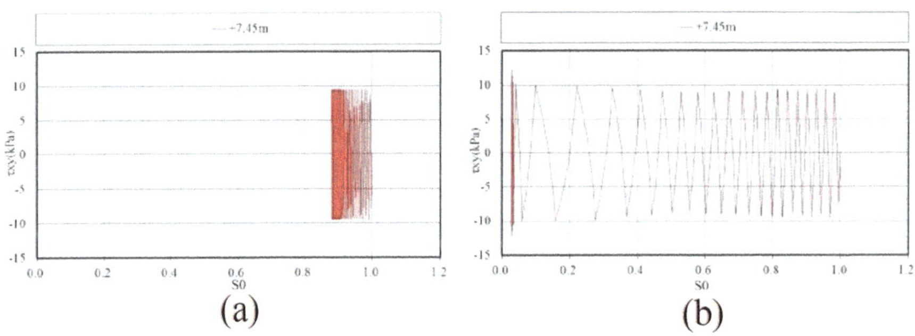

Fig. 15.19 Effects of the parameter $r''_{\varepsilon_d^c}$ on controlling the upper limit of the shear strain amplitude. (**a**) Case1 $r''_{\varepsilon_d^c} = 1.0$. (**b**) Case2 $r''_{\varepsilon_d^c} = 5.0$

consolidation settlement analysis, can be used as the initial state in the seismic response analysis.

Employing the proposed model should help solve challenges in combined geotechnical hazards.

Acknowledgements The author gratefully acknowledges Professor H. Ohta and Professor A. Iizuka for use of the computer code DACSAR to conduct simulations based on the Sekiguchi-Ohta model (1997).

This paper is a compilation of the results of the collective efforts of the working group of the FLIP consortium to develop a new model of clayey materials within the framework of the strain space multiple mechanism model. The working group consists of the following members: Susumu Iai, Tetsuo Tobita, Kyohei Ueda, Kengo Kusunoki, Kunihiro Yamaguchi, Yasutaka Kimura, Sho Mizuno, Masaya Ishikawa, Akinori Suzuki, Takashi Nakama, Hosono Atsushi, Soichi Tashiro, Osamu Ozutsumi, Kazuki Uemura, Keisuke Kitade, Naoki Orai, Shingo Mihira, Takeko Mikami, Junichi Hyodo, Koji Ichii, Eiji Kohama, and Yousuke Ohya. As the coordinator of the working group, the author gratefully acknowledges the contributions of the group's members.

References

Iai S, Ueda K (2016) Modeling of clay through strain space multiple mechanism model. Ann Disaster Prev Res Inst, Kyoto Univ 59B:92–114. (in Japanese)

Iai S, Tobita T, Ozutsumi O, Ueda K (2011) Dilatancy of granular materials in a strain space multiple mechanism model. Int J Numer Anal Methods Geomech 35(3):360–392

Iai S, Ueda K, Ozutsumi O (2015) Strain space multiple mechanism model for clay under monotonic and cyclic loads. In: 6th international conference on earthquake geotechnical engineering. Christchurch, New Zealand, 264, CD-ROM

Iizuka A, Ohta H (1987) A determination procedure of input parameters in elasto-viscoplastic finite element analysis. Soils Found 27(3):71–87

Sekiguchi H, Ohta H (1977) Induced anisotropy and time dependency in clays. In: 9th international conference on soil mechanics and foundation engineering, Tokyo

Shibata T (1963) On the volume change of normally consolidated clays. Ann Disaster Prev Res Inst, Kyoto Univ 6:128–134. (in Japanese)

Takeyama T, Pipatpongsa T, Iizuka A, Ohta H (2013) Stress-strain relationship for the singular point on the yield surface of the elasto-plastic constitutive model and quantification of metastability. In: Chu J, Wardani SPR, Iizuka A (eds) Geotechnical predictions and practice in dealing with geohazards. Springer, pp 229–239

Tatsuoka F, Ishihara M, Di Benedetto H, Kuwano R (2002) Time-dependent shear deformation characteristics of geomaterials and their simulations. Soils Found 42(2):103–129

Chapter 16
Modeling of Cohesive Soils: Consolidation and Seismic Response

Osamu Ozutsumi

Abstract A new constitutive model (cookie model) for clay was proposed to establish a prediction method to evaluate damages from complex disasters. The model was developed as an extension of a model (cocktail glass model) for liquefiable sandy soil. The models were installed on a two-dimensional dynamic effective stress analysis program. The program was used to apply a constitutive model (cookie model) to a clay layer under a highway embankment as an example. We could perform the self-weight analysis, the long-term consolidation settlement analysis continuously, and seismic response analysis within process of the consolidation settlement analysis. The results analytically indicated that earthquakes caused additional instantaneous settlement, additional long-term settlement, and additional increase in pore water pressure of the clay layer. The settlement advanced rapidly during the consolidation process, and rising of pore water pressure was observed for the embankment. These observational results could be analytically reproduced by applying seismic waves during the consolidation settlement analysis.

A new constitutive model (cookie model) for clay was proposed to establish a prediction method to evaluate damages from complex disasters. The model was developed as an extension of a model (cocktail glass model) for liquefiable sandy soil. The models were installed on a two-dimensional dynamic effective stress analysis program. The program was used to apply a constitutive model (cookie model) to a clay layer under a highway embankment as an example. We could perform the self-weight analysis, the long-term consolidation settlement analysis continuously, and seismic response analysis within process of the consolidation settlement analysis. The results analytically indicated that earthquakes caused additional instantaneous settlement, additional long-term settlement, and additional increase in pore water pressure of the clay layer. The settlement advanced rapidly during the consolidation process, and rising of pore water pressure was observed for

---

O. Ozutsumi (✉)
Meisosha Corporation, Tokyo, Japan
e-mail: ozutsumi@meisosha.co.jp

S. Iai (ed.), *Developments in Earthquake Geotechnics*, Geotechnical, Geological and Earthquake Engineering 43, DOI 10.1007/978-3-319-62069-5_16

the embankment. These observational results could be analytically reproduced by applying seismic waves during the consolidation settlement analysis.

16.1 Introduction

16.1.1 Summary

Several important facilities, such as oil storage stations or electric power plants, are often constructed in coastal zones. Such important facilities are often constructed on reclaimed land. Thus, the foundation grounds of these facilities often include sedimentary layers of clay and sand.

Following landfilling, the settlement of ground caused by consolidation of the clay layers occurs, and this is accompanied by the accumulation of strain in the structures including shore protection structures. Thus, in the event of an earthquake, seismic responses of the structures could differ from the responses of the structures not affected by the settlement. Additionally, instantaneous settlement and long-term settlement of clay layers due to an earthquake could occur. Furthermore, damages caused by liquefaction of sand layers, including landfill soil, may occur. However, prediction methods to evaluate damages to the structures due to these combined factors have not yet been established.

In order to establish this type of a prediction method, a constitutive model (cookie model) for clay was proposed by Iai et al. (2015), Iai and Ueda (2016) based on the framework of a strain space multiple mechanism model for granular materials. The model was installed on a two-dimensional dynamic effective stress analysis program "FLIP" based on the finite element method. Using the program, the constitutive model (cookie model) was applied to a clay layer under a highway embankment as an example of a problem. We performed the self-weight analysis, the long-term consolidation settlement analysis continuously, and seismic response analysis within process of the consolidation settlement analysis. In this chapter, the results of the analysis were discussed to show the applicability of the proposed constitutive model and the performance of the program including the constitutive model.

16.1.2 Computer Code for Complex Disaster Prediction

The constitutive model (cookie model) for clay as previously mentioned above was developed as an extension of the constitutive model (cocktail glass model) for sandy soil already proposed by Iai et al. (2011). First, the cocktail glass model was installed on the program "FLIP". The program could perform seismic response analysis, including liquefaction analysis, considering the permeability of the ground. This was followed by installing the cookie model on the program. The program could perform the seismic response analysis during the consolidation settlement analysis.

The program "FLIP" provided the multi-stage analysis function that enabled an analysis considering the construction process to accurately reproduce the initial stress state. Furthermore, the restart function and the read function of the seismic acceleration data were also in place to be able to deal with the aftershocks over several times for several decades.

Using these functions of FLIP, it was possible to continuously perform self-weight analysis, consolidation settlement analysis, and seismic response analyses in the process of consolidation settlement analysis.

16.1.3 Examples

This chapter discusses long-term consolidation settlement analysis of a highway embankment constructed on foundation ground including clay layers, and seismic response analysis during consolidation analysis by FLIP, as examples to show the ability of the cookie model to evaluate damages from a combined disaster. The program could consider liquefaction phenomena of sandy soil during earthquakes. However, liquefaction phenomena were not mentioned and instead instantaneous settlement, long-term settlement, and increases in excess pore water pressure of the clay layer caused by the earthquake were discussed.

16.2 The Constitutive Model (Cookie Model) for Clay

As mentioned above, the constitutive model (cookie model) was applied to the clay layer. The characteristics of this model were detailed in the previous chapter. This chapter examines Isotach damping (Tatsuoka et al. 2002) that was installed on the cookie model.

An aspect of the constitutive model includes the function reproducing prescribed coefficient of earth pressure at rest in the initial self-weight analysis and this is examined in the explanation of examples.

16.2.1 Isotach Damping

Isotach damping was proposed as damping based on the velocity of volumetric strain or shear strain. With respect to the cookie model, it is introduced as damping based on the shear strain velocity.

The results of dynamic element simulation using FLIP indicated that parameters for both Rayleigh damping and Isotach damping could be adjusted to show the similar damping effect for a motion of approximately 1 Hz. In the simulation, the distributed force (shear force) that varied sinusoidally with an amplitude of 30 kPa and a frequency of 1 Hz on the upper side of the element was loaded. Figure 16.1 shows the procedure for the dynamic element simulation. As an example, the results of the dynamic element simulation were shown using the cookie model element in Fig. 16.2. Shear force–shear strain relations in Fig. 16.2 are indicated by red lines. Additionally, shear stress of the element – shear strain relations are shown in the same figures and denoted by green lines. In this case, the difference between the share force and the shear stress corresponded to the damping force given that the effects of mass were not considered. A comparison of the upper figure of Rayleigh damping and the lower figure of Isotach damping indicated that both damping characteristics were similar. When other results of element simulation were considered, the findings indicated that both models of damping show similar effects on the decrease in the amplitude of displacement after unloading and on the phase shift between the element force and the damping force. However, they have different characteristics with respect to their dependence on the shear strain rate.

16.2.2 Dependency of Isotach Damping on the Shear Strain Rate

Figure 16.3 shows the dependency of Isotach damping on the shear strain rate in the case of kaolin clay. The horizontal axis in the figure denotes a shear strain rate, while the vertical axis corresponds to r_q that denotes the relative magnitude of the damping in the full range of the shear strain rate. As shown in the figure, r_q varies from 0.1 to 0.9, while the shear strain rate varies from 10^{-5} (1/s) to 10^{+7} (1/s). The parameter r_q had very low sensitivity with respect to the shear strain rate when compared to that of linear damping such as Rayleigh damping.

As discussed later in the study, consolidation analysis of a clay layer was performed in a foundation ground of a highway embankment, and seismic motions were loaded during the consolidation settlement analysis. The consolidation settlement analysis examined long-term phenomena over 50 years, and thus, the corresponding time integration interval (denoted as Δt) was set to 20 days as a standard. However, Δt was set to 0.01 s at the time of the earthquake response analysis. In the consolidation process after the end of the earthquake, Δt gradually increased from 0.01 s to 20 days (1,728,000 s). This increase in Δt sometimes causes an instability in the analysis. The results indicated that the use of Isotach damping was very effective in these types of problems. In this study, the maximum amount of the shear strain velocity was approximately 10^{-8} (1/s) when the consolidation settlement analysis was stable. Accordingly, the shear strain velocity was below the lower limit value 10^{-6} (1/s) as shown in Fig. 16.3, and then the Isotach

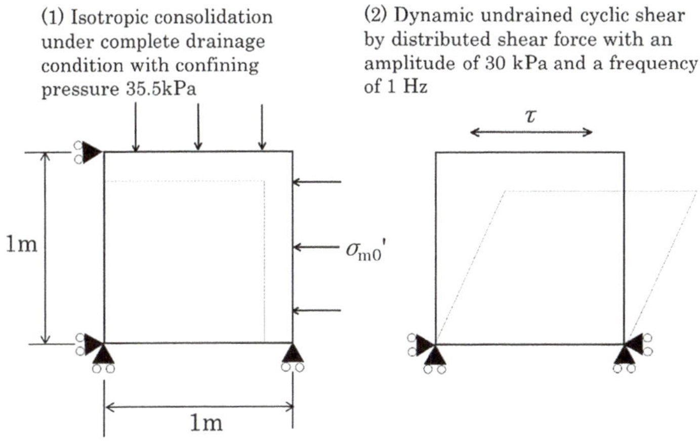

Fig. 16.1 Procedure for dynamic element simulation

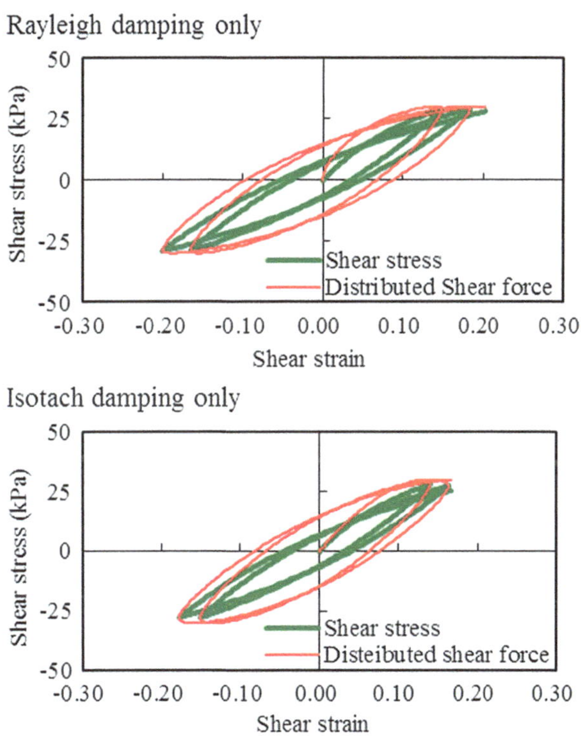

Fig. 16.2 Results of the dynamic element simulation

damping had no effect on the results. However, once the system became unstable, the shear strain rate appeared to increase beyond this lower limit. If this was the case, then proper damping worked, and the system was stable. Additionally, when

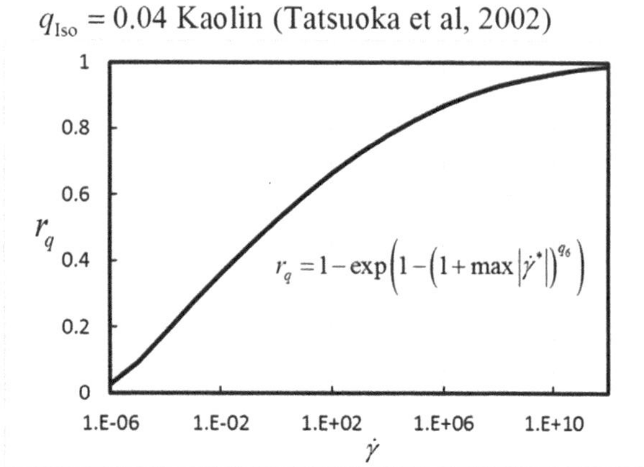

Fig. 16.3 Dependency of Isotach damping on shear strain rate (Based on Tatsuoka et al. 2002)

Rayleigh damping adjusted with 1 Hz instead of Isotach damping was applied in the process of increasing Δt, the analysis diverged frequently.

16.3 Examples

In the current study, consolidation settlement analysis of the highway embankment and seismic response analysis in the process of consolidation were performed.

16.3.1 Sectional View of the Highway Embankment

The sectional view of the highway embankment is shown in Fig. 16.4. It is located in Hitachi City, Japan. This sectional view was presented by a Committee of the Society of Geotechnical Engineering (Ohta et al. 2005). The committee organized concurrent analysis of the consolidation of the clayey ground using computer programs that employed the Sekiguchi-Ohta model (Sekiguchi and Ohta 1977).

As shown in the figure, a soft marine clay layer Ac2 exists with a thickness of 15 m in the foundation ground. Furthermore, Ac2 was sandwiched by sandy layers. At first, the highway embankment was constructed as a test embankment using dredge sand. This test embankment was then removed later, and the present embankment (hereafter referred to "the permanent embankment" for the purpose of distinguishing from the test embankment) was constructed using rock waste with the larger unit weight derived from the tunnel construction.

The settlement of the embankment lasted even after a period of 20 years from the construction of the test embankment. Therefore, overlaying of the pavement was repeated, and widening of the embankment accompanying the pavement was also

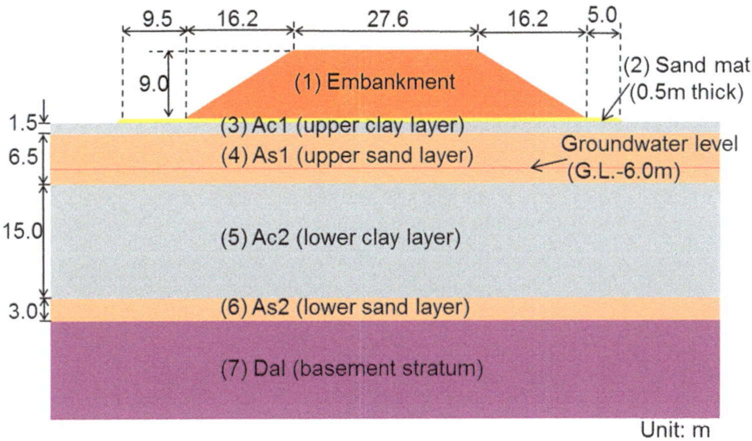

Fig. 16.4 Sectional view of the highway embankment (Ohta et al. 2005)

performed. However, the effects of the overlaying and the widening were ignored in the current study because it was not possible to refer to the detailed data.

16.3.2 Mesh View and the Output Locations

Figure 16.5 shows the finite element mesh (main area) used in the study. The figure shows the points where the analytical results were obtained. The time histories of the settlement due to the consolidation and the earthquakes are shown at point A on the ground level just below the embankment. Additionally, the response horizontal acceleration time histories and the response horizontal displacement time histories during earthquakes are shown at point A. The time history of excess pore water pressure due to the consolidation and earthquakes are shown at point B in the center of the Ac2 layer. Similarly, hysteresis curves of the element (effective stress path etc.) are shown at element B at the center of the Ac2 layer due to the consolidation and the earthquakes. The horizontal displacement distributions at the completion of the test embankment are shown at line C immediately under the toe of right slope of the embankment.

16.3.3 Side Boundary Conditions on the Excess Pore Water Pressure

Figure 16.6 shows the distribution of excess pore water pressure at the completion of the permanent embankment (on the 208th day from the starting date of the

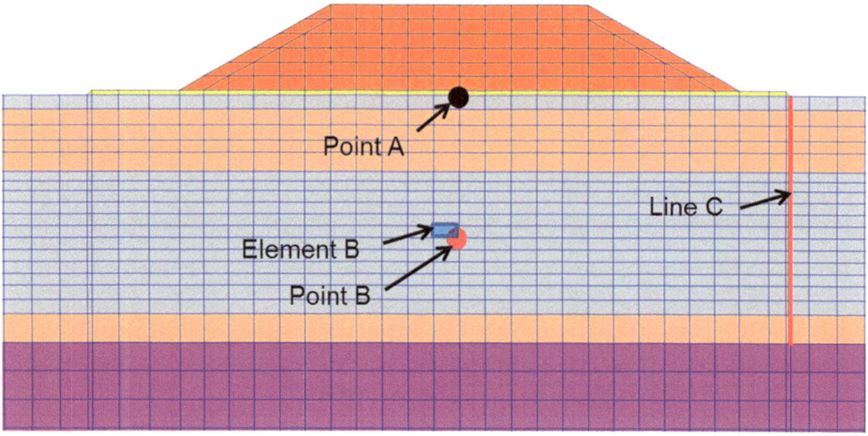

Fig. 16.5 Finite element mesh (main area) and the output locations

construction of the permanent embankment) wherein the effect of the test embankment is ignored. The side boundary conditions were permeable in this case. That is, excess pore water pressure was set as 0 at the boundaries. As observed in Fig. 16.6, the source of the excess pore water pressure was located in Ac2 directly under the embankment, and thus the excess pore water pressure that was set as 0 along the side boundaries was appropriate. In contrast, if the side boundary conditions were set as impermeable, then in the model, the increase in the excess pore water pressure due to the construction of the embankment remained high even after 50 years. This could be because the horizontally layered stratum specified at this site including the almost impermeable clay layers infinitely spread in its current state. However, this assumption is not realistic.

Figure 16.7 shows the distribution of the excess pore water pressure 50 s after the earthquake. The earthquake occurred 3200 days (8.8 years) after the start date of the permanent embankment construction. In this case, side boundary conditions were permeable before earthquake, but after the earthquake the conditions were set as impermeable. The distribution of the excess pore water pressure as shown in the figure was mainly caused by the seismic motion because the excess pore water pressure increased by the embankment loading showed a considerable decrease prior to the earthquake. The excess pore water pressure increased all over the Ac2 layer and did not disappear on the side boundaries. Furthermore, if the excess pore water pressure was set as 0 at the side boundaries, then the horizontal distribution of the excess pore water pressure had a sharp inclination near the side boundaries, and this was followed by a disturbance in the pressure near the side boundaries as shown in Fig. 16.8.

Accordingly, it was not sufficient to set the side boundary condition to be impermeable as noted above. Conversely, the disturbance in pressure near the side boundaries of the Ac2 layer occurred if the side boundary condition was set to permeable. Typically, it is appropriate to set the side boundary condition of a clay

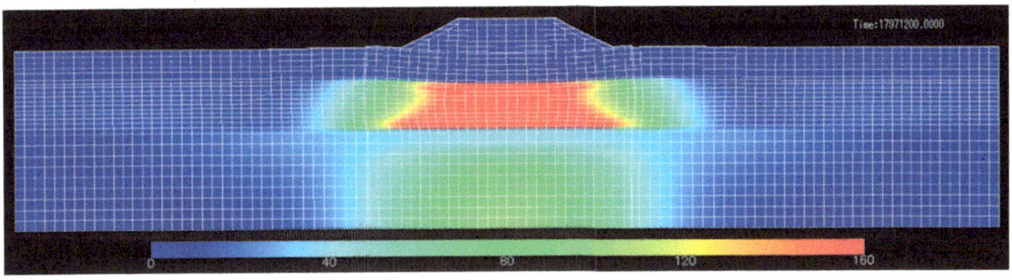

Fig. 16.6 Excess pore water pressure distribution with permeable side boundaries at the completion of the permanent embankment

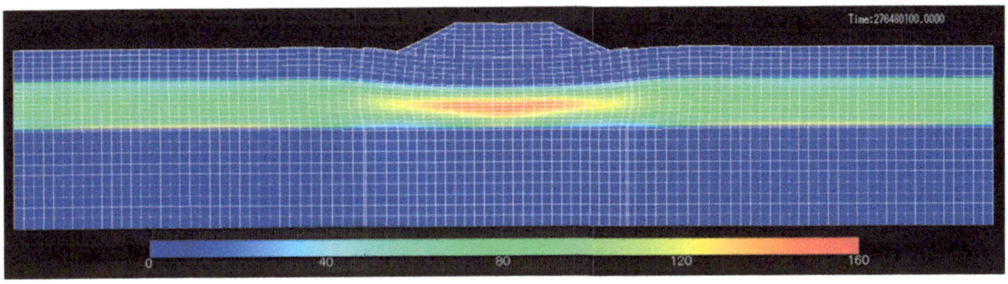

Fig. 16.7 Excess pore water pressure distribution with impermeable side boundaries after the first earthquake

layer with a low permeability coefficient as impermeable and that for a sandy soil layer with a large permeability coefficient as permeable. However, in this study, in the case of the sandy soil layer, a few items, such as considering the rise of excess pore water pressure even at the time of the earthquake, were missing from the model, and hence, the verification on these boundary conditions would be an issue in the future.

16.3.4 Construction Time Schedule of the Embankment

Table 16.1 shows the time schedule of the embankment construction. This table shows the days elapsed between the start date of the construction of the test embankment and the events. The start date of the construction of the test embankment was set as day 0. The test embankment was completed on the 208th day. It was then removed on the 520th day, and the construction of the permanent embankment commenced on the 936th day.

However, these dates were based on the figures listed in various materials and thus were not accurate.

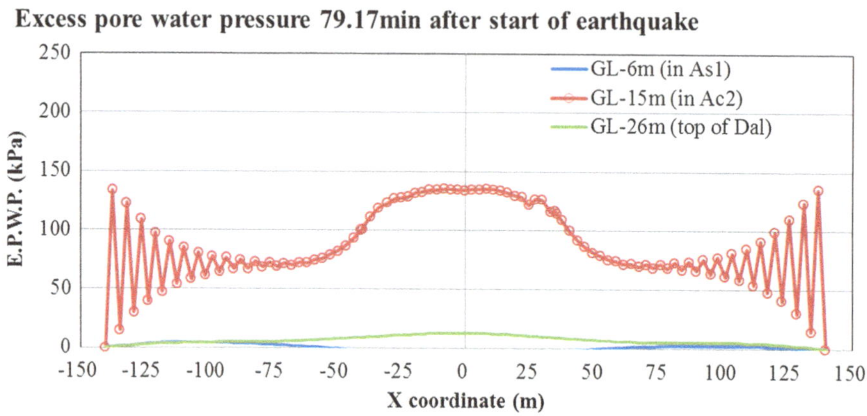

Fig. 16.8 Excess pore water pressure distribution at the center of Ac2 after the earthquake

Table 16.1 Construction time schedule of the embankment (Ohta et al. 2005)

Events	Time elapsed (day)
1. Beginning of test embankment construction	0
2. Completion of the test embankment	208
3. Beginning of test embankment removal	430
4. Completion of test embankment removal	520
5. Beginning of permanent embankment construction	936
6. Completion of permanent embankment	1144
7. Completion time of the simulation	18,250(50 years)

16.3.5 Model Parameters for Ac2 Used in the Concurrent Analysis

The committee for the concurrent analysis (Ohta et al. 2005) as mentioned previously, specified model parameter values for each layer as shown in Fig. 16.4. Among these soil layers, the model parameters as elasto-visco plastic material (Sekiguchi-Ota model) were specified for the two clay layers Ac1 and Ac2. With respect to each remaining soil layer, model parameters were specified as linear-elastic material. In a manner similar to the concurrent analysis, with respect to the FLIP analysis, the sandy soil layers were treated as a linear elastic solid, and only the Ac2 layer was modeled by a cookie model. The Ac1 layer exhibited low thickness, and thus, the Ac1 layer was also treated as a linear elastic solid to simplify the analysis. The permeability coefficient designated for each soil layer by the committee was also followed similar to that in the FLIP analysis.

Table 16.2 Model parameters for Ac2 specified in the concurrent analysis (Ohta et al. 2005)

Parameters	Symbol	Specified value
Void ratio	e_0	2.299
Compression index	λ	0.596
Swelling index	κ	0.272
Effective overburden pressure	σ_{vi}'	135.5 kPa
Over-consolidation ratio	OCR	1.00
Coefficient of earth pressure at rest	K_i	0.66
Coefficient of permeability	k	6.462E-6 m/day/(kN/m^3)
Effective Poisson's ratio	ν'	0.40
Secondary consolidation coefficient	α	9.03E-03
Initial volumetric creep strain rate	$\dot{v}_0$	6.02E-05 1/day

Table 16.2 shows part of model parameters for the elasto-visco plastic material as specified for Ac2 in the concurrent analysis. According to the table, Ac2 corresponded to normally consolidated clay and its coefficient of earth pressure at rest was 0.66. In the initial self-weight analysis, it was important to reproduce the coefficient of earth pressure at rest. This was performed to prevent effective confining pressure in the initial state from deviating from the assumed value.

16.3.6 Parameters for the Cookie Model for Ac2

Table 16.3 shows the parameters for the cookie model and the corresponding values for Ac2. Three sets of parameter values denoted by Cases A, B, and C were tested. Parameter values for rigidity were set in Cases A and B mainly based on both compression index (denoted as λ) and swelling index (denoted as κ) as shown in Table 16.2. Conversely, parameter values for rigidity were set in Case C based on the results of PS logging (Inagaki et al. 2002) conducted near the site. The secondary consolidation in Case A was not considered. However, this was considered in Cases B and C. The parameter values for the secondary consolidation were set based on the corresponding parameters listed in Table 16.2. Isotach damping was considered for all cases, and parameter values were used for kaolin clay (Tatsuoka et al. 2002).

In Cases A and B, bulk modulus (denoted as K_{Ua}) was calculated based on swelling index κ, void ratio (denoted as e_0) and effective overburden pressure (denoted as σ_{vi}') as shown in Table 16.2 using Eq. 16.1. The relation between bulk modulus and shear modulus for linear elastic material (see Eq. 16.2) was then used to calculate shear modulus (denoted as G_{ma}), and the value of the shear modulus corresponded to 352 kPa. In order to reproduce the coefficient of earth pressure at rest K_i as shown in Table 16.2 as a result of the initial self-weight analysis, the bulk modulus at reference state K_{La} was set as 2155 kPa using Eq. 16.3. In the Eq. 16.3, ϕ_f indicates internal friction angle. Nevertheless, it was

Table 16.3 Parameters of the cookie model for Ac2 in the analysis using "FLIP"

Parameters	Symbol	Specified value		
		Case A	Case B	Case C
Reference effective stress	P_a	135.5 kPa		135.5 kPa
Bulk modulus at reference state	K_{LA}	2155 kPa		90,481 kPa
Power index for bulk/shear moduli at static gravity analysis	m	0.5		
Reduction factor of bulk modulus for consolidation analysis	r_K	0.348		0.00844
Over-consolidation state	r_{pn0}	1.00		
Shear modulus at reference state	G_{ma}	352 kPa		14,780 kPa
Internal friction angle ($M = \sin\phi_f$)	ϕ_f	33.3°		
Coefficient of permeability	k	7.48E-11 m/s		
Parameter controlling contractive component	$r\varepsilon_{dc}$	0.85		5.0
Parameter controlling dilatancy at the steady state	Λ	0.544		
Secondary consolidation coefficient	α_C	0.0	9.03E-03	
Initial volumetric creep strain rate	$\dot{v}_0$	0.0 1/s	6.97E-10 1/s	
Isotach parameter representing max amplitude	riso	0.5		
Isotach parameter for normalizing strain rate	rgdot	1.0E+06 s		
Isotach parameter for representing strain rate dependency	qiso	0.04		

necessary to use the bulk modulus corresponding to the compression index λ in the consolidation settlement analysis. For this purpose, the reduction factor for bulk modulus $r_k$was set as 0.348 using Eq. 16.4 as follows:

$$K_{Ua} = \frac{1 + e_0}{\kappa} \sigma_{vi}' \quad (16.1)$$

$$G_{ma} = \frac{3K_{Ua}(1 - 2\nu')}{2(1 + \nu')} \quad (16.2)$$

$$(1 - m)K_{La} = \frac{(1 + \sin\phi_f)K_i - (1 - \sin\phi_f)}{(1 - K_i)\sin\phi_f} G_{ma} \quad (16.3)$$

$$r_k = \frac{1}{K_{La}} \frac{1 + e_0}{\lambda} \sigma_{vi}' \quad (16.4)$$

Shear modulus ($G_{ma} = 352$ kPa) in Cases A and B was equivalent to the shear wave velocity of 15 m/s. Conversely, shear wave velocity (denoted as *Vs*) corresponded to 100 m/s in the Ac2 layer based on PS logging conducted near the site (Inagaki et al. 2002). In Case C, shear modulus was calculated as 14,780 kPa from Vs = 100 m/s. The subsequent parameter setting procedure was the same as that in Cases A and B.

The parameter $r\varepsilon_{dc}$ that controlled the generation of contractive dilatancy was an important parameter and was set originally by referring to the effective stress path based on the results of the undrained shear test. However, given that it was not possible to refer to these test results, $r\varepsilon_{dc}$ was set such that it could reproduce the tendency of the settlement of the embankment and the excess pore water pressure generation in Ac2 that were already obtained by the concurrent analysis.

16.3.7 Cases Analyzed in the Study

Analyses were performed on the eight cases (Case 01–Case23) as shown in Table 16.4. These cases involved the combination of the parameter value sets in Cases A, B, and C for Ac2 and the analysis objects (1), (2), and (3) detailed as follows:

1. Consolidation settlement analysis under the test embankment loading
2. Consolidation settlement analysis under the permanent embankment loading
3. Analysis to investigate the influence of earthquakes during the consolidation settlement analysis under the permanent embankment loading

With respect to points (2) and (3) as listed above, the influences of the test embankment were ignored.

16.3.8 Analysis Procedure

The analyses of all the cases were performed in two stages, namely, the initial self-weight analysis of the ground and consolidation settlement analysis by loading the test embankment or permanent embankment. With respect to Cases 22 and 23, the earthquake response analyses was performed by applying seismic motions during consolidation settlement analysis.

The initial self-weight analysis, including steady permeation flow analysis, was performed. In the initial self-weight analysis, the groundwater level was set as −6 m, and the corresponding excess pore water pressure was set as 0. Both the sides and bottom of the analytical model constituted impermeable boundaries. In the subsequent consolidation settlement analysis, the embankment load was expressed by applying a distribution force equivalent to the embankment weight that increased at each construction step. This distribution force was applied to the upper surface of the portion of the embankment that was banked up in the previous step. In the consolidation settlement analysis, the excess pore water pressure was set as 0 at the ground surface. The bottom of the analytical model corresponded to an impermeable boundary, and both sides are described in Sect. 16.3.3.

In the consolidation settlement analysis, earthquake response analysis was sometimes performed during the consolidation settlement analysis. Thus, saturation/wet mass density was set to the soil elements, pore water elements, and embankment elements of FLIP in order to consider the effect of the inertial force.

Table 16.4 List of Cases analyzed in the study

Parameters of Ac2					
Cases	Initial shear modulus	Secondary consolidation	(1) Consolidation analysis for Test embankment	(2) Consolidation analysis for Permanent embankment	(3) Consolidation and seismic response analysis for Permanent embankment
Case A	By λ and κ	Ignore	Case01	Case11	–
Case B	By λ and κ	Apply	Case02	Case12	Case22
Case C	By PS logging	Apply	Case03	Case13	Case23

During the consolidation settlement analysis, the weight of the embankment was considered as the distributed load as described above, so gravity acceleration was not applied. Thus, although mass density was set to each element, this did not cause the weight to double count.

16.4 Consolidation Settlement Analysis Under the Test Embankment Loading

The results of consolidation settlement analyses under the test embankment loading (Cases 01, 02, and 03) were shown.

16.4.1 Schedule for Banking up the Test Embankment

Table 16.5 shows the schedule for banking up the test embankment. In the analysis, the embankment load was increased according to the progress schedule.

16.4.2 Time History of the Settlement and Excess Pore Water Pressure

Figure 16.9 shows the vertical displacement time history at point A on the ground level directly below the center of the embankment. The vertical axis represents vertical displacement (the settlement is negative), and the horizontal axis represents time (number of days since the commencement of the test embankment construction). The green, red, and blue lines correspond to Case 01, Case 02, and Case 03, respectively. The black line indicates the measured values (Tatta et al. 2003). Typically, the trend in Cases 01 and 02 generally coincides with the trend of the measured values. The measured values were similar to the settlement in Case 03 until the 165th day, after which the settlement progressed rapidly. There was a possibility that an event such as an earthquake occurred at this point (approximately February 1981), nevertheless, this was not considered at the moment.

Figure 16.10 shows the time history of excess pore water pressure at point B in the center of Ac2. In a manner similar to the vertical displacement time history, the green, red, and blue lines correspond to Case 01, Case 02, and Case 03, respectively. The black lines indicate measured values. Cases 01, 02, and 03 were generally consistent with the measured values as a tendency.

Table 16.5 Schedule for banking up the test embankment

Step	Events	Height (m)	Elapsed time (day)
1	Only foundation ground	0	0
2	Sand mat laying	0.5	90
3	Banking up (up to +3.5 m)	3.5	123
4	Leaving to stand	3.5	148
5	Banking up (up to +6.5 m)	6.5	183
6	Banking up (up to +9.5 m)	9.5	208
7	Leaving to stand	9.5	–

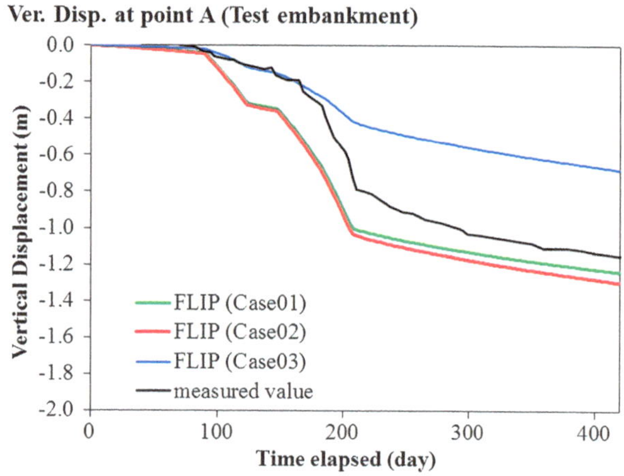

Fig. 16.9 Behavior of test embankment: vertical displacement time history at Point A

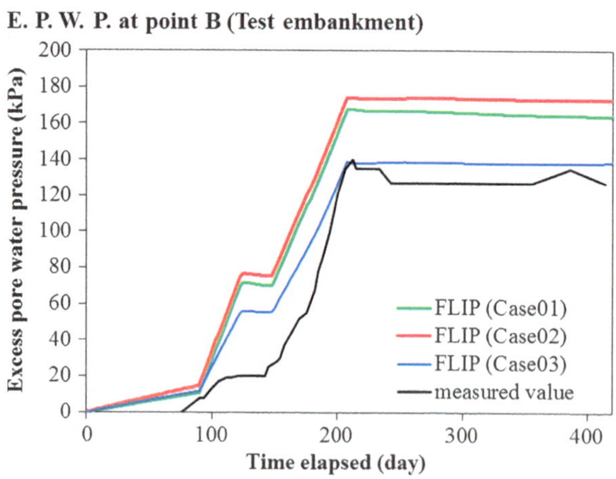

Fig. 16.10 Behavior of test embankment: excess pore water pressure at Point B

16.4.3 Distribution of Horizontal Displacement at the Time of Completion of Test Embankment

Figure 16.11 shows the distribution of horizontal displacement along Line C at the time of completion of the test embankment (the 208th day). The horizontal and vertical axes correspond to the horizontal displacement and depth, respectively.

The horizontal displacement distribution of Case 03 with the rigidity set based on the results of the PS logging was consistent with the measured value (Ota et al. 2005). Conversely, Cases 01 and 02 in which rigidity was set based on λ and κ as shown in Table 16.2 indicated a horizontal displacement of 0.8 m around GL – 14 m, which considerably exceeded the measured value. Each program participating in the concurrent analysis (Ota et al. 2005) indicated a horizontal displacement of approximately 1.2 m around G.L.-14 m, which considerably exceeded the measured value.

16.5 Consolidation Settlement Analysis Under Permanent Embankment Loading

This section describes the results of the consolidation settlement analysis under the permanent embankment loading (Cases 11, 12, and 13).

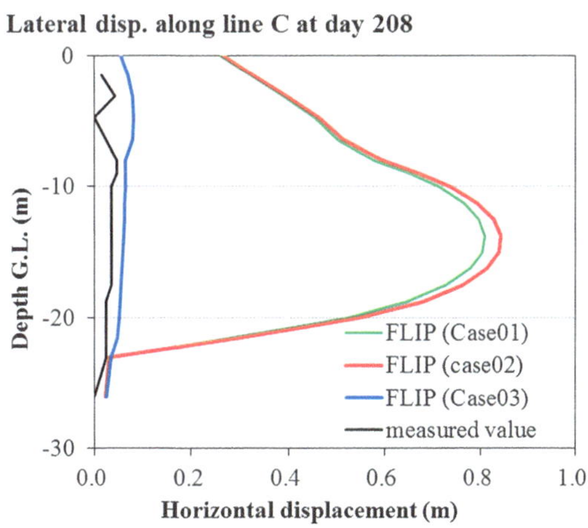

Fig. 16.11 Behavior of test embankment: lateral displacement along line C at the completion of the test embankment

16.5.1 Time History of the Settlement and Excess Pore Water Pressure

Figure 16.12 shows the vertical displacement time history at point A on the ground surface level just below the center of the embankment. The horizontal axis corresponds to time (the number of days elapsed from the start of the test embankment construction). The range of the horizontal axis extends from the 936th day when the permanent embankment construction commenced to the 18250th day (50 years later). The green, red, and blue lines correspond to Case 11, Case 12, and Case 13, respectively. The black line denotes measured values (Noda et al. 2004). The measured values indicate that settlement already reached approximately 1.07 m on the start date of the permanent embankment construction (936th day) due to the influence of the test embankment. The analysis was performed based on the assumption that the construction of the permanent embankment commenced afresh on the 936th day, and the influence of the test embankment was ignored. That is, in the analysis, the settlement on the 936th day was set as 0. Therefore, it was not possible to simply compare the measured values and analytical values.

First, Case 11, in which secondary consolidation was not considered, and Case 12, in which secondary consolidation was considered, were compared, and the comparison indicated that the settlement of Case 12 after 50 years exceeded that of Case 11 by approximately 0.7 m. This was followed by comparing the settlement between measured values and Cases 12 and 13, and the comparison indicated that both settlements progressed at approximately the same speed until approximately the 4000th day. Following this, the settlement in the measured values progressed rapidly (the part surrounded by the black broken circle).

Figure 16.13 shows the time history of excess pore water pressure at point B in the center of Ac2. In a manner similar to the figure for the time history of settlement, the green, red, and blue lines corresponded to Case 11, Case 12, and Case 13, respectively. Additionally, the marks ■ denoted the measured values

Fig. 16.12 Behavior of permanent embankment: vertical displacement time history at Point A

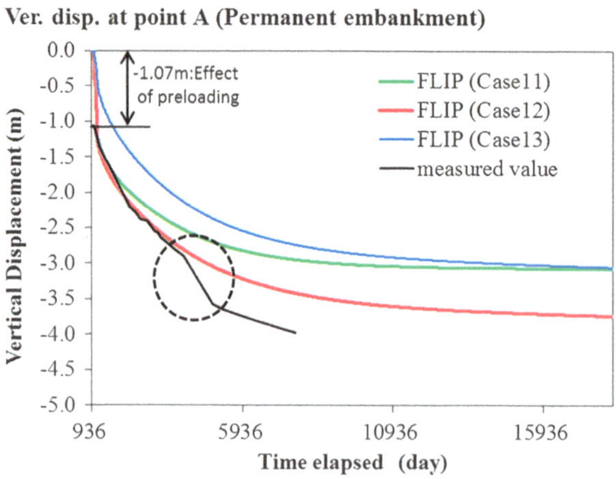

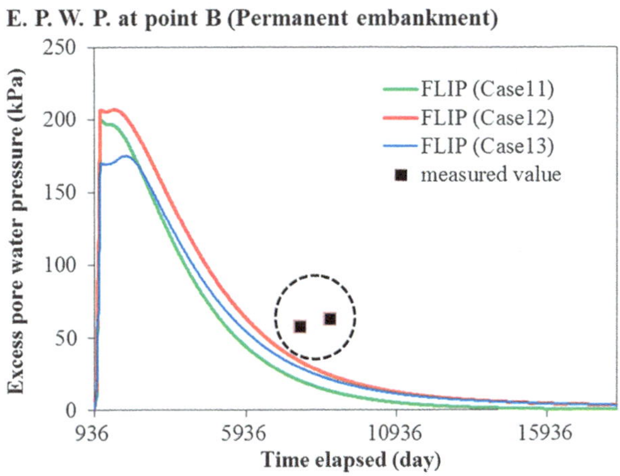

Fig. 16.13 Behavior of permanent embankment: excess pore water pressure at Point B

(Tatta et al. 2003). The measured values were measured at the 7718th day and the 8721st day with the same pore water pressure gauge. The problem was that the excess pore water pressure (measured value) increased with time (as denoted by the part surrounded by the black dashed circle). Conversely, this phenomenon was not observed in the results of analyses.

With respect to the two phenomena in the measured values (that is, the phenomenon in which the settlement progressed rapidly and the excess pore water pressure increased during consolidation), it is important to attempt to reproduce the same tendency through the seismic response analyses during consolidation settlement analysis.

16.6 Understanding the Influence of Earthquakes in the Consolidation Process

Seismic ground motion was applied on the 4136th day and the 7336th day (that is, the 3200th day and the 6400th day respectively from the commencement of permanent embankment construction) (Case 22, 23) to understand the effect of the earthquake in the process of consolidation settlement due to permanent embankment loading.

16.6.1 Seismic Wave

Earthquake response analyses were performed by applying the seismic wave as shown in Fig. 16.14 to the base ground surface at the 4136th day and the 7336th day

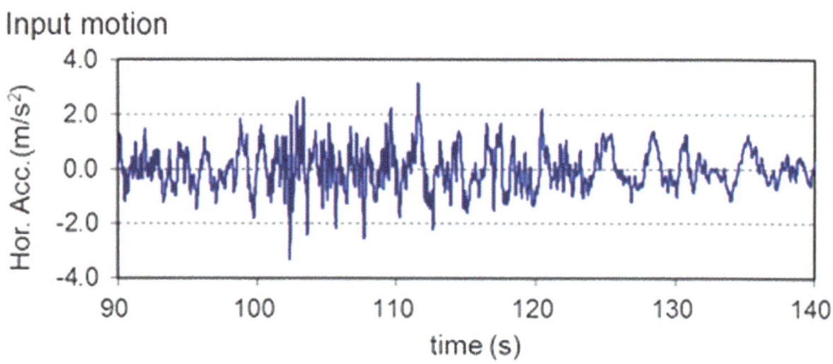

Fig. 16.14 Input seismic motion

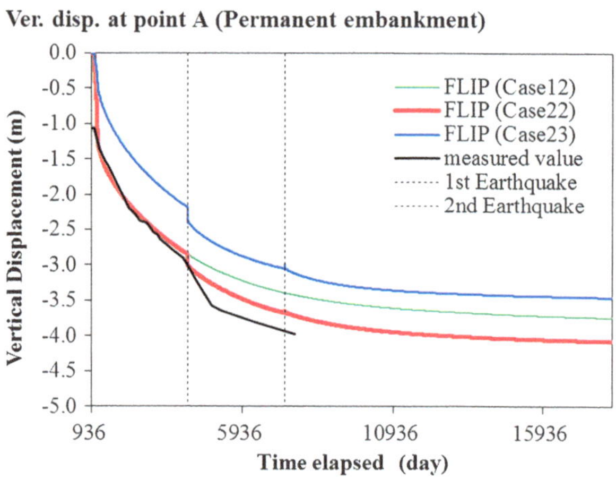

Fig. 16.15 Earthquake response: vertical displacement time history at Point A

from the commencement of the test embankment construction. These corresponded to imaginary earthquakes.

The seismic waveform as shown in Fig. 16.14 was obtained by taking out the principal motion (50 s) from the seismic waveform observed on the engineering base of the Shinagawa site during the great 2011 East Japan earthquake and by multiplying its amplitude by five times.

16.6.2 Time History of the Settlement and the Excess Pore Water Pressure

Figure 16.15 shows the vertical displacement time history at point A on the ground surface level immediately below the center of the embankment. Each line of red and blue color corresponds to Case 22 and Case 23, respectively. The vertical displacement history in Case 12 (redisplayed) was depicted by a green colored line to

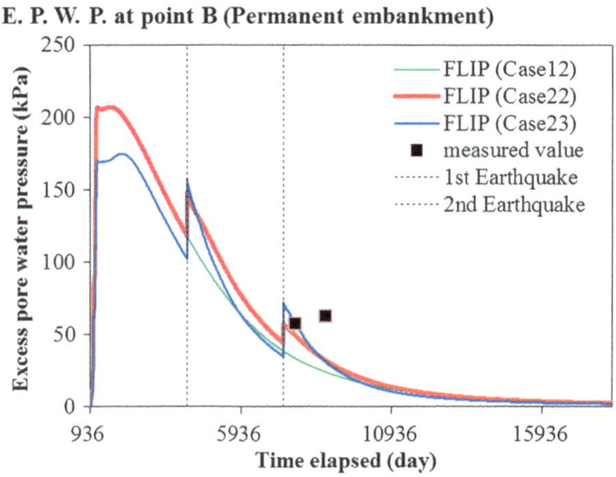

Fig. 16.16 Earthquake response: excess pore water pressure time history at Point B

facilitate a comparison. Case 12 was essentially the same as Case 22, but seismic motions were not applied in case 12. The black line indicates the measured value (redisplayed). The vertical dot lines indicate the time when the seismic motions were applied.

The vertical displacement time histories of Case 22 and Case 23 indicated that instantaneous settlement of 0.15 m in Case 22 and 0.20 m in Case 23 occurred on the 4136th day earthquake (the first earthquake). Case 12 that was not affected by seismic ground motion was compared with Case 22, and the results indicated a difference of 0.15 m in settlement at the time of the first earthquake, and after 50 years this difference expanded to 0.33 m.

Figure 16.16 shows the excess pore water pressure time history at Point B in the center of Ac2. In a manner similar to the vertical displacement time history chart, the time history in Case 12 was also shown for comparison purposes.

At the time of the first earthquake on the 4136th day, the excess pore water pressure increased by 30 kPa in Case 22 and 56 kPa in Case 23 for the first several days after the earthquake. At the time of the second earthquake on the 7336th day, the excess pore water pressure increased by 13 kPa in Case 22 and 37 kPa in Case 23 for the first several days after the earthquake.

Therefore, it was possible to analytically reproduce two types of observed phenomena, namely the rapid progress of the settlement and the increase of excess pore water pressure during consolidation, by applying the seismic motions during the consolidation settlement analysis.

16.6.3 Various Time Histories During the First Earthquake (Case 22)

Figures 16.17, 16.18, 16.19, and 16.20 show various time histories during 50 s of the first earthquake in Case 22. Fig. 16.17 shows the response acceleration time history at point A, Fig. 16.18 shows the response horizontal displacement time history at point A, Fig. 16.19 shows the response vertical displacement time history

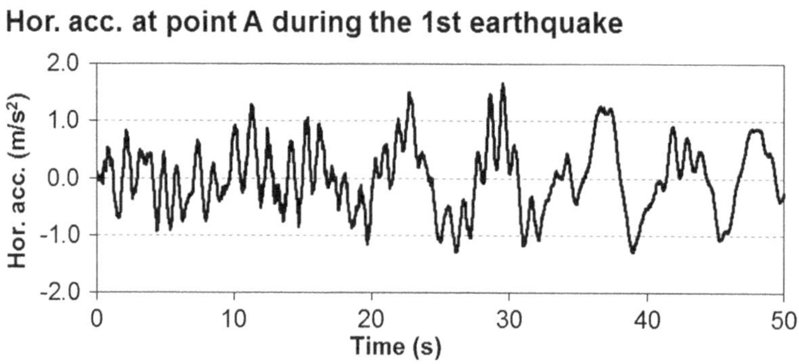

Fig. 16.17 Horizontal acceleration at Point A during the first earthquake (Case 22)

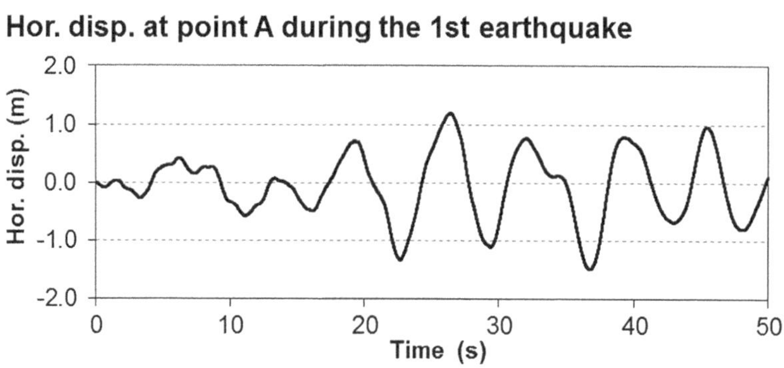

Fig. 16.18 Horizontal displacement at Point A during the first earthquake (Case 22)

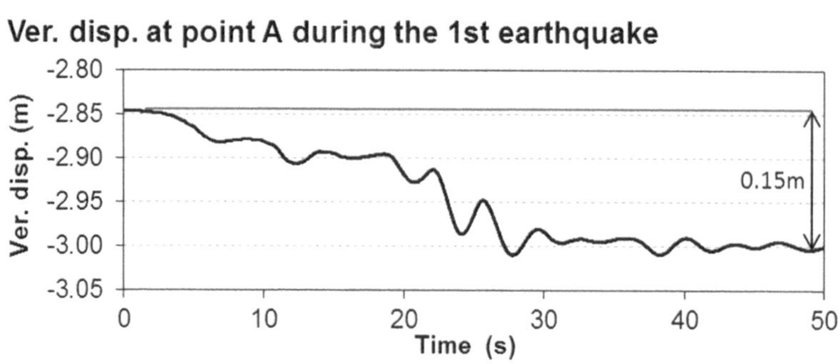

Fig. 16.19 Vertical displacement at Point A during the first earthquake (Case 22)

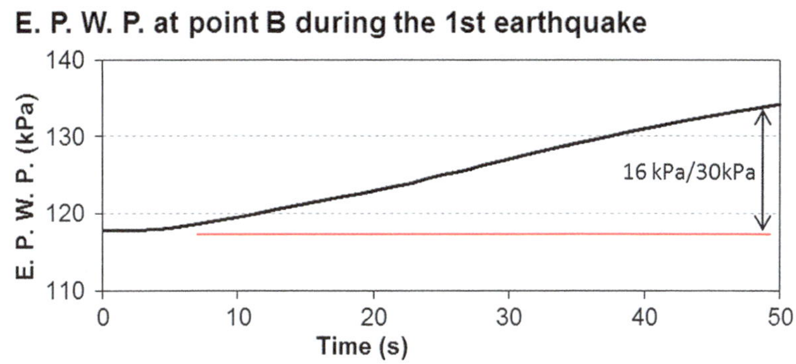

Fig. 16.20 Excess pore water pressure at Point B during the first earthquake (Case 22)

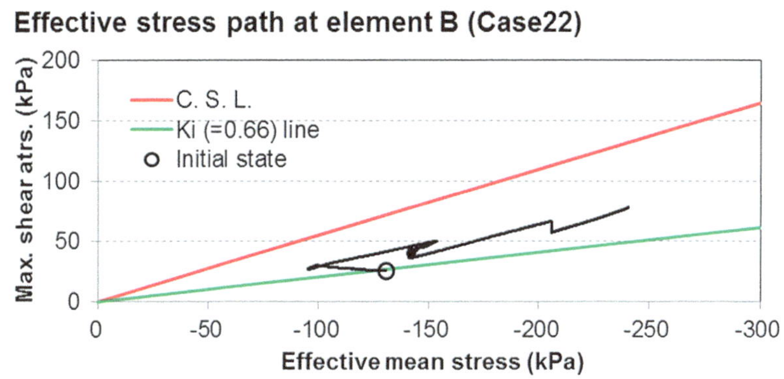

Fig. 16.21 Effective stress path in element B over 50 years (Case 22)

at point A, and Fig. 16.20 shows the excess pore water pressure time history at point B in the center of Ac2. The results suggested that a settlement of 0.15 m occurred at Point A for 50 s when the first seismic motion was applied. In contrast, the findings indicated that the pore water pressure increased by 16 kPa to a total of 30 kPa at Point B when the seismic motion was applied.

16.6.4 Hysteresis Curves at the Center of Ac2 (Case 22)

Figure 16.21 shows the effective stress path for 50 years in Element B at the center of Ac2, and Fig. 16.22 shows the relation between the volumetric strain and the mean effective stress.

According to the effective stress path, the initial state lies on the prescribed line of Ki (=0.66). Both figures indicated that the mean effective stress decreased once due to the permanent embankment loading. Following this, the mean effective stress increased and was accompanied with volume contraction. The mean effective

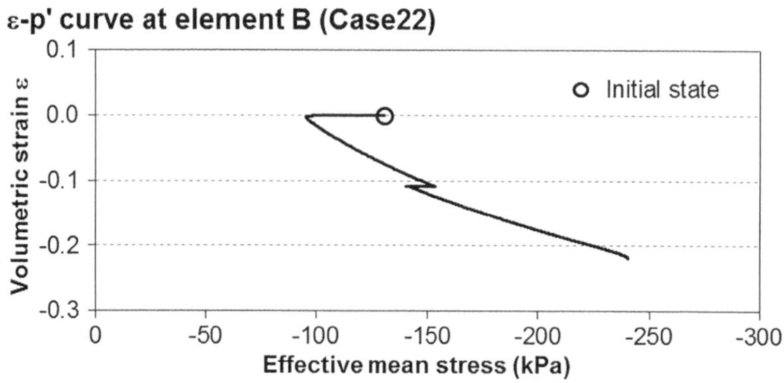

Fig. 16.22 Relation between volumetric strain and effective mean stress in Element B over 50 years (Case 22)

stress decreased again, and the maximum shear stress also exhibited a slight decrease at the time of the first earthquake. This was followed again by the consolidation process.

16.7 Conclusions

1. Applicability of the constitutive model for clay (cookie model) based on the strain space multiple mechanism model was demonstrated through a series of initial self-weight analysis, consolidation settlement analysis, and seismic response analysis within process of the consolidation settlement analysis.
2. Evaluation of the initial shear stiffness based on the PS logging proves better computational results, especially for lateral spreading under the toe of the slope of the embankment in the consolidation settlement analysis.
3. An instantaneous settlement of 0.15–0.20 m occurred when the seismic motion was applied during the consolidation settlement analysis. Additionally, the settlement amount was 0.33 m larger for 50 y than that in the case without the application of seismic motion.
4. The excess pore water pressure increased from 30 to 56 kPa with one earthquake.
5. The observed phenomena in which settlement progressed rapidly and excess pore water pressure increased during the consolidation process could be analytically reproduced by applying seismic waves during the consolidation settlement analysis.

Acknowledgement This study was conducted as a part of activities of the FLIP Consortium Association. The author gratefully acknowledges the members of the FLIP Consortium.

References

Iai S, Ueda K (2016) Modeling of clay through strain space multiple mechanism model. Ann Disaster Prev Res Inst, Kyoto Univ 59B:92–114. (in Japanese)

Iai S, Tobita T, Ozutsumi O, Ueda K (2011) Dilatancy of granular materials in a strain space multiple mechanism model. Int J Numer Anal Methods Geomech 35(3):360–392

Iai S, Ueda K, Ozutsumi O (2015) Strain space multiple mechanism model for clay under monotonic and cyclic loads. In: Proceedings of 6th international conference on earthquake geotechnical engineering, Christchurch, New Zealand, 264, CD-ROM

Inagaki M et al (2002) Consolidation of soft soil due to fill load. In: Proceedings of 37th research presentation conference of the Japanese Geotechnical Society (in Japanese)

Noda T et al (2004) Influences on progressive consolidation settlement of highly structured clayey soil. In: Proceedings of 39th research presentation conference of the Japanese Geotechnical Society (in Japanese)

Ohta H (chairman) et al (2005) Committee for study on application of FEM to a design in the geotechnical engineering, activity report, Japanese Geotechnical Society (in Japanese)

Sekiguchi H, Ohta H (1977) Induced anisotropy and time dependency in clays. In: Proceedings of 9th international conference on soil mechanics and foundation engineering, Tokyo

Tatsuoka F, Ishihara M, Di Benedetto H, Kuwano R (2002) Time-dependent shear deformation characteristics of geomaterials and their simulation. Soils Found 42(2):103–129

Tatta N et al (2003) Prediction of long-term deformation of road embankment on soft ground by fem and its application to performance design. J Jpn Soc Civ Eng No.743/III-64, 173–187 (in Japanese)

Chapter 17
Large Deformation (Finite Strain) Analysis: Theory

Kyohei Ueda

Abstract This chapter presents the finite strain formulation of a strain space multiple mechanism model for granular materials. Since the strain space multiple mechanism model has an appropriate micromechanical background in which the branch and complementary vectors are defined in the material (or referential) coordinate, the finite strain formulation is carried out by following the change in these vectors, both in direction and magnitude, associated with deformation in the material. By applying the methodology for compressible materials established in the finite strain continuum mechanics, decoupled formulation that decomposes the kinematic mechanisms into volumetric and isochoric components is adopted. Material (Lagrangian) description of the integrated form is given by a relation between the second Piola-Kirchhoff effective stress and the Green-Lagrange strain tensors; spatial (Eulerian) description by a relation between the Cauchy effective stress and the Euler-Almansi strain tensors. Material description of the incremental form is derived through the material time derivative of the integrated form. The counterpart in the spatial description is derived through the Lie time derivative, given as a relation between the Oldroyd stress rate of Kirchhoff stress and the rate of deformation tensor.

17.1 Introduction

The granular materials consist of an assemblage of particles that move along the macroscopic deformation. The motion of individual particle can be different from the macroscopic motion, with contacts newly formed or disappeared during the macroscopic deformation. However, the motion of particles, on average, may be consistent with the macroscopic deformation of the granular materials. This fact forms the basis for modeling the behavior of granular materials in terms of a continuum. The continuum may be characterized by a specific structure formed by the assemblage of particles through a unit vector representing direction of branch

K. Ueda (✉)
Disaster Prevention Research Institute, Kyoto University, Kyoto, Japan
e-mail: ueda.kyohei.2v@kyoto-u.ac.jp

S. Iai (ed.), *Developments in Earthquake Geotechnics*, Geotechnical, Geological and Earthquake Engineering 43, DOI 10.1007/978-3-319-62069-5_17

between the particles in contact and a complementary unit vector representing direction normal to the branch vector. These vectors characterizing the structure in the original configuration rotate along with the macroscopic rotation of the granular materials.

Among the various constitutive models proposed for granular materials, a strain space multiple mechanism model is a promising one to describe the specific structure of granular materials (Iai et al. 1992, 2011; Iai and Ozutsumi 2005). The model consists of a multitude of simple shear mechanisms with each oriented in an arbitrary direction and can describe the behavior of granular materials under complex loading paths, including the effect of principal stress axes rotation (Towhata and Ishihara 1985). The model has been implemented in a finite element program, called "FLIP ROSE (Finite element analysis of LIquefaction Process/ Response Of Soil-structure systems during Earthquakes)", and used in the analysis of numerous problems in practice for evaluating seismic performance of geotechnical works including level ground, retaining structures, embankments, and underground structures (Iai et al. 1998; Ozutsumi et al. 2002).

In the previous studies through the strain space multiple mechanism model, formulation was based on the infinitesimal strain theory. Although this formulation works relatively well in engineering practice, the application, strictly speaking, should be limited to phenomena under small strain, small displacement and small rotation. Therefore, the strain space multiple mechanism model has been extended within the finite strain framework that allows an analysis of the behavior of granular materials in the regime including relatively large strain and rotation (Ueda 2009; Iai et al. 2013a), and a large deformation (finite strain) analysis program, called "FLIP TULIP (Total and Updated Lagrangian program of LIquefaction Process)", has been developed by incorporating the extended model. This paper presents the finite strain formulation in terms of material (Lagrangian) and spatial (Eulerian) description.

17.2 Infinitesimal Strain Formulation

As a prelude to the finite strain formulation, the main components of the strain space multiple mechanism model are reviewed within the infinitesimal strain framework (Iai et al. 2011). Effective stress in granular materials as defined for continuum is given by a certain average of contact forces between the particles. In an assemblage of spherical particles, the contact force **P** can be partitioned into the direction of contact normal (or along the branch connecting the particle centers) **n** and tangential direction **t** as (see Fig. 17.1).

$$\mathbf{P} = F\mathbf{n} + S\mathbf{t} \tag{17.1}$$

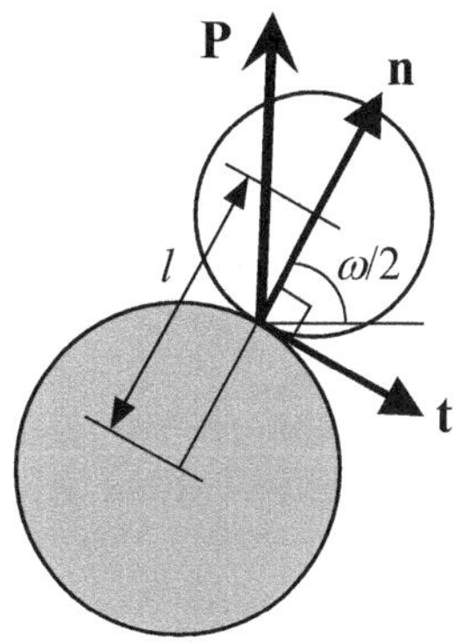

Fig. 17.1 Contact normal **n**, tangential direction **t** and contact force **P**

Macroscopic effective stress is given by taking an average over the contact forces within the representative volume element having volume V as (e.g. Thornton 1989)

$$\sigma' = \frac{1}{V}\sum l(F\mathbf{n}\otimes\mathbf{n} + S\mathbf{t}\otimes\mathbf{n}) \tag{17.2}$$

where l denotes length of the branch.

Before taking the average over all the contact forces in a representative volume element, the contact forces can be classified according to the orientation. Let us think of a class of contact forces of which direction is at an angle $\omega/2$ relative to the x axis in two dimensions and take the appropriate average to form a partial contribution to the macroscopic stress. The stress can then be given by combining all the partial stress contributions over the entire angles of ω.

By systematically sorting out the isotropic and deviator components of the second order tensors in Eq. 17.2 and by taking the number of summation to infinity, Eq. 17.2 is rewritten in two dimensions as follows (Mehrabadi et al. 1982; Iai et al. 2011):

$$\sigma' = -p'\mathbf{I} + \int_0^{\pi} (q_F\langle\mathbf{n}\otimes\mathbf{n}\rangle + q_S\langle\mathbf{t}\otimes\mathbf{n}\rangle)\mathrm{d}\omega \tag{17.3}$$

$$\langle\mathbf{n}\otimes\mathbf{n}\rangle = \mathbf{n}\otimes\mathbf{n} - \mathbf{t}\otimes\mathbf{t}, \langle\mathbf{t}\otimes\mathbf{n}\rangle = \mathbf{t}\otimes\mathbf{n} + \mathbf{n}\otimes\mathbf{t} \tag{17.4}$$

where p' denotes mean effective stress (compression positive), $\mathbf{I}$ denotes second order identity tensor, q_F, q_S denote micromechanical stress contributions to macroscopic deviator stress due to normal and tangential components of contact forces.

Equation 17.3 represents the mechanisms with the combination of biaxial shear $\langle\mathbf{n}\otimes\mathbf{n}\rangle$ and simple shear $\langle\mathbf{t}\otimes\mathbf{n}\rangle$. However, once they are idealized in terms of the second order tensors, they become indistinguishable except for the difference in the orientation with an angle of $\pi/4$ (Iai et al. 1992). Consequently, the term $\langle\mathbf{t}\otimes\mathbf{n}\rangle$, i. e. virtual simple shear mechanism, will be kept in use throughout this paper. Thus, Eq. 17.3 is rewritten as:

$$\sigma' = -p'\mathbf{I} + \int_0^{\pi} q\langle \mathbf{t} \otimes \mathbf{n} \rangle \mathrm{d}\omega \tag{17.5}$$

More details on the derivation of p' and q in Eq. 17.5 from the contact force components F and S may be found in another article (Iai et al. 2013b).

In order to derive the integrated form of the strain space multiple mechanism model, i.e. direct stress strain relationship, let us define the volumetric strain ε (extension positive) and the virtual simple shear strains γ as the projections of the macroscopic strain field to the second order tensors as follows:

$$\varepsilon = \mathbf{I} : \boldsymbol{\varepsilon} \tag{17.6}$$

$$\gamma = \langle \mathbf{t} \otimes \mathbf{n} \rangle : \boldsymbol{\varepsilon} \tag{17.7}$$

where the double dot symbol denotes double contraction. In addition, the effective volumetric strain ε' and virtual effective volumetric strain ε'' are introduced as follows:

$$\varepsilon' = \varepsilon - \varepsilon_{\mathrm{d}} \tag{17.8}$$

$$\varepsilon'' = \varepsilon - \varepsilon_{\mathrm{d}}^{\mathrm{c}} \tag{17.9}$$

where the dilatancy ε_{d} in Eq. 17.8 is decomposed into contractive component ${\varepsilon^{\mathrm{c}}}_{\mathrm{d}}$ and dilative components ${\varepsilon^{\mathrm{d}}}_{\mathrm{d}}$ as follows:

$$\varepsilon_{\mathrm{d}} = \varepsilon_{\mathrm{d}}^{\mathrm{c}} + \varepsilon_{\mathrm{d}}^{\mathrm{d}} \tag{17.10}$$

The contractive component of dilatancy is formulated, using a parameter $\mathrm{M_v}^*$ determined by the phase transformation angle, as follows:

$$\dot{\varepsilon}_{\mathrm{d}}^{\mathrm{c}} = \mathbf{I}_{\mathrm{d}}^{\mathrm{c}} : \dot{\boldsymbol{\varepsilon}} \tag{17.11}$$

where

$$\mathbf{I}_{\mathrm{d}}^{\mathrm{c}} = -\int_0^{\pi} \mathrm{M_v}^* |\langle \mathbf{t} \otimes \mathbf{n} \rangle|^* \mathrm{d}\omega \tag{17.12}$$

$$|\langle \mathbf{t} \otimes \mathbf{n} \rangle|^* = \begin{cases} \langle \mathbf{t} \otimes \mathbf{n} \rangle & \mathit{for}\ \langle \mathbf{t} \otimes \mathbf{n} \rangle : \dot{\boldsymbol{\varepsilon}} \geq 0 \\ -\langle \mathbf{t} \otimes \mathbf{n} \rangle & \mathit{for}\ \langle \mathbf{t} \otimes \mathbf{n} \rangle : \dot{\boldsymbol{\varepsilon}} < 0 \end{cases} \tag{17.13}$$

The formulation of dilative component of dilatancy will be given after the definition of the function q at the end of this section.

Using these variables defined in Eqs. 17.8 and 17.9, the mean effective stress p' and virtual simple shear stress q in Eq. 17.5 are defined through path dependent functions as follows:

$$p' = p'(\varepsilon') \tag{17.14}$$

$$q = q(\gamma, \varepsilon', \varepsilon'') \tag{17.15}$$

In particular, the mean effective stress in Eq. 17.14 is formulated with the power index l_K and the bulk modulus $K_{L/Ua}$ at the mean effective stress of $P_a{}'$ as

$$-\frac{dp'}{d\varepsilon'} = K_{L/U} = K_{L/Ua}\left(\frac{p'}{p_a{}'}\right)^{l_K} \tag{17.16}$$

$$K_{L/U} = \begin{cases} K_L & \text{if } \dot{\varepsilon}' \geq 0 \quad \text{(loading/neutral)} \\ K_U & \text{if } \dot{\varepsilon}' < 0 \quad \text{(unloading)} \end{cases} \tag{17.17}$$

whereas the virtual simple shear stress in Eq. 17.15 is formulated as a non-linear hysteretic function, where a back-bone curve is given by the following hyperbolic function (Hardin and Drnevich 1972);

$$q = \frac{\gamma/\gamma_v}{1 + |\gamma/\gamma_v|} q_v \tag{17.18}$$

The terms $q_v = q_v(\varepsilon')$ and $\gamma_v = \gamma_v(\varepsilon', \varepsilon'')$ defining the hyperbolic function are the shear strength and the reference strain of the virtual simple shear mechanism, respectively.

By applying the classic postulate of interlocking by Taylor (1948) (i.e. portion of shear strain energy is absorbed by dilative component of dilatancy in order to set free the assemblage of particles in granular materials from interlocking) to the backbone curve in Eq. 17.18, the dilative component of dilatancy is given by

$$\dot{\varepsilon}_d^d = \mathbf{I}_d^d : \dot{\varepsilon} \tag{17.19}$$

$$\mathbf{I}_d^d = r_{\varepsilon_d} \int_0^\pi \frac{\gamma/\gamma_v}{1 + |\gamma/\gamma_v|} M_{fv} \langle \mathbf{t} \otimes \mathbf{n} \rangle d\omega \tag{17.20}$$

where $r_{\varepsilon d}$ is a material parameter and M_{fv} is another parameter determined by the internal friction angle (Iai and Ueda 2016).

17.3 Finite Strain Formulation in Reference Configuration

17.3.1 Material Description of Integrated Form

In the finite strain formulation of the strain space multiple mechanism model, the unit vector **N** along the direction of the branch between the particles in contact and the unit vector **T** normal to **N** are defined in the reference (or initial) configuration

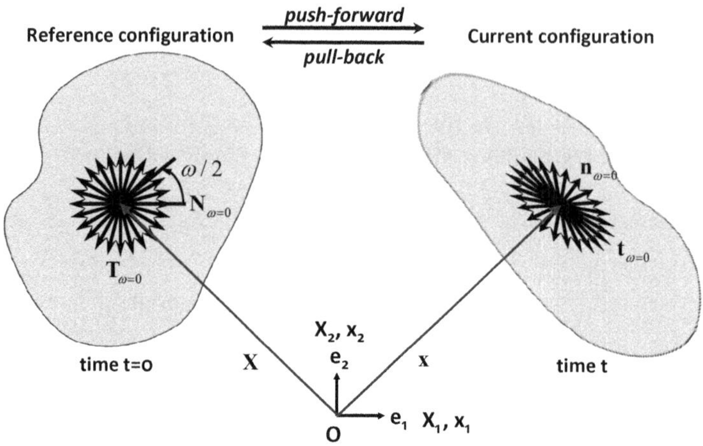

Fig. 17.2 Direction vectors along branches and position vectors in reference and current configurations

of the material. These vectors are assumed to change direction and magnitude into **n** and **t** in the current configuration through the deformation gradient **F** as follows:

$$\mathbf{n} = \mathbf{F}\mathbf{N}, \mathbf{t} = \mathbf{F}\mathbf{T} \tag{17.21}$$

$$\mathbf{F} = \frac{\partial \mathbf{x}}{\partial \mathbf{X}} \tag{17.22}$$

where the motion of the material transforms the reference position vector **X** defined on the material in the initial configuration to the current position vector **x** as shown in Fig. 17.2. By assuming that the motion of particles, on average, are consistent with the macroscopic deformation of the granular materials, the mapping through **F** in Eq. 17.21 is considered similar to the mapping through **F** for the particle centroids in discrete element modeling as schematically shown in Fig. 17.3.

The preliminary approach toward the finite strain formulation, but modified later because of a certain drawback as will be described, was to adopt the infinitesimal strain formulation in Eq. 17.5 directly to the spatial description of the strain space multiple mechanism model in the current configuration.

The pull-back operation (Holzapfel, 2000) of Eq. 17.5 to the reference configuration with multiplication of Jacobian determinant J results in the material description of the strain space multiple mechanism model as follows:

$$\mathbf{S}' = \mathbf{F}^{-1} J \boldsymbol{\sigma}' \mathbf{F}^{-\mathrm{T}} = -Jp'\mathbf{C}^{-1} + \int_0^{\pi} Jq \langle \mathbf{T} \otimes \mathbf{N} \rangle \mathrm{d}\omega \tag{17.23}$$

where $\mathbf{S}'$ denotes the second Piola-Kirchhoff effective stress and

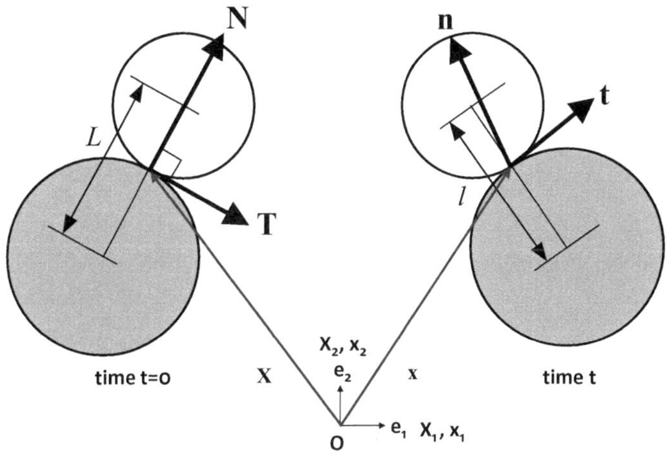

Fig. 17.3 Direction vectors along branches **N**, **n**, the complementary vectors **T**, **t**, and position vectors in reference and current configurations **X**, **x**

$$\mathbf{C} = \mathbf{F}^{\mathrm{T}}\mathbf{F} \tag{17.24}$$

$$J = \det\mathbf{F} \tag{17.25}$$

Although there is no flaw as a theory to formulate Eqs. 17.5 and 17.23, this type of formulation poses a certain drawback in practice of computation because the spatial direction vector **n** may no longer remain normal to **t** and, thus, the second term $\int q\langle\mathbf{t}\otimes\mathbf{n}\rangle d\omega$ in Eq. 17.5 and its counterpart in Eq. 17.23 can include volumetric component.

In order to avoid this drawback, the formulation through Eqs. 17.5 and 17.23 is modified by applying the approach well established for compressible materials (Holzapfel 2000) where the deformation is decomposed into volumetric and isochoric parts. By applying this approach to the strain space multiple mechanism model, the second Piola-Kirchhoff effective stress $\mathbf{S}'$ is decomposed as follows:

$$\mathbf{S}' = \mathbf{S}'_p + \mathbf{S}'_q \tag{17.26}$$

$$\mathbf{S}'_p = -Jp'\mathbf{C}^{-1} \tag{17.27}$$

$$\mathbf{S}'_q = J^{-1}\mathbb{Q} : \bar{\mathbf{S}}' \tag{17.28}$$

$$\bar{\mathbf{S}}' = \int_0^{\pi} Jq\langle\mathbf{T}\otimes\mathbf{N}\rangle d\omega \tag{17.29}$$

where the fourth-order tensor $\mathbb{Q}$, called projection tensor, extracts the isochoric component of the second order tensor $\mathbf{S}'$ in the reference configuration, being defined by using the fourth-order unit tensor $\mathbb{N}$ as follows:

$$\mathbb{Q} = \mathbb{N} - \frac{1}{2}\mathbf{C}^{-1} \otimes \mathbf{C} \tag{17.30}$$

Substitution of Eqs. 17.29 and 17.30 into Eq. 17.28 yields

$$\mathbf{S}'_q = \int_0^{\pi} q\left(\langle \mathbf{T} \otimes \mathbf{N} \rangle - \gamma \mathbf{C}^{-1}\right) \mathrm{d}\omega \tag{17.31}$$

with the following property:

$$2\gamma = \mathbf{C} : \langle \mathbf{T} \otimes \mathbf{N} \rangle \tag{17.32}$$

By taking the double contraction with $\mathbf{C}$, the stress components $\mathbf{S'}_p$ and $\mathbf{S'}_q$ in Eqs. 17.27 and 17.28 can be confirmed being indeed the volumetric and isochoric components as follows:

$$\mathbf{S}'_p : \mathbf{C} = -Jp'\mathbf{C}^{-1} : \mathbf{C} = -2Jp' \tag{17.33}$$

$$\mathbf{S}'_q : \mathbf{C} = \left(J^{-1}\mathbb{Q} : \bar{\mathbf{S}}'\right) : \mathbf{C} = J^{-1}\left\{\bar{\mathbf{S}}' - \frac{1}{2}(\mathbf{C} : \bar{\mathbf{S}}')\mathbf{C}^{-1}\right\} : \mathbf{C} = 0 \tag{17.34}$$

Although the approach described in Holzapfel (2000) refers to hyperelastic modeling, the constitutive model in this study is based on the mechanics of an assemblage of particles defined in Eq. 17.2 and differs from the hyperelastic models. If the strain energy should be evaluated for the thermodynamics consideration, it is readily computed through the equations that will be derived in Sect. 17.5.

The integrated form of the constitutive equation, i.e. direct stress strain relationship, in the reference configuration is derived by relating the macroscopic strain tensor to the macroscopic effective stress through the structure defined by Eqs. 17.26 through 17.29. The macroscopic strain tensor defined in the reference configuration is called the Green-Lagrange strain tensor, defined as

$$\mathbf{E} = \frac{1}{2}(\mathbf{C} - \mathbf{I}) = \frac{1}{2}\left(\mathbf{F}^{\mathrm{T}}\mathbf{F} - \mathbf{I}\right) \tag{17.35}$$

The first step to derive this relationship is to define the volumetric strain ε and the virtual simple shear strains γ within the finite strain framework. The volumetric strain is defined based on the following identity on the rate of change in volume

$$\dot{J} = J\mathrm{tr}\mathbf{d} \tag{17.36}$$

The term $\mathbf{d}$ in Eq. 17.36 denotes the rate of deformation tensor, defined as the symmetric part of the following spatial velocity gradient

$$\mathbf{l} = \mathrm{grad}\dot{\mathbf{u}} = \frac{\partial \dot{\mathbf{u}}}{\partial \mathbf{x}} \tag{17.37}$$

where $\mathbf{u} = \mathbf{x}–\mathbf{X}$ is the displacement vector. The integration of Eq. 17.36 with the initial condition of $\varepsilon = 0$ at $t = 0$ yields the definition of volumetric strain in the context of finite strain formulation as follows:

$$\varepsilon = \ln J \tag{17.38}$$

The virtual simple shear strain γ within the finite strain framework is given as a projection of the Green-Lagrange strain tensor $\mathbf{E}$ to the tensor representing the virtual simple shear component $\langle \mathbf{T} \otimes \mathbf{N} \rangle$ in the reference configuration as follows:

$$\gamma = \langle \mathbf{T} \otimes \mathbf{N} \rangle : \mathbf{E} \tag{17.39}$$

This definition is consistent with Eq. 17.32 by applying Eq. 17.35 because of $\langle \mathbf{T} \otimes \mathbf{N} \rangle : \mathbf{I} = \mathrm{tr} \langle \mathbf{T} \otimes \mathbf{N} \rangle = T_A N_A = 0$ where the vector $\mathbf{T}$ is normal to $\mathbf{N}$.

With the introduction of scalar variables defined in Eqs. 17.38 and 17.39 for the finite strain formulation, the scalar functions defining p' and q remain the same as those given in Eqs. 17.14 through 17.18 for the infinitesimal strain formulation.

17.3.2 Material Description of Incremental Form

Incremental form of the strain space multiple mechanism model in the reference configuration is derived by taking the material time derivative of both sides of Eq. 17.26 as

$$\dot{\mathbf{S}}' = \dot{\mathbf{S}}'_p + \dot{\mathbf{S}}'_q \tag{17.40}$$

$$\dot{\mathbf{S}}'_p = -\frac{\mathrm{D}}{\mathrm{D}t}\left(Jp'\mathbf{C}^{-1}\right) \tag{17.41}$$

$$\dot{\mathbf{S}}'_q = \frac{\mathrm{D}}{\mathrm{D}t}\left(J^{-1}\right)\left(\mathbb{Q} : \bar{\mathbf{S}}'\right) + J^{-1}\frac{\mathrm{D}}{\mathrm{D}t}\left(\mathbb{Q} : \bar{\mathbf{S}}'\right) \tag{17.42}$$

where the symbol D/Dt denotes the material time derivative, that reduces to the differentiation with respect to time $(\dot{\ })$ if the functions and variables are defined in the reference configuration. The incremental form of the constitutive equation is derived by relating this stress rate with the rate of Green-Lagrange strain tensor given by

$$\dot{\mathbf{E}} = \mathrm{sym}\left(\mathbf{F}^{\mathrm{T}}\mathrm{Grad}\dot{\mathbf{u}}\right) = \mathrm{sym}\left(\mathbf{F}^{\mathrm{T}}\partial\dot{\mathbf{u}} / \partial\mathbf{X}\right) \tag{17.43}$$

17.3.2.1 Volumetric Component

The volumetric component of stress rate in Eq. 17.41 may be computed as follows:

$$\begin{aligned}-\frac{\mathrm{D}}{\mathrm{D}t}(Jp'\mathbf{C}^{-1}) &= -\dot{J}p'\mathbf{C}^{-1} - J\dot{p}'\mathbf{C}^{-1} - Jp'\frac{\mathrm{D}}{\mathrm{D}t}(\mathbf{C}^{-1}) \\ &= -Jp'(\mathbf{C}^{-1} : \dot{\mathbf{E}})\mathbf{C}^{-1} - J\frac{\mathrm{d}p'}{\mathrm{d}\varepsilon'}\dot{\varepsilon}'\mathbf{C}^{-1} + 2Jp'\mathbf{C}^{-1} \odot \mathbf{C}^{-1} : \dot{\mathbf{E}}\end{aligned} \tag{17.44}$$

where the symbol $\odot$ denotes the tensor product defined as follows (Holzapfel 2000):

$$\left(\mathbf{C}^{-1} \odot \mathbf{C}^{-1}\right)_{ABCD} = \frac{1}{2}\left(C_{AC}^{-1}C_{BD}^{-1} + C_{AD}^{-1}C_{BC}^{-1}\right) \tag{17.45}$$

The rate of dilatancy and relevant strain components are given from the rate of Green-Lagrange strain tensor as follows:

$$\dot{\varepsilon} = \mathbf{C}^{-1} : \dot{\mathbf{E}} \tag{17.46}$$

$$\dot{\varepsilon}_{\mathrm{d}}^{\mathrm{c}} = \mathbf{C}_{\mathrm{dc}}^{-1} : \dot{\mathbf{E}} \tag{17.47}$$

$$\dot{\varepsilon}_{\mathrm{d}}^{\mathrm{d}} = \mathbf{C}_{\mathrm{dd}}^{-1} : \dot{\mathbf{E}} \tag{17.48}$$

$$\dot{\varepsilon}_{\mathrm{d}} = \mathbf{C}_{\mathrm{d}}^{-1} : \dot{\mathbf{E}} \tag{17.49}$$

$$\dot{\gamma} = \langle \mathbf{T} \otimes \mathbf{N} \rangle : \dot{\mathbf{E}} \tag{17.50}$$

where

$$\mathbf{C}_{\mathrm{d}}^{-1} = \mathbf{C}_{\mathrm{dc}}^{-1} + \mathbf{C}_{\mathrm{dd}}^{-1} \tag{17.51}$$

By referring to the infinitesimal strain formulation in Eq. 17.12, the tensor for defining the contractive component of dilatancy in Eq. 17.47 is given by

$$\mathbf{C}_{\mathrm{dc}}^{-1} = -\int_0^{\pi} \mathrm{M}_{\mathrm{v}}{}^{*}\left|\langle \mathbf{T} \otimes \mathbf{N} \rangle\right|^{*}\mathrm{d}\omega \tag{17.52}$$

Similarly, the tensor $\mathbf{C}_{\mathrm{dd}}^{-1}$ for defining the dilative component of dilatancy in Eq. 17.48 is given by

$$\mathbf{C}_{\mathrm{dd}}^{-1} = r_{\varepsilon_{\mathrm{d}}}J^{-1}\int_0^{\pi}\left(\frac{\gamma/\gamma_{\mathrm{v}}}{1 + |\gamma/\gamma_{\mathrm{v}}|}\right)\mathrm{M}_{\mathrm{fv}}(\langle \mathbf{T} \otimes \mathbf{N} \rangle - \gamma\mathbf{C}^{-1})\mathrm{d}\omega \tag{17.53}$$

The derivation of $\mathbf{C}_{\mathrm{dd}}^{-1}$, however, needs a preparation for defining the strain energy within the finite strain framework and thus will be discussed in Sect. 17.5.

Substitution of these equations into Eq. 17.44 yields the volumetric component of the incremental form as follows:

$$\dot{\mathbf{S}}'_p = \mathbb{C}_p : \dot{\mathbf{E}} \tag{17.54}$$

where

$$\mathbb{C}_p = J\left(K_{\mathrm{L/U}} - p'\right)\mathbf{C}^{-1} \otimes \mathbf{C}^{-1} + 2Jp'\mathbf{C}^{-1} \odot \mathbf{C}^{-1} - JK_{\mathrm{L/U}}\mathbf{C}^{-1} \otimes \mathbf{C}_{\mathbf{d}}^{-1} \tag{17.55}$$

$$K_{\mathrm{L/U}} = -\frac{\mathrm{d}p'}{\mathrm{d}\varepsilon'} \tag{17.56}$$

17.3.2.2 Isochoric Component

The first term of the isochoric component in Eq. 17.42 is given by

$$\frac{\mathrm{D}}{\mathrm{D}t}\left(J^{-1}\right)\left(\mathbb{Q} : \bar{\mathbf{S}}'\right) = -J^{-2}\dot{J}\left(\mathbb{Q} : \bar{\mathbf{S}}'\right) = -J^{-1}\left(\mathbb{Q} : \bar{\mathbf{S}}'\right)\mathbf{C}^{-1} : \dot{\mathbf{E}} = -\mathbf{S}'_q \otimes \mathbf{C}^{-1} : \dot{\mathbf{E}} \tag{17.57}$$

The derivation of the second term in Eq. 17.42 needs somewhat lengthy procedure. As a preparation, the material time derivative is rewritten as follows:

$$\frac{\mathrm{D}}{\mathrm{D}t}\left(\mathbb{Q} : \bar{\mathbf{S}}'\right) = \dot{\bar{\mathbf{S}}}' - \frac{1}{2}\frac{\mathrm{D}}{\mathrm{D}t}\left[\left(\mathbf{C} : \bar{\mathbf{S}}'\right)\mathbf{C}^{-1}\right] \tag{17.58}$$

Since the term $\langle \mathbf{T} \otimes \mathbf{N} \rangle$ is independent of time, the first term in Eq. 17.58, i.e., the material time derivative of $\bar{\mathbf{S}}'$, is written by

$$\dot{\bar{\mathbf{S}}}' = \int_0^{\pi} \frac{\mathrm{D}}{\mathrm{D}t}(Jq)\langle \mathbf{T} \otimes \mathbf{N} \rangle \mathrm{d}\omega \tag{17.59}$$

$$\frac{\mathrm{D}}{\mathrm{D}t}(Jq) = \dot{J}\,q + J\dot{q} = \dot{J}\,q + J\left(\frac{\partial q}{\partial \gamma}\dot{\gamma} + \frac{\partial q}{\partial \varepsilon'}\dot{\varepsilon}' + \frac{\partial q}{\partial \varepsilon''}\dot{\varepsilon}''\right) \tag{17.60}$$

By using Eqs. 17.46 through 17.50, Eq. 17.59 is computed as follows:

$$\dot{\bar{\mathbf{S}}}' = \overline{\mathbb{C}}_q : \dot{\mathbf{E}} \tag{17.61}$$

$$\overline{\mathbb{C}}_q = \int_0^{\pi} J\langle \mathbf{T} \otimes \mathbf{N} \rangle \otimes \mathbf{C}_q^{-1}\mathrm{d}\omega \tag{17.62}$$

$$\mathbf{C}_q^{-1} = q\mathbf{C}^{-1} + G_{\mathrm{L/U}}\langle \mathbf{T} \otimes \mathbf{N} \rangle + H_{\mathrm{L/U}}\left(\mathbf{C}^{-1} - \mathbf{C}_{\mathrm{d}}^{-1}\right) + L_{\mathrm{L/U}}\left(\mathbf{C}^{-1} - \mathbf{C}_{\mathrm{dc}}^{-1}\right) \quad (17.63)$$

$$G_{\mathrm{L/U}} = \frac{\partial q}{\partial \gamma} \quad (17.64)$$

$$H_{\mathrm{L/U}} = \frac{\partial q}{\partial \varepsilon'} \quad (17.65)$$

$$L_{\mathrm{L/U}} = \frac{\partial q}{\partial \varepsilon''} \quad (17.66)$$

The second term in the right side of Eq. 17.58 can be decomposed into three parts:

$$\frac{\mathrm{D}}{\mathrm{D}t}\left[(\mathbf{C} : \bar{\mathbf{S}}')\mathbf{C}^{-1}\right] = (\dot{\mathbf{C}} : \bar{\mathbf{S}}')\mathbf{C}^{-1} + (\mathbf{C} : \dot{\bar{\mathbf{S}}}')\mathbf{C}^{-1} + (\mathbf{C} : \bar{\mathbf{S}}')\frac{\mathrm{D}}{\mathrm{D}t}(\mathbf{C}^{-1}) \quad (17.67)$$

Since the terms in this equation may be computed by using Eqs. 17.28, 17.30, 17.45, and 17.61, Eq. 17.67 is computed as follows (Ueda 2009; Iai et al. 2013a):

$$\frac{\mathrm{D}}{\mathrm{D}t}\left[(\mathbf{C} : \bar{\mathbf{S}}')\mathbf{C}^{-1}\right] = \mathbf{C}^{-1} \otimes \mathbf{C} : \overline{\mathbb{C}}_q : \dot{\mathbf{E}} - 2(\mathbf{C} : \bar{\mathbf{S}}')\widetilde{\mathbb{Q}} : \dot{\mathbf{E}} + 2J\mathbf{C}^{-1} \otimes \mathbf{S}_q' : \dot{\mathbf{E}} \quad (17.68)$$

$$\widetilde{\mathbb{Q}} = \mathbf{C}^{-1} \odot \mathbf{C}^{-1} - \frac{1}{2}\mathbf{C}^{-1} \otimes \mathbf{C}^{-1} \quad (17.69)$$

Based on the above results, the isochoric component of the incremental form is computed as follows:

$$\dot{\mathbf{S}}_q' = \mathbb{C}_q : \dot{\mathbf{E}} \quad (17.70)$$

$$\mathbb{C}_q = J^{-1}\left[\mathbb{Q} : \overline{\mathbb{C}}_q + (\mathbf{C} : \bar{\mathbf{S}}')\widetilde{\mathbb{Q}}\right] - \left(\mathbf{C}^{-1} \otimes \mathbf{S}_q' + \mathbf{S}_q' \otimes \mathbf{C}^{-1}\right) \quad (17.71)$$

where $\mathbb{Q}$ is given in Eq. 17.30. The first term in Eq. 17.71 may be written down as follows:

$$\mathbb{Q} : \overline{\mathbb{C}}_q = \int_0^{\pi} J\left(\langle \mathbf{T} \otimes \mathbf{N} \rangle - \gamma \mathbf{C}^{-1}\right) \otimes \mathbf{C}_q^{-1} \mathrm{d}\omega \quad (17.72)$$

17.3.2.3 Material Description of Incremental Form

By summarizing the results derived above, the incremental form of the strain space multiple mechanism model in the reference configuration is given as follows:

$$\dot{\mathbf{S}}' = \frac{\partial \mathbf{S}'}{\partial \mathbf{E}} : \dot{\mathbf{E}} = \mathbb{C} : \dot{\mathbf{E}} \tag{17.73}$$

$$\mathbb{C} = \mathbb{C}_p + \mathbb{C}_q \tag{17.74}$$

where $\mathbb{C}_p$ and $\mathbb{C}_q$ are given in Eqs. 17.55 and 17.71.

Although the finite strain formulation looks more complicated than the infinitesimal strain counterpart, the formulation is basically the same except for the appearance of stress in the tangential stiffness tensor $\mathbb{C}$ because of the geometrical non-linearity (e.g., the change in the configuration, and the Jacobian determinant for taking into account the volume change in the configuration).

17.4 Finite Strain Formulation in Current Configuration

17.4.1 Spatial Description of Integrated Form

Integrated form of the strain space multiple mechanism model in the current configuration is obtained through the push-forward operation (Holzapfel 2000) of the material description in Eq. 17.26 as follows:

$$\sigma' = J^{-1}\mathbf{F}\mathbf{S}'\mathbf{F}^{\mathrm{T}} = \sigma'_p + \sigma'_q \tag{17.75}$$

where the volumetric and isochoric components are given by

$$\sigma'_p = J^{-1}\mathbf{F}\mathbf{S}'_p\mathbf{F}^{\mathrm{T}} = -J^{-1}\mathbf{F}(Jp'\mathbf{C}^{-1})\mathbf{F}^{\mathrm{T}} = -p'\mathbf{I} \tag{17.76}$$

$$\begin{aligned}\sigma'_q &= J^{-1}\mathbf{F}\mathbf{S}'_q\mathbf{F}^{\mathrm{T}} = J^{-1}\mathbf{F}(J^{-1}\mathbb{Q} : \bar{\mathbf{S}}')\mathbf{F}^{\mathrm{T}} \\ &= J^{-1}\left\{J^{-1}\mathbf{F}\bar{\mathbf{S}}'\mathbf{F}^{\mathrm{T}} - \frac{1}{2}J^{-1}(\mathbf{C} : \bar{\mathbf{S}}')\mathbf{I}\right\}\end{aligned} \tag{17.77}$$

The first and second terms in Eq. 17.77 are written by

$$\bar{\sigma}' = J^{-1}\mathbf{F}\bar{\mathbf{S}}'\mathbf{F}^{\mathrm{T}} = \int_0^{\pi} q\langle\mathbf{t} \otimes \mathbf{n}\rangle \mathrm{d}\omega \tag{17.78}$$

$$\frac{1}{2}J^{-1}(\mathbf{C} : \bar{\mathbf{S}}') = \frac{1}{2}\mathbf{I} : \bar{\sigma}' \tag{17.79}$$

Substitution of these equations back to Eq. 17.77 yields

$$\boldsymbol{\sigma}'_q = J^{-1}\mathbb{Z} : \bar{\boldsymbol{\sigma}}' \tag{17.80}$$

$$\mathbb{Z} = \mathbb{N} - \frac{1}{2}\mathbf{I} \otimes \mathbf{I} \tag{17.81}$$

Thus, the spatial description of integrated form is given by

$$\sigma' = \sigma'_p + \sigma'_q = -p'\mathbf{I} + J^{-1}\mathbb{Z} : \overline{\sigma}' \tag{17.82}$$

The second term in Eq. 17.82 may be written down as follows:

$$\sigma'_q = J^{-1} \int_0^{\pi} q(\langle \mathbf{t} \otimes \mathbf{n} \rangle - \gamma \mathbf{I}) \mathrm{d}\omega \tag{17.83}$$

The macroscopic strain field in the current configuration is defined by the Euler-Almansi strain tensor

$$\mathbf{e} = \frac{1}{2}\left(\mathbf{I} - \mathbf{F}^{-\mathrm{T}}\mathbf{F}^{-1}\right) \tag{17.84}$$

The projection of this strain tensor to the tensor of current virtual simple shear mechanism defined by

$$\gamma = \langle \mathbf{t} \otimes \mathbf{n} \rangle : \mathbf{e} \tag{17.85}$$

coincides with the virtual simple shear strain defined in the reference configuration by Eq. 17.39. The volumetric strain is given by Eq. 17.38 as is the case in the material description.

17.4.2 Spatial Description of Incremental Form

Following the well-established theory in continuum mechanics (Holzapfel 2000), the incremental form of the strain space multiple mechanism model described in the current configuration is derived through the Lie time derivative of the integrated form in Eq. 17.82 multiplied by the Jacobian determinant J. The Lie time derivative is based on the following three step procedure; (1) the pull-back operation to the reference configuration, (2) the material time derivative, and (3) the push-forward operation back to the current configuration. In this study, the results of the steps (1) and (2) have already been derived in Sect. 17.3, the rest is to apply the push-forward operation to these results.

The push-forward operation of the stress and strain rates defined in the reference configurations results in the following stress and strain rates in the current configuration:

$$\hat{\tau}' = \mathbf{F}\dot{\mathbf{S}}'\mathbf{F}^{\mathrm{T}} \tag{17.86}$$

$$\mathbf{d} = \mathbf{F}^{-\mathrm{T}}\dot{\mathbf{E}}\,\mathbf{F}^{-1} \tag{17.87}$$

where $\boldsymbol{\tau}' = J\boldsymbol{\sigma}'$ denotes the Kirchhoff effective stress and the symbol $(\hat{\ })$ denotes the Oldroyd stress rate defined by

$$\hat{\tau}' = \mathrm{Oldr}(\tau') = \dot{\tau}' - \mathbf{l}\tau' - \tau'\mathbf{l}^{\mathrm{T}} \tag{17.88}$$

Some literatures adopt Jaumann and Green-Naghdi stress rates that coincide with the Oldroyd stress rate under a specific condition such as with $\mathbf{d} = \mathbf{0}$ for Jaumann rate and $\mathbf{l} = \mathbf{R}\mathbf{R}^{\mathrm{T}}$ for Green-Naghdi rate, where $\mathbf{R}$ denotes rotation tensor defined through a polar decomposition of the deformation gradient $\mathbf{F}$. The rate of deformation tensor in Eq. 17.87 is given as the Lie time derivative of the Euler-Almansi strain tensor, written in parallel to Eq. 17.88 as

$$\mathbf{d} = \dot{\mathbf{e}} + \mathbf{l}^{\mathrm{T}}\mathbf{e} + \mathbf{e}\mathbf{l} \tag{17.89}$$

With these brief reviews on the well-established theory in the continuum mechanics, the incremental form of the strain space multiple mechanism model in the current configuration is given as follows:

$$\mathrm{Oldr}(J\sigma') = J\mathbb{c} : \mathbf{d} \tag{17.90}$$

$$c_{abcd} = J^{-1}F_{aA}F_{bB}F_{cC}F_{dD}C_{ABCD} \tag{17.91}$$

where c_{abcd} and C_{ABCD} denote components of the fourth-order tensors $\mathbb{c}$ and $\mathbb{C}$ defined for the current and reference configurations, respectively. In particular, the fourth-order tensor $\mathbb{C} = \frac{\partial \mathbf{S}'}{\partial \mathbf{E}}$ has been given in Eq. 17.74.

The push-forward operation of the fundamental fourth-order tensors in $\mathbb{C}$ yields the following results:

$$\begin{aligned} &J^{-1}F_{aA}F_{bB}F_{cC}F_{dD}\left(\mathbf{C}^{-1}\otimes\mathbf{C}^{-1}\right)_{ABCD} = J^{-1}F_{aA}F_{bB}F_{cC}F_{dD}C^{-1}_{AB}C^{-1}_{CD} \\ &= J^{-1}\left(F_{aA}C^{-1}_{AB}F_{bB}\right)\left(F_{cC}C^{-1}_{CD}F_{dD}\right) = J^{-1}\delta_{ab}\delta_{cd} = J^{-1}(\mathbf{I}\otimes\mathbf{I})_{abcd} \end{aligned} \tag{17.92}$$

$$\begin{aligned} &J^{-1}F_{aA}F_{bB}F_{cC}F_{dD}\left(\mathbf{C}^{-1}\odot\mathbf{C}^{-1}\right)_{ABCD} \\ &= \frac{J^{-1}}{2}\left\{\left(F_{aA}C^{-1}_{AC}F_{cC}\right)\left(F_{bB}C^{-1}_{BD}F_{dD}\right) + \left(F_{aA}C^{-1}_{AD}F_{dD}\right)\left(F_{bB}C^{-1}_{BC}F_{cC}\right)\right\} \\ &= \frac{J^{-1}}{2}(\delta_{ac}\delta_{bd} + \delta_{ad}\delta_{bc}) = \frac{J^{-1}}{2}\left(\mathbb{N} + \overline{\mathbb{N}}\right)_{abcd} \end{aligned} \tag{17.93}$$

$$\begin{aligned} &J^{-1}F_{aA}F_{bB}F_{cC}F_{dD}(\langle\mathbf{T}\otimes\mathbf{N}\rangle\otimes(\langle\mathbf{T}\otimes\mathbf{N}\rangle))_{ABCD} \\ &= J^{-1}F_{aA}(T_AN_B + T_BN_A)F_{bB}F_{cC}(T_CN_D + T_DN_C)F_{dD} \\ &= J^{-1}(t_an_b + t_bn_a)(t_cn_d + t_dn_c) = J^{-1}(\langle\mathbf{t}\otimes\mathbf{n}\rangle\otimes\langle\mathbf{t}\otimes\mathbf{n}\rangle)_{abcd} \end{aligned} \tag{17.94}$$

With these supplementary derivations, the push-forward operation on $\mathbb{C}$ in Eq. 17.73 yields the fourth-order tensor $\mathbb{c}$ in Eq. 17.90 as follows:

$$\mathbb{c} = \mathbb{c}_p + \mathbb{c}_q \tag{17.95}$$

$$\mathbb{c}_p = \left(K_{\mathrm{L/U}} - p'\right)\mathbf{I} \otimes \mathbf{I} + 2p'\mathbb{N} - K_{\mathrm{L/U}}\mathbf{I} \otimes \mathbf{I}_{\mathrm{d}} \tag{17.96}$$

$$\mathbb{C}_q = J^{-1}[\mathbb{Z} : \mathbb{c}_q + \mathrm{tr}\overline{\sigma}'\mathbb{Z}] - (\mathbf{I} \otimes \sigma'_q + \sigma'_q \otimes \mathbf{I}) \tag{17.97}$$

$$\overline{\mathbb{c}}_q = \int_0^{\pi} \langle \mathbf{t} \otimes \mathbf{n} \rangle \otimes \mathbf{I}_q \mathrm{d}\omega \tag{17.98}$$

$$\mathbf{I}_q = q\mathbf{I} + G_{\mathrm{L/U}} \langle \mathbf{t} \otimes \mathbf{n} \rangle + H_{\mathrm{L/U}}(\mathbf{I} - \mathbf{I}_{\mathrm{d}}) + L_{\mathrm{L/U}}\left(\mathbf{I} - \mathbf{I}_{\mathrm{d}}^{\mathrm{c}}\right) \tag{17.99}$$

where the tensor for describing the dilatancy is given by

$$\mathbf{I}_{\mathrm{d}} = \mathbf{F}\mathbf{C}_{\mathrm{d}}^{-1}\mathbf{F}^{\mathrm{T}} \tag{17.100}$$

$$\mathbf{I}_{\mathrm{d}}^{\mathrm{c}} = \mathbf{F}\mathbf{C}_{\mathrm{dc}}^{-1}\mathbf{F}^{\mathrm{T}} \tag{17.101}$$

The first term in Eq. 17.97 may be written down as follows:

$$\mathbb{Z} : \overline{\mathbb{c}}_q = \int_0^{\pi} (\langle \mathbf{t} \otimes \mathbf{n} \rangle - \gamma\mathbf{I}) \otimes \mathbf{I}_q \mathrm{d}\omega \tag{17.102}$$

The rate of dilatancy and relevant strain components are given, in the current configuration, from the rate of deformation tensor as follows:

$$\dot{\varepsilon} = \mathbf{I} : \mathbf{d} \tag{17.103}$$

$$\dot{\varepsilon}_{\mathrm{d}}^{\mathrm{c}} = \mathbf{I}_{\mathrm{d}}^{\mathrm{c}} : \mathbf{d} \tag{17.104}$$

$$\dot{\varepsilon}_{\mathrm{d}} = \mathbf{I}_{\mathrm{d}} : \mathbf{d} \tag{17.105}$$

$$\dot{\gamma} = \langle \mathbf{t} \otimes \mathbf{n} \rangle : \mathbf{d} \tag{17.106}$$

17.5 Strain Energy and Dilative Component of Dilatancy

Since the effective stress fields $\mathbf{S}'$ and $J\boldsymbol{\sigma}$ are work conjugate to the strain rate fields $\dot{\mathbf{E}}$ and $\mathbf{d}$, respectively (Holzapfel 2000), the strain energy rate in the strain space multiple mechanism model is given by

$$\int_{\Omega_0} \mathbf{S}' : \dot{\mathbf{E}}\, dV = \int_{\Omega_0} J\sigma' : \mathbf{d}dV = \int_{\Omega} \sigma' : \mathbf{d}dv \tag{17.107}$$

where the volume in the reference configuration is related with that in the current configuration through $JdV = dv$ and the symbols $\mathbf{\Omega}_0$ and $\mathbf{\Omega}$ denote the reference and current configurations. This relation forms the basis for applying the variational principles in deriving the primary results required for the finite element method. In particular, the relation described in the reference configuration forms the basis for the total Lagrangian method while the relation in the current configuration for the updated Lagrangian method.

In the reference configuration, the strain energy rate in terms of energy rate density per unit volume of the strain space multiple mechanism model defined in Eqs. 17.26, 17.27 and 17.31 is given from Eqs. 17.46 and 17.50 as follows:

$$\mathbf{S}' : \dot{\mathbf{E}} = -Jp'\dot{\varepsilon} + \int_0^{\pi} q(\dot{\gamma} - \gamma\dot{\varepsilon})\mathrm{d}\omega \tag{17.108}$$

In the current configuration, the strain energy rate in terms of energy rate density per unit volume is given from Eqs. 17.82, 17.83, 17.103, and 17.106 as follows:

$$\sigma' : \mathbf{d} = -p'\dot{\varepsilon} + J^{-1} \int_0^{\pi} q(\dot{\gamma} - \gamma\dot{\varepsilon})\mathrm{d}\omega \tag{17.109}$$

Alternatively, Eqs. 17.107 and 17.108 give the same result. Equation 17.109 (or its counterpart in Eq. 17.108 except for the multiplication of J) can be rewritten as

$$\dot{W} = \sigma' : \mathbf{d} = \dot{W}_p + \int_0^{\pi} \dot{W}_{q\text{v}}\mathrm{d}\omega \tag{17.110}$$

where

$$\dot{W}_p = -p'\dot{\varepsilon} \tag{17.111}$$

$$\dot{W}_{q\text{v}} = J^{-1}q(\dot{\gamma} - \gamma\dot{\varepsilon}) \tag{17.112}$$

Equation 17.110 defines the fundamental relationship between the macroscopic strain energy rate $\dot{W}$ and the micromechanical strain energy rate $\dot{W}_{q\text{v}}$ given by the individual virtual simple shear mechanism that constitutes the strain space multiple mechanism model. Except for the correction term in the virtual simple shear strain $-\gamma\dot{\varepsilon}$ and the correction for the Jacobian determinant, the relation remains the same as that for the infinitesimal strain formulation.

From this energy relationship, the dilative component of dilatancy is given for the current configuration through a derivation analogous to that for the infinitesimal strain formulation as

$$\dot{\varepsilon}_d^d = \mathbf{I}_d^d : \mathbf{d} \tag{17.113}$$

$$\mathbf{I}_d^d = r_{\varepsilon_d} J^{-1} \int_0^{\pi} \left(\frac{\gamma/\gamma_v}{1 + |\gamma/\gamma_v|} \right) M_{fv} (\langle \mathbf{t} \otimes \mathbf{n} \rangle - \gamma \mathbf{I}) d\omega \tag{17.114}$$

The counterpart in the reference configuration is given by the pull-back operation $\mathbf{C}_{dd}^{-1} = \mathbf{F}^{-1}\mathbf{I}_{dd}\mathbf{F}^{-T}$ to give the result shown earlier in Eq. 17.53.

17.6 Governing Equations of Porous Saturated Materials

In order to complete the description of the finite strain formulation of granular materials, equilibrium and mass balance of the porous saturated materials are briefly reviewed within the finite strain framework. The spatial description of the governing equations of the boundary value problems of porous materials are given in terms of the Cauchy total stress tensor $\boldsymbol{\sigma}$, displacement $\mathbf{u}$ and pore water pressure p as follows (Zienkiewicz and Bettess 1982):

$$\begin{aligned}
&\mathrm{div}\sigma(\mathbf{x}, t) + \rho(\mathbf{x}, t)\mathbf{g}(\mathbf{x}) = \rho(\mathbf{x}, t)\ddot{\mathbf{u}}(\mathbf{x}, t) \\
&\mathrm{div}(\mathbf{k}(\mathbf{x})\mathrm{grad}(p(\mathbf{x}, t))) - \mathrm{div}\dot{\mathbf{u}}\,(\mathbf{x}, t) - \mathrm{div}(\mathbf{k}(\mathbf{x})\rho_f(\mathbf{x}, t)\mathbf{g}(\mathbf{x})) \\
&\qquad = -\mathrm{div}(\mathbf{k}(\mathbf{x})\rho_f(\mathbf{x}, t)\ddot{\mathbf{u}}(\mathbf{x}, t)) + n\dot{p}\,(\mathbf{x}, t)/K_f \\
&\mathbf{u}(\mathbf{x}, t) = \bar{\mathbf{u}}(\mathbf{x}, t) \qquad \text{on } \partial\boldsymbol{\Omega}_{\mathbf{u}} \\
&\mathbf{t}(\mathbf{x}, t) = \sigma(\mathbf{x}, t)\mathbf{n} = \bar{\mathbf{t}}(\mathbf{x}, t) \qquad \text{on } \partial\boldsymbol{\Omega}_{\sigma} \\
&p(\mathbf{x}, t) = \bar{p}(\mathbf{x}, t) \qquad \text{on } \partial\boldsymbol{\Omega}_{p} \\
&\mathbf{q}(\mathbf{x}, t) = \mathbf{k}(\mathrm{grad}\,p - \rho_f\mathbf{g} + \rho_f\ddot{\mathbf{u}})\mathbf{n} = \bar{\mathbf{q}}(\mathbf{x}, t) \quad \text{on } \partial\boldsymbol{\Omega}_{\mathbf{q}} \\
&\mathbf{u}(\mathbf{x}, 0) = \mathbf{u}_0(\mathbf{X}), \dot{\mathbf{u}}\,(\mathbf{x}, 0) = \dot{\mathbf{u}}_0(\mathbf{X}) \\
&p(\mathbf{x}, 0) = p_0(\mathbf{X}), \dot{p}\,(\mathbf{x}, 0) = \dot{p}_0(\mathbf{X})
\end{aligned} \tag{17.115}$$

where ρ and ρ_f denote bulk and fluid density; $\mathbf{g}$ denotes acceleration of gravity; $\mathbf{k}$ denotes permeability tensor; K_f denotes bulk modulus of pore fluid; n denotes porosity; $\partial\boldsymbol{\Omega}_{\mathbf{u}}$, $\partial\boldsymbol{\Omega}_{\boldsymbol{\sigma}}$, $\partial\boldsymbol{\Omega}_p$, and $\partial\boldsymbol{\Omega}_{\mathbf{q}}$ denote boundaries constrained with displacement, stress, pore water pressure, and pore water inflow, respectively. The mass balance with respect to the density is given by

$$\rho(\mathbf{x}, t)J = \rho_0 \tag{17.116}$$

By using the definition of effective stress (i.e., $\boldsymbol{\sigma} = \boldsymbol{\sigma} - p\mathbf{I}$), the equilibrium equation in Eq. 17.115 is rewritten as follows:

$$\mathrm{div}(\sigma' - p\mathbf{I}) + \rho\mathbf{g} = \rho\ddot{\mathbf{u}} \tag{17.117}$$

This equation is written, together with the boundary conditions on $\partial\boldsymbol{\Omega}_{\boldsymbol{\sigma}}$, through a virtual work principle as

$$\int_{\Omega} [(\sigma' - p\mathbf{I}) : \delta\mathbf{e} - (\rho\mathbf{g} - \rho\ddot{\mathbf{u}}) \cdot \delta\mathbf{u}]\mathrm{d}v - \int_{\partial\Omega_{\sigma}} \bar{\mathbf{t}} \cdot \delta\mathbf{u}\mathrm{d}s = 0 \tag{17.118}$$

Equation 17.118 representing equilibrium in the current configuration is transformed into the reference configuration through the pull-back operation (Holzapfel 2000) as follows:

$$\int_{\Omega_0} \left[\left(\mathbf{S}' - Jp\mathbf{C}^{-1}\right) : \delta\mathbf{E} - \left(\rho_0\mathbf{g} - \rho_0\ddot{\mathbf{u}}\right) \cdot \delta\mathbf{u}\right]\mathrm{d}V - \int_{\partial\Omega_{0\sigma}} \bar{\mathbf{T}} \cdot \delta\mathbf{u}\mathrm{d}S = 0 \tag{17.119}$$

Similarly the second equation in Eq. 17.115 representing mass balance of fluid is written through the Galerkin method, by using arbitrary function of $\tilde{h}$ that becomes zero at the pore water pressure boundary $\partial\boldsymbol{\Omega}_p$, as follows:

$$\begin{aligned}&\int_{\Omega} \tilde{h}\mathrm{div}(\mathbf{k}(\mathrm{grad}p))\mathrm{d}v - \int_{\Omega} \tilde{h}\mathrm{div}\dot{\mathbf{u}}\,\mathrm{d}v - \int_{\Omega} \tilde{h}\mathrm{div}(\mathbf{k}\rho_{\mathrm{f}}\mathbf{g})\mathrm{d}v\\&+ \int_{\Omega} \tilde{h}\mathrm{div}\left(\mathbf{k}\rho_{\mathrm{f}}\ddot{\mathbf{u}}\right)\mathrm{d}v - \int_{\Omega} \tilde{h}n\dot{p}\,/K_{\mathrm{f}}\mathrm{d}v + \int_{\partial\Omega} \tilde{h}\left(\bar{\mathbf{q}} - \mathbf{q}\right)\mathrm{d}s = 0\end{aligned} \tag{17.120}$$

By applying the Gauss theorem, the following equation is obtained

$$\begin{aligned}&\int_{\Omega} \left(\mathrm{grad}\tilde{h}\right)^{\mathrm{T}}\mathbf{k}(\mathrm{grad}p)\mathrm{d}v + \int_{\Omega} \tilde{h}\mathrm{div}\dot{\mathbf{u}}\,\mathrm{d}v - \int_{\Omega} \left(\mathrm{grad}\tilde{h}\right)^{\mathrm{T}}\left(\mathbf{k}\rho_{\mathrm{f}}\left(\mathbf{g} - \ddot{\mathbf{u}}\right)\right)\mathrm{d}v\\&+ \int_{\Omega} \tilde{h}n\dot{p}\,/K_{\mathrm{f}}\mathrm{d}v - \int_{\partial\Omega_{\mathrm{q}}} \tilde{h}\bar{\mathbf{q}}\mathrm{d}s = 0\end{aligned} \tag{17.121}$$

In Eq. 17.121, the integration over a surface on the last term is restricted to $\partial\boldsymbol{\Omega}_{\mathbf{q}}$ because of $\partial\boldsymbol{\Omega}_p \cap \partial\boldsymbol{\Omega}_{\mathbf{q}} = \varnothing$. The pull-back operation of this equation gives

$$\begin{aligned}&\int_{\Omega_0} J\left(\mathbf{F}^{-T}\mathrm{Grad}\tilde{h}\right)^{\mathrm{T}}\mathbf{k}\left(\mathbf{F}^{-T}\mathrm{Grad}p\right)\mathrm{d}V + \int_{\Omega_0} J\tilde{h}\mathbf{F}^{-\mathrm{T}} : \mathrm{Grad}\dot{\mathbf{u}}\,\mathrm{d}V\\&- \int_{\Omega_0} J\left(\mathbf{F}^{-\mathrm{T}}\mathrm{Grad}\tilde{h}\right)^{\mathrm{T}}\left(\mathbf{k}\rho_{\mathrm{f}}\left(\mathbf{g} - \ddot{\mathbf{u}}\right)\right)\mathrm{d}V + \int_{\Omega} J\tilde{h}n\dot{p}\,/K_{\mathrm{f}}\mathrm{d}V - \int_{\partial\Omega_{0\mathrm{q}}} \tilde{h}\bar{\mathbf{Q}}\mathrm{d}S = 0\end{aligned} \tag{17.122}$$

The rest is the discretization by using the interpolation functions, the linearization for deriving the tangential stiffness matrix for the Newton type iteration procedure, and the vector-matrix description of tensors. In the linearization, the fourth order tangential stiffness tensor $\mathbb{C}$ or $\mathbbm{c}$ play a primary role. Various geometrically non-linear terms appear in the equilibrium and mass balance equations through the derivation (the details are omitted). For the equations on the equilibrium of solid, refer to text books on finite element method (e.g. Bathe 1996). As mentioned earlier, the equations for the current configuration give the formulation of updated Lagrangian method, while those for the reference configuration give the formulation of the total Lagrangian method.

17.7 Model Parameters

The macroscopic shear strength τ_{m} and the shear modulus G_{m} are defined for the current configuration so that the virtual simple shear strength q_{v} and the virtual reference strain γ_{v} are calibrated for two dimensional analysis through Eq. 17.83. The assumption of hyperbolic function for the backbone curve on the virtual simple shear mechanism as shown in Eq. 17.18 gives the virtual simple shear stress, for small strain $\gamma \approx 0$, as

$$q = \frac{q_{\mathrm{v}}}{\gamma_{\mathrm{v}}}\gamma \tag{17.123}$$

and, for large strain $\gamma \approx \infty$, as

$$q = \mathrm{sgn}(\gamma) q_{\mathrm{v}} \tag{17.124}$$

Consequently the parameters q_{v} and γ_{v} are calibrated with respect to the macroscopic shear strength τ_{m} and elastic shear modulus G_{m} as follows:

$$q_{\mathrm{v}} = \frac{J}{\sqrt{\frac{1}{4}\left[\int_0^{\pi} \mathrm{sgn}(\gamma)(n_{11} - n_{22})\mathrm{d}\omega\right]^2 + \left(\int_0^{\pi} \mathrm{sgn}(\gamma) n_{12} \mathrm{d}\omega\right)^2}}\tau_{\mathrm{m}} \tag{17.125}$$

$$\gamma_{\mathrm{v}} = \frac{\int_0^{\pi} (n_{12})^2 \mathrm{d}\omega}{\sqrt{\frac{1}{4}\left[\int_0^{\pi} \mathrm{sgn}(\gamma)(n_{11} - n_{22})\mathrm{d}\omega\right]^2 + \left(\int_0^{\pi} \mathrm{sgn}(\gamma) n_{12} \mathrm{d}\omega\right)^2}}\gamma_{\mathrm{m}} \tag{17.126}$$

where

$$\langle \mathbf{t} \otimes \mathbf{n} \rangle = \begin{pmatrix} n_{11} & n_{12} \\ n_{21} & n_{22} \end{pmatrix}, \quad n_{12} = n_{21} \tag{17.127}$$

and $\gamma_{\mathrm{m}} = \tau_{\mathrm{m}}/G_{\mathrm{m}}$ is the reference strain often referred to in the discipline of soil dynamics.

The model parameters and the rest of the formulation to complete the description of the strain space multiple mechanism are the same as those given for the infinitesimal strain formulation (Iai et al. 2011). As described earlier, the strain space multiple mechanism model based on the finite strain theory has been incorporated into the large deformation analysis program called "FLIP TULIP". Examples of the application of the program with the extended model to geotechnical works including embankments, port structures are covered in the following chapter and the references (Ueda et al. 2015, 2016; Ueda and Iai 2016).

17.8 Conclusions

In this study, finite strain formulation is given to a strain space multiple mechanism model for granular materials. The formulation is carried out by following the change, both in direction and magnitude, in the unit vector defined along the branch and the complementary vector normal to it originally defined in the reference configuration. The formulation proposed in this study may be characterized as follows:

1. Adoption of the infinitesimal strain formulation directly to the spatial description of the strain space multiple mechanism model poses a certain drawback because the virtual simple shear term can include volumetric component. In order to overcome this drawback, decoupled formulation, that decomposes the kinematic mechanisms into volumetric and isochoric components, is adopted by using the fourth order projection tensor.
2. Material (Lagrangian) description of integrated form is given by a relation between the second Piola-Kirchhoff effective stress tensor and the Green-Lagrange strain tensor; spatial (Eulerian) description by a relation between the Cauchy effective stress tensor and the Euler-Almansi strain tensor. In particular, the volumetric strain is defined as a logarithm of Jacobian determinant, whereas the virtual simple shear strain is defined as a projection of relevant strains to the tensors representing the virtual simple shear mechanism.
3. Material description of incremental form is derived through the material time derivative of the integrated form. The counterpart in the spatial description is derived through the Lie time derivative, given as a relation between the Oldroyd stress rate of the Kirchhoff stress and the rate of deformation tensor. Although the finite strain formulation looks more complicated than the infinitesimal strain counterpart, the formulation is basically the same except for the appearance of stress in the tangential stiffness tensor because of the geometrical nonlinearity.
4. The relation between the macroscopic and micromechanical strain energies is derived in the decoupled form. Except for the correction term in the virtual simple shear strain for volumetric component and the correction for the Jacobian determinant, the relation remains the same as that for the infinitesimal strain formulation.

References

Bathe KJ (1996) Finite element procedures. Prentice Hall, p 1037
Hardin BO, Drnevich VP (1972) Shear modulus and damping in soils: design equation and curves. J Soil Mech Found Div, ASCE 98(SM7):667–692
Holzapfel GA (2000) Nonlinear solid mechanics: a continuum approach for engineering. Wiley, West Sussex, p 455
Iai S, Ozutsumi O (2005) Yield and cyclic behaviour of a strain space multiple mechanism model for granular materials. Int J Numer Anal Methods Geomech 29(4):417–442
Iai S, Ueda K (2016) Energy-less strain in granular materials – micromechanical background and modelling. Soils Found 56(3):391–398
Iai S, Matsunaga Y, Kameoka T (1992) Strain space plasticity model for cyclic mobility. Soils Found 32(2):1–15
Iai S, Ichii K, Liu H, Morita T (1998) Effective stress analyses of port structures. Soils Found., Special Issue on Geotechnical Aspects of the January 17 1995 Hyogoken-Nambu Earthquake 2:97–114
Iai S, Tobita T, Ozutsumi O, Ueda K (2011) Dilatancy of granular materials in a strain space multiple mechanism model. Int J Numer Anal Methods Geomech 35(3):360–392
Iai S, Ueda K, Tobita T, Ozutsumi O (2013a) Finite strain formulation of a strain space multiple mechanism model for granular materials. Int J Numer Anal Methods Geomech 37 (9):1189–1212
Iai S, Tobita T, Ozutsumi O (2013b) Evolution of induced fabric in a strain space multiple mechanism model for granular materials. Int J Numer Anal Methods Geomech 37 (10):1326–1336
Mehrabadi MM, Nemat-Nasser S, Oda M (1982) On statistical description of stress and fabric in granular materials. Int J Numer Anal Methods Geomech 6(1):95–108
Ozutsumi O, Sawada S, Iai S, Takeshima Y, Sugiyama W, Shimazu T (2002) Effective stress analyses of liquefaction-induced deformation in river dikes. Soil Dyn Earthq Eng 22:1075–1082
Taylor DW (1948) Fundamentals of soil mechanics. Wiley, New York. 700 p
Thornton C (1989) A direct approach to micromechanically based continuum models for granular material. In: Mechanics of granular materials JSSMFE: XII ICSMFE, pp 145–150
Towhata I, Ishihara K (1985) Modelling soil behaviour under principal stress axes rotation. In: Proceedings of 5th international conference on numerical methods in geomechanics. Balkema, Nagoya, pp 523–530
Ueda K (2009) Finite strain formulation of a strain space multiple mechanism model for granular materials and its application. Doctoral thesis, Kyoto University, Japan (in Japanese)
Ueda K, Iai S (2016). Seismic response analysis of caisson type composite breakwater based on finite strain theory. In: The 1st international conference on natural hazards and infrastructure, Chania
Ueda K, Iai S, Ozutsumi O (2015) Finite deformation analysis of dynamic behavior of embankment on liquefiable sand deposit considering pore water flow and migration. In: The 6th international conference on earthquake geotechnical engineering. Christchurch; Paper No. 215
Ueda K, Iai S, Tobita T (2016) Centrifuge model tests and large deformation analyses of a breakwater subject to combined effects of tsunami. Soil Dyn Earthq Eng 91:294–303
Zienkiewicz OC, Bettess P (1982) Soils and other saturated media under transient, dynamic conditions. In: Soil mechanics – transient and cyclic loads. Wiley, pp 1–16

Chapter 18
Large Deformation (Finite Strain) Analysis: Application

Noriyuki Fujii

Abstract To accurately estimate the damage of soil-structure systems during earthquakes, reliable analytical methods and appropriate modeling of soils and structures are necessary. Herein seismic response analyses are performed on an embankment, a caisson-type composite breakwater, and a caisson-type quay wall to verify the applicability of the strain space multiple mechanism model, where both the total and updated Lagrangian (TL/UL) formulations are introduced based on the large deformation (finite strain) theory. Both infinitesimal and large deformation analyses are performed to examine the effect of geometrical nonlinearity. All of the computed results (e.g., deformation and excess pore pressure ratio) indicate that the TL and UL formulations are theoretically and numerically equivalent, validating the computer program for large deformation analysis. In addition, the large deformation analyses decrease the amount of deformations compared to the infinitesimal deformation analysis in these three cases. This tendency becomes more significant as the amplitude of the input ground motions increases. We also perform a centrifuge model test to investigate the large deformation mechanism of liquefiable sloping ground and compare the experimental results with numerical simulations. The comparison indicates that the large deformation analysis by the UL formulation reasonably simulates the experiment, whereas the infinitesimal deformation analysis overestimates the deformation.

18.1 Introduction

In the 1995 Hyogoken-Nanbu Earthquake, the Kobe Port was hit by strong motions with a peak ground acceleration of 0.54 g in the horizontal direction and 0.45 g in the vertical direction (Inagaki et al. 1996). The ground motion inflicted significant damage to many port structures, especially those constructed from rigid blocks such as concrete caissons (The Great Hanshin-Awaji Earthquake Survey Editorial Board 1998). Historically, Japan has been repeatedly hit by major earthquakes. Other

N. Fujii (✉)
OYO Corporation, Tokyo, Japan
e-mail: fujii-noriyuki@oyonet.oyo.co.jp

S. Iai (ed.), *Developments in Earthquake Geotechnics*, Geotechnical, Geological and Earthquake Engineering 43, DOI 10.1007/978-3-319-62069-5_18

recent earthquakes causing serious damage to social infrastructure facilities include the 2003 Tokachi-Oki, the 2004 Niigata-ken Chuetsu, and the 2011 Great East Japan earthquakes. In the near future, the Nankai Trough earthquake is forecasted to occur off the southern Japan coast. If the Tokai, Tonankai, and Nankai earthquakes occur simultaneously, many soil-structure systems would be seriously damaged. Therefore, it is crucial to predict the structural damage during major earthquakes as accurately as possible and to improve the seismic performance by implementing appropriate remedies as necessary.

To estimate the seismic damage to soil-structure systems such as port structures, various techniques from simplified methods to detailed ones have been proposed. Among these, effective stress analysis using the finite element method is very effective because it can take the nonlinearity of soils and the dynamic interactions between ground and structures into account using an appropriate constitutive model for soils. As a constitutive model, a strain space multiple mechanism model (Iai et al. 1992, 2011), which has been implemented in the finite element program called "FLIP ROSE", is particularly promising. FLIP ROSE can consider the effect of rotating the principal stress axis directions, which is known to play an important role in the cyclic behavior of soils, and is classified into two types (i.e., the multi-spring model and the cocktail glass model) depending upon the difference in the dilatancy formulation. The FLIP ROSE program has been used in the seismic design of soil-structure systems.

However, models have been formulated based on the infinitesimal deformation theory. Strictly speaking, the applications are limited to phenomena with a small deformation and rotation. Consequently, the model has been extended based on two types of large deformation (finite strain) formulations (Iai et al. 2013) to take the effect of geometrical nonlinearity into account. To more accurately estimate the damage to soil-structure systems during an earthquake, the extended model has been implemented in a large deformation analysis program called "FLIP TULIP".

The objectives of this study are to examine how well the program can simulate the induced damage to an embankment in centrifuge model tests and a caisson type breakwater/quay wall in Kobe Port during the 1995 Hyogoken-Nanbu earthquake using the extended strain space multiple mechanism model and to verify the applicability of the model to the seismic response of soil-structure systems.

18.2 Application to Embankments

18.2.1 Centrifuge Model Test on the Seismic Behavior of Embankments

A series of centrifuge model tests was performed in the P.W.R.I. (2000) to study the dynamic behavior of an embankment on a liquefiable sand deposit. The applied centrifugal acceleration was 50 G. In the experiments, various remedies were adopted to prevent the settlement of an embankment. In this research, the simulation target is the result of a test case without remedies.

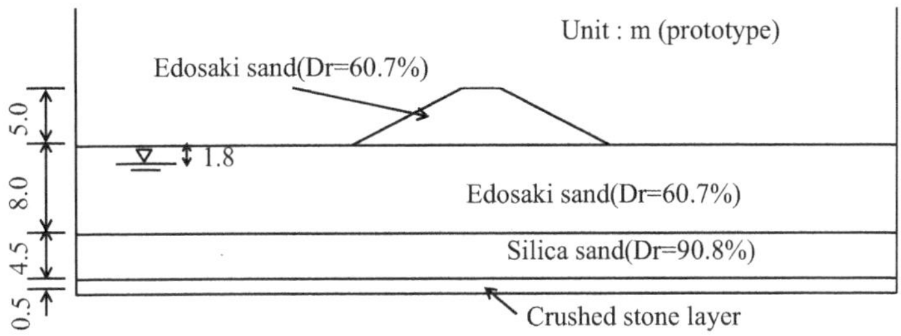

Fig. 18.1 Cross-section diagram in the centrifuge model test (Modified after P.W.R.I. 2000)

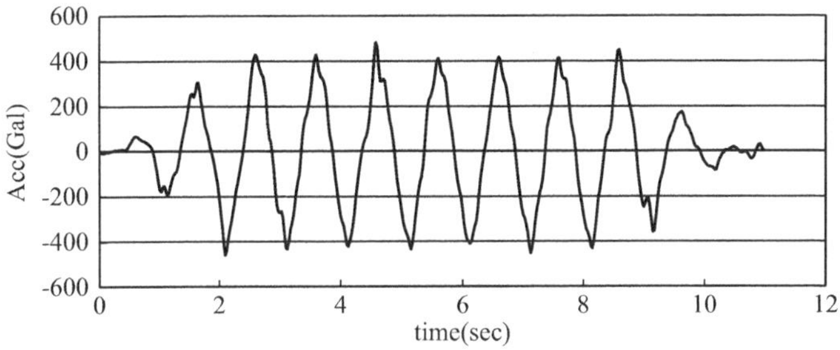

Fig. 18.2 Time history of the input acceleration (P.W.R.I. 2000)

Figure 18.1 shows a model of the cross section of an embankment on the sand deposit with a 13.0-m depth. The crest height is 5.0 m, and the lateral length at the top and bottom are 3.0 m and 23.0 m, respectively. The groundwater level is 1.8 m below the ground's surface. The relative densities of the upper layer and the lower layer of the saturated sand deposit are approximately 60% and 90%, respectively.

Figure 18.2 shows the input motion, which is composed of seven sinusoidal waves (main shock) with a peak acceleration of approximately 0.4 g and subsequent aftershocks with acceleration amplitudes less than 0.1 g. Figure 18.3 shows the measured residual deformation of the embankment after the excess pore water pressure dissipates. The solid lines express the deformed configuration with reference to the original configuration (broken lines). The induced crest settlements are 2.04 m and 2.30 m at the end of the main shock and the aftershocks, respectively. The deformation mechanism involves lateral expansion of soils near the toes of the embankment in the direction opposite to the embankment and vertical compression and shearing of soils under the embankment.

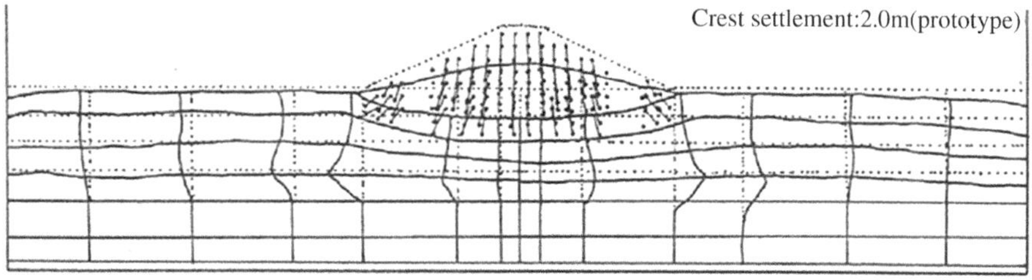

Fig. 18.3 Measured residual deformation in the centrifuge model test (P.W.R.I. 2000)

Table 18.1 Soil parameters (Ueda 2010)

Soil layer	ρ t/m^3	σ_{ma} kN/m^2	m_g, m_K	G_{ma} kN/m^2	K_{ma} kN/m^2	n	h_{max}	ϕ_f (°)	ϕ_p (°)	ε_d^{cm}
Bank (Edosaki sand)	1.70	98	0.5	84,000	218,000	0.49	0.26	34	–	–
Edosaki sand	1.68	98	0.5	41,000	107,000	0.49	0.26	34	–	–
Edosaki sand (Liq)	1.86	98	0.5	41,000	107,000	0.49	0.26	34	28	0.10
No.7 silica sand	1.98	98	0.5	86,000	224,000	0.40	0.26	48	33	0.10

Soil layer	$r_{\varepsilon dc}$	$r_{\varepsilon d}$	q_1	q_2	l_K	r_K	Sl	c_1	qus kN/m^2
Bank (Edosaki sand)	–	–	–	–	–	–	–	–	–
Edosaki sand	–	–	–	–	–	–	–	–	–
Edosaki sand (Liq)	18.00	0.20	1.00	2.50	2.00	0.50	0.005	1.00	17
No.7 silica sand	6.00	0.06	1.00	3.50	2.00	0.50	0.005	1.00	–

18.2.2 Outline of Finite Element Analysis

Both the embankment and the foundation ground were modeled by the cocktail glass model and the model parameters shown in Table 18.1. These parameters were set based on a previous study (Ueda 2010), in which the model parameters specifying the characteristics of liquefaction and dilatancy were determined by referring to the liquefaction resistance curves obtained from the cyclic triaxial tests under the undrained condition. The parameter q_{us} in Table 18.1, which is the undrained shear strength to describe the steady state of sand, was determined for Edosaki sand from the void ratio and the fines content following Motoshima et al. (2008). The parameter is not defined for the silica sand, which has a fines content of only 0.7%, because the undrained shear strength for clean sand generally becomes very large.

The numerical analysis was carried out with the same prototype dimensions as the centrifuge model test. To replicate the boundary conditions of the rigid container in the test, the degrees of freedom for displacements at the base were fixed both horizontally and vertically, but only the horizontal displacements were fixed at the side boundaries. Seismic response analysis was performed for 11 s under the undrained conditions following a self-weight analysis. The recorded main shock shown in Fig. 18.2 at the bottom of the container was used for the input motion. The numerical time integration was done by the Wilson θ method ($\theta = 1.4$) with a time step of 0.01 s. Rayleigh damping ($\alpha = 0.0$, $\beta = 0.0002$) was used to ensure the stability of the numerical solution process. Additionally, we also carried out a series of analyses by changing the amplitude of the input acceleration to study the difference in the computed deformation between the infinitesimal and large deformation analyses as a function of the magnitude of the input acceleration.

18.2.3 Numerical Results

Figure 18.4 shows the computed residual deformation of the embankment with the distribution of the maximum excess pore water pressure ratio. The deformed configuration obtained in the infinitesimal deformation analysis differs from the measurement shown in Fig. 18.3. The shape of the embankment becomes flat and

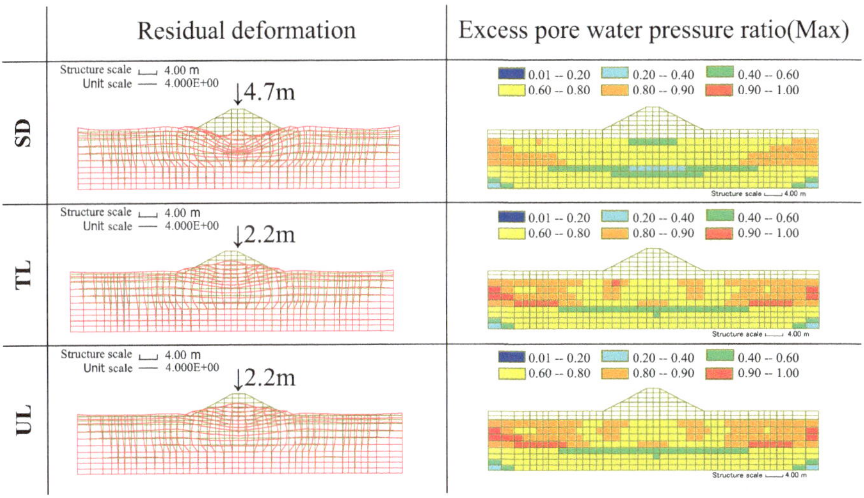

SD: Small (infinitesimal) deformation analysis

TL: Large deformation analysis (TL formulation)

UL: Large deformation analysis (UL formulation)

Fig. 18.4 Computed residual deformation with the distribution of the excess pore water pressure ratio. *SD* Small (infinitesimal) deformation analysis, *TL* Large deformation analysis (TL formulation), *UL* Large deformation analysis (UL formulation)

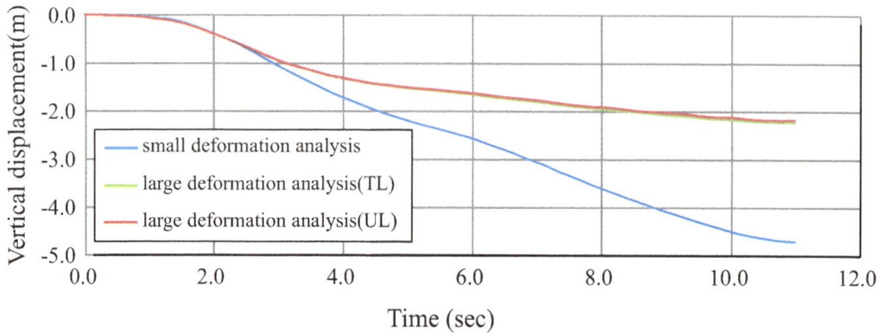

Fig. 18.5 Time history of the computed vertical displacement of the embankment crest

the residual height is almost equal to that of the surrounding ground. On the other hand, the results obtained from the large deformation analyses agree well with the measurement. The computed deformation mode is characterized by a crest settlement associated with lateral spread in the foundation soils similar to the measured one. Additionally, a significant difference is not recognized in the deformed shape and the distribution of the excess pore water pressure ratio between the TL and UL formulations. These results indicate that the cocktail glass model, which has been extended based on the finite strain formulation, shows a reasonable capability to reproduce the deformation of an embankment irrespective of the formulation.

Figure 18.5 shows the time history of the crest settlement. A comparison between the infinitesimal large deformation analyses indicates that the effect of geometrical nonlinearity becomes obvious after the computed settlement in the infinitesimal deformation analysis exceeds one-fifth of the embankment height. Similar to the deformed shape in Fig. 18.4, the TL and UL formulations do not differ.

Figure 18.6 shows the relationship between the peak acceleration of the input motion and the amount of residual settlement of the embankment crest. The difference (i.e., the effect of geometrical nonlinearity) of the settlement between the infinitesimal and the large deformation analyses becomes more significant as the peak acceleration increases. This result indicates that the infinitesimal deformation analysis is inappropriate to evaluate the seismic performance of embankments and the use of the large deformation analyses (TL or UL formulation) is inevitable in the case of large earthquakes (e.g., the 1995 Kobe earthquake).

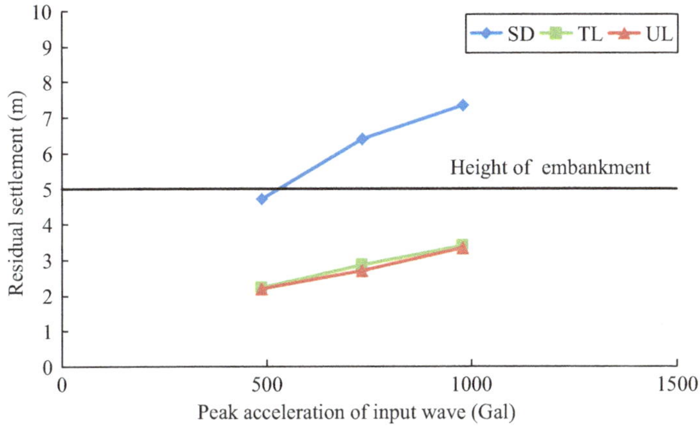

Fig. 18.6 Relationship between the peak acceleration of the input wave and the residual settlement

18.3 Application to Caisson-Type Breakwaters

18.3.1 Summary of Damaged Breakwater

The target of numerical simulation is Breakwater No. 7, which is located south of Rokko Island in Kobe Port. Figure 18.7 shows a typical cross section and deformed configuration of the breakwater. The breakwater is a composite type, which consists of a concrete caisson installed on foundation rubble without backfilling and embedment. It is constructed on decomposed granite soils, which are backfilled into the area excavated from the original alluvial clay layer. The clay layer reaches a thickness of up to 25 m at the southern front where Breakwater No. 7 is located. During the 1995 Hyogoken-Nanbu earthquake, the breakwater induced a settlement of 1.4–2.6 m. However, the horizontal displacements of the breakwater were less than tens of centimeters regardless of the PGA of 0.54 g in the horizontal direction.

The caisson settled into the foundation rubble due to liquefaction or degradation of the loose deposit and the gravity force of the caisson itself. Additionally, the settlement also dragged down the rubble and pushed it into the loose deposit below. The dynamic behavior of the breakwater contrasts that of a caisson-type quay wall, which is mainly caused by liquefaction in the reclaimed areas behind the quay wall due to strong ground motions. (See the next section). In other words, while quay walls are strongly influenced by the backfilled rubble and the reclaimed soils behind them, the behavior of a breakwater is governed largely by the foundation rubble and backfilled soils below them and the dynamic water pressure. This is why to accurately evaluate the behavior of the foundation rubble, backfilled soils play a key role in predicting the damage (e.g., settlement) of the breakwater.

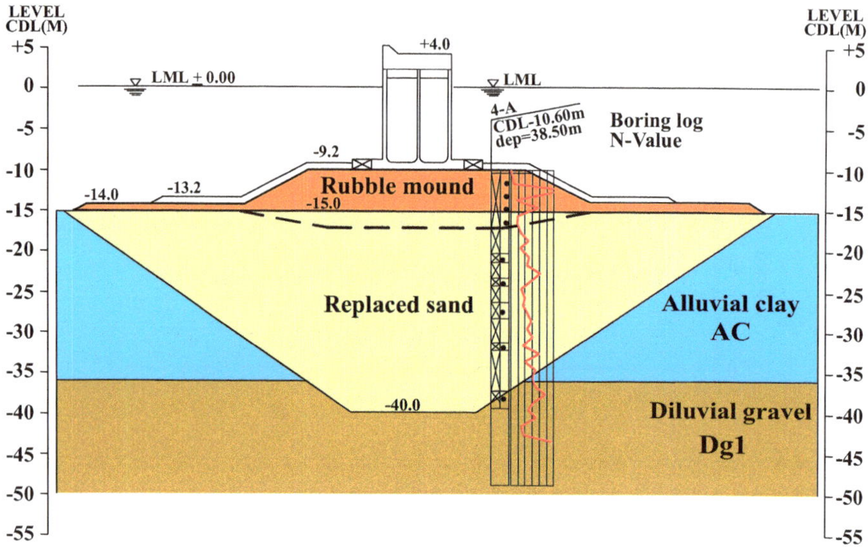

Fig. 18.7 Cross-section of the No. 7 breakwater at the Kobe Port (Iai 1996)

Table 18.2 Soil parameters (Iai et al. 1992)

Soil layer	ρ t/m^3	σ_{ma} kN/m^2	m_g,m_k	G_{ma} kN/m^2	K_{ma} kN/m^2	ν	n	h_{max}
Clay	1.70	143	0.5	74,970	195,500	0.33	0.45	0.21
Replacet sand	1.80	98	0.5	84,695	220,872	0.33	0.45	0.24
Rubble mound	2.00	98	0.5	180,000	469,000	0.33	0.45	0.24

Soil layer	ϕ_f (°)	ϕ_p (°)	$s1$	$w1$	$p1$	$p2$	$c1$	qus kN/m^2
Clay	34	–	–	–	–	–	–	–
Replacet sand	34	28	0.005	4.70	0.50	0.70	1.82	100
Rubble mound	34	–	–	–	–	–	–	–

18.3.2 Outline of Finite Element Analysis

The replaced sand, alluvial clay, and rubble mound were modeled by the multi-spring model. Table 18.2 shows the parameters, which were set in accordance to a previous study (Ozutsumi 2003), in which the dilatancy parameters for the replaced sand were determined by referring to the liquefaction resistance curves obtained from undrained cyclic triaxial tests (The Third District Port Construction Bureau 1997) of decomposed granite soil (Dr = 67–77%) taken from the reclaimed ground on Rokko Island. The parameter q_{us} in Table 18.2, which is the undrained shear strength to describe the steady state of sand, was determined from undrained monotonic triaxial compression tests.

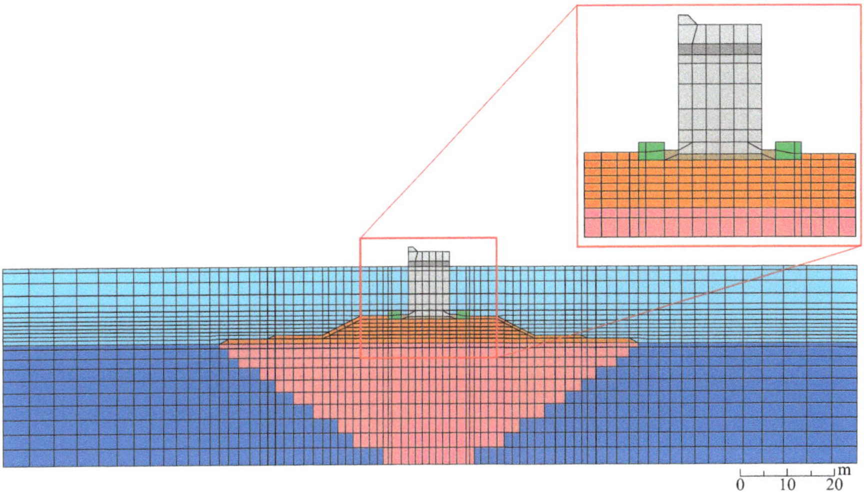

Fig. 18.8 Finite element mesh for the analysis (Iai et al. 1992)

The concrete caisson was idealized by linear plane elements, which were extended for the finite strain analysis based on the concept of the multiple shear mechanism (Ueda et al. 2009). The relative sliding and separation between the foundation rubble and the concrete caisson were taken into account by introducing joint elements. The seawater was modeled as an incompressible fluid and formulated as an added mass matrix considering the equilibrium and continuity of the fluid at the solid–fluid interface (Zienkiewicz et al. 2000).

The finite element mesh for numerical analyses is depicted in Fig. 18.8. Before the seismic response analysis, a static (self-weight) analysis was performed with a gravity force under drained conditions to simulate the initial stress acting in situ before an earthquake. Following the static analysis, seismic response analyses were performed under undrained conditions with viscous boundaries on the bottom and both sides. As an input motion, both the horizontal (NS) and the vertical (UD) components of the earthquake motion recorded at Port Island during the 1995 Hyogoken-Nanbu earthquake, shown in Fig. 18.9, were adopted. The EW component of the horizontal waves, whose direction was parallel to the longitudinal direction of Breakwater No. 7, was not considered in the analysis. To investigate the relationship between the amplitude of the input motions and the induced settlement of the caisson, the observed motions shown in Fig. 18.9 were used through an amplitude adjustment in this study.

The numerical time integration was implemented by the Wilson θ method ($\theta = 1.4$) using a time step of 0.01 s. Rayleigh damping ($\alpha = 0.0$ and $\beta = 0.002$) was used to ensure the stability of the numerical solution process. In the computation, both infinitesimal and large deformation analyses were performed to compare the effect of geometrical nonlinearity. In the large deformation analyses, we

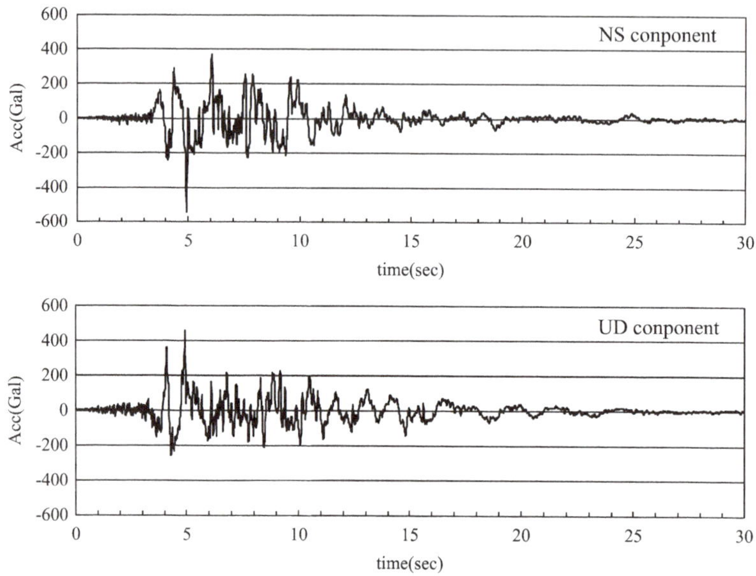

Fig. 18.9 Time history of input wave

Fig. 18.10 Residual deformation and the excess pore water pressure ratio (Cases 1–5)

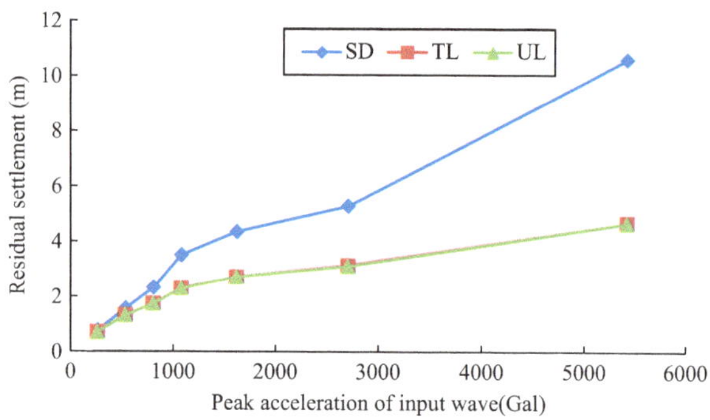

employed both the TL formulation with a material description and the UL formulation with a spatial description.

18.3.3 Numerical Results

Figure 18.10 shows the relationship between the peak acceleration of the input motion and the amount of residual settlement of the caisson top. When the peak acceleration is less than or equal to 500 Gal, the infinitesimal and large deformation analyses are essentially the same. However, the difference (i.e., the effect of geometrical nonlinearity) of the settlement becomes more significant as the peak

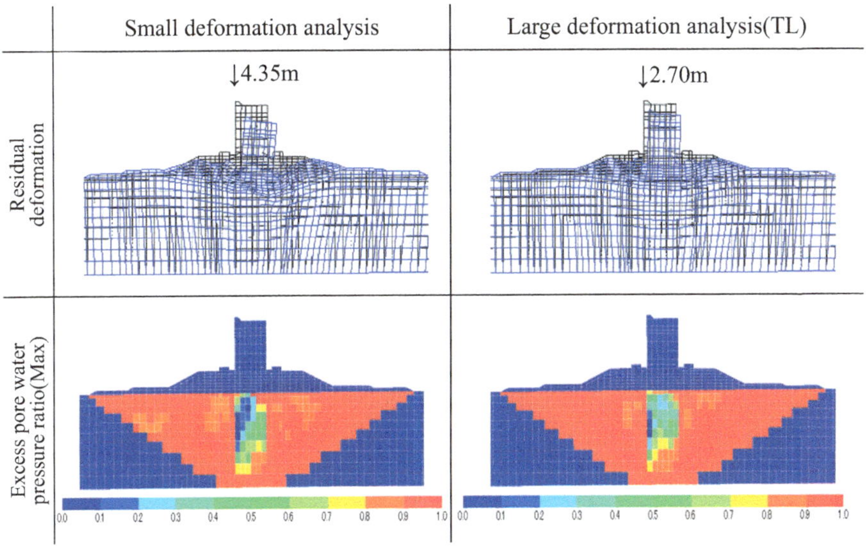

Fig. 18.11 Residual deformation and the excess pore water pressure ratio (1630 Gal)

acceleration increases when the acceleration is over 500 Gal or the amount of the induced settlement in the infinitesimal deformation analysis exceeds 15% (i.e., approximately 2 m) of the wall height. This means that the infinitesimal deformation analysis is inappropriate if the computed settlement of the caisson top is larger than 15% of the wall height. It is necessary to switch from an analytical technique based on the infinitesimal strain theory to one based on the finite strain theory. Figure 18.10 also indicates that the relationship in the TL formulation coincides with that in the UL formulation.

The computed residual deformation and the distribution of the maximum excess pore water pressure ratio under a peak acceleration of 1630 Gal are depicted in Fig. 18.11. In addition to the settlement, the degree of inclination of the caisson also differs between the infinitesimal and large deformation analyses. This result indicates that the use of the large deformation analyses (TL or UL formulation) is recommended, especially if the caisson shape is asymmetric, to evaluate the seismic performance of a caisson-type breakwater.

18.4 Application to Caisson-Type Quay Walls

18.4.1 Summary of the Damaged Quay Wall

The target of the numerical simulation is a caisson-type quay wall, called the RF3 quay wall, which is located on the northeast side of Rokko Island in Kobe Port. The RF3 quay wall was damaged during the 1995 Hyogoken-Nanbu earthquake.

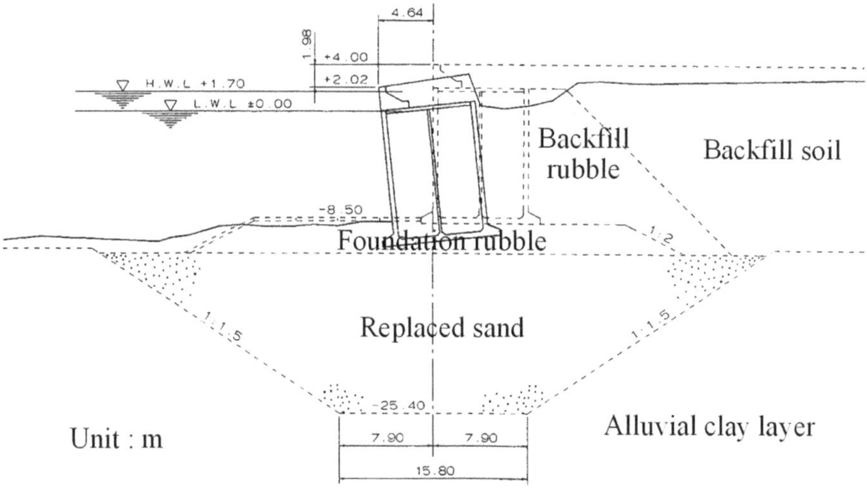

Fig. 18.12 Residual deformation observed in the Kobe earthquake (Inatomi et al. 1995)

Fig. 18.12 shows the cross-section and deformed configuration of the quay wall (Inatomi et al. 1995). The height and width of the caisson are 11.5 and 8.0 m, respectively. In addition to the crest settlement of the caisson (158 cm on average), large lateral displacements of the caisson toward the seaside were observed (370 cm at the caisson top on average). The lateral movement was attributed to liquefaction in the reclaimed areas behind the quay wall due to the strong ground motions (Fig. 18.9). The residual inclined angle of the caisson was 3.1° after shaking.

18.4.2 Outline of Finite Element Analysis

The replaced sand, alluvial clay, backfill, and foundation rubble were modeled by the multi-spring model. Table 18.3 shows the parameters, which were set based on a previous study (Motoshima et al. 2008), in which the dilatancy parameters for the replaced sand and backfilled soil were determined by referring to the liquefaction resistance curves obtained from undrained cyclic triaxial tests (The Third District Port Construction Bureau 1997). The parameter q_{us} in Table 18.3, which is the undrained shear strength to describe the steady state of sand, was determined from the undrained monotonic triaxial compression tests.

To model the concrete caisson, linear plane elements based on the concept of the multiple shear mechanism within the framework of the finite strain theory (Ueda et al. 2009) were used. Joints elements were also used in order to consider the relative sliding and separation between the concrete caisson and the foundation or backfilled rubble. The dynamic water pressure acting on the caisson during shaking was taken into account by the fluid-structure interface elements as well as the fluid elements.

Table 18.3 Soil parameters (Motoshima et al. 2008)

Soil layer	ρ t/m^3	σ_{ma} kN/m^2	m_g, m_k	G_{ma} kN/m^2	K_{ma} kN/m^2	ν	n	h_{max}
Backfill soil	1.8	63	0.5	79,380	207,000	0.33	0.45	0.3
Backfill soil (Liq)	1.8	63	0.5	79,380	207,000	0.33	0.45	0.3
Replaced sand	1.8	44	0.5	58,320	152,000	0.33	0.45	0.3
Clay	1.7	143	0.5	74,970	195,500	0.33	0.45	0.3
Backfill rubble Rubble mound	2.0	98	0.5	180,000	469,000	0.33	0.45	0.3

Soil layer	ϕ_f (°)	ϕ_p (°)	$s1$	$w1$	$p1$	$p2$	$c1$	qus kN/m^2
Backfill soil	36	–	–	–	–	–	–	–
Backfill soil (Liq)	36	28	0.005	6.00	0.50	0.80	2.43	100
Replaced sand	37	28	0.005	4.00	0.60	0.60	2.35	100
Clay	30	–	–	–	–	–	–	–
Backfill rubble rubble mound	40	–	–	–	–	–	–	–

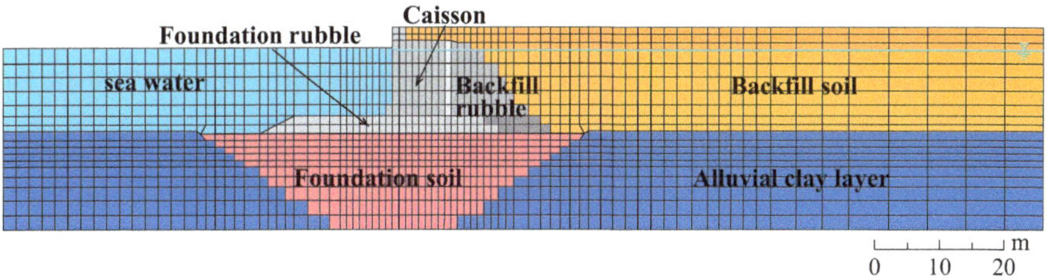

Fig. 18.13 Finite element mesh for analysis (Ozutsumi 2003)

Figure 18.13 shows the finite element mesh for numerical analyses. Following the static (self-weight) analysis, seismic response analyses were performed under undrained conditions with viscous boundaries on the bottom and both sides. The recorded earthquake motions shown in Fig. 18.9 were adopted as the input motions. Furthermore, to investigate the relationship between the amplitude of the input motions and the induced deformation of the caisson, the observed motions shown in Fig. 18.9 were used through amplitude adjustment in this study.

The numerical time integration was implemented by the Wilson θ method ($\theta = 1.4$) using a time step of 0.01 s with Rayleigh damping ($\alpha = 0.0$ and $\beta = 0.002$). In the simulation, both infinitesimal and large deformation (both the TL and UL formulations) analyses were performed to compare the effect of geometrical nonlinearity.

Table 18.4 Comparison of the computed displacements and the measurements

		Analysis results		
	Actual damage	SD	TL	UL
Residual horizontal displacement (m)	3.70	2.62	2.60	2.60
Residual vertical displacement (m)	−1.58	−1.44	−1.35	−1.35
Residual inclination angle (°)	3.10	1.55	1.54	1.46

18.4.3 Numerical Results

Table 18.4 compares the computed displacements with the measurements. With respect to the computed results, the infinitesimal and large deformation analyses do not produce a distinctive difference. This may be because the effect of geometrical nonlinearity is insignificant under the given motions (Fig. 18.9). Table 18.4 indicates that the computed lateral displacement and inclined angle are smaller than the observations although the measured vertical displacement is well simulated. However, considering the fact that the measured values shown in Table 18.4 are the average over all caissons belonging to the RF3 quay wall and vary by location, the constitutive model and program is considered to be an effective way to evaluate the seismic performance of caisson-type quay walls.

Figure 18.14 shows the relationship between the peak acceleration of the input motion and the amount of residual deformation (i.e., lateral and vertical displacements, inclined angle). When the peak acceleration is less than or equal to 500 Gal, there is not a significant difference between the infinitesimal and large deformation analyses. However, the difference (i.e., the effect of geometrical nonlinearity) of the deformation becomes more significant as the peak acceleration increases when the acceleration is over 500 Gal or the amount of the induced lateral and vertical displacements in the infinitesimal deformation analysis exceeds 2.5 and 1.5 m, respectively. The tendency is remarkable in the case for the vertical displacement (i.e., settlement). On the other hand, the difference becomes smaller beyond 1630 Gal for the inclined angle, which may be because sinking of the caisson becomes a dominant deformation mode as the input acceleration increases. Figure 18.14 also indicates that the relationship in the TL formulation coincides with that in the UL formulation. However, the simulation based on the TL formulation becomes unstable over 1630 Gal due to mesh distortion.

As a reference, the computed residual deformation and distribution of the maximum excess pore water pressure ratio under the peak acceleration of 1630 Gal are depicted in Fig. 18.15. The infinitesimal deformation analysis induces a larger settlement of the caisson compared to the large deformation analyses.

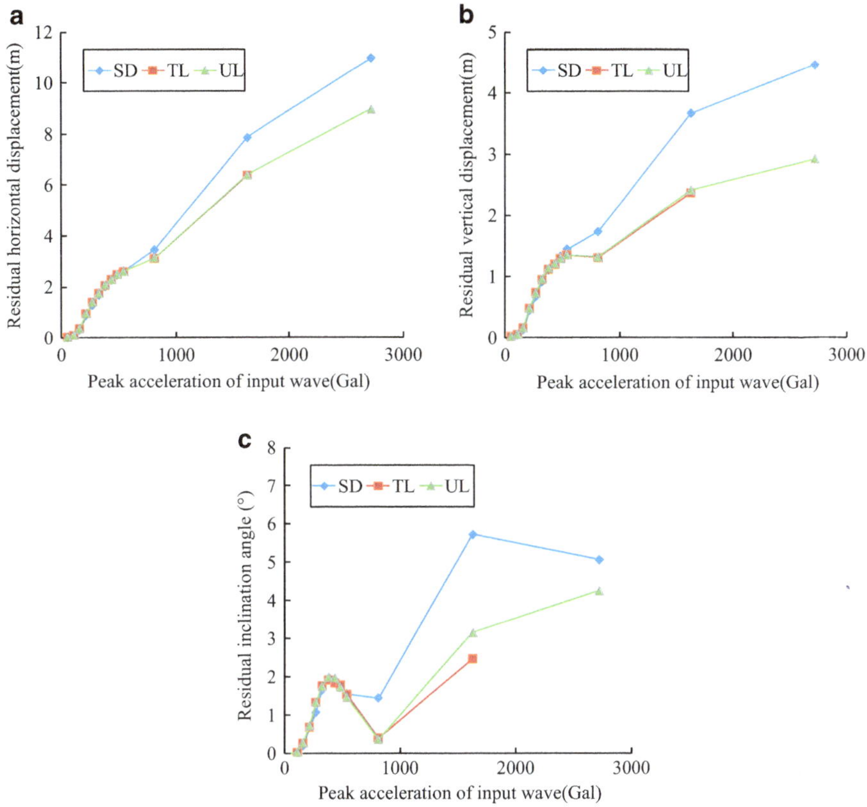

Fig. 18.14 Analysis results. (**a**) Horizontal displacement. (**b**) Vertical displacement. (**c**) Inclination angle

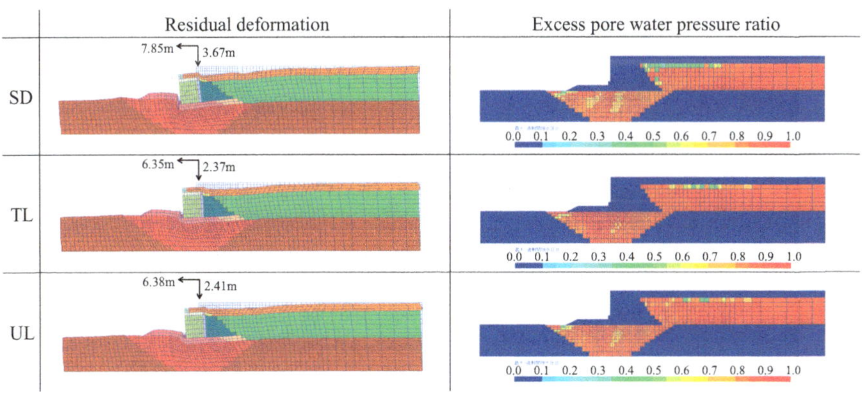

Fig. 18.15 Residual deformation and the excess pore water pressure ratio (1630 Gal)

18.5 Centrifuge Model Tests and Numerical Simulations for Liquefiable Sloping Grounds

18.5.1 Summary of the Centrifuge Model Test

To investigate the large deformation mechanism of a liquefiable sloping ground, a centrifuge model test was carried out in the geotechnical centrifuge of the Disaster Prevention Research Institute, Kyoto University. An objective of the model test is to provide experimental data that can be used to verify the applicability of the large deformation analysis program (i.e., FLIP TULIP). The centrifuge machine had an arm length of 2.5 m and was equipped with a shaking table, allowing the model containers to be exposed to a dynamic excitation in the tangential direction of flight.

Figure 18.16 shows the cross-section view of a liquefiable sloping ground. The water level was set above the ground, and the response acceleration was measured at three different points (i.e., A2–A4) within the ground. Nine excess pore water pressure transducers were installed at the locations specified in Fig. 18.16. A rigid container with inner dimensions of 45 cm (width) × 30 cm (height) × 15 cm (depth) was filled with Silica sand No. 7 ($\rho_s = 2.641$ t/m^3, $e_{max} = 1.219$, $e_{min} = 0.675$) by the air pluviation method to create a sloping ground with a relative density of 50%. In the experiments for saturated ground, the viscosity of the pore fluid was scaled with a factor of μ, which corresponds to the centrifugal acceleration relative to water. To produce water with a specific viscosity, a methylcellulose solution (Metolose) was employed.

The model test was performed under a centrifugal acceleration of 50 G. After spinning up the centrifuge, the waveform (six-step excitation with sinusoidal waves) shown in Fig. 18.17 was given to the shaking table as an input motion.

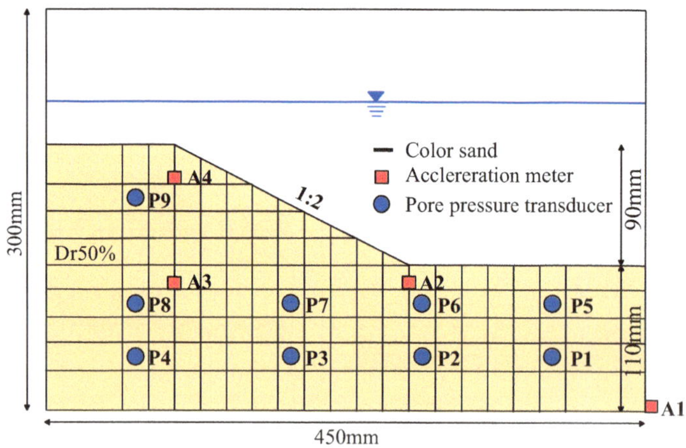

Fig. 18.16 Cross-section view of the model (on the model scale)

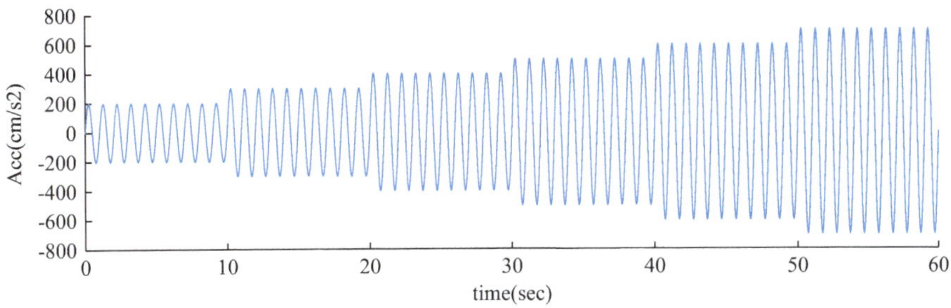

Fig. 18.17 Input waveform (on a prototype scale)

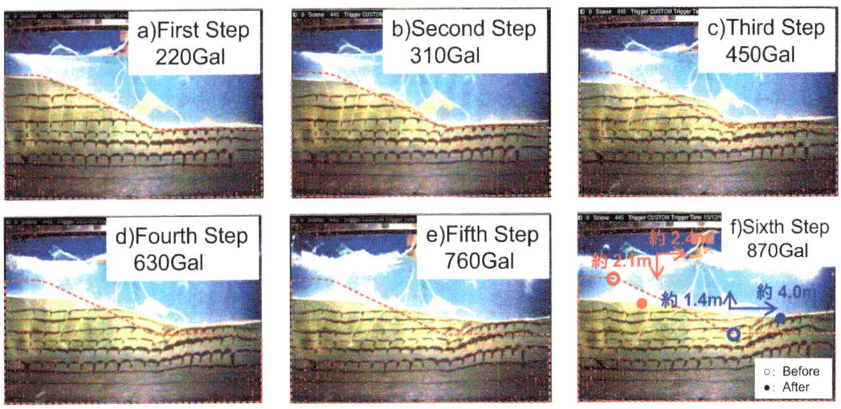

Fig. 18.18 Deformed configuration captured by a high-speed camera

18.5.2 Results of the Centrifuge Model Tests

Figure 18.18 shows the deformed configuration captured by a high-speed camera installed on the centrifuge arm. Although no significant deformation is observed during the first step with an acceleration amplitude of 220 Gal (Fig. 18.18a), the upper part of the sloping ground starts to move downstream during the second step with an acceleration of 310 Gal (Fig. 18.18b). The amount of the lateral movement and settlement increase as the amplitude of the acceleration increases (Fig. 18.18c–f). Finally, the toe of the slope moves 1.4 (upheaval) and 4.0 m in the vertical and lateral directions, respectively, due to liquefaction (Fig. 18.18f). The vertical and lateral displacements at the top of the slope are 2.1 (settlement) and 2.4 m, respectively.

Figure 18.19 shows the time histories of the measured excess pore water pressure and the initial effective overburden pressure. At the lower side of the ground (P5, P6), the excess pore water pressure reaches and then exceeds the initial effective overburden pressure, indicating that the ground is fully liquefied due to the strong motions. The excess pore water pressure does not reach the initial effective

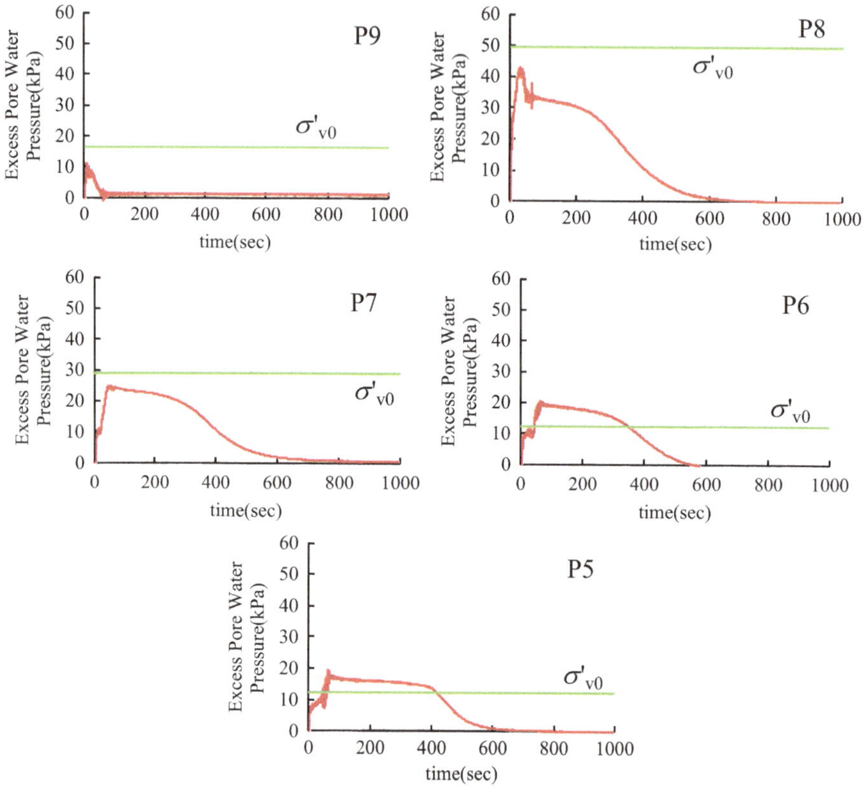

Fig. 18.19 Measured time histories of the excess pore water pressure (on a prototype scale)

overburden pressure at other points, but the stiffness and the strength of the ground significantly decrease due to the build-up of the excess pore water pressure. Figure 18.19 also indicates that the excess pore water pressure perfectly dissipates about 600 s after the shaking ends.

18.5.3 Numerical Simulation

To verify the applicability of the large deformation analysis program (i.e., FLIP TULIP) to the behavior of liquefiable sloping ground, infinitesimal and large deformation analyses were carried out using the multi-spring model for the soil layers. Because the simulation was implemented under the undrained conditions, the drainage phase is beyond the scope of this study. The model parameters were set by referring to a previous study (Ozutsumi 2003).

Figure 18.20 shows the computed residual deformation with the distribution of the excess pore water pressure ratio. The residual deformation in the infinitesimal

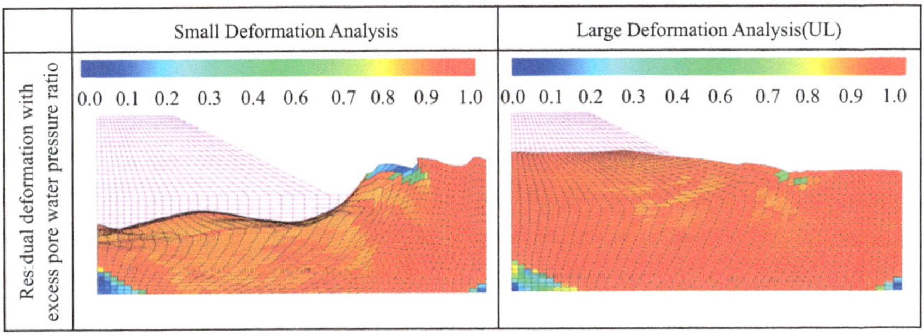

Fig. 18.20 Computed residual deformation with the excess pore water pressure ratio

deformation analysis does not reproduce the observed deformation, and the lateral and vertical displacements are overestimated. On the other hand, the observed deformation is reasonably simulated in the large deformation analysis.

18.6 Conclusions

In this paper, seismic response analyses were performed on an embankment, a caisson-type breakwater, and a caisson-type quay wall in order to verify the applicability of the strain space multiple mechanism model where both the TL and UL formulations are introduced based on a large deformation (finite strain) theory. The primary conclusions are summarized as follows:

1. Infinitesimal deformation analysis overestimates the observed crest settlement of the embankment, but large deformation analyses (both the TL and UL formulations) well simulate the observed deformation. As the peak acceleration of the input motions increases, the difference between the infinitesimal and large deformation analyses (i.e., the effect of geometrical nonlinearity) in the settlement becomes more significant. The results indicate that infinitesimal deformation analysis is inappropriate and large deformation analysis is inevitable to evaluate the seismic performance of embankments in large earthquakes.
2. The numerical simulation of the caisson-type breakwater indicates that the effect of geometrical nonlinearity on the crest settlement of the caisson becomes more significant as the peak acceleration increases when the amount of the induced settlement in the infinitesimal deformation analysis exceeds 15% of the wall height. This means that infinitesimal deformation analysis is inappropriate if the computed settlement of the caisson top is larger than 15% of the wall height. It is necessary to adopt the finite strain theory.
3. The numerical simulation of the caisson-type quay wall indicates that the effect of geometrical nonlinearity on the residual deformation of the caisson becomes more significant as the peak acceleration increases. The tendency is remarkable in the case of vertical displacement (i.e., settlement). On the other hand, the

difference between the infinitesimal and large deformation analyses becomes smaller beyond the threshold value of the acceleration for an inclined angle due to the change in the dominant deformation mode.

4. All of the computed results (e.g., deformation, excess pore pressure ratio) indicate that the TL and UL formulations are not only theoretically but also numerically equivalent, validating the computer program for large deformation analysis.

Acknowledgements This study was conducted as part of the activities of the large deformation analysis working group of the FLIP consortium. The working group consists of the following members: Susumu Iai, Tetsuo Tobita, Kyohei Ueda, Akira Yuasa, Junichi Hyodo, Kyohei Sato, Tsuguhiro Shibamata, Naonori Kuwabara, Tatsuru Yamamoto, Daisuke Shibata, Katsuya Uno, Yoshiro Shiozaki, Kenji Hayashi, Keiichi Sumiya, Tomonari Imono, Shigeru Sato, Masumi Sekino, Osamu Ozutsumi, Koichi Masuda, Tsuyoshi Nakata, Fumitaka Ohsawa, Takashi Nakama, Viradeth Phommachanh, Hironobu Murakami, Takeko Mikami, Kenji Ebisu, Kazuaki Uemura, Koji Ichii, Eiji Kohama, Yousuke Ohya, Yukio Tamari, and Tomohiro Nakahara. The author, as the coordinator of the working group, hereby gratefully acknowledges their contributions.

References

Dynamic Soil Laboratory, Earthquake Engineering Center, Ministry of Construction Public Works Research Institute (2000) Dynamic centrifuge model experiment report on earthquake response effect of embankment by solidification improvement of law. Research Materials for Public Works Research Institute No. 3688 (in Japanese)

Iai S (1996) Deformation analysis of ground and structure system during large earthquake. Lecture Collection of Port and Harbor Research Institute Lecture meeting in 1996 (in Japanese)

Iai S, Matsunaga Y, Kameoka T (1992) Strain space plasticity model for cyclic mobility. Soils Found 32(2):1–15

Iai S, Tobita T, Ozutsumi O, Ueda K (2011) Dilatancy of granular materials in a strain space multiple mechanism model. Int J Numer Anal Methods Geomech 35(3):360–392

Iai S, Ueda K, Tobita T, Ozutsumi O (2013) Finite strain formulation of a strain space multiple mechanism model for granular materials. Int J Numer Anal Methods Geomech 37 (9):1189–1212

Inagaki H, Iai S, Sugano T, Yamazaki H, Inatomi T (1996) Performance of caisson type quay walls at Kobe Port: soils and foundations. Special Issue, pp 119–136

Inatomi T, Zen K, Toyama S, Uwabe T, Iai et al (1995) Damage to port and port-related facilities by the 1995 Hyogoken-nanbu earthquake. Technical Note of the Port and, Harbor Research Institute Ministry of Transport, No.857

Motoshima K, Iai S, Yokoyama N, Sawada S (2008) Influence of fine fraction content, void ratio, etc. to the steady state of sandy soil. In: The 43rd JGS national conference, 1799–1800 (in Japanese)

Ozutsumi O (2003) Study on numerical methods for estimating earthquake induced damage of soil-structure systems on liquefiable ground. Doctoral thesis, Kyoto University (in Japanese)

The Great Hanshin-Awaji Earthquake Survey Editorial Board (1998) The Great Hanshin-Awaji earthquake survey, analysis of the cause of damage. Maruzen Co., Ltd. (in Japanese)

The Third District Port Construction Bureau of Transport Ministry and Kiso-Jiban Consultants Co., Ltd. (1997) Soil test report of Kobe Port (in Japanese)

Ueda K, Iai S, Tobita T, Ozutsumi O (2009) Finite strain formulation of elastic body based on multiple shear mechanism. Ann Disaster Prev Res Inst 52(B):383–394. (in Japanese)

Ueda K (2010) Formulation of large deformation analysis of multiple shear model as sand dynamics model and its applicability. Kyoto University Doctoral Thesis (in Japanese)
Zienkiewicz OC, Taylor RL, Zhu JZ (2000) The finite element method: its basis and fundamentals, 6th edn. Elsevier

Index

S. Iai (ed.), *Developments in Earthquake Geotechnics*, Geotechnical, Geological and Earthquake Engineering 43, DOI 10.1007/978-3-319-62069-5

Printed by Printforce, the Netherlands